Joining textiles

The Textile Institute and Woodhead Publishing

The Textile Institute is a unique organisation in textiles, clothing and footwear. Incorporated in England by a Royal Charter granted in 1925, the Institute has individual and corporate members in over 90 countries. The aim of the Institute is to facilitate learning, recognise achievement, reward excellence and disseminate information within the global textiles, clothing and footwear industries.

Historically, The Textile Institute has published books of interest to its members and the textile industry. To maintain this policy, the Institute has entered into partnership with Woodhead Publishing Limited to ensure that Institute members and the textile industry continue to have access to high calibre titles on textile science and technology.

Most Woodhead titles on textiles are now published in collaboration with The Textile Institute. Through this arrangement, the Institute provides an Editorial Board which advises Woodhead on appropriate titles for future publication and suggests possible editors and authors for these books. Each book published under this arrangement carries the Institute's logo.

Woodhead books published in collaboration with The Textile Institute are offered to Textile Institute members at a substantial discount. These books, together with those published by The Textile Institute that are still in print, are offered on the Woodhead web site at: www.woodheadpublishing.com. Textile Institute books still in print are also available directly from the Institute's web site at: www.textileinstitutebooks.com

A list of Woodhead books on textiles science and technology, most of which have been published in collaboration with The Textile Institute, can be found towards the end of the contents pages.

We are always happy to receive suggestions for new books from potential editors. To enquire about contributing to our textiles series, please send your name, contact address and details of the topic(s) you are interested in to francis.dodds@woodheadpublishing.com. We look forward to hearing from you.

The Woodhead team responsible for publishing this book:
Commissioning Editor: Kathryn Picking
Publications Co-ordinator: Lynsey Gathercole
Project Editor: Cathryn Freear
Editorial and Production Manager: Mary Campbell
Production Editor: Richard Fairclough
Cover designer: Terry Callanan

Woodhead Publishing Series in Textiles: Number 110

Joining textiles

Principles and applications

Edited by

I. Jones and G. K. Stylios

The Textile Institute

WOODHEAD PUBLISHING

Oxford Cambridge Philadelphia New Delhi

Published by Woodhead Publishing Limited in association with The Textile Institute
Woodhead Publishing Limited, 80 High Street, Sawston, Cambridge CB22 3HJ, UK
www.woodheadpublishing.com
www.woodheadpublishingonline.com

Woodhead Publishing, 1518 Walnut Street, Suite 1100, Philadelphia, PA 19102-3406, USA

Woodhead Publishing India Private Limited, G-2, Vardaan House, 7/28 Ansari Road, Daryaganj, New Delhi – 110002, India
www.woodheadpublishingindia.com

First published 2013, Woodhead Publishing Limited

British Library Cataloguing in Publication Data
A catalogue record for this book is available from the British Library.

Library of Congress Control Number: 2012953127

ISBN 978-1-84569-627-6 (print)
ISBN 978-0-85709-396-7 (online)
ISSN 2042-0803 Woodhead Publishing Series in Textiles (print)
ISSN 2042-0811 Woodhead Publishing Series in Textiles (online)

The publishers' policy is to use permanent paper from mills that operate a sustainable forestry policy, and which has been manufactured from pulp which is processed using acid-free and elemental chlorine-free practices. Furthermore, the publishers ensure that the text paper and cover board used have met acceptable environmental accreditation standards.

Typeset by Replika Press Pvt Ltd, India
Printed and bound in the UK by the MPG Books Group

Contents

Contributor contact details

(* = main contact)

Editors and Chapter 20

Ian Jones
TWI Ltd
Granta Park
Great Abington
Cambridge CB21 6AL
UK

E-mail: ian.jones@twi.co.uk

Professor George K. Stylios
Heriot-Watt University
Research Institute for Flexible Materials
School of Textiles and Design
Galashiels TD1 3HF
UK

E-mail: G.Stylios@hw.ac.uk

Chapters 1 and 3

John McLoughlin* and Steven Hayes
Manchester Metropolitan University
Manchester
UK

E-mail: j.mcloughlin@mmu.ac.uk

Chapters 2, 7 and 15

Professor George K. Stylios
Heriot-Watt University
Research Institute for Flexible Materials
School of Textiles and Design
Galashiels TD1 3HF
UK

E-mail: G.Stylios@hw.ac.uk

Chapter 4

John McLoughlin* and Anita Mitchell
Manchester Metropolitan University
Manchester
UK

E-mail: j.mcloughlin@mmu.ac.uk

Chapter 5

Miguel Carvalho* and Helder Carvalho
University of Minho – Department of Textile Engineering
Campus de Azurém
4800-058 Guimarães
Portugal

E-mail: migcar@det.uminho.pt; helder@det.uminho.pt

Luís F. Silva
University of Minho – Department of Mechanical Engineering
Campus de Azurém
4800-058 Guimarães
Portugal

E-mail: lffsilva@dem.uminho.pt

Chapter 6

Professor Dr Arunangshu Mukhopadhyay*
Department of Textile Technology
National Institute of Technology Jalandhar
India - 144011

E-mail: arunangshu@nitj.ac.in; mukhopadhyay.arunangshu@gmail.com

Associate Professor Dr Vinay Kumar Midha
Department of Textile Technology
National Institute of Technology Jalandhar
India - 144011

E-mail: midhav@rediffmail.com

Chapter 8

Edward M. Petrie
Independent Consultant
407 Whisperwood Drive
Cary, NC 27518
USA

E-mail: empetrie@hotmail.com

Chapter 9

Elisabeth Stammen* and Professor Dr-Ing. Klaus Dilger
Institute of Joining and Welding
Technical University Braunschweig, branch office Aachen
Dennewartstr. 25
52068 Aachen
Germany

E-mail: e.stammen@tu-braunschweig.de

Chapter 10

Eunkyoung Shim
North Carolina State University
2401 Research Drive
Raleigh, NC 27695-8301
USA

E-mail: eshim@ncsu.edu

Chapters 11 and 12

Ian Jones
TWI Ltd
Granta Park
Great Abington
Cambridge CB21 6AL
UK

E-mail: ian.jones@twi.co.uk

Chapter 13

Ian Jones*
TWI Ltd
Granta Park
Great Abington
Cambridge CB21 6AL
UK

E-mail: ian.jones@twi.co.uk

Arvind Patil
Vascutek Ltd
Newmains Avenue
Inchinnan
Renfrewshire PA4 9RR
UK

E-mail: A.Patil@Vascutek.com

Chapter 14

Associate Professor Edita Vujasinovic*
Department of Material, Fibres and Textile Testing
Faculty of Textile Technology
University of Zagreb
Prilaz baruna Filipovica 28a
HR-10000 Zagreb
Croatia

E-mail: edita.vujasinovic@ttf.hr

Professor Dubravko Rogale
Department of Clothing Technology
Faculty of Textile Technology
University of Zagreb
Prilaz baruna Filipovica 28a
HR-10000 Zagreb
Croatia

E-mail: dubravko.rogale@ttf.hr

Chapters 16 and 18

Professor Stana Kovacevic*
Department of Textile Design and Management
Faculty of Textile Technology
University of Zagreb
Prilaz baruna Filipovica 28a
HR-10000 Zagreb
Croatia

E-mail: stana.kovacevic@ttf.hr

Professor Darko Ujevic
Department of Clothing Technology
Faculty of Textile Technology
University of Zagreb
Prilaz baruna Filipovica 28a
HR-10000 Zagreb
Croatia

E-mail: darko.ujevic@ttf.hr

Chapter 17

D. J. Tyler
Department of Clothing Design and Technology
Hollings Faculty
Manchester Metropolitan University
Manchester
UK

E-mail: d.tyler@mmu.ac.uk

Chapter 19

A. Pourmohammadi
University of Payame Noor
Parand
Iran

E-mail: apourmohammadi@baftineh.com

Woodhead Publishing Series in Textiles

1 **Watson's textile design and colour Seventh edition**
Edited by Z. Grosicki

2 **Watson's advanced textile design**
Edited by Z. Grosicki

3 **Weaving Second edition**
P. R. Lord and M. H. Mohamed

4 **Handbook of textile fibres Volume 1: Natural fibres**
J. Gordon Cook

5 **Handbook of textile fibres Volume 2: Man-made fibres**
J. Gordon Cook

6 **Recycling textile and plastic waste**
Edited by A. R. Horrocks

7 **New fibers Second edition**
T. Hongu and G. O. Phillips

8 **Atlas of fibre fracture and damage to textiles Second edition**
J. W. S. Hearle, B. Lomas and W. D. Cooke

9 **Ecotextile '98**
Edited by A. R. Horrocks

10 **Physical testing of textiles**
B. P. Saville

11 **Geometric symmetry in patterns and tilings**
C. E. Horne

12 **Handbook of technical textiles**
Edited by A. R. Horrocks and S. C. Anand

13 **Textiles in automotive engineering**
W. Fung and J. M. Hardcastle

48 **Medical textiles and biomaterials for healthcare**
Edited by S. C. Anand, M. Miraftab, S. Rajendran and J. F. Kennedy

49 **Total colour management in textiles**
Edited by J. Xin

50 **Recycling in textiles**
Edited by Y. Wang

51 **Clothing biosensory engineering**
Y. Li and A. S. W. Wong

52 **Biomechanical engineering of textiles and clothing**
Edited by Y. Li and D. X-Q. Dai

53 **Digital printing of textiles**
Edited by H. Ujiie

54 **Intelligent textiles and clothing**
Edited by H. R. Mattila

55 **Innovation and technology of women's intimate apparel**
W. Yu, J. Fan, S. C. Harlock and S. P. Ng

56 **Thermal and moisture transport in fibrous materials**
Edited by N. Pan and P. Gibson

57 **Geosynthetics in civil engineering**
Edited by R. W. Sarsby

58 **Handbook of nonwovens**
Edited by S. Russell

59 **Cotton: Science and technology**
Edited by S. Gordon and Y-L. Hsieh

60 **Ecotextiles**
Edited by M. Miraftab and A. R. Horrocks

61 **Composite forming technologies**
Edited by A. C. Long

62 **Plasma technology for textiles**
Edited by R. Shishoo

63 **Smart textiles for medicine and healthcare**
Edited by L. Van Langenhove

64 **Sizing in clothing**
Edited by S. Ashdown

81 **Engineering textiles: Integrating the design and manufacture of textile products**
Edited by Y. E. El-Mogahzy

82 **Polyolefin fibres: Industrial and medical applications**
Edited by S. C. O. Ugbolue

83 **Smart clothes and wearable technology**
Edited by J. McCann and D. Bryson

84 **Identification of textile fibres**
Edited by M. Houck

85 **Advanced textiles for wound care**
Edited by S. Rajendran

86 **Fatigue failure of textile fibres**
Edited by M. Miraftab

87 **Advances in carpet technology**
Edited by K. Goswami

88 **Handbook of textile fibre structure Volume 1 and Volume 2**
Edited by S. J. Eichhorn, J. W. S. Hearle, M. Jaffe and T. Kikutani

89 **Advances in knitting technology**
Edited by K-F. Au

90 **Smart textile coatings and laminates**
Edited by W. C. Smith

91 **Handbook of tensile properties of textile and technical fibres**
Edited by A. R. Bunsell

92 **Interior textiles: Design and developments**
Edited by T. Rowe

93 **Textiles for cold weather apparel**
Edited by J. T. Williams

94 **Modelling and predicting textile behaviour**
Edited by X. Chen

95 **Textiles, polymers and composites for buildings**
Edited by G. Pohl

96 **Engineering apparel fabrics and garments**
J. Fan and L. Hunter

97 **Surface modification of textiles**
Edited by Q. Wei

98 **Sustainable textiles**
Edited by R. S. Blackburn

99 **Advances in yarn spinning technology**
Edited by C. A. Lawrence

100 **Handbook of medical textiles**
Edited by V. T. Bartels

101 **Technical textile yarns**
Edited by R. Alagirusamy and A. Das

102 **Applications of nonwovens in technical textiles**
Edited by R. A. Chapman

103 **Colour measurement: Principles, advances and industrial applications**
Edited by M. L. Gulrajani

104 **Fibrous and composite materials for civil engineering applications**
Edited by R. Fangueiro

105 **New product development in textiles: Innovation and production**
Edited by L. Horne

106 **Improving comfort in clothing**
Edited by G. Song

107 **Advances in textile biotechnology**
Edited by V. A. Nierstrasz and A. Cavaco-Paulo

108 **Textiles for hygiene and infection control**
Edited by B. McCarthy

109 **Nanofunctional textiles**
Edited by Y. Li

110 **Joining textiles: Principles and applications**
Edited by I. Jones and G. Stylios

111 **Soft computing in textile engineering**
Edited by A. Majumdar

112 **Textile design**
Edited by A. Briggs-Goode and K. Townsend

113 **Biotextiles as medical implants**
Edited by M. King and B. Gupta

114 **Textile thermal bioengineering**
Edited by Y. Li

115 **Woven textile structure**
B. K. Behera and P. K. Hari

116 **Handbook of textile and industrial dyeing Volume 1: Principles, processes and types of dyes**
Edited by M. Clark

117 **Handbook of textile and industrial dyeing Volume 2: Applications of dyes**
Edited by M. Clark

118 **Handbook of natural fibres Volume 1: Types, properties and factors affecting breeding and cultivation**
Edited by R. Kozłowski

119 **Handbook of natural fibres Volume 2: Processing and applications**
Edited by R. Kozłowski

120 **Functional textiles for improved performance, protection and health**
Edited by N. Pan and G. Sun

121 **Computer technology for textiles and apparel**
Edited by J. Hu

122 **Advances in military textiles and personal equipment**
Edited by E. Sparks

123 **Specialist yarn and fabric structures**
Edited by R. H. Gong

124 **Handbook of sustainable textile production**
M. I. Tobler-Rohr

125 **Woven textiles: Principles, developments and applications**
Edited by K. Gandhi

126 **Textiles and fashion: Materials design and technology**
Edited by R. Sinclair

127 **Industrial cutting of textile materials**
I. Viļumsone-Nemes

128 **Colour design: Theories and applications**
Edited by J. Best

129 **False twist textured yarns**
C. Atkinson

130 **Modelling, simulation and control of the dyeing process**
R. Shamey and X. Zhao

131 **Process control in textile manufacturing**
Edited by A. Majumdar, A. Das, R. Alagirusamy and V. K. Kothari

132 **Understanding and improving the durability of textiles**
Edited by P. A. Annis

133 **Smart textiles for protection**
Edited by R. A. Chapman

134 **Functional nanofibers and applications**
Edited by Q. Wei

135 **The global textile and clothing industry: Technological advances and future challenges**
Edited by R. Shishoo

136 **Simulation in textile technology: Theory and applications**
Edited by D. Veit

137 **Pattern cutting for clothing using CAD: How to use Lectra Modaris pattern cutting software**
M. Stott

138 **Advances in the dyeing and finishing of technical textiles**
M. Gulrajani

139 **Multidisciplinary know-how for smart textile development**
T. Kirstein

140 **Handbook of fire resistant textiles**
F-S. Kilinc-Balci

141 **Handbook of footwear design and manufacture**
A. Luximon

142 **Textile-led design for the active ageing population**
J. McCann and D. Bryson

143 **Optimising decision making in the apparel supply chain using artificial intelligence (AI)**
W. K. Wong, Z. X. Guo and S. Y. S. Leung

144 **Mechanisms of flat weaving technology**
V. Choogin, P. Bandara and E. Chepelyuk

145 **Innovative jacquard textile design using digital technologies**
F. Ng and J. Zhou

146 **Advances in shape memory polymers**
J. Hu

147 **Clothing manufacture management: A systematic approach to planning, scheduling and control**
J. Gersak

148 **Anthropometry, apparel sizing and design**
D. Gupta and N. Zakaria

Introduction

In one reference, this book provides up-to-date information on sewing, adhesive bonding and welding methods for joining textiles. Despite the importance of joining processes in product manufacture of all kinds, before now there has never been a single volume dedicated to covering all the major textile joining methods.

The methods for joining textiles have been practised for longer than fibre-based textiles have been available. Based on exploration in Russia, archaeologists believe that Stone Age people (30 000 years ago) used sewing needles of bone and ivory to join animal skins together into clothing. Much later, some 12 000 years ago in Neolithic times, weaving was developed. This timeline has some logic given that threads with useful strength for sewing needed to be available before they could be woven into cloth. Now textiles are probably the most ubiquitous materials on earth. Everyone comes into contact with them in many forms every day; and almost every textile product contains a joint of some kind.

The earliest sewing machines were developed by Frenchman Barthélemy Thimonnier in the 1750s, constructed of wood (see Chapter 4 for further discussion of the history of sewing). This and subsequent developments significantly improved the working efficiency and consistency for seam and product making over manual sewing techniques. Developments, including the introduction of the rotary shuttle in the early twentieth century, provided the opportunity for lockstitch, chainstitch and overlock stitching in home and industrial machines. More recent industrial machines achieve over 5 000 stitches per minute and incorporate monitoring and control mechanisms to counter seaming problems such as pucker and thread breakage. So-called intelligent sewing machines have become available which can set up themselves and dynamically control their mechanisms as a function of the material properties. Even so, mass production of textile products has largely been reliant on hand manipulation of the fabrics through the sewing machines, which is very labour intensive. Unlike the revolution in other manufacturing industries that has seen the introduction of automated and robotic systems on production lines, this level of automation has largely not been seen in the textile industry. This is not least because of the difficulties in handling very flexible materials. The concepts of robotic automation, recent progress and application areas are discussed.

As the range of fabric types and demands from the seam and textile products have increased, the range of methods available for making seams in textiles is increasing. New joining techniques have been developed as new technologies arise. These have included heat sealing, dielectric welding (since 1940s), ultrasonic welding (1960s) and laser welding (1980s), and more recently machines introducing new variations on the processes, combining the benefits of more than one technique. These systems bring much wider opportunities for greater automation, different performance capability and new aesthetic potential. For example, in contrast to stitching, welding or bonding typically provide a continuous seam suitable for fluid or gas sealing. Welding methods are used for synthetic fabrics, woven or non-woven, coated and uncoated, and for joining fabrics to rigid plastic parts. Adhesive bonding is widely used in laminating fabrics, sealing previously stitched seams, and is being used increasingly in various forms to provide a seaming method in its own right. Adhesive methods are also applicable to natural and synthetic fabrics and dissimilar material combinations. The range of techniques available increases the flexibility of the designer, particularly in working with new materials and multicomponent systems such as laminates and composites products. Some of the welding and bonding methods also bring new opportunities for automation. Many of the processes require access to only one side of the fabric, which brings the possibility for different manipulation techniques and fabric holding methods using vacuum beds, for example.

The future will bring challenges and will drive automation. The broadening scope of textile products, and the introduction of more complex and multifunctional materials in areas such as medical dressings, wearable electronics and industrial fabrics, will bring a need for joining methods that provide the same performance and productivity without adversely affecting the new functions of the textiles.

Following an overview of fibre and fabric types, covering natural and synthetic materials and their joining requirements, the book is split into four main parts. Part I deals with the fundamental theory of sewing and sewing machines. This provides an understanding of the mechanics of the seam before and after sewing and the dynamic forces generated on the needle and fabric. The practical requirements of operating sewing machines are described along with the main stitch types. The recent developments in mechanised, monitored and automated equipment are described in fundamental and practical terms. This section also raises the problems that can occur related to sewing, and provides solutions and methods for their avoidance.

Part II introduces adhesive bonding of textiles: the adhesive types used, the bonding methods and application to seaming and textile lamination. There is certainly overlap between adhesive bonding and welding in that the same equipment can often be used for both purposes. By definition, adhesive bonding makes use of a third material at the joint to provide attachment between the two fabrics. The adhesive is commonly one that is melted or re-melted in the process, but the fabrics themselves are not melted. Welding occurs when one or both of the original fabrics are melted at the joint line. Welding does not require the use of a third material between the fabrics, yet also does not require the full thickness of the fabric to be melted.

Part III provides details of the equipment and processing methods used in welding: heat sealing, hot air, hot wedge, dielectric (also known as RF or HF), ultrasonic and laser welding. The variations in these processes are discussed, including their use for bonding where relevant. This section also introduces the performance of welded seams including the joint strength and leak-tightness developed using the different techniques.

The final part discusses the application of textile joining in some typical product areas, including non-iron shirts, car seats, tarpaulins, tents, protective clothing and in fabrics with electronic components.

This book is intended as a valuable reference for manufacturing and production engineers, designers and researchers working with products using textiles or combinations of textiles and other materials. It also provides a basis for researchers working to advance the development of the joining methods mentioned here and for the inevitable development of new methods. Scientific and practical insight is provided in the use of sewing, bonding and welding techniques for application to all weights and types of textiles materials with uses in manufacturing from garment making and medical devices through to tarpaulins and inflatables.

It would have been impossible to complete this unique detailed review of the aspects of textiles joining without the insightful contributions from all the collaborating authors. They have provided content worthy of the leaders in their fields that they are. We are also very grateful to the staff at Woodhead Publishing Limited for their co-ordination of this project, and at TWI Ltd and Heriot-Watt University for their support throughout the development of the book and to those involved with the editors in the many years of research that made this book possible.

Ian Jones
George K. Stylios

1
Types of fabric and their joining requirements

J. MCLOUGHLIN and S. HAYES, Manchester Metropolitan University, UK

DOI: 10.1533/9780857093967.1

Abstract: The first part of this chapter describes the most common fibres used in fabric production and how materials have evolved into high-tech smart materials. The second part describes the categories of fabrics, the applications and joining requirements in the construction of these materials into products. The stitches and seams used in the manufacture of these textiles are described and discussed. It is also important to mention that wherever these products are made, the principles outlined in this chapter apply.

Key words: textiles, fabrics, joining, seams, stitches, sewability, manufacturing textiles, high performance fabrics.

1.1 Introduction

Textiles have been synonymous with human activity for thousands of years and as the years have progressed, their usage has become wider and more varied. Many people's perception of textiles is of traditional textiles such as apparel, clothing and home furnishings, such as curtains and bedding. However, textiles have diversified and advanced into areas that affect our lives daily, from automotive textiles to even more high-tech end uses such as smart materials that are able to react to the environment. High performance products for outdoor pursuits such as camping, walking and hiking are a particularly large and important area of the textile market. Materials have been developed from the 1930s up to the present day that are capable of producing high water repellence whilst, at the same time, being breathable and comfortable against the skin. Garments have been designed with fabric properties that are able to save or prolong life in dangerous situations. All of these high performance textiles are constructed using a similar combination of stitched seams and welded joints.

The creation of a fabric begins with the fibre from which the yarn is spun and then woven or knitted into a material. This process is a lengthy one and incorporates a highly developed form of technology. The textile industry began by relying on a range of natural fibres. The invention of synthetic fibres created a new generation of high performance yarns and fabrics resulting in products of very high tensile strength, fully waterproof whilst being comfortable and breathable. Examples of these are given later in this chapter. Figure 1.1 highlights the different categories of fibres.

Of the many diverse fibres used in textiles, only the most commonly used fibres and fabrics for apparel and outerwear products are considered here. The sections in the first part of this chapter describe the most common fibres used in production of these fabrics and how materials have evolved into high-tech smart materials. The sections in the second part of the chapter describe the categories of fabrics, the applications and joining requirements in the construction of these materials into products. The stitches and seams used in the manufacture of these textiles are described and discussed. It is also important to mention that wherever these products are made, the principles outlined in this chapter apply.

1.2 The main types of natural fibre

Prehistoric men and women used animal skins and plants to protect themselves from the cold and against the sun, sand and dust, etc. This can be described as the birth of textiles and became the start of the revolution that has become a multibillion pound global industry. We now exploit four main sources of natural materials which are:

- bast fibre such as flax, jute, ramie, hemp, etc.
- wool and other animal hairs
- cotton
- silk.

1.2.1 Cellulose fibres

Cellulose fibres are the fibres that are produced naturally from plants. The most important of these fibres, cotton, is the most widely used natural fibre in the world and accounts for over 50% of the world's production (Figs 1.2 and 1.3). It is used extensively in apparel, homeware and outerwear and is recognised for its comfort and breathability and also for its ability to absorb moisture, thus making it comfortable against the skin. It is blended extensively with polyester and with other synthetic fibres, particularly spandex, and is used in many applications, particularly in the manufacture of shirting. The crease resistance properties of the polyester combined with the comfort and moisture regain of cotton produce what is called 'easy care' on the label on shirts, the most common blend ratio being 65% polyester and 35% cotton. This makes the ironing of this type of garment much easier. It has also been developed into high performance products such as tent and ventile fabrics which will be discussed later in this chapter.

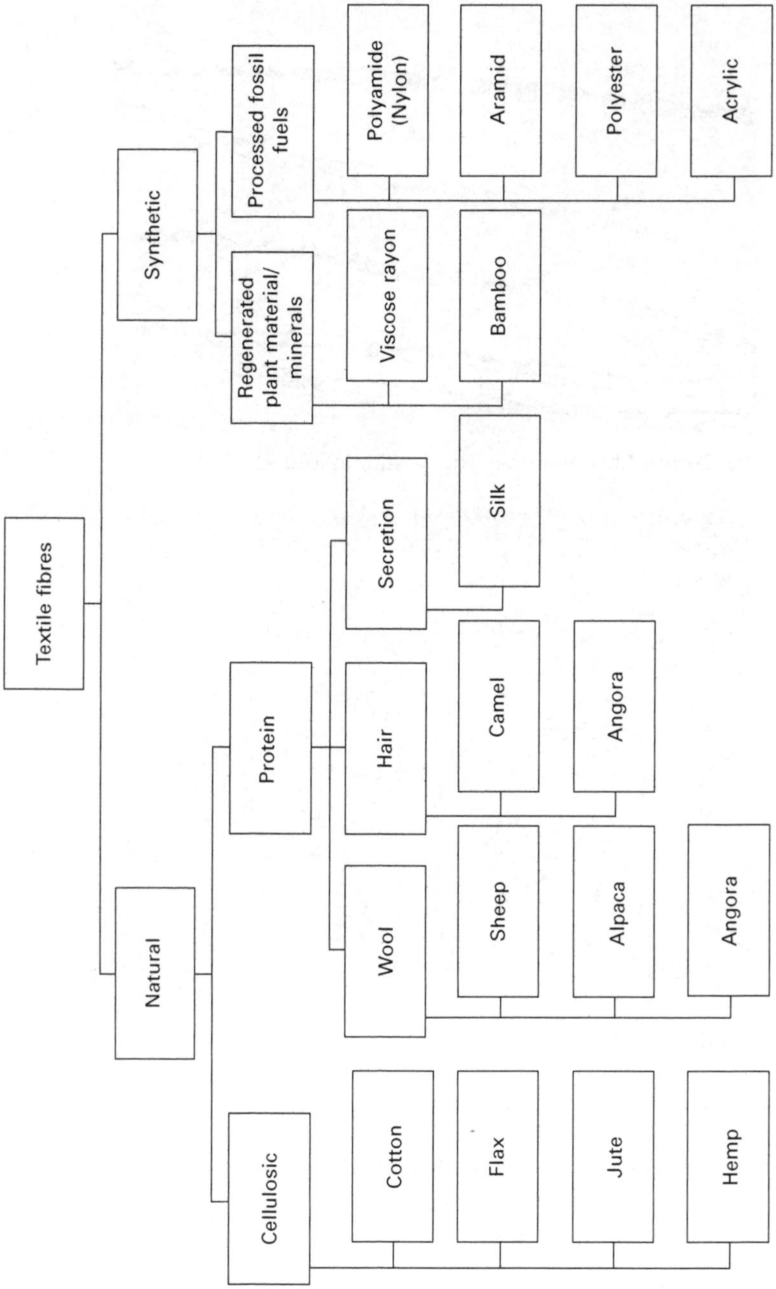

1.1 Categorisation of fibre types.

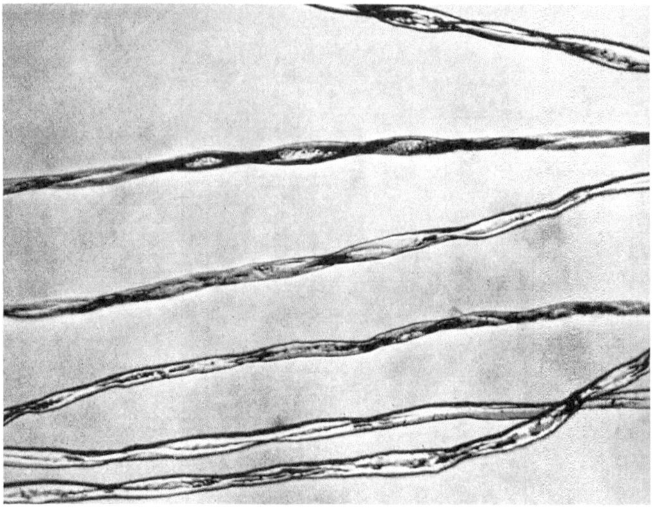

1.2 Cotton fibre (source: The Textile Institute).

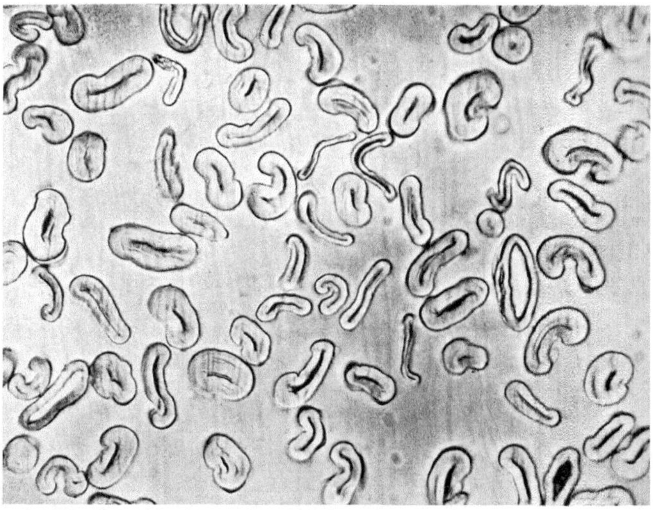

1.3 Cotton fibre cross section.

Under a microscope, the cotton fibre appears as a very fine regular fibre and can range in length from 10 to 65 mm depending upon the quality of the fibre. The longer the fibre is, then the better quality yarn can be spun resulting in a more comfortable and better quality product. Key properties of cotton include the following:

- It is one of the only fibres that gets stronger when it is wet, enabling a smoother manufacturing process when weaving into fabric because of its ability to withstand the rigours of the weaving process.

- It is a relatively strong fibre due to its polymer structure and its crystalline nature.
- It is relatively inelastic due to its crystalline nature.
- Cotton is a hydrophilic fibre due to its amorphous regions, making it able to absorb up to 50% moisture when wet.
- It has the ability to conduct heat energy, minimising any destructive heat accumulation, making it able to withstand very hot ironing temperatures.
- Cotton fibres are weakened and destroyed by acids.

1.2.2 Flax

Another fibre commonly used in the production of apparel is flax, normally known as linen (Figs 1.4 and 1.5). The word is derived from the old English 'fleax'. Linen is the term applied to yarn spun from the flax fibres. It is a heavier fibre than cotton and it is for this reason that most linen materials are of light construction. Heavier linen materials would be uncomfortably heavy to wear. The fibre morphology (appearance) of linen is that of a much thicker fibre than cotton and it is much longer, ranging in length from 10 to 100 cm. The colour of flax varies from light blond to grey blond, and the differences in shade result from the agricultural and climatic conditions in which it was grown.

Other properties of flax include the following:

- It is a very strong fibre due to its high crystalline nature.
- It is very inelastic, again due to its polymer structure and crystalline nature.
- It has the same water absorbency properties as cotton.
- Like cotton, flax fibres are weakened and destroyed by acids.

1.4 Flax fibre.

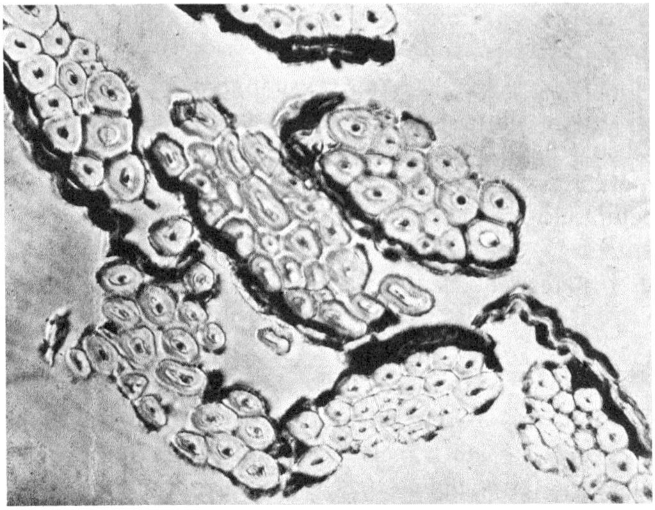

1.5 Flax fibre cross section.

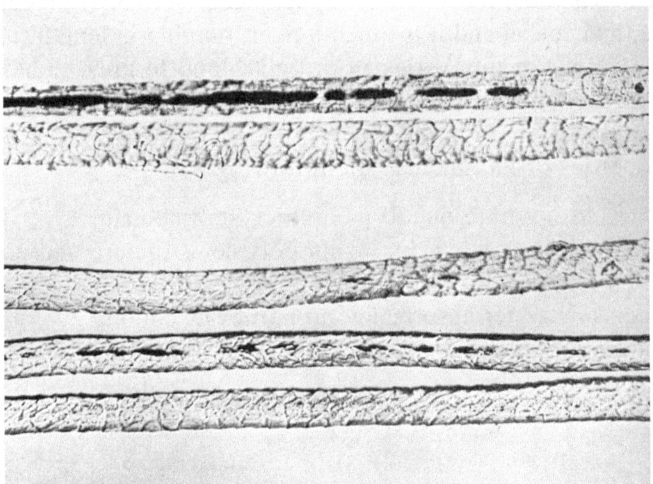

1.6 Wool fibre.

1.2.3 Protein fibres

Protein fibres are the resultant product from an animal, and these include wool, silk, mohair, alpaca, angora, cashmere, etc. Only the most commonly used protein fibres, namely wool and silk, are considered here. The word wool is derived from the term 'wull' in old English and is the fibre from the fleece of domesticated sheep. It is a natural, protein and multi-cellular staple fibre (Figs 1.6 and 1.7). The fibre is a crimped fine to thick regular fibre that

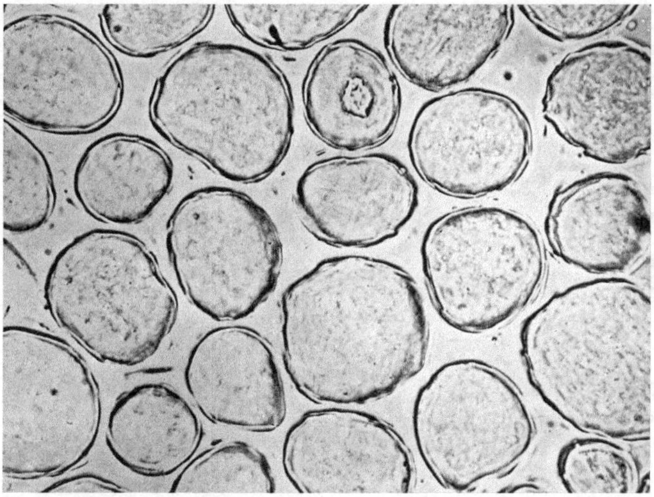

1.7 Wool fibre cross section.

has natural serrations (scales) along its length. This fibre has a very high natural crease resistance and very good resilience, partly attributable to the elastic nature of the wool fibre. The warmth of wool fabrics is due more to the air spaces in the material than to the fibres themselves.

Other properties of wool include the following:

- It is a relatively weak fibre with a low tensile strength.
- Wool has a very high absorbency, approximately 100%.
- Wool has very good elastic recovery and excellent resilience due to its crimped configuration.
- Wool is more resistant to acids than to alkalis.
- Exposure to sunlight and weather tends to yellow or dull coloured wool textile materials.

1.2.4 Silk fibres

In the old English, the name silk was 'sioloc' which is thought to have originated from the Greek interpretation of 'seres' meaning the people from Eastern Asia, namely the Chinese. This is a very fine fibre with translucent filaments and is used extensively in the manufacture of apparel, particularly shirts, dresses and ties. The raw silk strands are encased by a protein called sericin which enables the fibre to withstand prolonged exposure to weather, making the fibre very weather resistant due to this coating of sericin (Figs 1.8 and 1.9).

The silk filament has a high degree of strength associated with it, but when the fibre gets wet, it loses strength and the fibre is more sensitive to

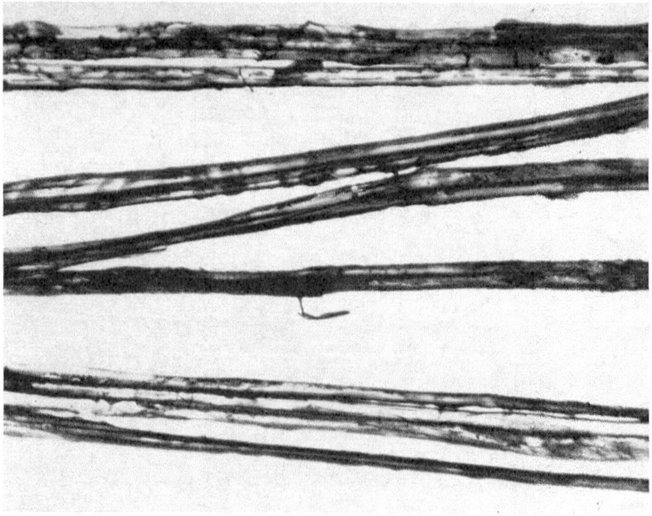

1.8 The silk fibre coated with the gum sericin.

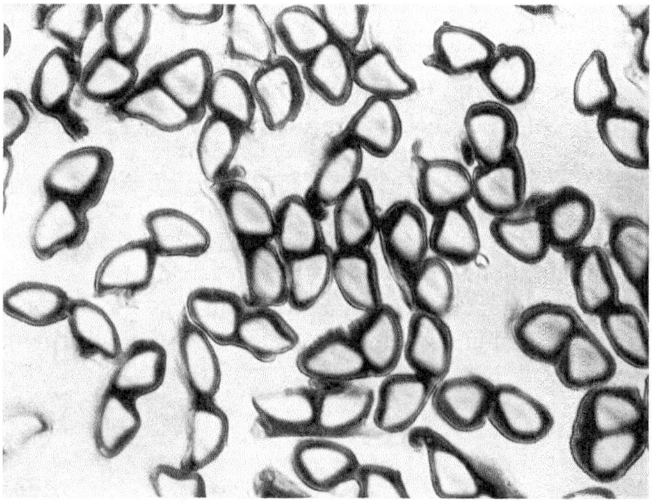

1.9 Silk fibre cross section.

heat than wool due to chemical interactions within the fibre. The fibre is more degraded by acids than wool and the resistance to the environment is not as good as that of wool.

1.3 Synthetic fibres

Because of the many and diverse numbers of synthetic fibres, only three of the most common ones are briefly considered here. Synthetic fibres are man-made fibres made from materials such as coal and other substances, such as chemicals, etc. The most common of the fibres used are polyamide, polyester and viscose.

1.3.1 Viscose

Viscose is a man-made fibre made from natural fibre materials such as tree bark (Figs 1.10 and 1.11). It is usually described as a regenerated cellulose fibre. The morphology of the fibre is as follows.

- It is a fine regular filament or staple fibre.
- Its cross section is that of a kidney been shape.
- It has a harsh bright lustrous appearance.

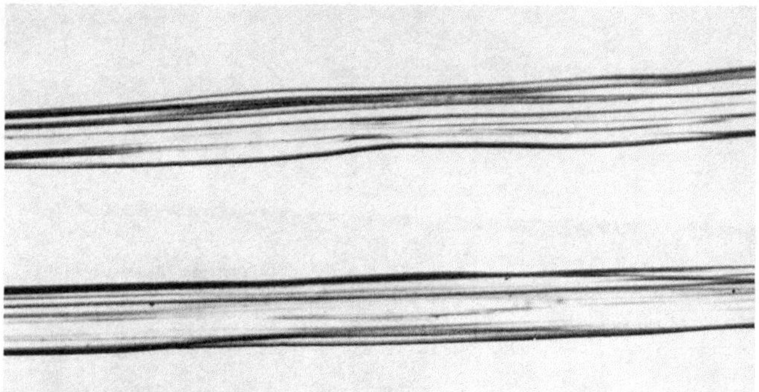

1.10 Viscose fibre.

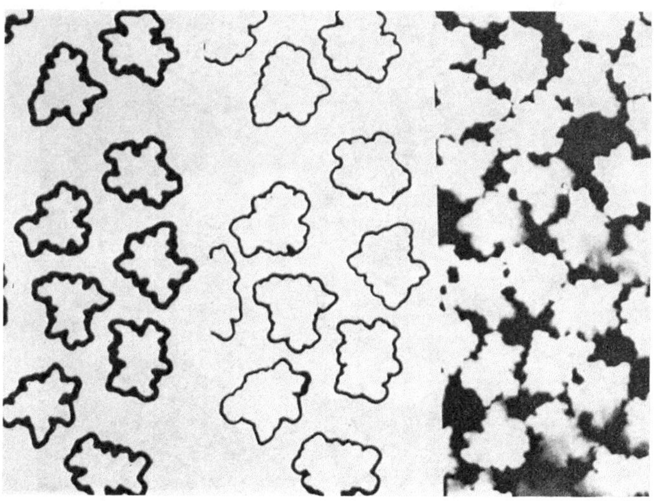

1.11 Viscose fibre cross section.

- It is treated with chemicals to make the fibre appear to be white.
- It is very amorphous, meaning it absorbs water very easily.
- It is a weaker fibre than cotton.
- When wet, viscose is only half its strength when dry.

1.3.2 Polyester

Polyester is the most common of all the man-made fibres (Figs 1.12 and 1.13). It is considered to be a rather heavy fibre and for this reason is manufactured as lightweight or thin fabrics, since thick polyester materials would be too heavy. This fibre is used extensively in the production of apparel, blended with cotton to manufacture shirts, blouses, underwear, etc., and suiting fabrics, trousers and many other products such as high performance outwear jackets, ski wear, tents and car seat covers. Some of the properties of polyester are as follows.

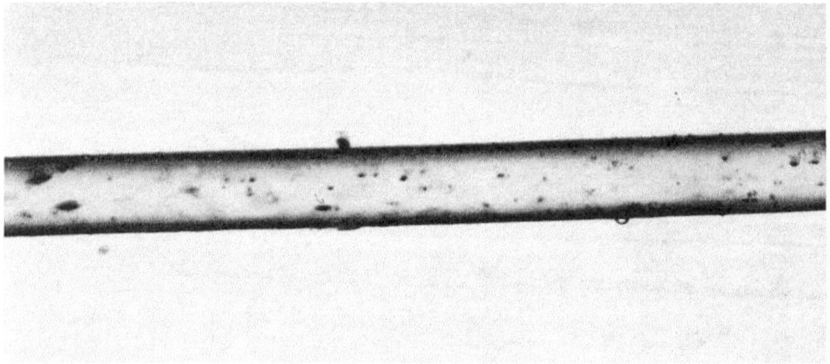

1.12 Polyester fibre.

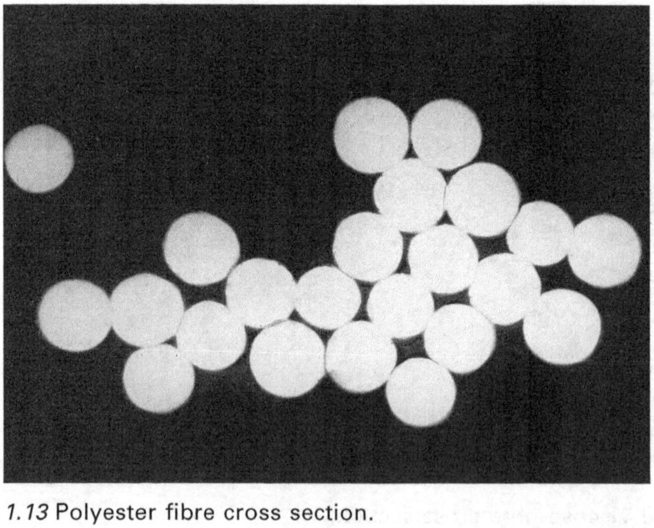

1.13 Polyester fibre cross section.

- It is a fine, regular and translucent fibre.
- It can be crimped and textured.
- It has good easy care properties, especially when blended with cotton.
- It has a highly crystalline structure, making it hydrophobic (highly resistant to water penetration).
- It has high tenacity (strength).
- It possesses very good colour fastness.

1.3.3 Nylon

Nylon, derived from polyamide, is a very regular, translucent fibre. The two most important products from polyamide are nylon 6,6 and nylon 6, both of which are used extensively in producing fabrics for outerwear and tent materials (Figs 1.14 and 1.15). The nylon filament may be likened to a glass rod due to its longitudinal cross section, being essentially featureless along its length. Textile materials composed of nylon tend to be lightweight which makes it the ideal material for the outdoor enthusiast, particularly with regard to hiking and backpacking. The specific viscosity of nylon is lower than polyester meaning that the fibre is less dense thus making it a lighter weight fibre.

Other properties of nylon include the following.

- Nylon is a strong, elastic and durable fibre, due to its crystalline polymer system and its chemical structure.
- Nylon 6,6 is stronger than nylon 6 and also has less extensibility, which

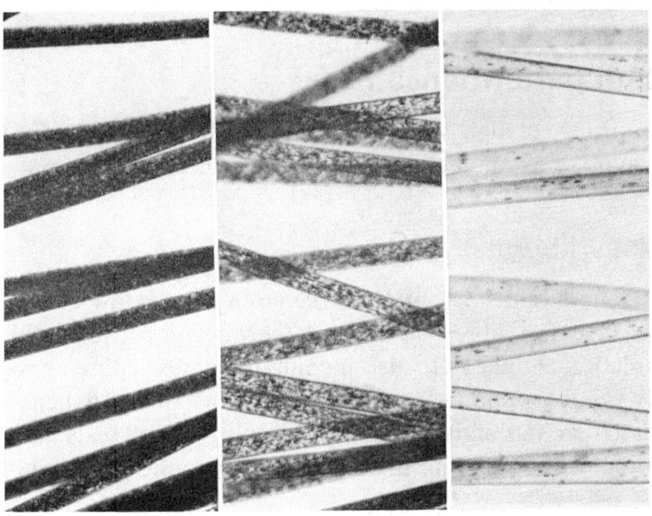

1.14 Nylon fibre 6.6; showing the fibres with three levels of content.

1.15 Nylon fibre 6.6 cross section.

is useful for tent materials as the fabric flysheet stretches less when the tent is erected.

- Nylon suffers a loss of tenacity when wet.
- High performance fabrics such as tent materials are impregnated with silicone treatments in order to improve the hydrostatic head of the fabric, making them highly waterproof.
- Nylon has poor heat conductivity.
- Nylon has only a fair resistance to sunlight and weather, particularly ultraviolet light.

1.4 High performance fibres

There is a huge range of high performance fibres. This section briefly discusses three representative examples: glass fibres, Kevlar and Dyneema.

1.4.1 Glass fibres

These fibres generally display high tensile strength at a low elongation and very low density. The fibres have good thermal properties making them ideal for insulation and they are also incombustible by nature as well as not releasing any vapours or poisonous materials when exposed to heat. They are also resistant to oils, fats and solutions and display a good resistance to acids and alkalisation. The following list of properties applies to this fibre:

- excellent strength to weight ratio
- excellent elastic recovery

- good thermal properties
- excellent resistance to sunlight
- good resistance to acids, although can be attacked by some forms of acid, e.g., concentrated hydrofluoric and sulphuric acid, and also hot phosphoric acids
- highly resistance to most chemicals in general use
- highly resistance to organic solvents
- excellent acoustic properties.

End uses for this fibre include mats, thermal insulation, electrical insulation, sound insulation, and reinforcement of various materials, e.g., tent poles, sound absorption, heat and corrosion resistant fabrics. Other uses are high strength fabrics, pole vault poles, arrows, bows and Irish step dance shoes.

1.4.2 Kevlar

Kevlar comes from the polyamide family. They are derived from benzene resulting in a fibre with extremely long molecular chains giving it an almost ridge-like structure. Properties of this fibre are:

- excellent strength to weight ratio
- excellent thermal stability
- tensile fatigue is high
- high resistance to cutting
- good flex resistance
- poor abrasion resistance
- poor resistance to ultraviolet radiation.

Kevlar is a well-known component of personal armour, including combat helmets, ballistic face masks and ballistic vests. It is also used to manufacture gloves, sleeves, jacket chaps and other articles of clothing designed to protect users from cuts, abrasions and heat. Motor cycle safety clothing is another area where Kevlar is used extensively.

1.4.3 Dyneema

Dyneema, invented some years ago by the Company DSM, is reputed to be the world's strongest fibre, 15 times stronger than steel on a weight for weight basis. It is highly flexible and offers high protection against mechanical hazards. Some properties of this fibre are:

- high strength to low weight ratio
- low specific gravity; it will float on water
- excellent chemical resistance
- highly UV resistant

- moisture resistant
- highly durable.

End uses for this fibre are: automotive, glass, paper, food, household equipment, steel, construction and emergency services industries.

1.5 The process of creating fabrics from fibres

Before they can be turned into a fabric, the natural and synthetic fibres discussed above need first to be converted to yarn. Fibre for yarns is supplied either in short staple lengths (staple yarn) or long filaments (filament yarn). There are different manufacturing processes for the various types of fibre. Yarns can be manufactured in a wide range of structures such as single, ply and cord structures. More complex structures are known as fancy yarns or core-spun yarns. Yarns can also be manufactured with differing amounts of twist to hold the fibres together. When woven or knitted into a fabric, yarn type and structure has a significant impact on the properties of the fabric.

A suggested categorisation of fabrics is given in Fig. 1.16. The earliest fabrics were made exclusively from plant fibre. Protein fibres such as wool were not developed until around 2500 BC and silk from around 1500 BC. These first-generation textiles were colourless and used principally for protection against the environment. The next revolutionary step was the discovery of how to colour these fibres and fabrics. Initially these garments were expensive and only the very rich could afford them. The representation of status through clothing became the basis for fashion. Today fashion is available at very reasonable prices (Figs 1.17 and 1.18).

The change in the functionality of textiles thus evolved from the first generation of textile materials, those that provided a barrier to the elements, to the evolution of the second generation, providing a barrier and projecting an image. The industry can now be seen as evolving a third generation of 'active' or 'smart' textiles. The previous generations of fabrics can be described as passive, not able to adapt to environmental changes. However, materials are being developed that are able to adapt to the environment, with

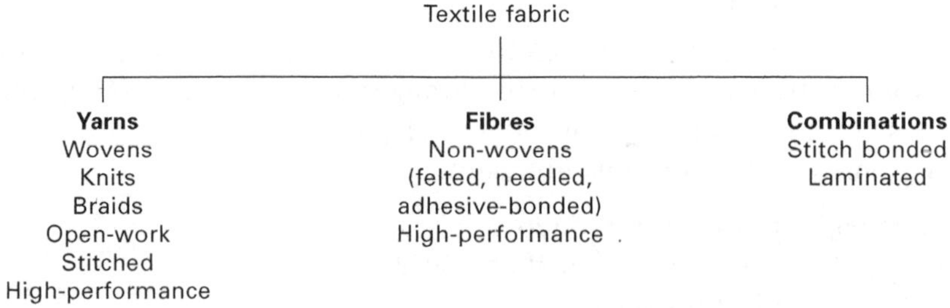

1.16 Categorisation of textile fabric.

1.17 Fashion from the 1950s.

1.18 Fashion from the 1990s to the present decade.

fibres, yarns and fabrics being combined with electronics to produce smart materials. These electronic components include:

- strain gauges

- flexible switches
- displacement transducers
- electrodes
- temperature transducers.

These produce electrically active fibre structures with significant electrical properties in order to incorporate conductive elements into the fabric structure.

Garments have been designed and developed to measure respiratory measurements (Fig. 1.19), heart beat and blood pressure. Work is being carried out to design garments that can incorporate the SIM from a mobile phone so you can have a telephone call without having to use a handset.

1.6 Woven fabric structures

Derived from basket making, weaving is the earliest form of fabric manufacture, going back to 7000 BC. By the end of the Roman period weaving was a highly developed form of technology. There are hundreds of different fabric constructions, the most common being plain-woven fabrics to more complex structures such as twill, herringbone, satin and sateen weaves. Fabrics are designed on point paper similar to graph paper. The design method is given as:

- pattern draft of point paper design
- it is marked and read from bottom left to top right
- the vertical columns represent the warp yarns and the horizontal rows represent the weft yarns

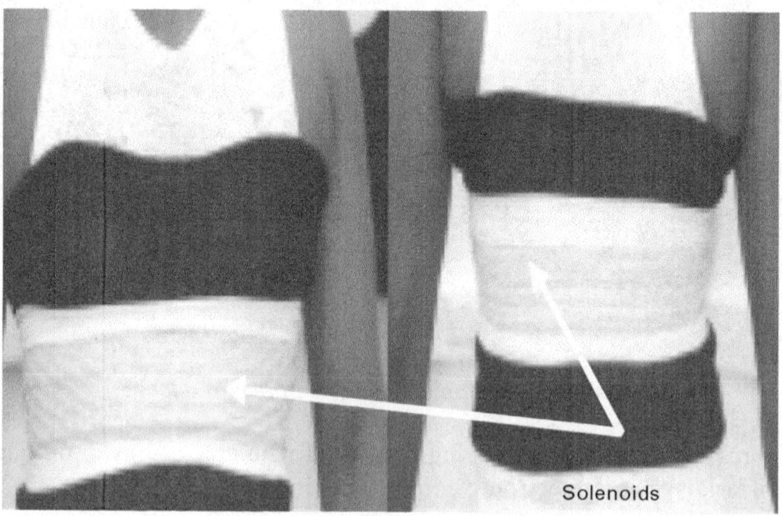

1.19 The future of textiles.

- a mark placed over any weft cell indicates that the warp is lifted over the weft in that intersection
- An absence of a mark indicates the opposite, i.e. warp remains beneath the weft yarn.

1.6.1 Plain-woven fabric

An example of a point paper design for a plain-woven fabric is given in Fig. 1.20 and an example of the fabric is given in Figs 1.21 and 1.22. Properties of a plain-woven fabric are:

- The fabric is the simplest and tightest of interlacing warp and weft yarns.
- Opposite sides of the fabrics are the same.
- These fabrics have high abrasion and are resistant to yarn slippage.
- Types of fabrics produced include batiste, cambric, Donegal, fresco and voile.

1.6.2 Twill weaves

A twill weave is a fabric composed of yarns in a diagonal line where each warp yarn lifts over (and/or remains over) more than one weft. Examples of these types of fabrics are given in Figs 1.23 and 1.24.

1.20 A plain woven fabric drafted on point paper.

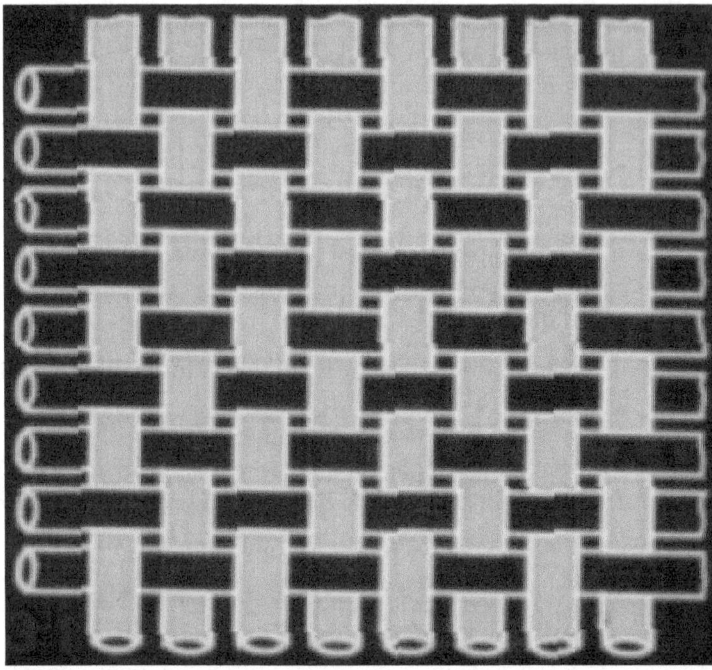

1.21 Diagram of fabric design.

1.22 Plain woven fabric.

1.23 Pattern draft of a twill fabric.

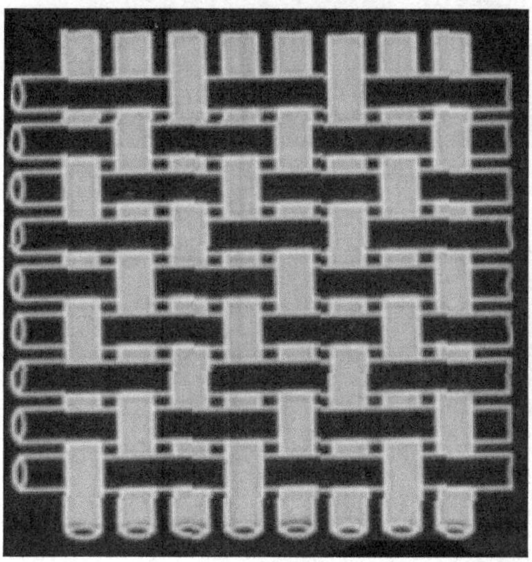

1.24 Interlacement of the yarns of the twill fabric.

Properties of the twill fabric are:

- The twill fabric order of interlacing causes diagonal lines to appear in the fabric.
- These lines can either run in the Z direction or the S direction.

- Warp-faced twills show a predominance of warp yarns on the face of the fabric.
- Weft-faced twills show a predominance of weft yarns on the face of the fabric.

Examples of twill warp-faced fabrics are given in Figs 1.25 and 1.26 and weft-faced fabrics in Figs 1.27 and 1.28.

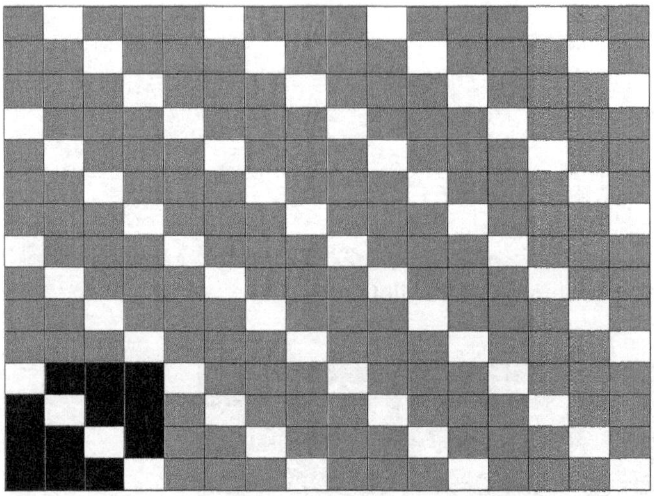

1.25 Warp-faced twill pattern.

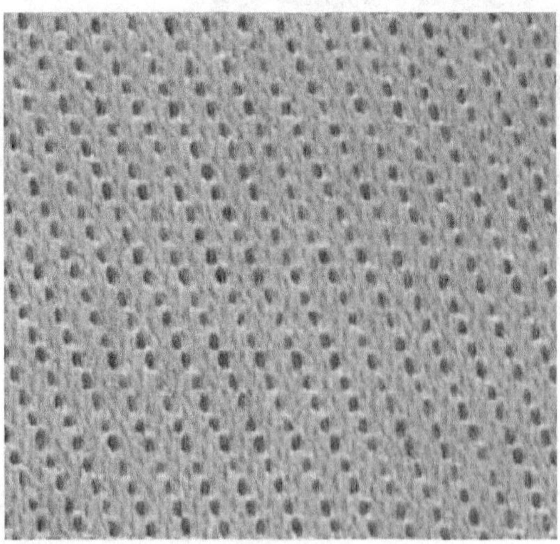

1.26 Warp-faced twill fabric.

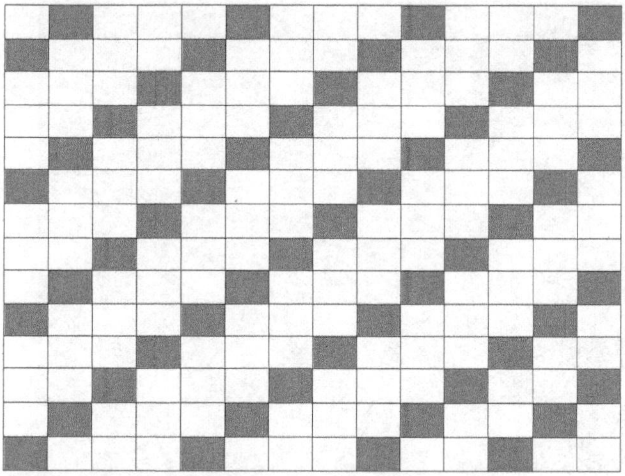

1.27 Weft-faced twill pattern.

1.28 Weft-faced twill fabric.

1.6.3 Satin and sateen fabrics

Satin and sateen fabrics use a high density of warp and weft-faced yarns to produce materials. They are smooth, uniform and lustrous due to the scarcity of interlacings and the density of threads. Examples of these fabrics are shown in Figs 1.29–1.32.

1.7 Knitted fabric structures

Knitted fabrics have advanced to a stage of high performance materials capable of reacting to the environment (Fig. 1.19). These materials have conductive

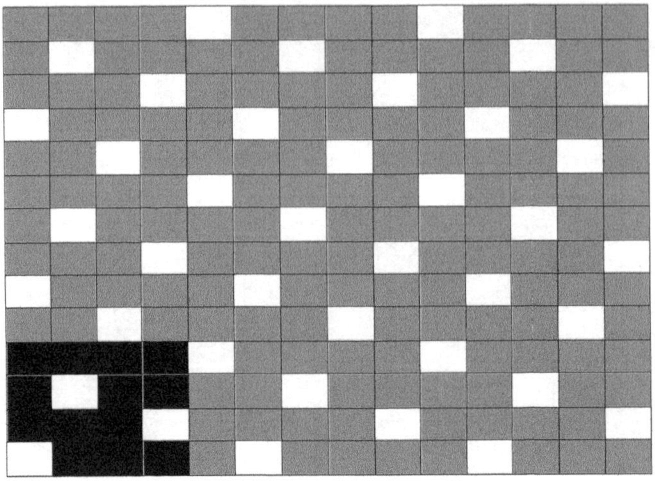

1.29 Satin pattern.

1.30 Satin fabric.

elements knitted into the garment or product and are termed 'smart fabrics'. However, most knitted products perform similar functions to their woven cousins, namely in apparel and the automotive industry. There are also many uses for knitwear in the medical industry. There are many types and structures of knitted fabrics; therefore, only the most common are described here. The most common fibres used in the manufacture of knitwear are:

- cotton
- wools
- cashmere

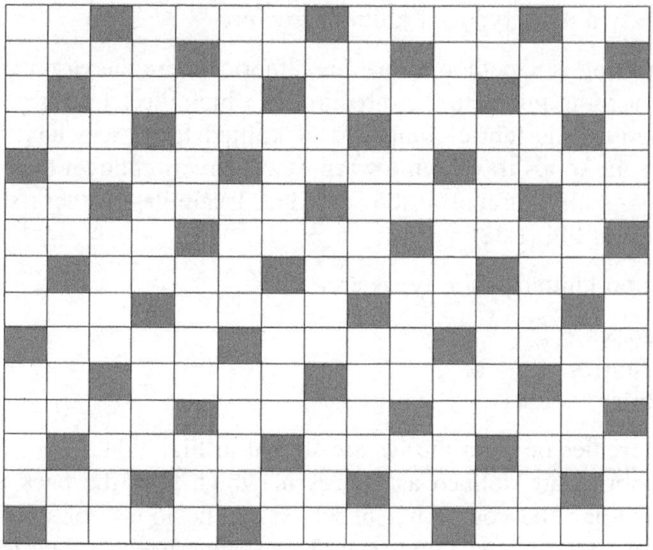

1.31 Sateen pattern.

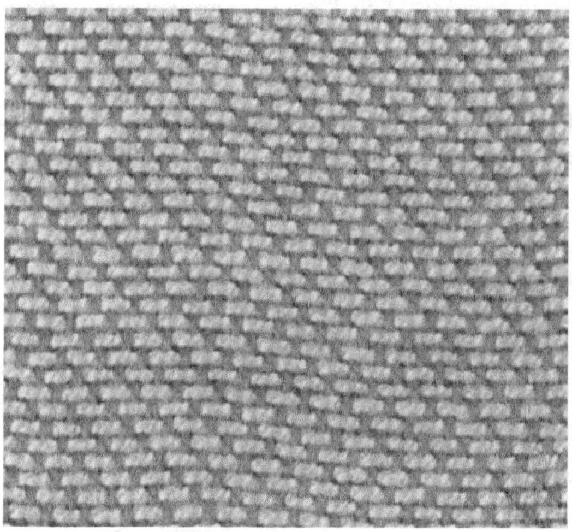

1.32 Sateen fabric.

- silk
- polyester
- acrylics
- polyamide/nylon
- acetate
- triacetate.

There are two main types of knitted structure:

- Weft knitting is a method of making a fabric where the yarn is fed more or less at right angles to the direction in which the fabric is produced.
- Warp knitting is fabric composed of knitted loops in which the yarn forming the loops travels in a warp-faced direction down the length of fabric generally parallel to the selvedge. Examples of these structures are given in Fig. 1.33.

Some common knitted fabric types are:

- plain fabrics
- ribbed fabrics
- purl fabrics.

The characteristics of plain fabrics are shown in Fig. 1.34.

Ribbed fabrics are defined as fabrics in which both the back and face loops occur along the course but in which all the loops contained within any single wale are of the same type. This means back or face loops. An example of these materials is given in Fig. 1.35.

Purl fabrics are described as a fabric in which both back and face loops occur in some or all of the wales. The structure may be identified by the particular knitting sequence used. An example of these is given in Fig. 1.36.

In manufacturing apparel there are three main methods of manufacture. These are: cut and sew, shaping and linking and full fashioned/integral production.

- Cut and sew garments are where flat panels of knitted fabric are pressed

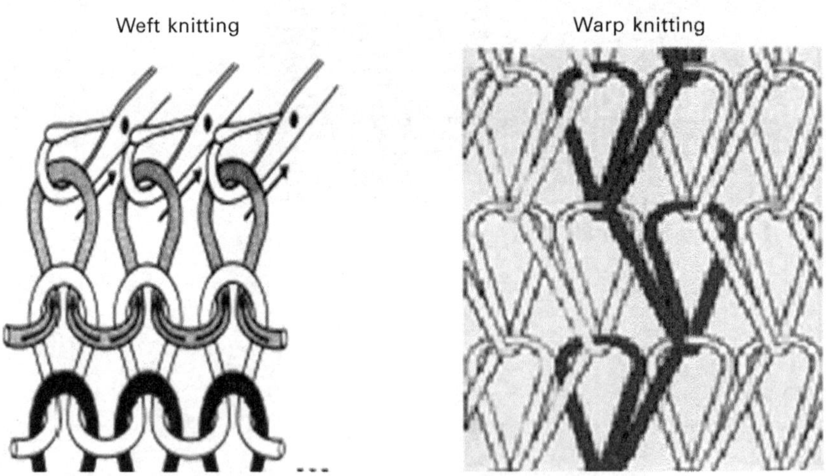

1.33 Weft and warp knitted structures.

- Technical face
- Produced on one needle bed
- Curls back at selvedges
- Curls forward at start and finish
- Unravels from top and bottom
- Low extensibility
- Lightweight

1.34 Characteristics of plain knitted fabrics.

- Face loop
- Two needle beds
- Stable (no edge curling)
- Unravels from end last knitted
- Very extensible
- Thermal properties

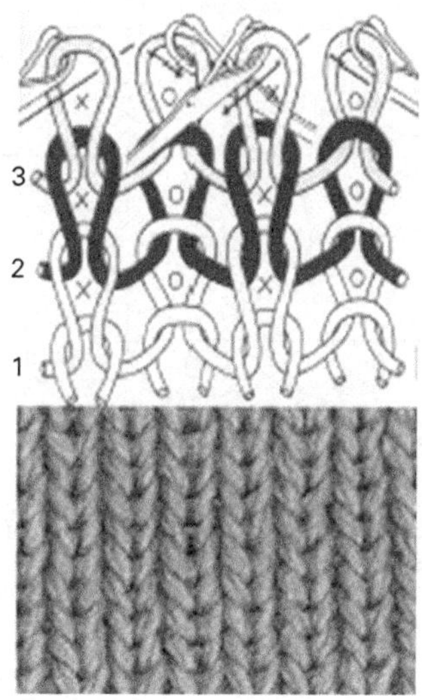

1.35 Characteristics of knitted rib fabrics.

on a flat press to stabilise and set the material. The fabric is then cut to shape for the garments to be sewn and these are processed in the sewing room. The process route for these garments is given in Fig. 1.37.

- Shaping and linking is where the garment panels are knitted to shape; no cutting takes place whatsoever. The panels are stitched together using linking machines providing much neater, smaller and nicer seams. An example of the production route is given in Fig. 1.38.

1.36 Characteristic of a purl knitted fabric.

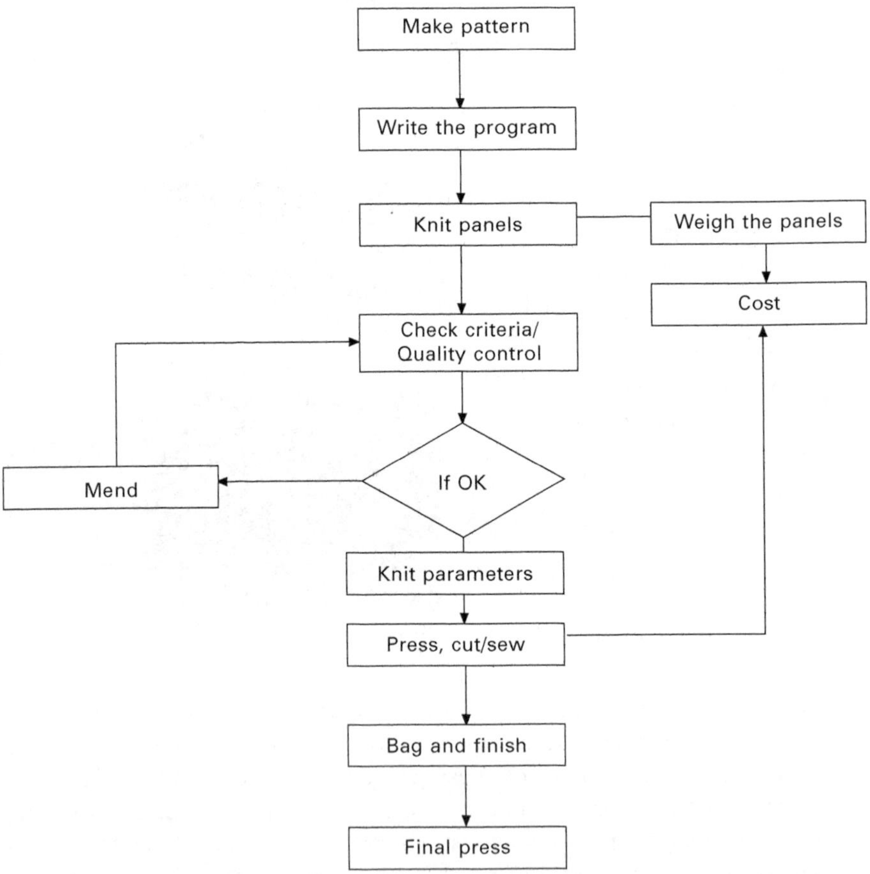

1.37 Flow diagram for cut and sew knitting process.

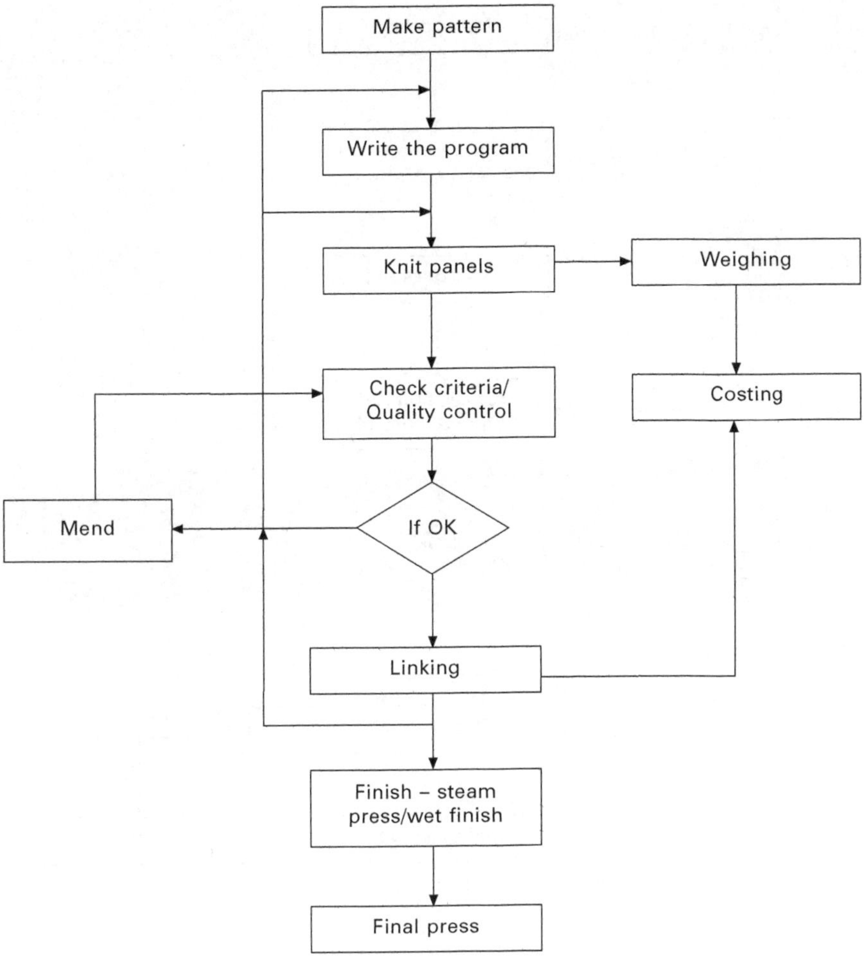

1.38 Flow diagram for shaping and linking knitting process.

- Fully fashioned/integral is where the garment is knitted whole in one piece with few or no seams on the face of the garment. Modern knitting machines are capable of knitting three-dimensional shapes. A production flow diagram is given in Fig. 1.39.

1.8 Non-woven fabrics

A final group of fabrics that are neither woven nor knitted are non-wovens. These are fabrics made from continuous filament yarns or from staple fibre webs or batts. The main types of web formation are dry laid, wet laid and spun laid. The webs are strengthened by bonding using various techniques which include adhesive bonding, mechanical interlocking by needling or by

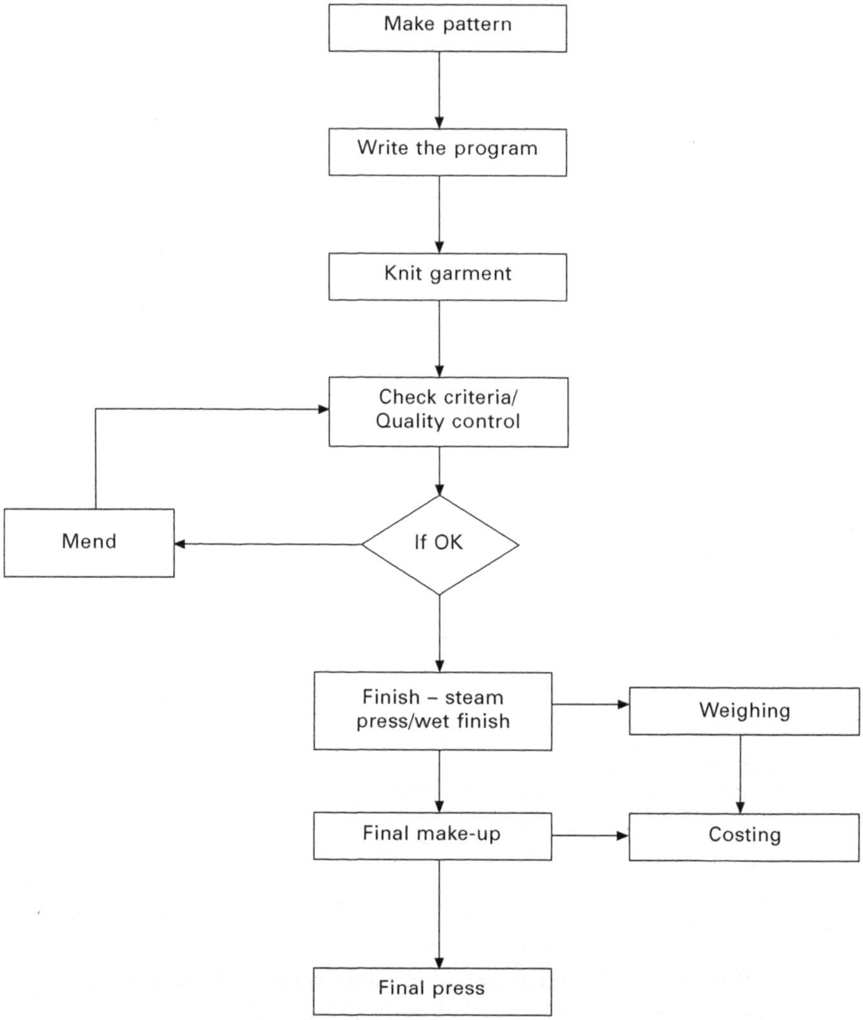

1.39 Flow diagram for fully fashioned production for knitwear.

fluid jet entanglement. Thermal bonding and stitch bonding are also methods commonly used. Non-wovens are used for a range of applications including diapers, sanitary wear, wipes, medical clothing, carpet backing and more specialised uses such as geotextiles.

1.9 Joining fabrics: seams and stitches

When joining materials together, the following areas need to be considered. These include:

- aesthetic appeal
- strength
- durability
- comfort in wear
- equipment availability
- cost.

1.9.1 Seams

The seam has to be suitable for the purpose for which it is intended and the seam type is dependent upon the product being sewn. It must combine the required standards of appearance and performance whilst ensuring economy of production. There are many types of seams categorised into specific groups. The most common and most used seams are considered here.

- Class 1, superimposed seams (Figs 1.40 and 1.41) are generally two or more plies of fabric laid upon one another in the same orientation. They are usually stitched together near the edge of the fabric with one or more

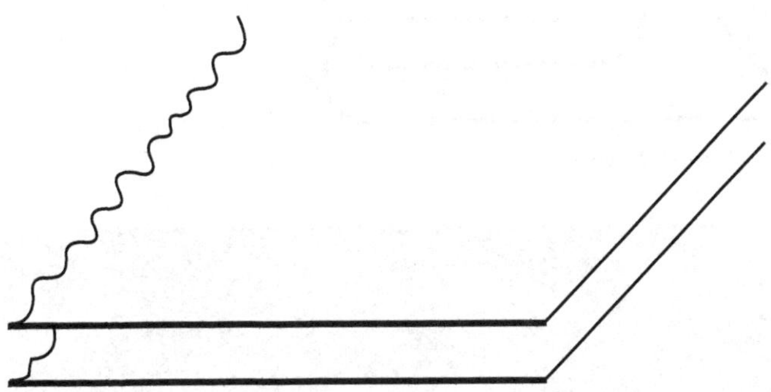

1.40 Superimposed seam.

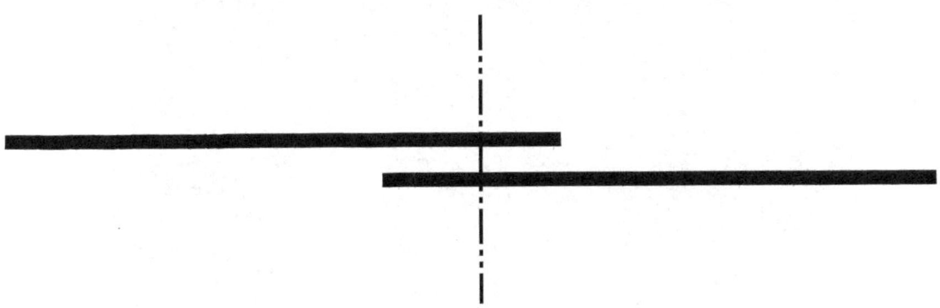

1.41 Superimposed seam cross section.

rows of stitching. The rows of stitching may be sewn simultaneously or consecutively and seams may be reinforced with fused or stay tapes along the length of the seam.

- Class 2, doubled lapped seams (Fig. 1.42) are very strong seams used extensively in many areas of apparel, outerwear and high performance sewn products.
- Class 3, bound seams (Fig. 1.43), are used in many areas of sewn product manufacture and produce a neat seam on the face of the fabric. They are used for many applications including luggage and tent manufacture.
- Class 4, flat seams (Fig. 1.44), are produced with a minimum of two plies of fabric butted together at the raw edge on the same level. A minimum

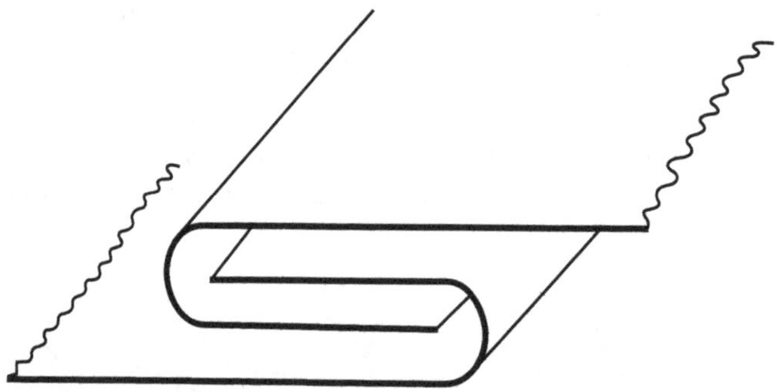

1.42 Double lapped seam.

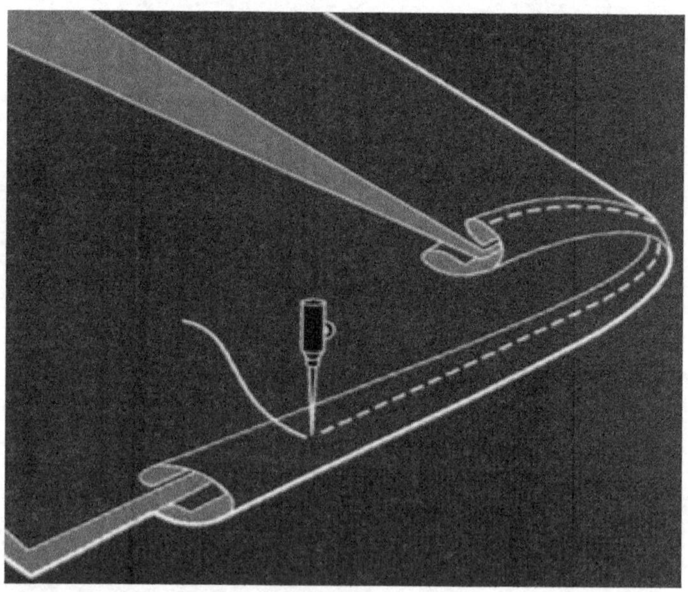

1.43 Example of a bound seam.

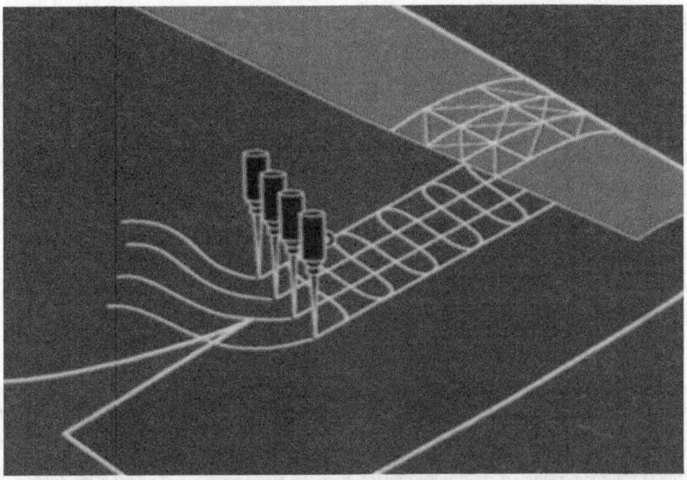

1.44 Example of a flat seam.

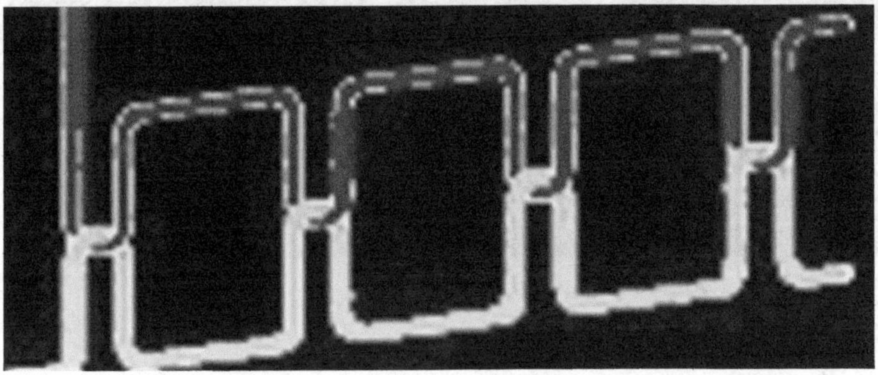

1.45 301 lock-stitch.

of two rows of stitching are introduced simultaneously which are joined on one or both surfaces. They are often used in knitwear to reduce bulk around the join especially in underwear.

1.9.2 Stitch types

The stitch types commonly used in the production of these types of products are:

- Stitch type 301
- Stitch type 401
- Stitch type 516.

Stitch type 301 (Fig. 1.45) looks the same on both sides of the fabric and consists of the interlacing of the passing of a thread around or over a separate thread supply or a loop of that supply.

Stitch type 401 (Fig. 1.46) is the passing of a thread through a loop formed by a separate thread supply.

Stitch type 516 (Fig. 1.47) consists of two independent stitch types produced on the same machine, namely a five-thread over-lock machine. The first stitch type, stitch 504 is produced as a three-thread over-lock stitch which provides a neatening on the edge of the fabric to eliminate fraying on the fabric edges. The second stitch type is stitch 401, providing the strength, elasticity and durability for the seam.

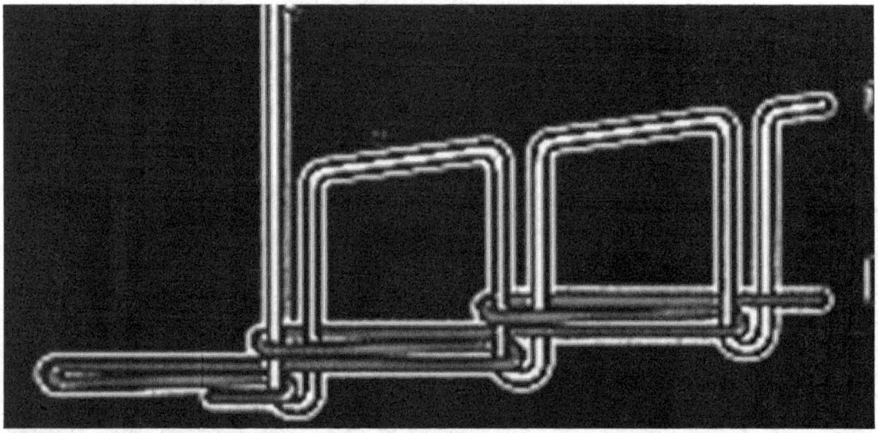

1.46 Two-thread chainstitch 401.

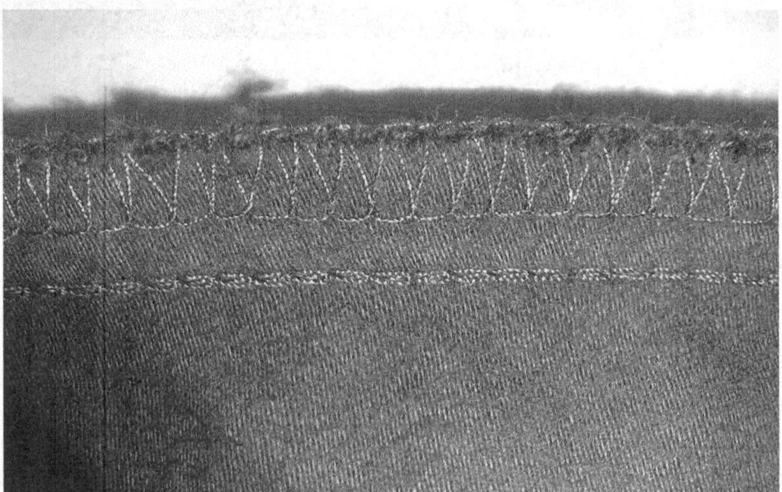

1.47 Example of 516 stitch type.

More detail of sewing and stitches of textile are given in Chapter 3. Many types of products consist of a combination of stitches and seams and some of these products are discussed below.

1.10 Stitching in practice: the case of high performance fabrics

As discussed earlier in the chapter, the properties of a fabric are dependent upon the properties of the fibre as well as the finishing treatments applied to the material. High performance fabrics are materials that are able to withstand extreme conditions. An example that was described above is Kevlar, which is a fibre derived from polyamide, used in the manufacture of ballistic protection garments for police and the armed services. Other high performance materials have been developed from fibres that have been around for thousands of years. The invention of synthetic fibres has expanded the use of these fabrics, creating materials that are very lightweight yet durable and very strong. This section discusses two examples of high performance fabrics and their joining requirements.

1.10.1 Ventile fabrics

The pedigree of this fabric goes back as far as the late 1930s, when war was threatening Europe. The British government was worried that there would be a shortage of flax which was principally used to make fire buckets and hoses. It was decided by the government that an alternative material was required; therefore research was commissioned into the use of cottons woven in such a way as to keep water in. A fabric was developed. Research continued and intensified during the Second World War as there was an urgent need to develop garments for pilots flying fighter escort missions over the sea. The fabric needed to be comfortable in the cockpit and be able to provide protection if they should ditch in the sea. Ventile fabrics that achieved this objective were developed by the Shirley Institute in Manchester. These garments could extend the life expectancy of a pilot from 2 minutes to 20 minutes, enabling a realistic chance of rescue.

Ventile fabrics for RAF clothing went into production in 1943. Garment designs have changed throughout the years but this material is still widely used by the RAF and as garments used in many outdoor activities. These fabrics are extremely durable and offer many years of protection and comfort. An example of a garment is given in Fig. 1.48.

Other end uses for ventile are:

- garments for hunting, fishing, climbing, hiking and skiing
- rainwear

1.48 A ventile garment as used by the RAF and for hiking.

Table 1.1 Fabric specifications of ventile materials

Qualities	L34	L28	L24	L19
Weight (gsm)	165–168	285–300	200–205	235–240
Threads/cm – Warp	95	71	81	66
Threads/cm – Weft	35	26	30	24
Hydrostatic head (mm)	750	900	750	750
Vapour permeability (%)	93–98			

- hats
- horse riding
- general garments for walking, etc.
- surgical gowns.

There are different grades/quality of ventile materials; these are given in Table 1.1.

The seams used in joining these garments together are double lapped seams (Fig. 1.42) which are one of the strongest seams used in garment manufacture. This type of seam consists of two or more plies lapped together secured with one or more rows of stitches. This provides a very strong seam with fabric edge protection often used on denim jeans, other denim products and tents, which will be discussed later. The machines commonly used for sewing this type of seam are feed-off-the-arm machines, using a double lap seam folder (Figs 1.49 and 1.50).

Needles recommended for sewing these fabrics are Schmetz SDI or SPI size 80s–90s depending upon the thickness of the seam. These needles

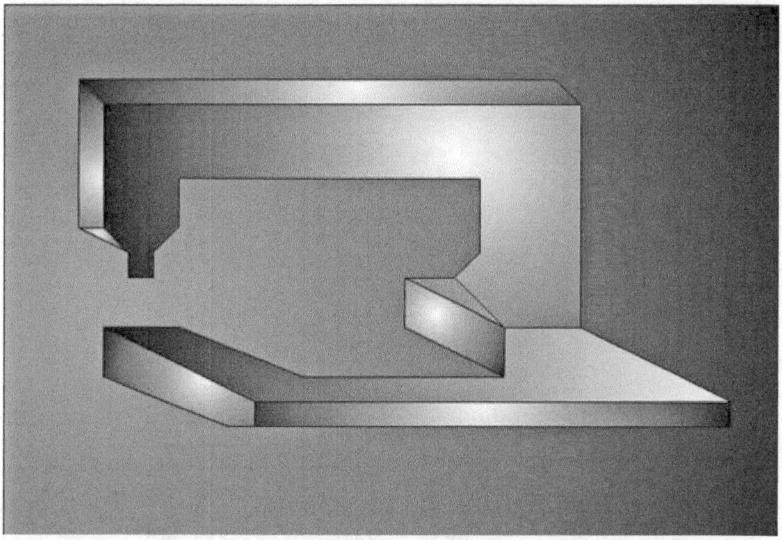

1.49 Feed-off-the-arm machine casting shape.

1.50 Feed-off-the-arm machine with double lap-seam folder.

have very acute angles and so minimise the amount of surface penetration, decreasing the amount of disruption to the fibres. An example of this point type is given in Fig. 1.51 compared to a round point needle in Fig. 1.52.

Sewing threads used for sewing ventile materials are core-spun polyester, size 100s–120s. These sewing threads are durable and maintain a high degree

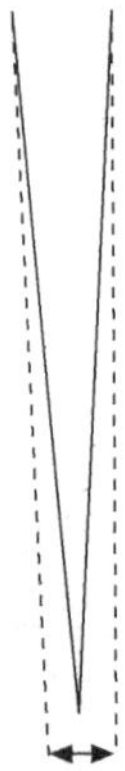

1.51 Acute round point needle.

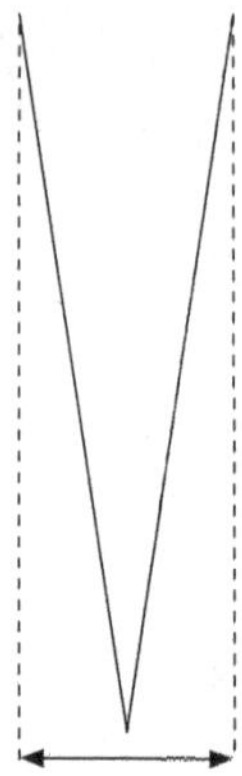

1.52 Normal round point needle.

1.53 Moleskin smock garment as used by leaders of East Lancashire Scouts.

of yarn strength to thread size ratio. Stitch density for these materials is given as 8–9 stitches per 2.5 cm.

1.10.2 High performance outerwear for hiking and tents

Most high performance outwear are plain woven fabrics using natural and synthetic fibres, namely cotton, polyester and nylon. Figure 1.53 shows an example of a garment worn by scout leaders and Fig. 1.54 gives an example of a garment used for outdoor pursuits, namely hiking and walking.

1.54 Garment with Gore-tex used for walking, hiking and trekking.

Modern synthetic garments use the material Gore-tex, which gives a garment the properties of being fully waterproof whilst at the same time enabling them to be breathable, improving the comfort to the wearer. Gore-Tex materials are typically based on thermo-mechanically expanded polytetrafluoroethylene and other fluoropolymer products. These materials are used in a wide variety of applications such as high performance fabrics, medical implants, filter media, insulation for wires and cables, gaskets and sealants. However, Gore-Tex is best known for its use in protective, yet breathable, rainwear.

Outerwear and in particular tents are susceptible not only to the elements of wind and rain but also to the degrading effects of sunlight, especially ultraviolet light (Fig. 1.55). This phenomenon affects products predominantly manufactured from nylon.

These products are usually joined using a combination of seams and stitches as mentioned above. The main stitch type used is stitch 301 and seams consist mainly of double lapped seams for strength on the outside of the fabric.

Tent fabrics use similar combinations of stitches and seams; the outer skin of the tent (the fly-sheet) uses the double lapped seam. Another seam extensively used in tents (on the ground sheet) is bound seams. This method of joining fabrics enables two or more fabrics to be joined together in one

1.55 Damage to a high performance tent fabric due to ultraviolet light.

1.56 Cotton canvas tent.

operation. Examples of tent designs and joining methods are given in Figs 1.56–1.59.

Tent fabrics also use a combination of taped and welded seams (Fig. 1.60) to guarantee that the tent will be fully waterproof. These are particularly applied to the groundsheet covering the stitches to prevent water from entering the bottom of the tent.

1.57 Mountain tent with a tunnel design.

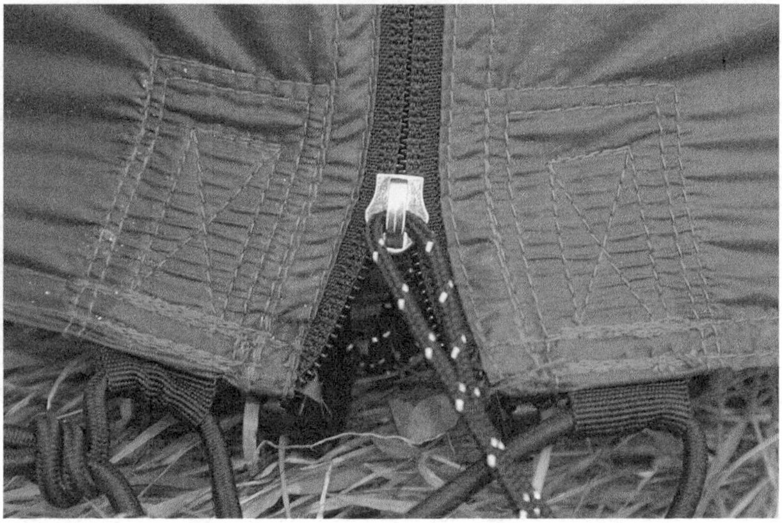

1.58 Reinforced flag stitching for anchoring the tent to the ground.

1.11 Alternative methods of joining fabrics: welded seams

Methods of joining textile materials to form garments other than sewing with needle and thread have existed for several decades and include bonding,

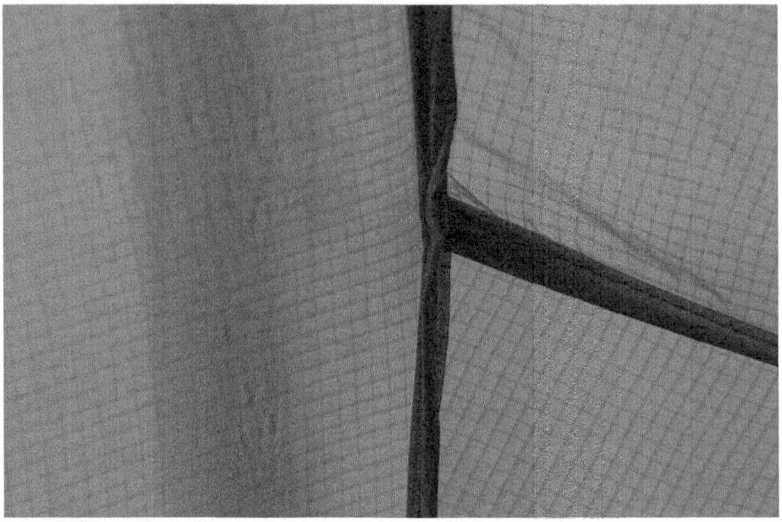

1.59 The flysheet of the tent joined together using bound seams as illustrated above.

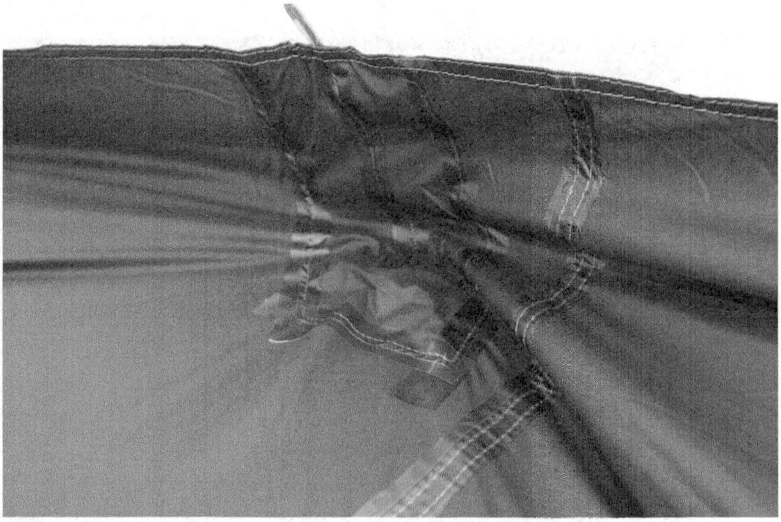

1.60 Taped seams over sewn seams.

welding or indeed seamless shaping. Until recently, however, the first and second of these approaches have been used sparingly in the creation of commercially viable garments (Cooklin *et al.*, 2006).

Developments in both production technology and textile materials have resulted in several brands of performance outdoor clothing offering garments predominantly constructed from welded seams with many beneficial claims being made about garment pliability and compliance, weight, waterproofing and strength. Initially welding was applied to garment manufacturing most effectively in the production of short seams or stampings (such as vents, button holes and eyelets) although continuous welding is capable of producing long seams and has seen recent acceptance as a method of joining garments.

Several methods of welding exist. In all methods, the thermoplastic material is joined by raising the temperature of the substrate to a level that melts the fabric at the interface of the two plies. Pressure applied in a controlled manner promotes the flow of one material into the other and upon cooling a weld bead is formed. Welding techniques include the following.

- Hot Air: A focused stream of heated compressed air is directed toward the join line of the interface between the fabric plies. A roller is placed directly behind this area to apply pressure once the material begins to flow and before/as it cools. The profile of this roller can define the weld impression.
- Hot wedge (thermal sealers): Electrically heated jaws are pressed into contact with the thermoplastic material for the duration required to transfer heat to the interface. The profile of the jaws dictates the weld impression.
- Dielectric welding (RF): Alternating waves from an electromagnetic generator excite the bipolar molecules of a dielectric material to such an extent that the molecular vibration causes friction between the molecules which results in the fabric interface melting. In seaming, the fabrics are placed between two electrodes, one live and the other a ground electrode.
- Ultrasonic welding: This is discussed in the next section.

1.12 Ultrasonic welding

Of the various welding techniques available, ultrasonic welding has become one of the most widely used for thermoplastics. Since they burn when heated, thermosets cannot be ultrasonically joined. Thermoplastics can be further categorised as amorphous or crystalline (Fig. 1.61). Amorphous resins exhibit random, spaghetti-like structure. They do not greatly dampen energy introduced into the material. As heat is applied, they soften and do not have a sharply defined melting temperature. Amorphous resins include ABS, acrylic, polycarbonate, polystyrene and polysulfone. Crystalline resins have an orderly pattern, like coiled springs. Just as metal springs dampen vibration, so do crystalline materials. They also have a well-defined melting temperature. Crystalline materials include acetal, nylon, polyester, polyethylene, polypropylene and polyphenylene sulfide.

Every ultrasonic unit contains the following five elements (Fig. 1.62):

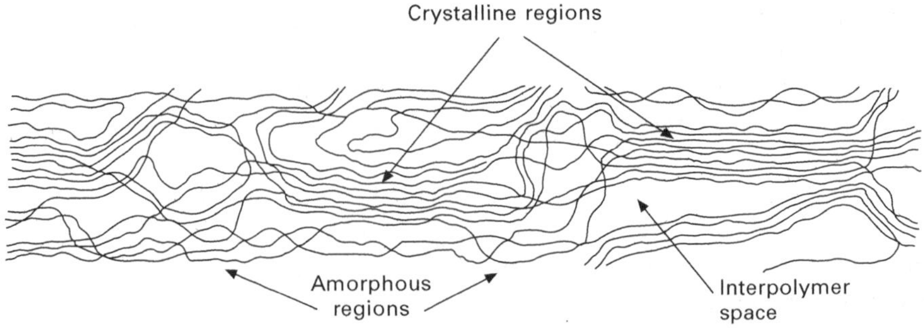

1.61 Polymer structure (source: Gohl and Vilensky, 1983).

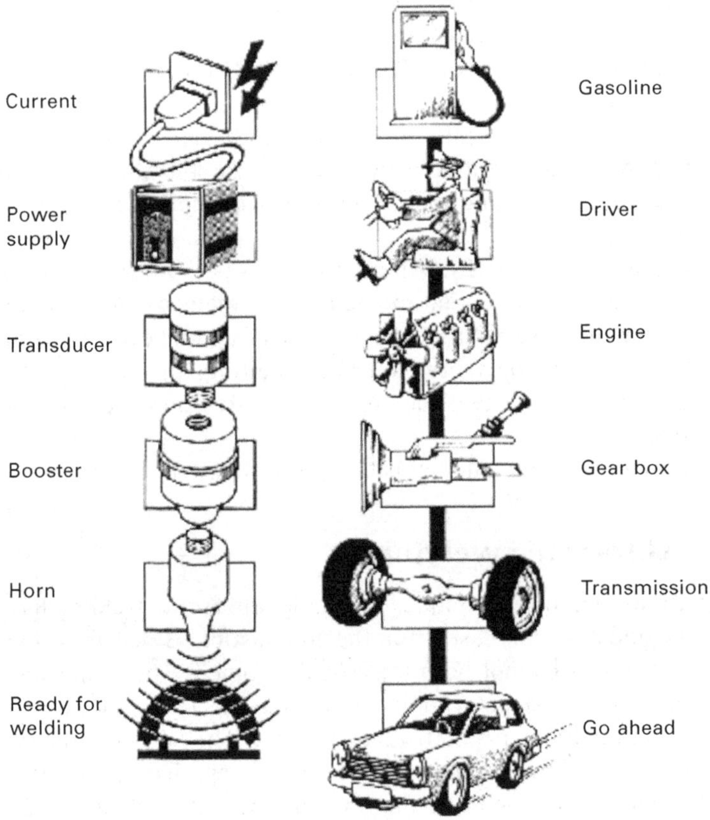

1.62 Welding elements (source: Rinco Ultrasonics).

- a power supply that takes line power at 50–60 cycles and changes it to a high ultrasonic frequency of 20,000–40,000 cycles per second (20–40 kHz)
- a converter or transducer that contains piezoelectric crystals that change

the incoming high-frequency electrical signal to mechanical vibration of the same frequency
- a booster that transmits the vibration energy and increases its amplitude (controllable)
- a horn to deliver the vibration energy by contact with the parts to be welded
- a pattern wheel to support the work piece.

There are two basic ultrasonic welding techniques: plunge welding and continuous welding. In plunge welding, the parts are placed under a tool or horn; the horn descends to the part under moderate pressure and the weld cycle is initiated. In the continuous welding process, the horn may 'scan' the part, or the material is passed over or under the horn on a continuous basis.

Basically, the following three parameters are essential for ultrasonic welding:

1. amplitude, measured in μm
2. pressure, measured in pa
3. weld time, measured in ms.

These three settings can vary and are determined for each application or part by trials. In order to be able to reproduce the once predetermined values, they are carefully registered. Because some areas to be welded can be very different in nature, geometry, material and surface, an experienced technician is usually needed to determine the settings of the machine through experimentation.

1.13 Conclusions

The development of textile materials is the oldest and most diverse industry in the world. The most basic materials originally designed to withstand cold and the environment have advanced to the point where we now have materials that are able to adapt to the environment. All the fabrics and products discussed use very similar combinations of stitches, seams and needles. More information on fabric sewability is given in Chapter 3. The science and experimentation into the development of textile fabrics and fabric joining methods looks set to continue for many years to come.

1.14 Acknowledgements

Special thanks to the Textile Institute for the pictures of the fibres and weave constructions. The work on welding and sewn seams was conducted with the technical assistance of Mr Neville Haynes and Mr Derek Hebdon.

1.15 References and further reading

Cooklin, G., Hayes, S. and McLoughlin, J. (2006), '*Introduction to Clothing Manufacture*', 2nd edn. Blackwell Publishing, Oxford.

Gohl, E.P.G. and Vilensky, L.D. (1983), '*Textile Science*'. Longman Cheshire, Melbourne.

Hayes, S.G. and McLoughlin, J. (2007), 'Welded and sewn seams: a comparative analysis of their mechanical behaviour', in Ariadurai, S.A. and Wimalaweera, W.A. (eds) *The Textile Institute 85th World Conference*, 1–3 May 2007, Colombo, pp. 131–142.

McLoughlin, J. (2000), Steps to improving quality at the needle point, *Industrial Focus*, October, Mauritius.

McLoughlin, J. (2008a) 'Technological advances in sewing garments', in Fairhurst, C. (ed.), *Advances in Apparel Production*, Woodhead Publishing, Cambridge, ch. 10.

McLoughlin, J. (2008b), 'Automated fabric inspection', in Fairhurst, C. (ed.), *Advances in Apparel Production*. Woodhead Publishing, Cambridge, ch. 13.

McLoughlin, J. and Hayes, S.G. (2008), 'An analysis on the relationship between sewing parameters and fabric parameters and of their impact on seam quality (part 1)', in: *The Textile Institute 86th World Conference*, 18–21 November 2008, Hong Kong.

McLoughlin, J., Sabir, T. and Hayes, S. (2010), 'Fabric parameter mapping for seam sewability', *International Journal of Fashion Design, Technology and Education*, 3(2), 77–88.

Tyler, D.J. (2008), *Carr & Latham's Technology of Clothing Manufacture*. Blackwell Publishing, Oxford.

Part I

Sewing technology

2
The mechanics of stitching

G. K. STYLIOS, Heriot-Watt University, UK

DOI: 10.1533/9780857093967.1.47

Abstract: An explanation of stitch formation and the interaction of the components and variables that constitute a 'good seam' are considered. It is shown how changes to those variables can affect the stitch equilibrium from a flat stable appearance to an unstable manifestation resulting in unbalanced seams, differential feed, bedding-in and seam puckering. This analysis considers the mechanics of the sewn assembly during and after stitching, highlighting the importance of fabric and thread properties and their relationships with the sewing foot pressure, feed dog and the relevant frictional forces during sewing and with sewing thread and its recovery after sewing. The important role of fabric behaviour determined by its bending, shear and other properties provides an explanation of sewing faults and how to eliminate them.

Key words: seam, stitch, sewing equilibrium, stitch formation, friction, fabric properties, differential feeding, seam pucker, dry relaxation, thread tension, bedding-in, load extension.

2.1 Introduction

The stitching formation for joining together two or more pieces of fabric, under any seam type, is a complicated process involving many components. During stitching these components react with each other under the influence of operator handling and at varied speed, rendering the process dynamic. Any variation of these factors will influence the feeding of the fabric, the forces of the machine mechanisms acting on the thread and fabric coupled with the subjective handling of the machine operator.

Many researchers have dealt with different stitch types; some have focused on the sewing needle,[1,2] the thread,[3–5] or the fabric[6–10] and others on the sewing machine mechanisms[11,12] with considerable success. Here an attempt will be made to establish a generic explanation of 'a good stitch' by showing that when the variables that constitute the formation of the stitch are at an optimum, the result of joining is producing a seam of good appearance. And that when one or more of these variables changes the desirable equilibrium, they ought to be adjusted to achieve a functional stitch assembly of good and undeformed appearance.

Let us therefore consider the formation of stitching and the factors that influence the stability of a stitch by examining the most popular of the stitches:

a single needle lockstitch. The manifestations of the instabilities that we are considering are the known problems associated with sewing deformations along the seam line, e.g., unbalanced seams, skip stitches, differential feed, bedding-in stitches and seam pucker, which are unacceptable in garment construction for fashion and retailing.

2.2 The principles of stitching

Let us examine each stitching component before we relate them together during the formation of a stitch.

'The stitch' has survived since the Stone Age[13] and is achieved by a needle and thread because we have not found a better method of joining two fabrics together that gives the wearer comfort of movement, durability and good aesthetics. The three primary stitching components are the sewing thread, the sewing needle and, of course, the fabric. Because we are increasingly moving away from cut and sew in knitted fabrics and because, even for knitted fabrics that are still being stitched, stitching is not considered as the main part in the garment, we shall concentrate on woven cloths. The sewing machine which has survived for centuries along with the handling of the operator provide the means of manufacturing and may be considered as secondary, not because they are unimportant, but because the stitch can be made by hand and without the machine, and that in some operations the handling is completely removed by automation. The sewing machine variables, as we call them, are the sewing thread tensions, the foot pressure, the feed dog height, along with the appropriate fitting of the needle size and shape and the type of feed dog. The speed of stitching should be as stable as possible, but piece working and productivity make this impossible to achieve in manufacturing. Speed fluctuations unfortunately affect the settings of sewing machine variables because they are made prior to stitching, named 'static settings'. Stitching, however, is a dynamic operation and needs to be compensated or adapted as machine speed fluctuations may affect the 'equilibrium' and this can cause additional difficulties. We will introduce some dynamic aspects of stitching later in this discussion. The sewing needle size and shape is an important aspect but has been dealt with elsewhere.[14]

2.2.1 The concept of sewing equilibrium and its mechanism

Let us consider two plain weave fabric pieces, fabrics A and B, being stitched by a single needle lockstitch sewing machine, as shown in Fig. 2.1, lockstitch being the most common and simple stitch found in garment manufacture, and let us try to express this basic stitching operation graphically by introducing simple mathematical relationships.

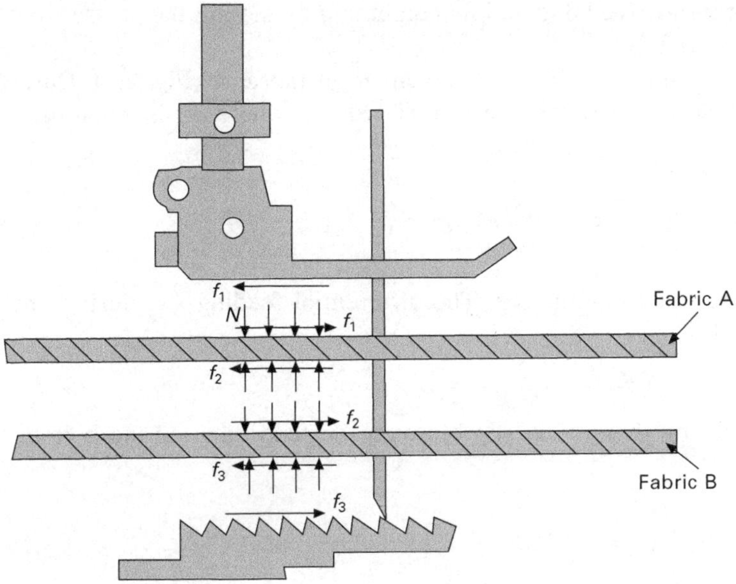

2.1 Single needle lockstitch sewing of fabrics A and B.

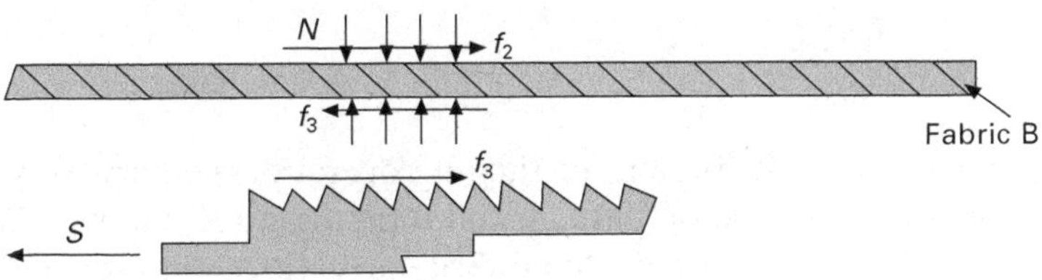

2.2 Movement of fabric B.

During stitching

During sewing, both fabrics A and B move forward by the action of the feed dog. Fabric B is directly driven forward by a frictional force f_3 acting between the feed dog teeth and fabric B (Fig. 2.2). Fabric A is moved forward by a frictional force f_2 acting between fabric A and fabric B, whilst f_1 is the frictional force between foot pressure and fabric A which prevents the top fabric A from moving forward.

Let's look at the movement of fabric B during any given time dt,

$$dl_B = Sdt - dl_{B3}$$

where $dl_{B3} \propto f_3$.

When k_3 is high enough, dl_{B3} is very small, which means that the grip is good and there is no slip between the feed dog and the bottom fabric B. In most cases this is true and mechanics try to achieve this by using the appropriate feed dog and its height and by setting the presser foot pressure appropriately.

Let's now consider the movement of fabric A (Fig. 2.3). During a given time dt, fabric A has moved forward dl_A, where dl_A is given by:

$$
\begin{aligned}
dl_A &= S\,dt \\
&= S\left(1 - \frac{cSk_1}{Nk_2}\right) dt
\end{aligned}
$$

where c is in N/mm/sec. The differential feeding $\Delta\ell_3$ during one stitch of length ℓ is

$$\Delta\ell_3 = \ell_B - \ell_A$$

supposing there is no slip between the feed dog and the bottom fabric B, i.e.

$$
\begin{aligned}
\ell_3 &= \int_{t_1}^{\ell/s} S\,dt - \int_{t_1}^{\ell/s} S\left(1 - \frac{cSk_1}{Nk_2}\right) dt \\
&= \int_{0}^{\ell/s} \frac{ck_1S^2}{Nk_2}\,dt \qquad [2.1] \\
&= \frac{cSk_1\ell}{Nk_2}
\end{aligned}
$$

It is important to have friction in stitching operations, otherwise we would not be able to feed the fabric through and form the stitch, but its relationship and the reaching of what we call 'equilibrium' is critical for producing a good seam. From Eq. [2.1] we can see that the differential feeding $\Delta\ell_3$ decreases as the frictional coefficient between fabrics A and B increases, we can see that it increases with the frictional coefficient between the presser foot and top fabric A, and with an increase of the stitch length and the speed of the machine, and that it decreases by increasing of the pressure of the foot.

If this 'equilibrium' is not achieved when the sewing machine is set up prior to starting the stitch formation, and if it is not compensated by operator handling and/or other sewing aids, potentially and depending on the fabric behavioural properties, will manifest itself as a sewing fault, usually in a

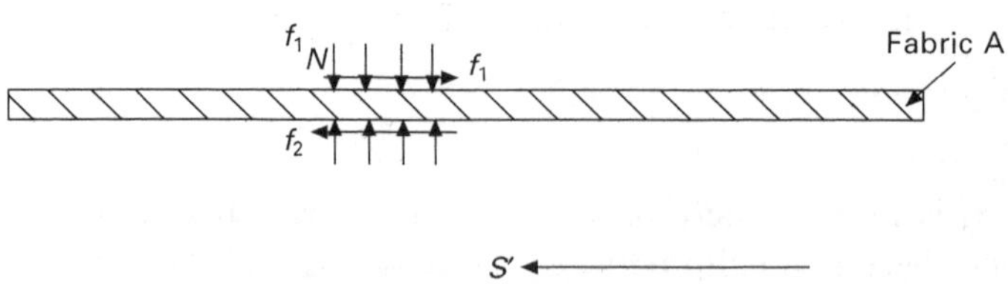

2.3 Movement of fabric A.

deformed and wrinkled seam. Hence during the sewing process, due to the foot pressure and friction forces among the fabrics and sewing machine, both fabrics A and B deform to a certain degree which will depend on the properties of both fabrics.

After stitching

Having performed the sewing operation, and having stitched our fabrics A and B, let us examine the behaviour of the stitching assembly by considering one single stitch along the sewing line.

2.2.2 Differential feeding

The deformation of fabric A, the top fabric in a stitch, and the forces acted upon the different stitching components before and after fabric deformation are shown in Figs 2.4 and 2.5 respectively, where

$$\ell_{\mathrm{A}} = \frac{f_1}{E_{\mathrm{TA}}} + \frac{h_{\mathrm{A}}N}{E_{\mathrm{CA}}}$$

$$= N\ \frac{k_1}{E_{\mathrm{TA}}} + \frac{h_{\mathrm{A}}}{E_{\mathrm{CA}}}$$

in one stitch length $\ell - \ell' + \Delta\ell_A$.

Similar relationships apply to fabric B, the bottom fabric in a stitch, as shown in Fig. 2.6, where

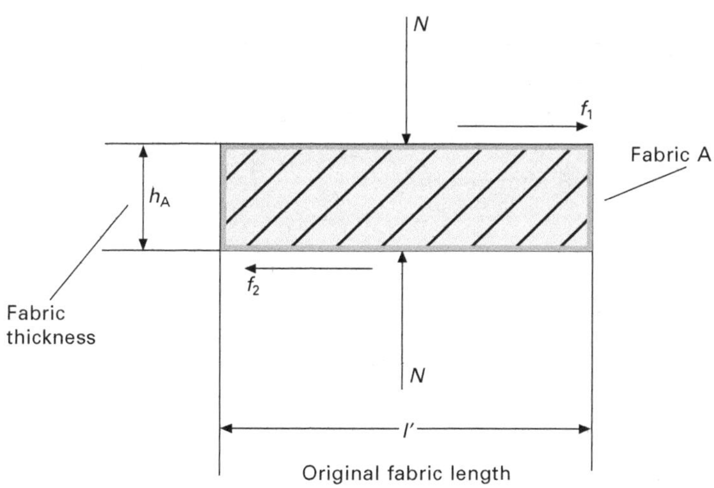

2.4 Fabric A before deformation.

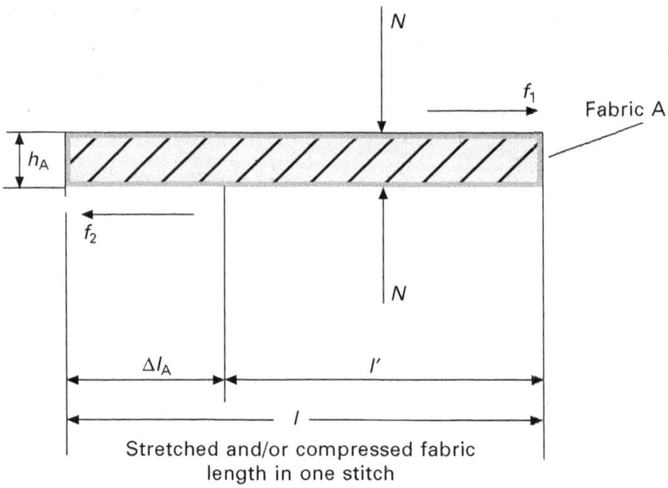

2.5 Fabric A after deformation.

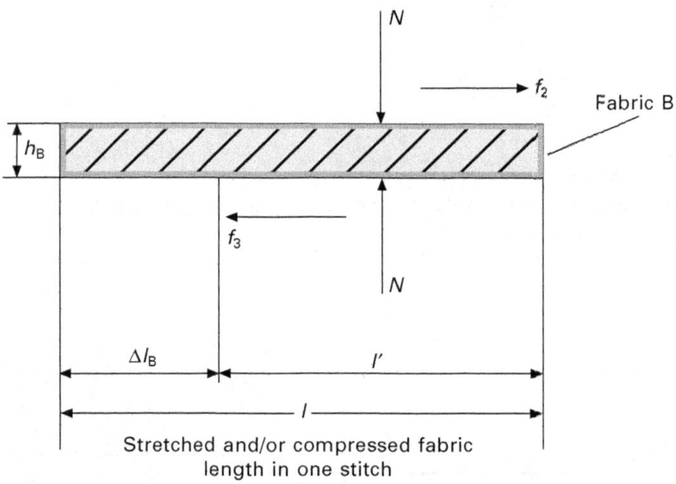

2.6 Fabric B after deformation.

$$\ell_B = \frac{f_2}{E_{TB}} + \frac{h_B N}{E_{CB}} - Ck_3$$

$$= N \; \frac{k_1}{E_{TB}} + \frac{h_B}{E_{CB}} \; - Ck_3$$

If Ck_3 is a frictional representation of the resistance of the bottom fabric B to deformation, then the differential deformation of fabrics A and B is as follows:

$$\ell_{AB} = \ell_A - \ell_B$$
$$= N\left(\frac{k_1}{E_{TA}} + \frac{h_A}{E_{CA}} - \frac{k_2}{E_{TB}} - \frac{h_B}{E_{CB}}\right) + Ck_3$$

Consequently, the total differential displacement deformation of fabric A and fabric B is

$$\ell = \ell_{AB} + \ell_3$$
$$= N\left(\frac{k_1}{E_{TA}} + \frac{h_A}{E_{CA}} - \frac{k_2}{E_{TB}} - \frac{h_B}{E_{CB}}\right) + Ck_3 + \frac{cSk_1\ell}{Nk_2} \quad [2.2]$$

It is evident from Eq. [2.2] that the higher the differential displacement of fabric A and fabric B, the more prone the stitch will be to becoming unstable and hence displacing the seam and/or deforming it out of the flat state and causing faults such as seam pucker. These simple relationships point towards fabric properties such as surface friction, fabric tensile and compression properties and thickness, coupled with the sewing machine's foot pressure and sewing speed.

2.2.3 Dry relaxation

After leaving the sewing machine and on completion of the sewing operation, the seam will relax. The relaxation force will depend on the magnitude of the displacement deformation of fabrics A and B as shown in Figs 2.7 and 2.8, and/or of the effect of sewing thread tensions, as we will see later. The

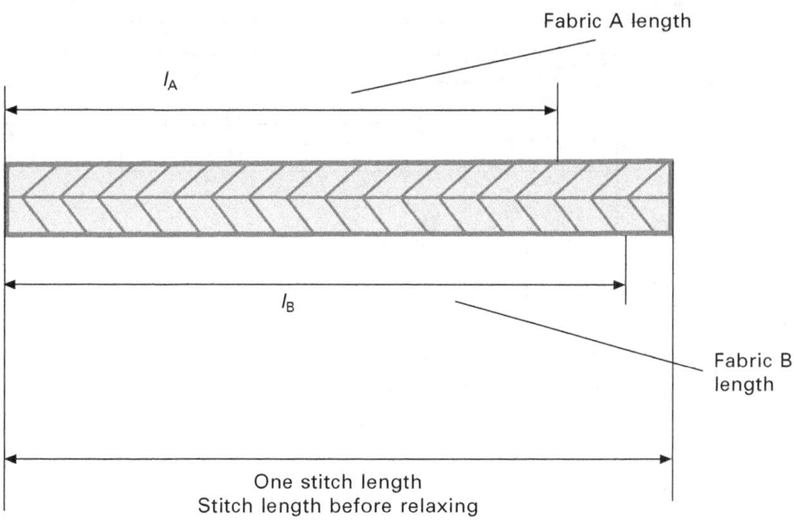

2.7 Displacement deformation of fabric A.

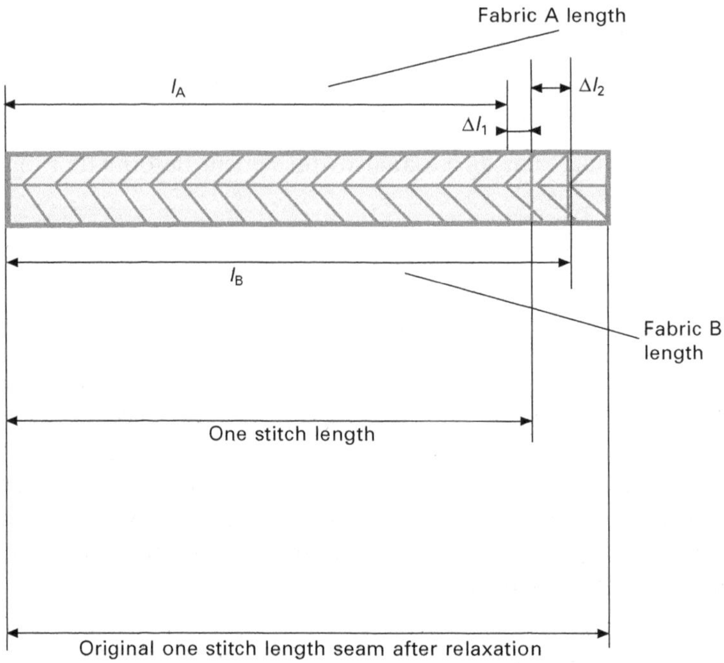

2.8 Displacement deformation of fabric B.

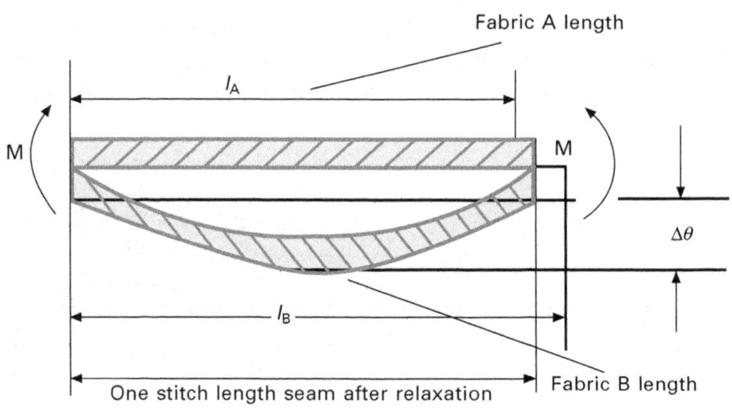

2.9 Deformation of fabric B due to curling couple M.

deformation of the bottom fabric B will depend on the curling couple M, Fig. 2.9, generated by the relaxation force and the resistance to curling, which are dependent on the bending rigidity of the fabric. Similarly the deformation of top fabric A due to the recovery of the needle thread after relaxation will depend on its resistance to curling.

Consequently, considering the above relationships we can construct the following equation:

$$\mu \frac{M}{B_B}$$

$$\mu \frac{\ell_1}{B_B} \mu \frac{\ell_2}{B_B} \qquad [2.3]$$

$$\frac{\ell_2}{B_B} = \frac{1}{B_B} \; N \; \frac{k_1}{E_{TA}} + \frac{h_A}{E_{CA}} - \frac{k_1}{E_{TB}} - \frac{h_B}{E_{CB}} \; + \; \ell_3 + C \, \frac{ck_1 \ell S}{k_2 N}$$

which may give us the equilibrium state that we need for a flat stitch. Equation [2.3] establishes the importance of fabric bending behaviour, saying that the higher the bending rigidity B_B of the fabric the smaller the deformation, verifying what we already experience in practice with very thin, delicate fabrics made of microfibers (shingosen in Japanese[15]) and the nanofabrics which are under development.[16] However, Eq. [2.3] also shows that if the fabric cannot withstand the forces acted upon it during the sewing operation, it will move away from a stable flat configuration to an unstable faulty manifestation such as of seam pucker. A particular case of such a fabric that is difficult to control is a structurally jamming one in which its constituent components in terms of yarn and thread diameters, stitch and yarn densities are governed by the following relationship:

$$d_s \quad \frac{1}{2n} \quad 1 - \frac{2d}{p}$$

where d_s is sewing thread diameter, n is number of stitches per length unit, d is yarn diameter and p is yarn spacing. In such cases, jamming and deformation manifestations such as seam pucker very much depend on the relative bending rigidities of the fabric and thread in a unit of stitch.

2.2.4 The effect of thread tension

The effect of thread tension on seam performance is paramount and this is another consideration which creates difficulties. This has led thread makers to develop sewing threads with low shrinkage and interlining companies to stabilise seams so that when the thread tries to recover and shrinks in the stitch, the fabric is locally stiffened so that it does not alter its flat state of equilibrium, hence improving the resistance to fabric curling in the seam only. Figure 2.10 shows this case of equilibrium, in which when the sewing thread either in the needle and/or in the bobbin T_n (or T_b) is high; its elongation $\Delta\ell_T$ is governed as follows. For the needle thread T_n

$$\ell_T = \frac{T_n}{E_T}$$

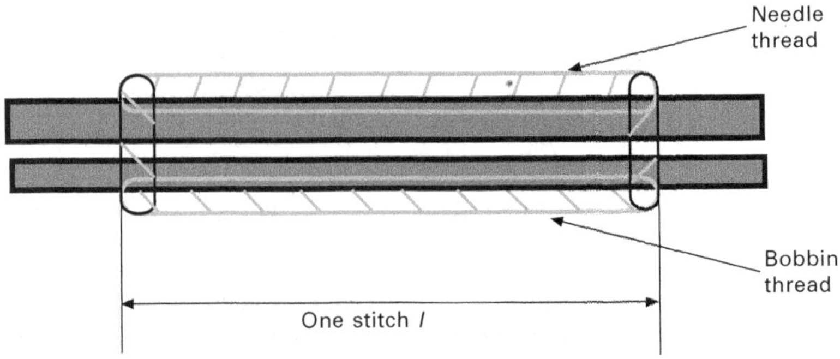

2.10 Thread tension equilibrium in a stitch, where *T*n and *T*b is the needle and bobbin thread tensions respectively.

For the bobbin thread T_b

$$\ell_T = \frac{T_b}{E_T}$$

or

$$L_T = \frac{T_n}{E_T} + \frac{T_b}{E_T}$$

as may be the case with the lockstitch stitch formation in which the combined thread elongation ΔL_T usually acts on top fabric A since the needle thread has to overcome the tension of the bobbin thread, by pulling the thread from the bobbin, to make the stitch. Unbalanced stitching is common where either the needle or the bobbin thread tensions are excessive and not equal, manifesting in the stitch not being set inside and in the mid-point of the body of the fabric assembly.

Extensibility of sewing threads and seam fault manifestation

Let us further illustrate the importance of thread elongation in sewing. During lockstitch formation the needle thread pulls the bobbin thread into the fabric. Thus the tension of the needle thread must be at least equal to the tension of the bobbin thread plus the frictional forces on both threads in order to produce a balanced seam. This means that certain minimum thread tensions are unavoidable. Let us now examine the extensibility of typical commercial spun polyester and polyester (61%)/cotton (31%) core spun sewing threads (T-180, T-120, T-100 and T-80) immediately after sewing and at dry recovery after 48 hours in standard atmospheric conditions, to highlight this condition. Figure 2.11 shows that there is a significant percentage of thread extension, averaging 5.7% to 9.6% immediately after sewing and 1.5% to 2.0% after

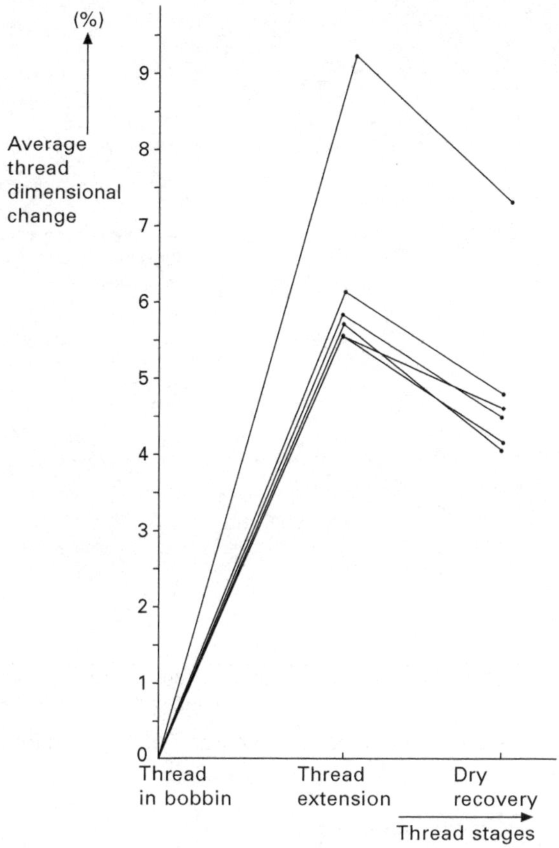

2.11 Average load-recovery of commercial sewing threads.

dry recovery. This suggests that after sewing, the thread may produce further seam deformation due to trying to recover from extension. The magnitude of this deformation will depend on the properties of the fabric.

This can be verified when the load extension of every sewing thread is examined according to British Standard 1932: Part 1:1965.[17] All mentioned threads have shown similar behaviours at loads less than 5 N. Figure 2.12 highlights this by showing a typical load extension diagram of T-120 100% spun polyester thread indicating that significant extension occurs even at low sewing speeds and, even when the load is as low as 2 N, the extension is more than 20%.

Consequently the higher the thread tension either of the needle and/or the bobbin, the higher the thread elongation, which will tend to compress the fabric, if the fabric has low bending stiffness, and will deform the fabric in the stitch during relaxation and especially after laundering. This case where stitches are bed-in rather that lay-on (Fig. 2.13) should be avoided. Again to

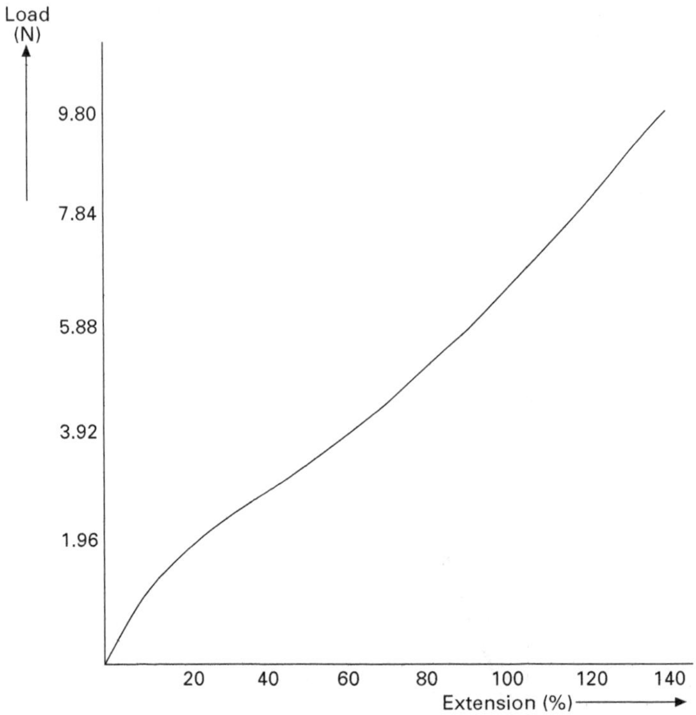

2.12 Load–extension of 100% spun polyester T-120 commercial sewing thread.

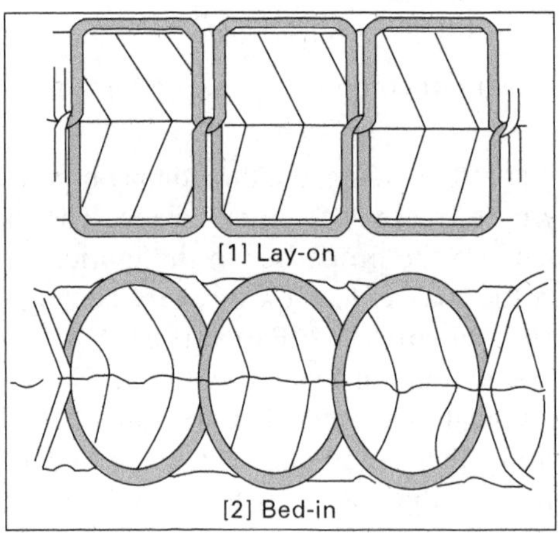

2.13 Representation of bedding-in and lay-on stitches.

achieve an undeformed seam of that appearance, the bending behaviour of the fabric is very important and the mechanism of this kind of deformation is shown in Fig. 2.14.

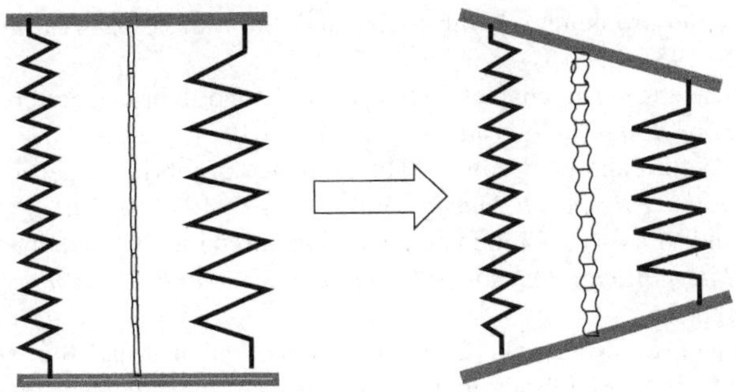

2.14 The importance of fabric stiffness in seam pucker.

2.3 Conclusion

Even in cases where you have the optimum grip on the fabrics during feeding, the minimum practically achievable thread tension and the best suited sewing thread, a fabric may have such a low bending stiffness and other properties that, whatever the remedy during sewing, the equilibrium will never be reached and hence a deformation or other fault will occur, usually of a seam pucker type. The case of non-iron shirts, considered in Chapter 15 of this book, is an important commercial application, highlighting the remedy of such difficulties and it illustrates very clearly 'the mechanics of stitching' presented in this discussion.

Whichever method we decide on to join textile fabrics, the interactions with the fabric mechanics are paramount, and if the industry wants to numerically automate the making up of garments, it must be able to define the unique behaviour of every fabric in relation to joining process parameters. 'Whether mass customisation or mass production, the character of the fabric always holds the label.'

2.4 References

1. S. Simmons, 'An analysis of forces in a fabric-needle sewing system', *Clothing Research Journal*, Vol. 7, No. 2, 1979, p. 51.
2. R.A. Khan, S.P. Hersh and L.P. Grady, 'Simulation of needle-fabric interactions in sewing operations', *Textile Research Journal*, Vol. 40, No. 6, 1970, p. 489.
3. G. Stylios and D.W. Lloyd, 'Prediction of seam pucker in garments by measuring fabric and thread mechanical properties and geometrical relationships', *International Journal of Clothing Science and Technology*, Vol. 2, No. 1, 1990, p. 6.
4. I. Ajiki and R. Postle, 'Viscoelastic properties of threads before and after sewing', *International Journal of Clothing Science and Technology*, Vol. 15, No. 1, 2003, p. 16.
5. M. Mori and M. Niwa, 'Investigation of the performance of sewing thread', *International Journal of Clothing Science and Technology*, 1994, Vol. 6, No. 2/3, 1994, p. 20.

6. L.H. Scott, 'Some problems relating to sewing', *Journal of the Textile Institute*, Vol. 42, No. 8, 1951, p. 653.
7. M. Cednas, 'The shrinkage characteristics of fabric and their effect upon garment pucker', *Clothing Research Journal*, Vol. 2, No. 3, 1974, p. 122.
8. E. Rosenblad-Wallin and M. Cednas, 'The influence of fabric properties on seam puckering', *Clothing Research Journal*, Vol. 1, No. 3, 1973, p. 20.
9. S. Kawabata and M. Niwa, 'Clothing engineering based on objective measurement technology', *International Journal of Clothing Science and Technology*, Vol. 10, No. 3/4, 1998, p. 263.
10. A.G. De Boos and A.F. Roczniok, 'Communications: "engineering" the extensibility and formability of wool fabrics to improve garment appearance', *International Journal of Clothing Science and Technology*, Vol. 8, No. 5, 1996, p. 51.
11. F.B.N. Ferreira, S.C. Harlock and P. Grosberg, 'A study of thread tensions on a lockstitch sewing machine (Part II)', *International Journal of Clothing Science and Technology*, Vol. 6, No. 5, 1994, p. 26.
12. D. Rogale, 'Garment sewing processing parameters: determination using numerical methods and computers', *International Journal of Clothing Science and Technology*, Vol. 7, No. 2/3, 1995, p. 56.
13. B. Odar, 'A Dufour Bladelet from Potočka zijalka (Slovenia)', *Arheološki vestnik*, Vol. 59, No. 13, 2008, p. 9.
14. G. Stylios and Y.M. Xu, 'An investigation of the penetration force profile of the sewing machine needle point', *Journal of the Textile Institute*, Vol. 86, No. 1, 1995, p. 148.
15. K. Yamazaki and M. Okamoto, 'A review of new directions in shingosen and synthetic-fiber textiles for sportswear', *Journal of the Textile Institute*, Vol. 88, No. 3, 1997; Special Issue: Issue 3, Parts 1 and 3.
16. M.B. Bazbouz and G.K. Stylios, 'A new mechanism for the electrospinning of nano yarns', *Journal of Applied Polymer Science*, Vol. 124, No. 1, 2012, p. 195.
17. British Standard 1932: Part 1: 1965, 'Determination of breaking load and extension', BS Handbook 11: 1974, Section 3/32 yarns.

2.5 Appendix: nomenclature and notation

E_T	fabric elastic modulus
E_C	fabric compression modulus
B_B	fabric bending rigidity
T	thread tension
$\Delta\ell_T$	thread elongation
S	feeding speed (sewing speed)
N	foot pressure
fabric A	top fabric
fabric B	bottom fabric
Ck_3	frictional resistance of the bottom fabric B to deformation
f_1	friction between presser foot and fabric A
f_2	friction between fabric A and fabric B
f_3	friction between fabric B and feed dog
k_1	friction coefficient between presser foot and fabric A

k_2	friction coefficient between fabric A and fabric B
k_3	friction coefficient between fabric B and feed dog
$\Delta\ell_1$	fabric A elongation after relaxing
$\Delta\ell_2$	fabric B compression after relaxing
$\Delta\ell_3$	differential feeding between top and bottom fabrics mainly due to the low friction coefficient between fabrics

3
The sewing of textiles

S. HAYES and J. MCLOUGHLIN, Manchester Metropolitan University, UK

DOI: 10.1533/9780857093967.1.62

Abstract: This chapter describes the techniques and intricacies used in the production of sewn products. It also explains the stitches and seams used in the production of apparel and focuses particularly on the lockstitch machine to give an intimate and detailed explanation of the settings for the machine to produce high-quality products. Stitch types, different machine usage and product quality are also discussed in detail. The chapter also contains checklists of problems that can occur whilst sewing garments and how to rectify these problems.

Key words: sewn products, clothing technology, clothing machine technology, stitching, sewing textiles, textiles, clothing engineering.

3.1 Introduction

The performance of a garment is affected by the quality of fabrics used in its manufacture as well as factors determined by the technology of the garment manufacturing process. These can involve some lengthy and complex processes. When joining fabrics together, there are a number of factors that need to be taken into account. These include:

- the machine settings
- needle size and point type
- stitch type
- the category of the seam to be used
- the properties of sewing thread
- operator handling
- fabric properties and parameters.

All of these factors are equally important in producing a quality seam. If one or more of them are incorrect, a poorly sewn and finished garment can result. Therefore well-organised companies have assessment structures in place in order to minimise production downtime. Garment technologists, quality managers and clothing machine engineers have a contributing role in minimising or eliminating production problems (McLoughlin, 1998, 1999, 2000), but in order to ensure a smooth production process, an understanding of these factors has to be achieved.

This chapter describes the techniques used in the production of sewn products. It also explains the stitches and seams used in the production of apparel and focuses particularly on the lockstitch machine to give an intimate and detailed explanation of the settings for the machine to produce high-quality products. Stitch types, different machine usage and product quality are also discussed in detail. The chapter also contains checklists of problems that can occur whilst sewing garments and how to rectify these problems.

3.2 Types of sewing machine

Early commercial sewing machines were made from wood. In 1790 Thomas Saint applied for a patent to build a wooden industrial sewing machine producing the chainstitch. Whilst the stitches invented hundreds of years ago from chain- to lockstitches remain the same today, the engineering has changed, enabling wooden machines to be replaced with metal ones and the introduction of modern day electronics. The introduction of pneumatics and hydraulics has also created much higher production speeds.

Machines are usually categorised into four main types. These are:

- the basic sewing machine
- the mechanised sewing machine
- semi-automatic machines
- automatic transfer lines.

Mobile hanging rail systems (Fig. 3.1) are examples of mechanised and automated systems. These types of systems have been known to be inflexible due to machine breakdown or situations that can create a bottleneck in production such as misplaced production items.

It is important to use the correct machine for the type of operation on the garment. Hundreds of machines have been developed for many different and varied operations in the production of apparel to tents, tarpaulins and bouncy castles. The machine bed casting, feed system and stitch formation can be applied to many different types of sewn product operations. Using the correct machine for the operation is important in order to produce the required amount of production and also the desired quality of seam. Chain-stitch machines, for example, are extensively used in industry for apparel applications such as:

- side seams on garments jeans, shirts, blouses, etc.
- lingerie
- high performance garments and tents
- trousers, suits and skirts
- hems on fleeces and knitwear plus many more.

The most common of all machines in industry, the industrial lockstitch

3.1 Hanging rail production system.

machine, can have numerous machine bed types, different feeding systems and different derivations of lockstitches associated with it. Using these variables, lockstitch machines have a particularly wide range of application. Lockstitch machines used for the production of apparel can be compared to the same type of machine used for sewing larger pieces of material such as homeware, automotive and outdoor products. Examples of these machine types are given in Figs 3.2–3.4.

As has been noted, the most used machine in industry is the basic ISU (integrated sewing unit) lockstitch sewing machine, commonly called a flat machine in the industry (see Figs 3.5 and 3.6). This type of machine is the most prolific and widely used machine in the world. Most companies manufacturing apparel and many other sewn products will have this type of machine in their production units. It is this for this reason that this machine and the settings that influence the seam quality are explained in the greatest detail here. The ISU machine is a fully integrated unit that usually consists of:

- automatic thread trimmers, enabling the threads to be cut without using scissors

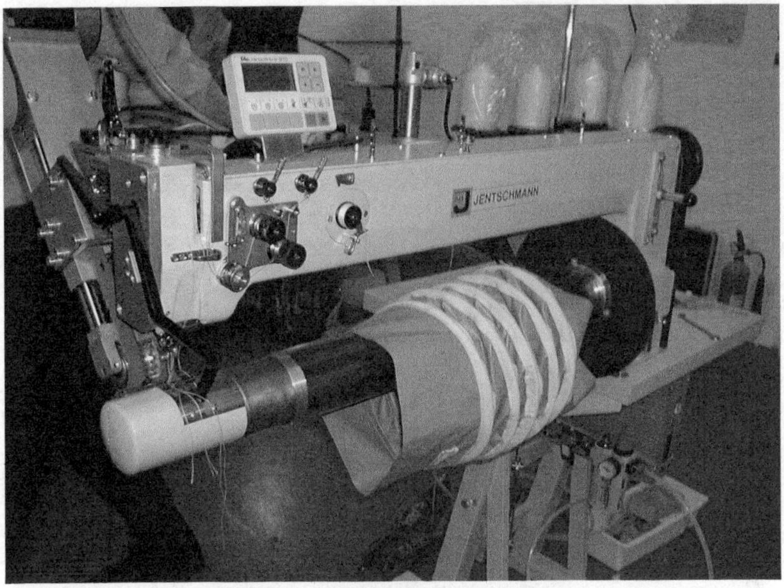

3.2 Long arm cylinder arm lockstitch machine.

3.3 Long arm lockstitch machine for curtains and bedding.

- automatic presser foot lift
- automatic back tacking
- stitch counting and ply sensing devices.

The machine described in detail in this chapter is a Juki flatbed (Fig. 3.7), drop feed, lockstitch (stitch type 301). The lockstitch type 301 is the commonest

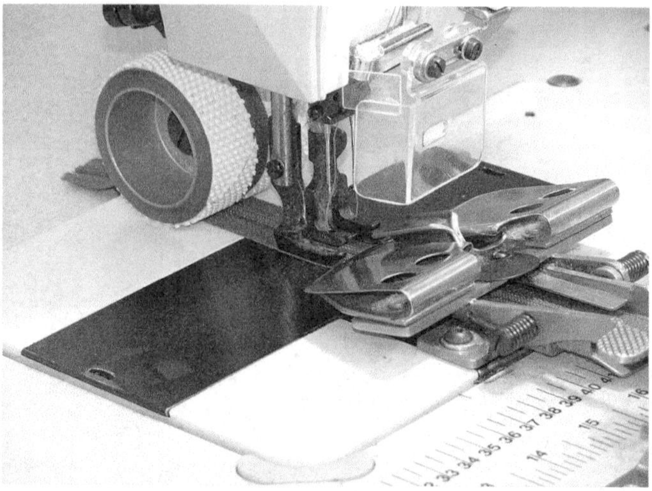

3.4 Lockstitch with puller and unison feed.

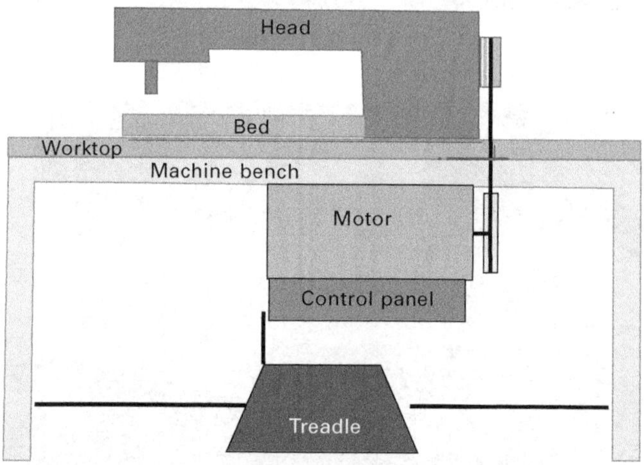

3.5 Schematic diagram of basic machine with all components.

of all the stitch types of all the machines used in the clothing industry. It is often referred to as a double lockstitch due to the way it locks together inside the material. This stitch type is formed by the interlacing of a needle thread supply with the bobbin thread supply underneath the machine bed. These stitches are very secure, as a break in one stitch will not cause the seam to completely unravel, although it will compromise the overall seam performance.

A common problem with the lockstitch machine is that the formation of the stitches within the material causes the yarns inside the fabric to be

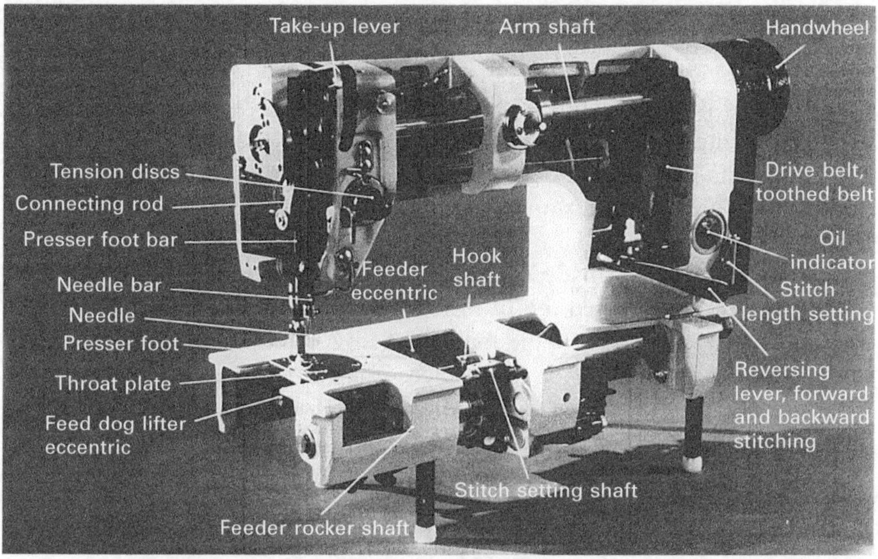

3.6 Diagram of a lockstitch sewing head.

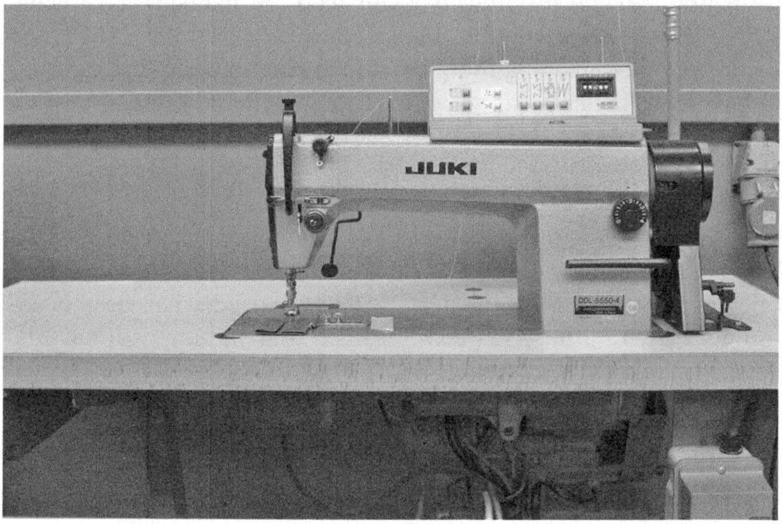

3.7 Juki single needle ISU lockstitch.

displaced. This phenomenon of displacement often causes seam pucker, a gathering of the material to produce an uneven, crinkled or distorted seam. Explanations of seam pucker are given in Section 3.10.

3.3 Machine feeding systems

Since they are a particularly important part of the machine, feeding systems are discussed in more detail in this section. There have been many feeding systems invented for the sewing of textile products. The principal role of the feeding mechanism is to move the material from one stitch position to the next over the prescribed distance. In doing so the material must be controlled precisely under the minimum pressure.

The feed system is made up of the throat plate, presser foot and feed dog. They are often called the fittings because they fit together. The throat plate is designed to support the material being stitched and allow the easy passage of the material over its surface. The throat plate is manufactured with a needle hole to allow the needle to penetrate the fabric. The needle hole must be large enough for the needle and the sewing thread and there are slots in the throat plate that allow the feeder to rise and feed the material. The slots should match the width of the rows of teeth on the feed dog without allowing contact.

There are various feed mechanisms and these have evolved over the last 40 years to encompass a wide range for many different types of operations on a product. In order to feed the material correctly as required for effective stitch formation, several different types of feed mechanisms have been adopted. These include: four motion drop feed, differential drop feed, needle feed, compound feed, feeding foot, variable top and bottom feed, alternating foot, unison feed, puller feed, wheel feed, cup feed and manual feed.

3.3.1 Four motion drop feed

This system (see Fig. 3.8) is the most common of all feed systems used. The feeder engages the underside of the fabric ply intermittently and is set up (timed) to engage the material when the needle has risen clear from the top ply of the fabric. The feed is named four-motion as it has four degrees of movement:

- Motion 1 – Rising above the plate to contact the fabric.
- Motion 2 – Feed the fabric the required stitch distance.
- Motion 3 – Descend beneath the plate releasing contact with the fabric.
- Motion 4 – Travelling back underneath the plate to the required distance to repeat the feeding process.

3.3.2 Differential drop feed

This feed system (Fig. 3.9) employs two feed dogs set in series which are driven in a similar manner to the single four motion drop feed dog. It is

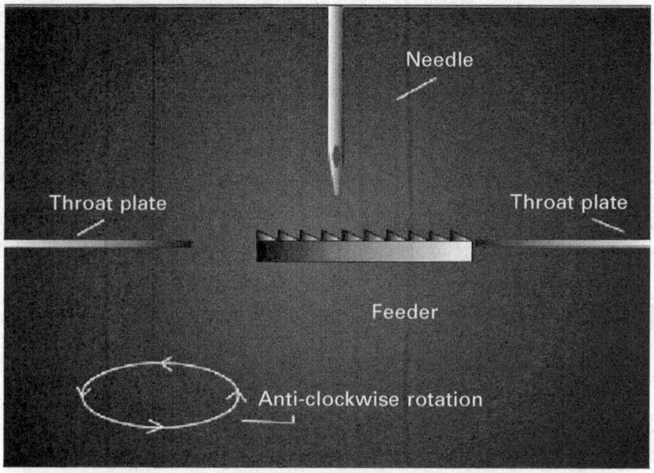

3.8 Four motion drop feed.

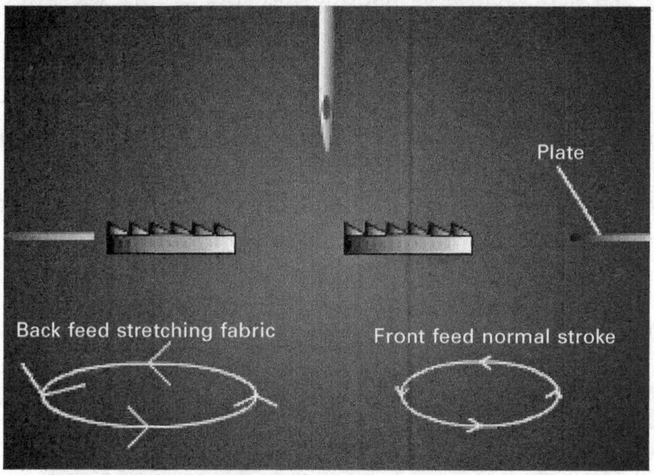

3.9 Differential drop feed.

possible with this feed system to control the ratio of feed between the two feed dogs. By feeding more with the back feed dog than the front feed dog (in the same stitch cycle), it is possible to stretch the fabric bottom ply. Conversely, by feeding more with the front feed dog than the back feed dog, it is possible to introduce fullness into the bottom ply.

3.3.3 Needle feed

The progression of the fabric is achieved by the longitudinal vibration of the needle bar alone. No feed dog is employed in feeding the material. This feed is rarely used in isolation. Today, it is mainly combined with a drop feed to give a compound feeding system.

3.3.4 Compound feed

This feed system (Fig. 3.10) combines the needle feed with the four motion drop feed system. The feed timing is such that the feed dog rises and engages the bottom fabric ply at the same time that the needle descends into the fabric. Both the feed dog and needle move the fabric through the prescribed stitch length. This can help to reduce feeding pucker and ply slippage during stitching because the fabric plies are pinned together during sewing.

3.3.5 Feeding foot

No feed dog is employed in this system. The action is mimicked by the presser foot descending and engaging the fabric, moving it through the machine then disengaging and rising before travelling forwards for the next stitch.

3.3.6 Variable top and bottom feed

This feed mechanism (Fig. 3.11) is a combination of a feeding foot synchronised with a bottom four motion drop feed system. These feeding mechanisms are often used for the sewing of high friction materials such as simulated leather and composites, where the use of a static presser foot is unsuitable.

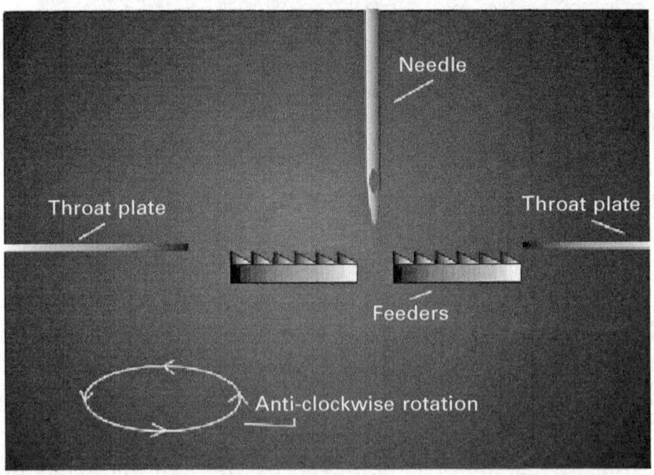

3.10 Compound feed.

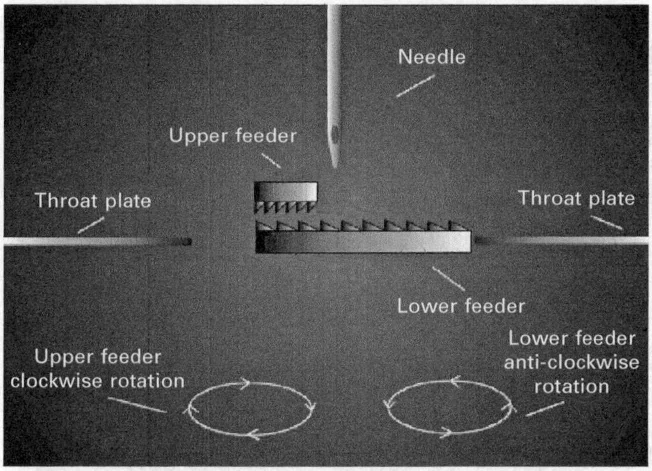

3.11 Variable top and bottom feed.

3.3.7 Alternating foot

These feed mechanisms consist of a feeding foot and a lifting foot in conjunction with a feed dog. The feed dog and the feeding foot transport the fabric whilst the lifting foot is clear of the work piece. The lifting foot descends to hold the fabric whilst the other two components return for the next stitch. These are excellent for the seaming of very bulky seams.

3.3.8 Unison feed

These are similar to alternating feed but provide more positive feed again. Both sections of the two-part presser foot contact the fabric and transport it through the machine sequentially. This means that one of the presser feet is in contact with the fabric at all times.

3.3.9 Puller feed

This is an auxiliary feed (additional to the main feed system) which takes the form of a continuously or intermittently turning weighted roller positioned at the rear of the needle (Fig. 3.12). It is used mainly in the seaming of heavy fabrics or to maintain tension in lighter weight materials.

3.3.10 Wheel feed

In this system, a driven roller, wheel or belt is employed above, below or in both positions to move the fabric instead of a feed dog and presser foot. The surface of the wheel is often knurled or coated to provide increased friction for feeding leather or plastics.

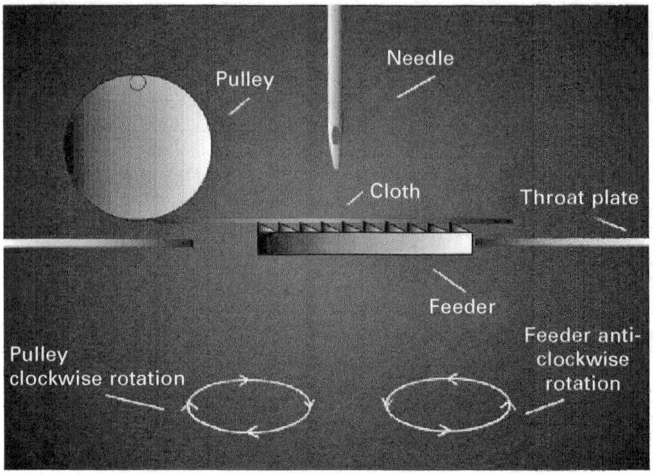

3.12 Puller feed system.

3.13 Cup seaming machine.

3.3.11 Cup feed

Two horizontally mounted feed wheels are driven simultaneously in order to move the fabric from right to left through machines with horizontal needles (Fig. 3.13). It is most often used in the seaming of knitted goods and skins. The positioning of the feed system allows greater visibility of both fabric plies.

3.3.12 Manual feed

The operator moves the fabric beneath the needle alone. No feed dog is employed, thus allowing variable stitch length at the operator's discretion.

Used principally for baisting operations on 101 single thread chain-stitch machines or for decorative 'embroidery' stitches, it has also been employed for jigging collars for men's shirts.

3.4 Machine settings and sewing quality

Optimum settings are the settings that give the best possible settings on the machine for the fabric to be sewn. Advanced sewing machine settings consist of settings that can only be made by hand either by using gauges or a subjective estimation by the engineer making the adjustment on the machine. It is important to reiterate that these settings have a significant impact upon the quality of the seam. Some examples of these alterations are given below.

3.4.1 Optimum presser foot force

The presser foot force should be reduced as much as possible in order to provide an absolute minimum pressure for the fabric to allow for correct feeding of the material without fabric ply slippage. A presser foot adjustment screw located on the top of the machine head adjusts this setting. The pressure of the presser foot can be measured using a presser foot measurement device (Fig. 3.14). Once the required pressure is established, a measurement of the presser foot adjusting screw can be obtained using a standard measurement rule (Fig. 3.15).

3.4.2 Optimum thread tensions

The thread tensions should be adjusted to be as slack as possible in order to produce a well-balanced stitch. The thread tension of the static (top thread

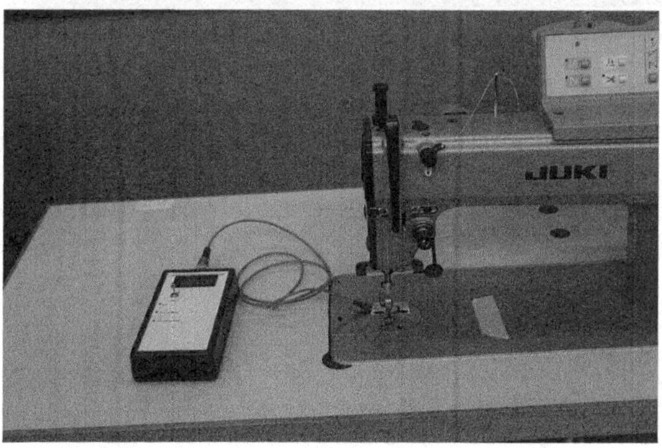

3.14 Presser foot force measurement gauge.

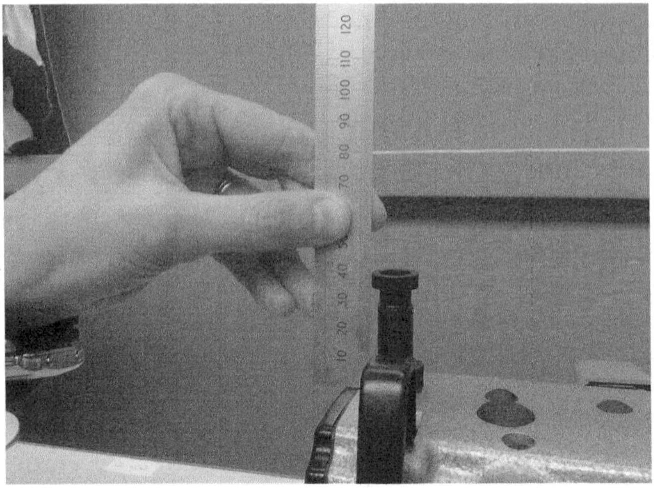

3.15 Presser foot force measurement adjustment screw.

3.16 COATS thread tension metering device.

tension) has the highest set tension and the spool tension is adjusted to be in harmony with the top thread tension. Devices used to measure these tensions are the COATS thread tension-metering device (Fig. 3.16) and a less expensive option is shown in Fig. 3.17. A bobbin thread tension measuring device is shown in Fig. 3.18.

3.4.3 Optimum type and height of the feeder

The metal feed dog is the component that feeds the material through the machine. There are many types available depending on the fabric to be sewn.

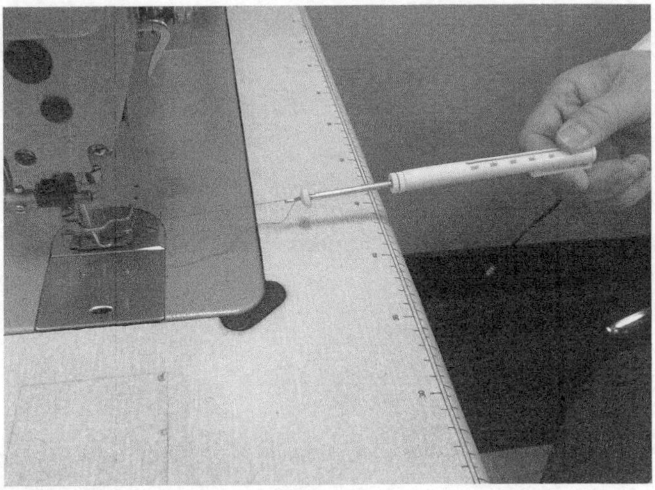

3.17 Pencil thread tensioning device.

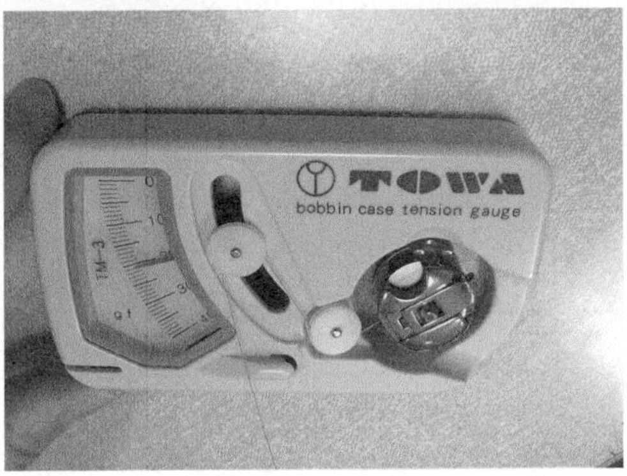

3.18 Bobbin tension measuring device.

When sewing denim, for example, a heavy duty feed with a course tooth set should be used, but when sewing fine fabrics such as shirting materials, a finer tooth set can be used. This is a feeder especially developed for sewing fine, lightweight fabrics. Examples of feeding components are shown in Fig. 3.19. The proper height setting for the feeder is that the feed should be set at one full tooth above the throat plate when the needle is at the top dead centre (its highest position). Examples of this adjustment are given in Figs 3.20 and 3.21.

3.4.4 Feed dog tilt

The tilting of the feed is an important adjustment that can have a dramatic impact on how the fabric performs through the machine. There are varied

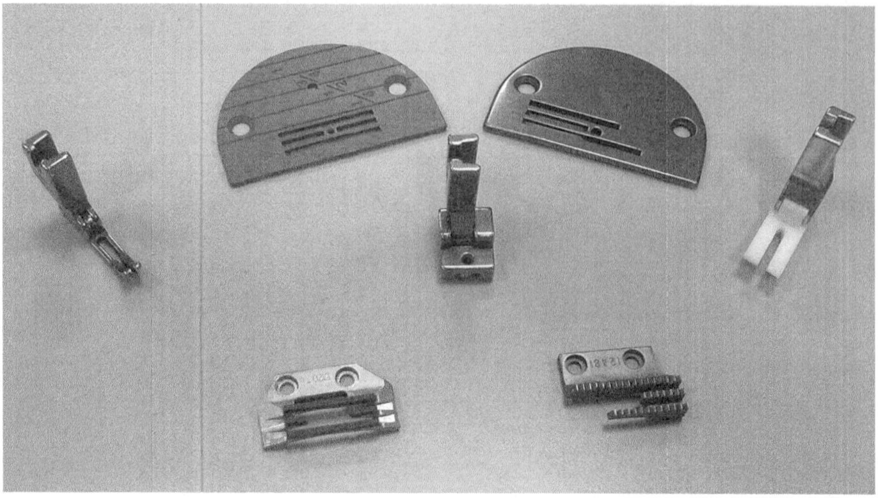

3.19 Examples of feeding components.

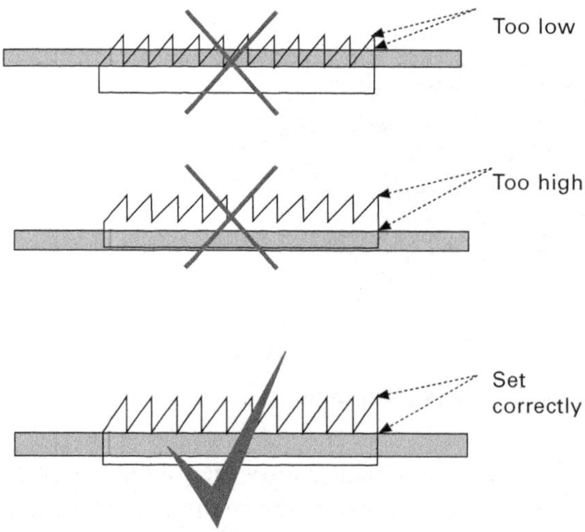

3.20 Correct setting of feed dog at one full tooth above the throat plate.

and important reasons for adjusting this setting. One can be when attaching a rigid material to a stretchy material. The difference in the extensibility of the two materials between the presser foot and the feeder causes a deformation of the material on the fabric with the greater extensibility thus causing feed pucker. A typical area on a garment can be when attaching a zip into a trouser where the zip is usually a rigid component and trousers are of a more extensible nature. Figures 3.22–3.26 give examples of pre and post

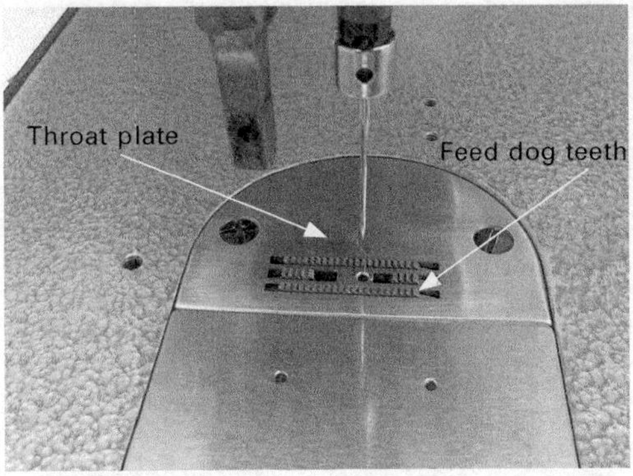

3.21 Feed dog teeth set to one full tooth above the stitching plate.

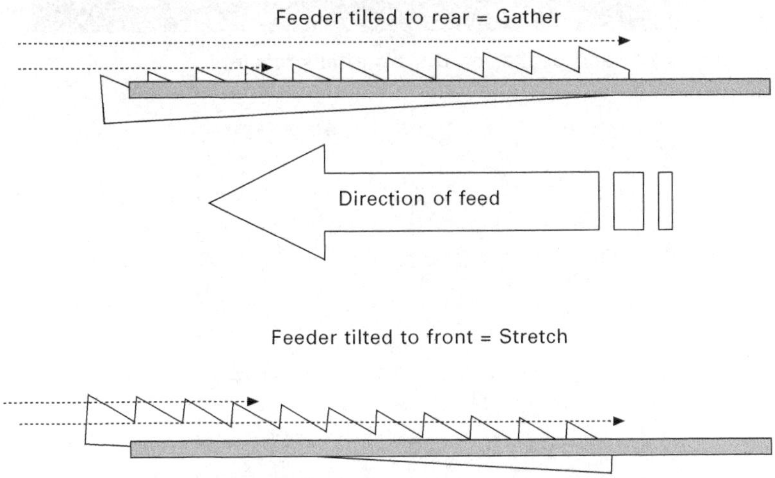

3.22 Feed tilted to gather and feed tilted to stretch.

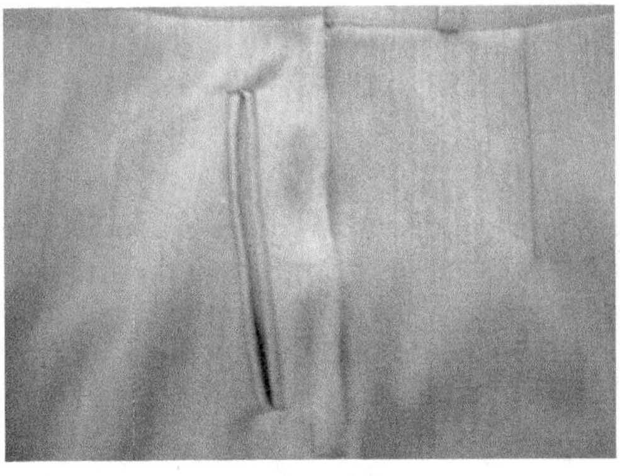

3.23 Feed pucker in the zip before feed tilt adjustment.

3.24 Feed pucker on the zip after pressing.

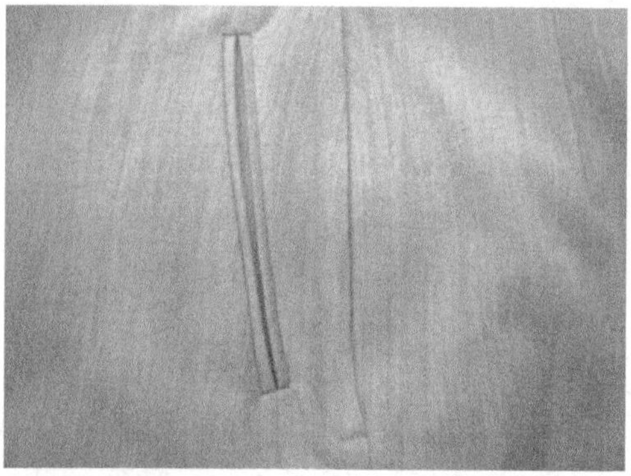

3.25 Feed pucker reduced by feed tilt.

feed adjustment where the feed has been set to gather in order to feed more fabric into the seam.

3.4.5 Feed timing

The adjustment of the feed timing is crucial in order to provide the smooth feeding of the material and this has a direct impact on the formation of the stitch. Fabric feeding can be a major cause of seam deformation. Kennon and Hayes (2000) investigated the fabric feed timing on a lockstitch sewing machine and concluded that by retarding the feed timing by 25 degrees, the tension in the stitch formation was reduced, therefore reducing the effect of seam pucker on the fabric.

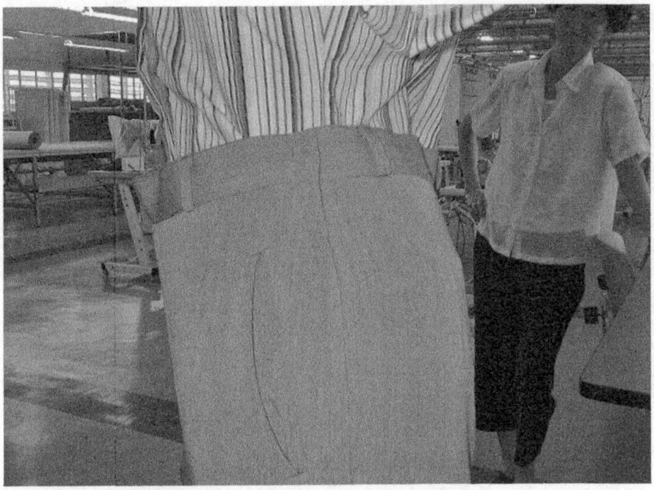

3.26 Finished garment on the zip after pressing.

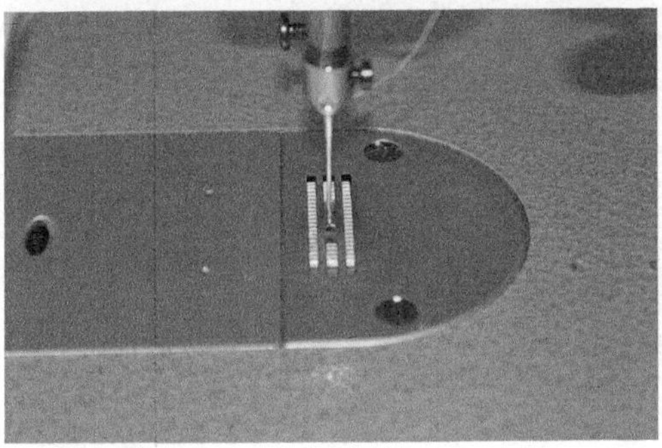

3.27 Feed timing setting.

Lockstitch machines can employ different feed systems depending on the types of fabric and the sewing operation on the product. The correct timing for the feeder on the drop feed lockstitch machine is that, when the point of the needle is just about to enter the throat plate hole, the teeth of the feed should be level with the top of the throat plate. An example of the feed timing setting is given in Fig. 3.27 and the adjustment for the feed timing is shown in Fig. 3.28.

3.4.6 Adjustment of the check spring

The check spring (Fig. 3.29) is a very important part of the sewing machine and it plays a fundamental role in stitch formation. This adjustment needs to be made under sewing conditions with fabric under the machine. The most

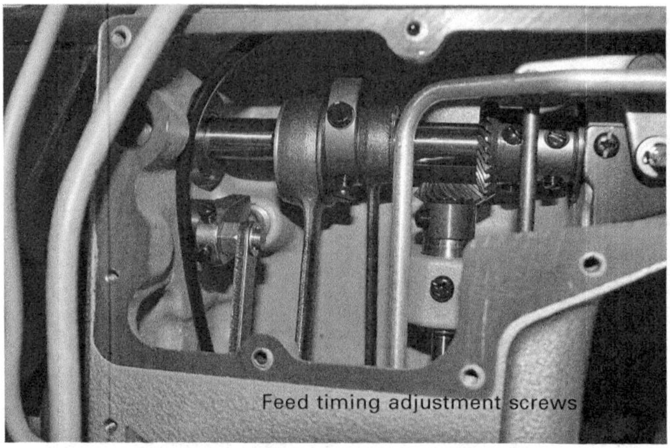

3.28 Feed timing adjustments.

3.29 Check spring set between ten and quarter to the hour.

common setting of the check spring is to set it so that the spring is between the hours of ten to and quarter to on a 24-hour clock.

When the sewing mechanism picks up the sewing thread, the thread is transported down and around the hook and base to form the stitch. At some point of its formation, the thread reaches what is termed its six o'clock position (Fig. 3.30) and at this position is subjected to the highest amount of strain. It is at this point that the thread needs as much support as possible. The check spring (Fig. 3.29) should wink or move from its stop position, releasing tension on the needle thread. Therefore, the main function of the check spring is to provide a means of reducing the strain of the sewing thread at critical points of sewing.

If the setting is incorrect, loosen screw 1 (Fig. 3.31) and turn the entire tension assembly until this condition is obtained. It may be necessary to sew a few inches between adjustments to get an accurate setting.

3.30 Sewing thread in the six o'clock position.

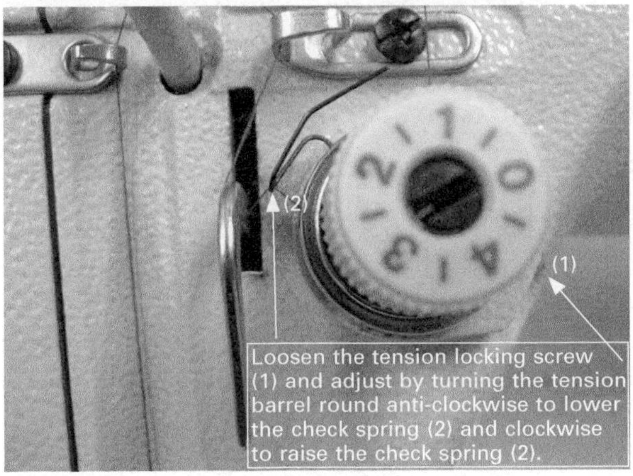

3.31 Check spring adjustment to reduce strain.

3.4.7 Floating of the presser foot

One setting that is little known about is the floating of the presser foot. This is usually only altered in exceptional circumstances, if all other settings have failed to produce the desired quality. The foot bar (Fig. 3.32) is adjusted in order to leave a small gap between the presser food and throat plate when the feed dog is at its lowest position. The space between this gap (Fig. 3.33) can vary depending on the thickness and composition of the material.

The reasons for performing this adjustment is that the surface of the presser foot skims the surface of the material enabling less pressure on the material whilst still maintaining adequate pressure to feed the fabric. The frictional forces on the fabric are also reduced. It is a very delicate adjustment and should only be performed by a highly skilled engineer.

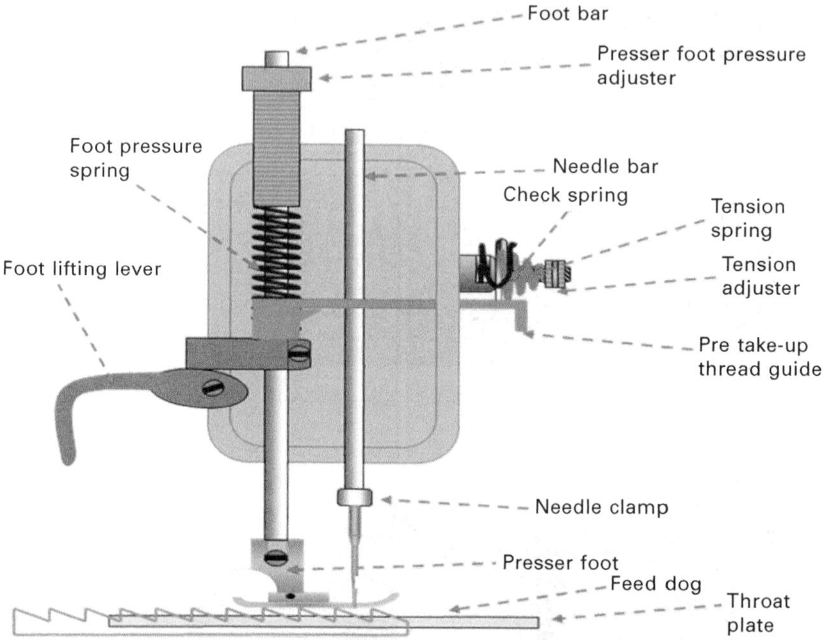

3.32 Behind the faceplate.

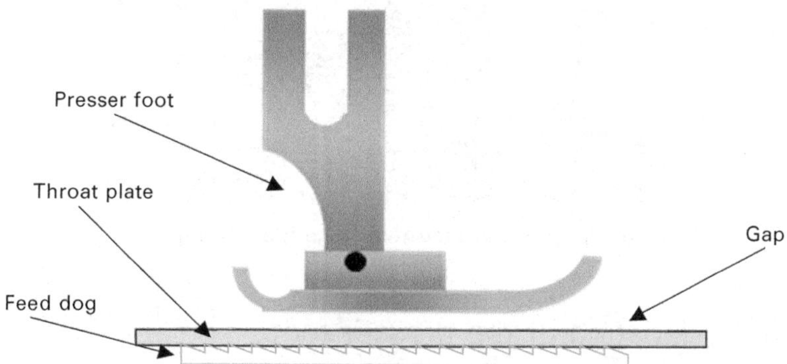

3.33 Gap between the bottom of the presser foot and throat plate.

3.4.8 Thread tension

As previously mentioned, the top thread tension and bobbin tension on a lockstitch machine should be set as slack as possible in order to form a good quality stitch. The bobbin thread should be wound with a low tension to enable the thread to be interlaced more easily with the top tension thread into the centre of the fabric. Examples of acceptable and unacceptable stitch quality caused by tension problems are given in Figs 3.34, 3.35 and 3.36 (Union Special, 1988).

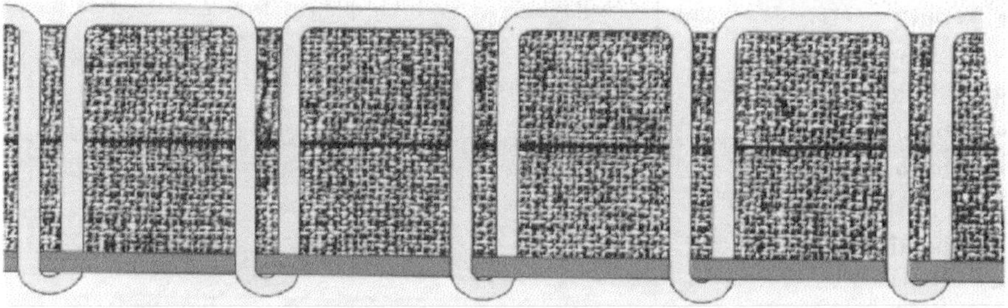

3.34 Stitch slack underneath caused by top thread tension too slack and bobbin thread tension too tight.

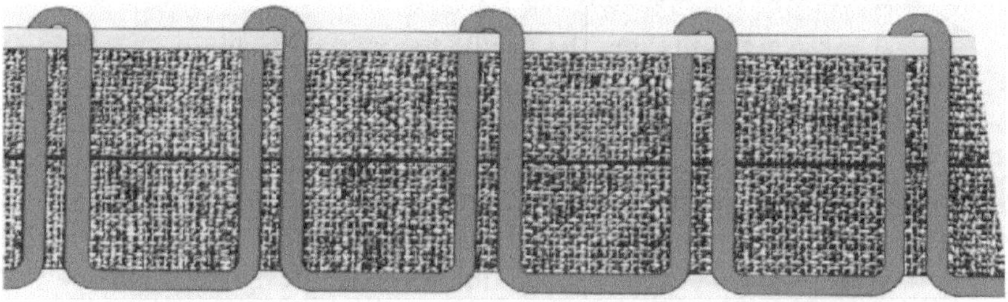

3.35 Stitch slack on top caused by the bobbin tension too slack and the top thread tension too tight.

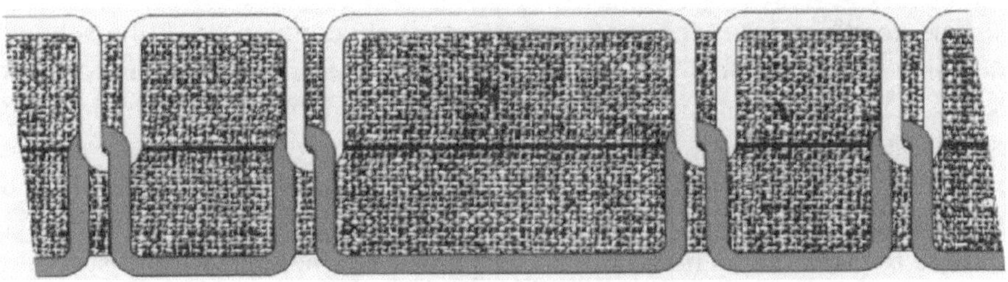

3.36 Stitch formed correctly with both threads interlaced in the centre of the material.

3.5 Needle size and point type

The factors considered when choosing the needle are:

- fabric type
- fabric density
- fabric composition
- seam thickness
- the type of machine.

The component parts of the needle also need to be taken into consideration. These are the:

- butt
- shank
- shoulder
- blade
- long groove
- short groove
- needle eye
- scarf
- needle point
- needle tip.

Some of these features are explained in more detail below.

- The shank is what fits into the needle bar of the machine.
- Some needle types have a short groove that runs from the scarf of the needle up to the shoulder.
- The grooves, which are channelled into the blade, are designed as a protective channel for the sewing thread.
- The needle eye is threaded with the sewing thread.
- The scarf is the flattened part of the needle so designed to enable the sewing mechanism (in the case of a lockstitch, the sewing hook) to pick up a loop of the sewing thread and thus form a stitch.
- The point and the tip are the first point of contact with the fabric.

The needle-point is the most important single component of the sewing machine due to the fact it is the carrier and deliverer of the sewing thread to the sewing mechanism. It has to penetrate the fabric whilst minimising damage to the material. If it is not changed regularly, it can be responsible for major quality problems. It is also subject to the most abuse of all the machine parts as it penetrates the material at speeds of 5,000–6,000 times per minute for lockstitch and 8,000–10,000 times per minute in chain-stitches. The friction caused by the penetration of the needle into the fabric causes extreme needle heating with temperatures in excess of 250°C (Schmetz, 2000).

There have been hundreds of needle-points developed, including needles for sewing knitwear, woven materials and leather. Examples of needle-points commonly used on fabrics are acute round point needles and round point needles. An example of this needle-point can be seen in Fig. 3.37 compared to a normal round point in Fig. 3.38. Examples of needlepoint designs and the component parts of the needle can be seen in Figs 3.39 and 3.40.

A smaller diameter needle reduces the mechanical forces exerted on the yarns. Needle diameters can range in thickness from size 50s (0.50 mm) to in excess of 150s (1.50 mm).

When fabrics are stitched together, the impact from the needle as it penetrates the fabric can cause buckling and distortion of the yarns and the

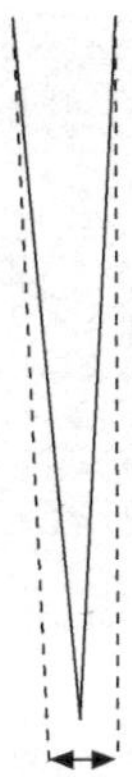

3.37 Acute round point.

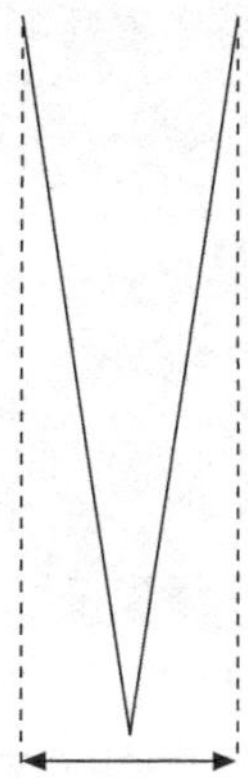

3.38 Normal round point.

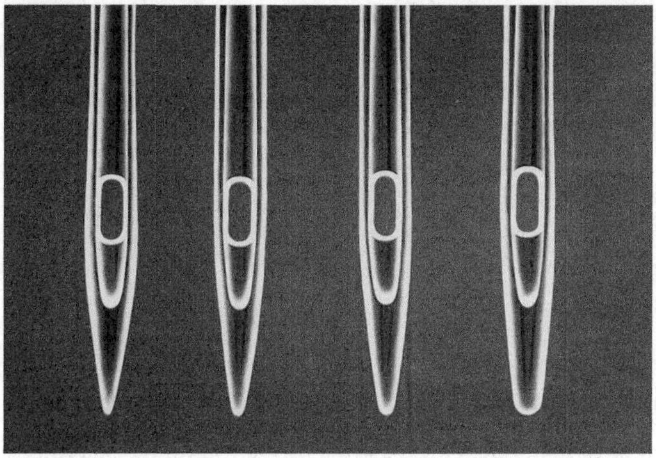

3.39 Examples of needle-point types, from right to left, acute round point, round point, light ballpoint, heavy ballpoint.

fibres. The mechanical strain on the yarns increases if the needle is damaged (Fig. 3.41), thereby causing the fibres to rupture thus reducing the seam strength significantly. The following factors need to be taken into account in order to help avoid this problem:

- Use a needle with a smaller diameter for the fabric and seam being sewn.
- Adapt the opening of the sewing plate to fit the needle size.
- Use a sewing thread with the correct diameter for the needle eye.
- Use the correct needle point for the type of fabric you are sewing.
- Consider whether the type of seam that you are using to construct the garment could be changed or use multiple seaming in order to divide the strain.

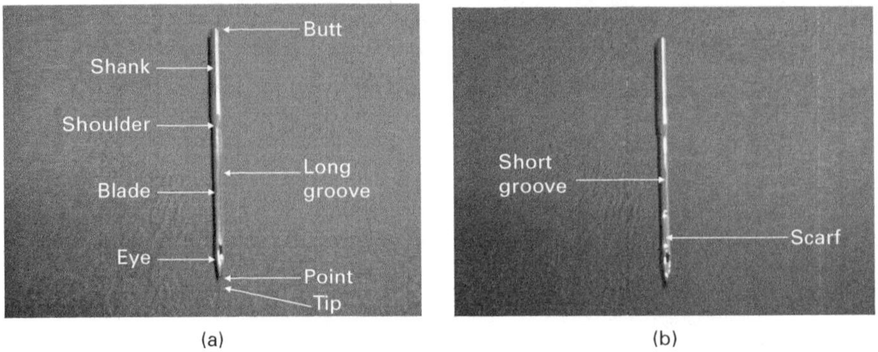

3.40 Needle component parts: (a) needle front; (b) needle back.

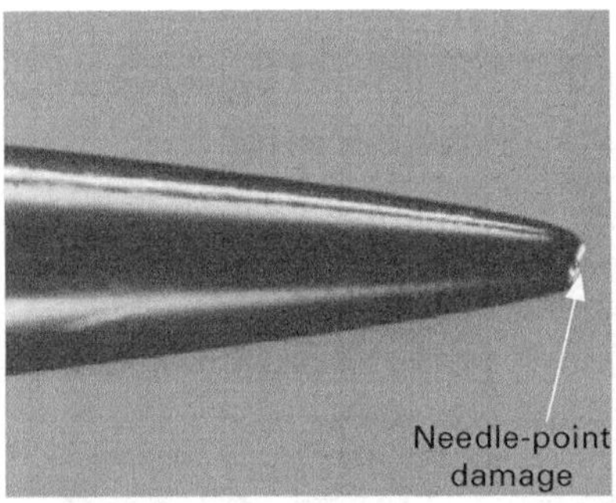

3.41 Damaged needle-point (Schmetz, 2000).

3.6 Stitch classification and applications

The importance of using the correct stitch formation for the seaming operation on the product cannot be understated. For example, you would not use a chain-stitch on seams used on parachutes as a break in the stitch would cause the seam to completely unravel. Therefore, careful selection of stitches is vital in order to make the product fit for its purpose and end use.

Stitches are defined by the method of formation and classified into six different categories. The method of formation falls into three different methods which are:

- Intralooping – The passing of a loop of thread through another loop of the same thread supply. Example: 101 single thread chain-stitch.
- Interlooping – The passing of one loop of thread through a loop formed by a separate thread supply. Example: 401 double lock chain-stitch.
- Interlacing – The passing of a thread around, or over, a separate thread supply or a loop of that supply. Example: 301 lockstitch.

The stitches are classified by stitch numbering systems of three digits

(International Organisation for Standardisation, 1991). These classifications are given as:

- Class 100 – Single thread chain-stitches
- Class 200 – Hand stitches
- Class 300 – Lockstitches
- Class 400 – Multi-thread chain-stitches
- Class 500 – Overedge/overlock stitches
- Class 600 – Covering chain-stitches.

Each specific stitch type is designated by the second and third digits of the number, hence the 301 stitch formation described above is a single needle lockstitch and is specifically identified by this number. There are hundreds of stitch types used in the manufacture of textile products, therefore only the most commonly used are discussed here.

Class 100 chain-stitches

Class 100 chain-stitches (Fig. 3.42) are formed by the intralooping of a needle thread supply on or around the fabric. Single thread chain-stitch seams are often used for temporary applications due to their ease of removal. This is because each successive loop is dependent on the previous loop for security. Applications of this stitch type include:

- baisting – Temporary holding of fabric pieces before final securing stitch is introduced
- sack closure – Ease of removal allows use for securing industrial sacking openings
- coinelli – Up to 65 needles can be employed to produce decorative seams with the chain on the face fabric
- can be used to produce button holes and to attach buttons
- end of seam must be secured in all cases to avoid the running back of seam.

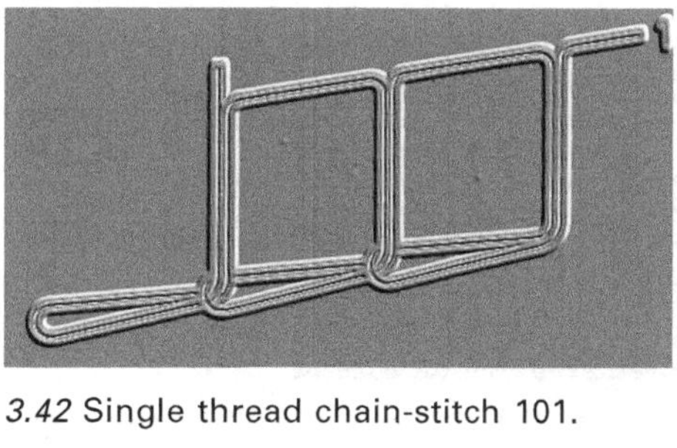

3.42 Single thread chain-stitch 101.

Class 300 lockstitches

Class 300 lockstitches (Fig. 3.43) are often referred to as double lockstitch. This stitch type is formed by interlacing a needle thread supply with the bobbin thread supply underneath. These stitches are very secure as a break in one stitch will not cause the seam to completely unravel, although it will compromise the overall seam performance. They are formed with a single thread and bobbin thread and are the most widely used stitch in low volume production. They exhibit great strength and resilience if correct thread types are used. They can provide adequate extension (up to 30%) for comfort stretch garments. They have the same appearance on both sides and it is possible to match threads to different fabric ply colours. This is the only stitch to reliably sew around 90° when pivoting the fabric at the needle point which is very important for top-stitching (collars and cuffs, etc.).

Class 400 multi-thread chain-stitches

Class 400 multi-thread chain-stitches (Fig. 3.44) are formed by the interlooping of a needle thread with a separate looper thread on the underside of the fabric. This stitch is often referred to as a double locked stitch because each needle thread loop is interconnected with two loops of the same single underthread. It exhibits good strength and increased extension/recovery properties due to lower static thread tension and interlooped threads. It is less prone to pucker, again due to lower static thread tension and interlooped threads on the underside. It is excellent for long seams because of continuous thread supplies.

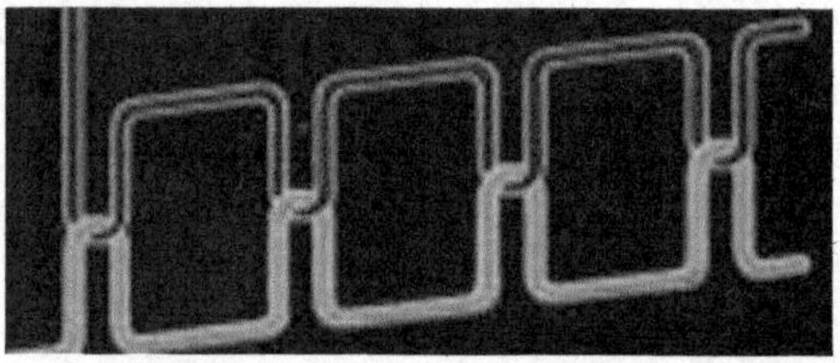

3.43 Single needle lock-stitch 301.

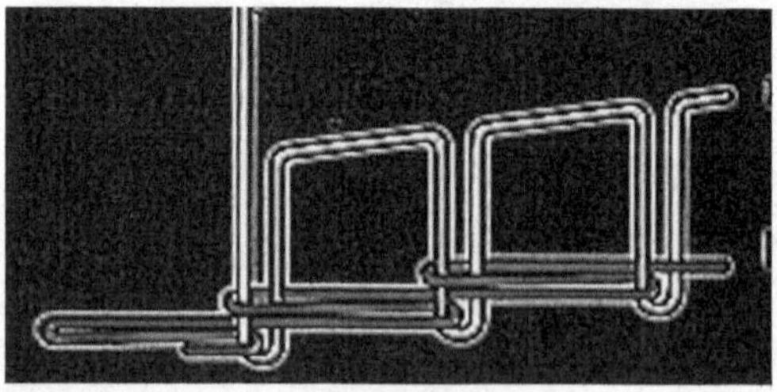

3.44 Two thread chain-stitch 401.

Class 500 over-edge stitches

Class 500 over-edge stitches (Fig. 3.45) are formed with at least one of the sewing threads passing around the fabric edge. Many variations exist employing from one to four threads only one of which is the needle thread. They are generally used to neaten the cut edge of fabric plies. The needle thread provides the seam strength and the looper threads give the stitch greater extensibility. Often the sewing threads used in the loopers can be chosen for softness and appearance.

3.7 Seam classification and applications

Some form of seam joining is used in virtually all sewn product manufacture involving the stitching of materials using sewing threads. This process is still by far the most important method used in the joining of textiles. A seam should be suitable for the purpose for which it is intended and the seam type is dependent on the product being sewn. It must meet the required standards of appearance and performance while ensuring the economy of production which is critical to a modern production environment. The main classes according to UK and US standards are shown in Table 3.1.

When choosing a seam for a garment operation, the following factors should be considered:

- aesthetic appeal
- strength

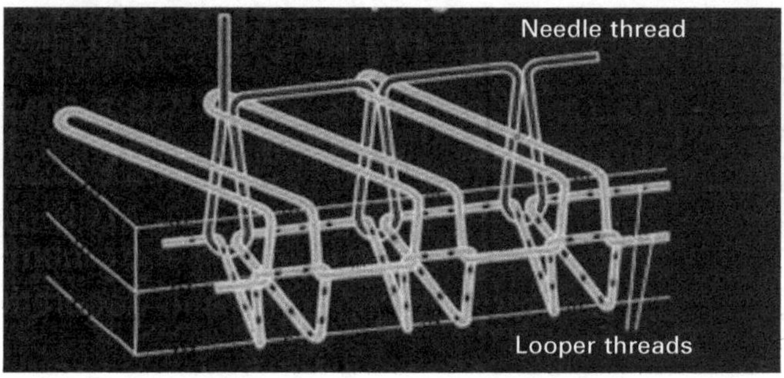

3.45 Over-edge stitch 504.

Table 3.1 Description of seam types

British Standard	Federal Standard	Description	Example
Class 1	SS	Superimposed	SSa – 1
Class 2	LS	Lapped	LSc – 1
Class 3	BS	Bound	BSa – 1
Class 4	FS	Flat	Efa – 1

- durability
- comfort in wear
- ease of assembly
- equipment availability
- cost.

These factors were mentioned in Chapter 1. However, it is important to consider the main areas where seams play an integral part in the garment, particularly in relation to economic production whilst fulfilling the requirements essential for the product. Therefore the seams that are of major importance in this area of garment manufacture are briefly considered here.

3.7.1 Class 1: Superimposed seams

These seam structures (Figs 3.46 and 3.47) are the most basic and easiest to produce of all the seam types and are frequently used in many areas of garment construction as well as many other sewn products on the market.

Typical areas of a garment where these seams are used are side seams on a pair of jeans, side seams on blouses or shirts and trousers, also attaching

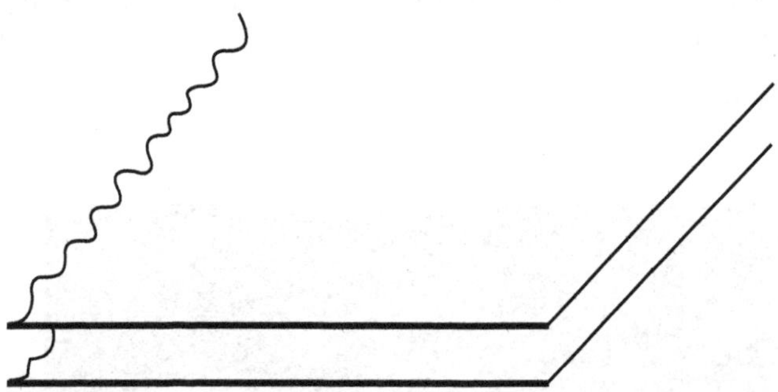

3.46 Superimposed seam.

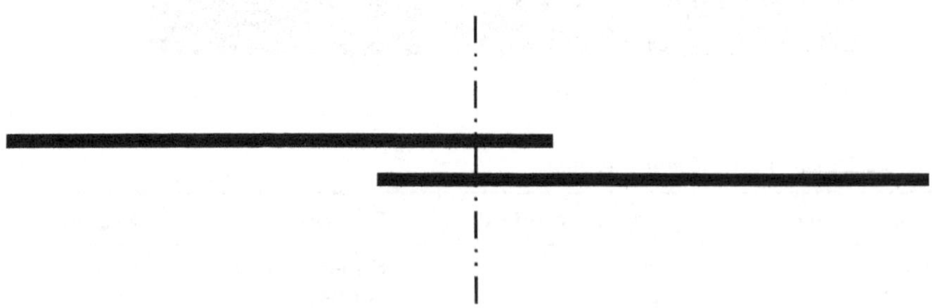

3.47 Superimposed seam cross section.

zips and trims, etc. The fabric edges would usually be overlocked in order to prevent frayed ends of the material. The seam is durable and is easy to produce as it only involves one fabric placed on top of another. However, a disadvantage of this seam is that it has strength limitations due to the fact that the integrity of the seam is dictated by the strength and type of sewing thread. You would not sew the side seams of a judo suit, for example, using this type of seam. Judo is a very physical form of martial arts, which involves hands on contact with your opponent. The physical pulling and pushing of the fabric would no doubt cause the stitches to break causing a rupture on the seam. Therefore in order to maintain the seam strength on this type of garment, it is desirable that another type of seam and construction is used, and this is discussed next.

3.7.2 Class 2: Double lapped seams

These seam types (Fig. 3.48 and 3.49) are regarded as amongst the strongest available and are used in many areas of sewn product manufacture, particularly side seams of denim jeans, parachutes, tent joining and the judo suit mentioned above.

3.7.3 Class 3: Bound seams

Bound seams (Figs 3.50–3.52) are commonly used on luggage, inside tent joining and also have many uses on apparel. Examples of these can be hemming and binding operations for fabric edge neatening.

3.7.4 Class 4: Flat seams

Flat seams (Fig. 3.53) are seams produced with a minimum of two plies of fabric with the raw edges butted together and usually stitched using a cover-seam machine. These seams are most commonly used in knitted products,

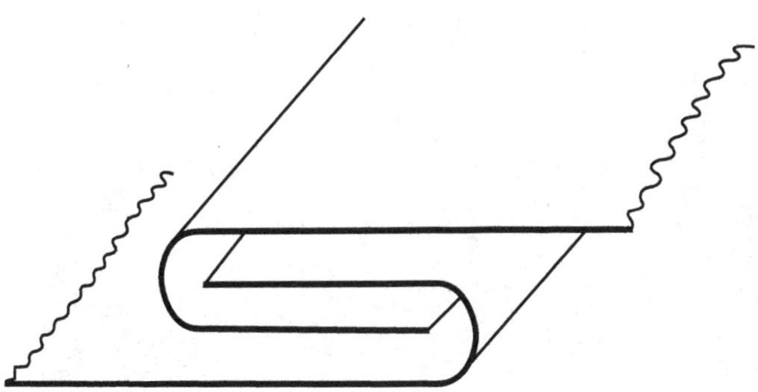

3.48 Double lapped seam.

3.49 An outside leg of a pair of jeans using a double lapped seam.

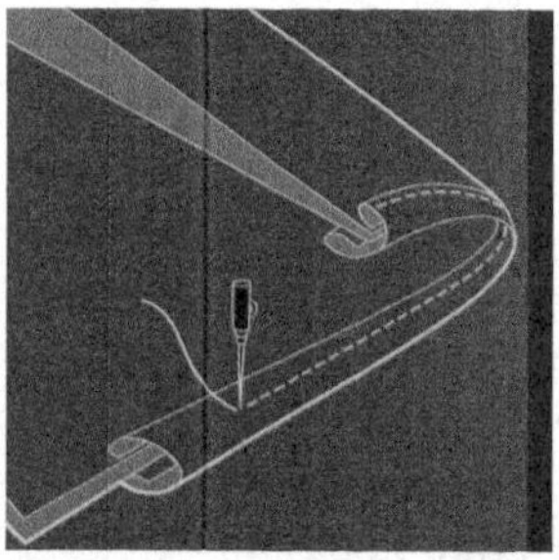

3.50 Bound seam.

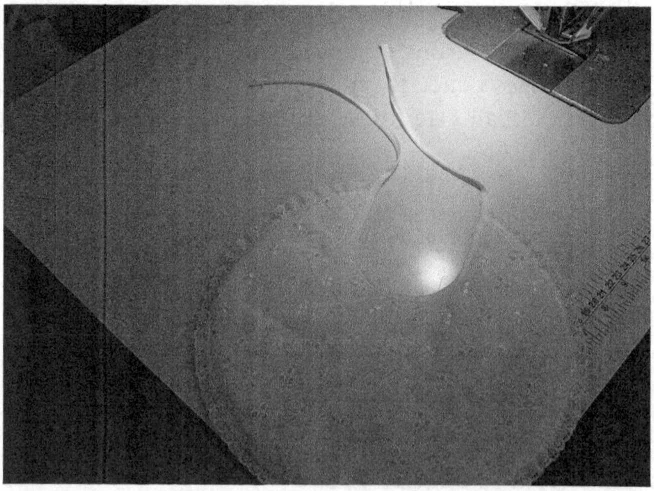

3.51 The binding of a necktie for a babies bib.

3.52 The binding attachment on the lockstitch machine.

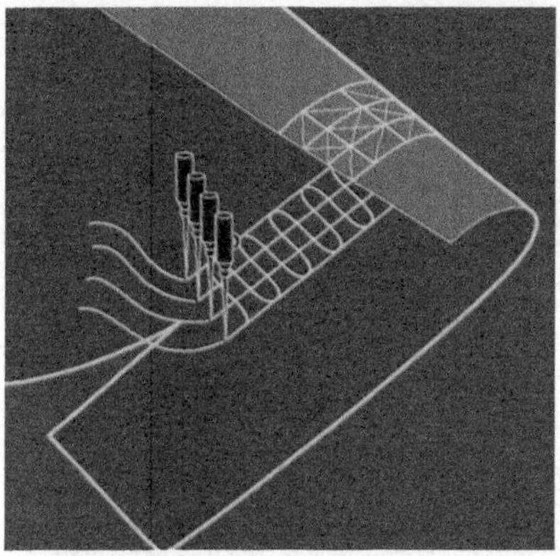

3.53 Flat seam.

particularly underwear. A major advantage in using this seam type is reducing the bulk of the seam (Figs 3.54–3.56).

3.8 Sewing threads

3.8.1 Factors affecting the choice of sewing thread

In choosing a sewing thread, both thread and product requirements must be taken into account. Thread requirements include:

- sewability
- loop strength
- elongation

3.54 Over locking machine with a puller feed.

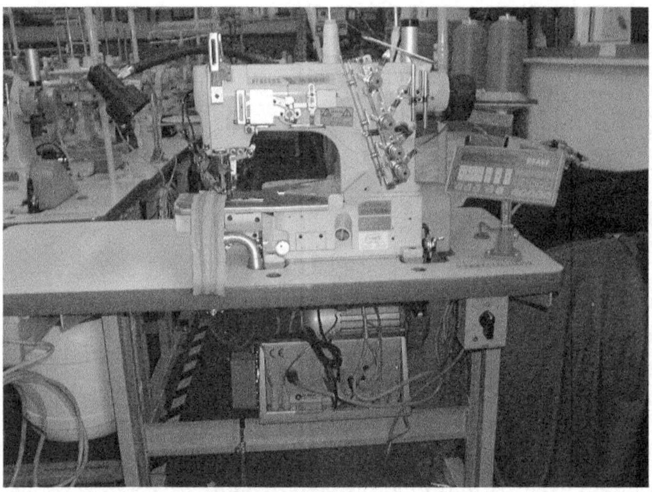

3.55 Cylinder arm cover-seam.

- shrinkage
- abrasion resistance
- colour fastness
- resistance to chemicals, heat, light, etc.

Product requirements consist of the following factors:
- required seam strength
- type of seam
- stitch type

3.56 Over-locking a cushion cover closing using a bound seam.

- stitch density (stitches per cm)
- type of material being sewn
- type of sewing machine and related equipment
- conditions under which the product must perform
- normal life of product
- cost effectiveness.

Different physical properties of the sewing thread are required in order to optimise the performance of any product.

3.8.2 Fibre types for sewing threads

The main cause of seam failure has been identified as thread breakage, hence selection of an appropriate and fit-for-purpose sewing thread is of great importance for product performance. Sewing thread fibres come in a variety of forms. Staple length fibres refer to fibres that either vary in length, or contain filaments cut to a specific length during the manufacturing process. Continuous filament fibres, on the other hand, refer to synthetic fibres of no definite length.

Cotton

The main characteristics of cotton thread are good wet strength and heat resistance, with adequate tenacity and elastic recovery. However, cotton exhibits poor resistance to sunlight, acid, abrasion and exposure to flames. Cotton thread comes in different finishes. Mercerised cotton is formed by passing threads under tension through a cold solution of 20% caustic soda. This causes swelling of the fibres and adds approximately 12% tensile strength when compared to non-mercerised equivalent. Mercerised cotton is used widely in industrial and domestic production. Glacé or polished cotton is cotton strengthened and protected by application of a mixture of starch and lubricants to the outer surface (leaving it non-resistant to wet treatment). To increase smoothness, the surface is polished by high-speed rotating brushes, possibly with the application of wax. This finish facilitates machine stitching, including that of heavy fabrics. Cotton has a number of applications in spun thread and core thread formats (see below). In spun-thread format these include uses as varied as women's lingerie and heavy leather gloves, and in core thread format, when wrapped around a continuous filament of polyester, in blouses, heavy overalls, and jeans, etc.

Silk

The characteristics of silk thread are high lustre and high extensibility (up to 20%). However, it has a poor resistance to abrasion and is suitable only for a restricted range of applications including buttons, button holes and decorative stitching.

Synthetic fibres

A number of synthetic fibres are used for sewing thread and can improve the performance of a product in certain specific applications. Polyester, for example, exhibits 100% greater seam strength than cotton thread of the same size. Both polyester and nylon are more resistant to abrasion and chemical degradation than cotton or silk. Nylon threads exhibits good resistance to alkalis but poor resistance to acids; conversely, polyester exhibits good resistance to acids and relatively poor resistance to alkalis. PPTA (poly *p*-phenylene terephthalamide) sewing thread retains strength at high temperatures. It is thus ideal for applications where such properties are critical, for example, flame resistant furnishings. Polypropylene can be used under different weather conditions, as the presence of moisture results in an average 8% increase in strength and a decrease in elongation resulting from a plasticising effect.

Fibre finishes include glazing, waxing and treating threads with silicone. Each of these treatments is designed to facilitate movement of the thread in the machine, enhancement of thread tenacity, and to protect the needle thread from the high needle temperatures that can be generated at higher sewing speeds (up to 300°C) to which natural fibres are insensitive. Textured thread is made from continuous filaments of polyester or nylon using a variety of techniques. Air-texturing is a mechanical process in which filaments are passed between a pair of rollers through an air jet, with the first pair of rollers set at a faster speed than the second, thus leading to length overfeed. The air jet blows the filaments apart, forming loops at random points both within, and on the surface of, the filament. Heat treatment sets the loop structure. The final stability of the sewing thread structure is the result of a combination of loop entanglement, twisting and heat-setting. Air-textured threads exhibit high elongation appropriate for the seaming of kitted garments and swimwear where seams are overlocked and covered. Stitch stability is excellent, making it suitable for shoe in-seaming. Air-textured threads are used for purposes such as seaming of flags to heavy denim jeans.

3.8.3 Sewing thread construction

The structure of sewing thread impacts seam behaviour, making an understanding of thread construction important in understanding performance.

Monofilament threads

These are formed from a single continuous filament, normally nylon, and are translucent, reflecting the colour of the textile into which they are stitched. Monofilaments are stiff, and wearer discomfort has been reported where monofilaments have been used in the manufacture of clothing. High shrinkage has also been reported, with some monofilaments shrinking as much as 10% on boiling with an average of 3–4%. Monofilaments must only be used at medium sewing speeds and low tension. Monofilament threads are typically used for interior textiles such as bedding, curtains, home furnishings etc.

Spun thread

Spun thread is made either from cotton staple or polyester fibres spun into single yarns with two or more of these subsequently plied together to make the sewing thread. Spun synthetic yarns are made by cutting continuous filaments into short sections and then spinning them in the same manner as natural fibres. Alternatively, continuous filaments may be broken by stretching and the resultant yarns may be twisted and/or plied.

Spun thread yarns have a fuzz on the surface giving them a soft handle. They have a high tensile strength and a high resistance to abrasion as well as good tenacity and are extensible. The extensibility of some polyester threads resembles that of cotton. Some nylon threads are extensible to a degree that makes them suitable for joining knitted products. As they are often thermoplastic in nature, spun threads are vulnerable, however, to the high temperatures that are generated at high stitching speeds. A machine operating at 4,000–5,000 stitches per minute may generate temperatures as high as 300–350°C, whereas a thermoplastic spun thread will melt at 240–260°C. Spun threads are used in a range of clothing applications, from lingerie to gloves, and, as stated above, are also suitable for knitted garments.

Twisted multifilament thread

Twisted multifilament threads are continuous filaments of polyester or nylon twisted together into a cohesive bundle and then plied to make a thread. They are then dyed, stretched and heat set to achieve the desired physical characteristics. Twisted multifilament threads are exceptionally strong for their size with three-ply threads demonstrating higher seam strength compared to two-ply threads. The strength of the thread increases with the increase in twist in the plied structure. A higher degree of twist also helps to gain circularity in the cross section of the thread, thus reducing the contact area between thread and needle, resulting in lower strength loss during sewing. Twisted multifilament threads are used for a variety of purposes, from swimming suits, to applications in the automobile industry.

Core-spun thread

Core-spun threads are made by spinning a wrap of staple cotton or polyester around a continuous filament of polyester or nylon, combining the desirable characteristics of each into a single yarn. Core-spun threads have a fuzz on the surface which gives them a good handle and makes them easy to handle during sewing. The filament adds high strength and durability, whilst the cotton fibres add bulk and contribute to the protection of the core from heat and abrasion. When wrapped with polyester, core-spun threads exhibit good resistance to chemical attack and excellent colour fastness. Core-spun threads are often used where higher seam strength is required. Core-spun sewing threads are usually soft finished and are then used in products such as blouses and other garments. They may also be polished for use in the manufacture of leather goods and footwear.

Monocord thread

Monocord thread is made from continuous filaments of nylon that have been bonded together, and exhibits high seam strength. Monocord thread is used for heavy duty applications such as furniture or shoes.

3.8.4 Factors affecting sewing thread performance

Twist

As yarn is twisted, its length contracts, a process that can be partially recovered by subjecting the resulting yarn to tension. As twist increases, extensibility first increases and then decreases beyond an optimum point. The same is true for yarn strength, which first increases and then decreases until breakage occurs. When the twist level is high, the fibres are less aligned to the yarn axis and thus contribute less to the yarn strength. When twist level is low, failure occurs primarily through fibres slipping past each other. Although twist is not essential to the strength of continuous filaments, they may still be twisted to allow for better cohesion performance in the sewing process. It is important to be aware of twist accumulation during sewing which may cause jamming at the eye of the needle during sewing. This results in a varying twist level of the thread and consequent varying tensile behaviour of the stitch. This potentially affects thread strength, extensibility and appearance.

Thread diameter

Thread diameter is an important property as it affects a number of parameters relating to seam performance. For example, coarser thread exhibits higher seam strength due to its higher breaking load, whereas a finer thread will embed itself into the fabric of the material making it less susceptible to abrasion. Thread diameter also affects the creation of undulations in dense fabrics. Thread diameter increases during stitch formation (by up to 30% for mercerised cotton), resulting from abrasion of the thread against elements of the sewing machine, cyclic stretching resulting from irregular tensioning during the stitch formation, and potential reduction in twist between the eye of the needle. A linear relationship exists between thread diameter and the number of stitch formation cycles for spun threads. This is probably due to a reduction in twist and abrasion which increases thread hairiness, resulting in loosening of the thread structure and accompanying expansion in diameter.

3.8.5 Sewing thread quality characteristics

Sewing thread quality characteristics include appearance, durability, extensibility and strength. The priority given to each of the characteristics will vary depending on the final product and its uses. Traditionally, the appearance of stitches and seams has been assessed using the subjective procedure of visual inspection, though instrumental techniques have also been developed. Thread instability can affect the appearance of a seam through the creation of surface undulations along the seam line. For example, threads used for shirts of a polyester/cotton blend need to display a shrinkage of less than 1% in order to avoid an undesirable effect on the appearance of the product. Topstitching is used to define design and construction seams and as such, thick and/or lustrous thread is typically used. Embroidery stitching requiring a smooth, satiny, appearance is achieved by using a fine mercerised cotton, or alternatively, a silk thread.

The durability of a seam may be affected by a variety of factors including ultraviolet (UV) radiation, chemical attack, mould and mildew, abrasion and temperature. Resistance to UV radiation is important, for example, in curtains, where seams are exposed to sunlight for long periods of time. Core-spun polyester or nylon threads may be appropriate for such application, as long as the covering is not cotton. A thread constructed from synthetic resins, acrylic and plasticisers in continuous filament form, plied and bonded, has been specifically developed for products where sensitivity to UV radiation is an issue.

Seams may also come under chemical attack either through use, or during the cleaning process. For example, chlorine, present in swimming pools or used as a household bleach, can attack threads, as can perspiration from close-fitting clothes. As has been noted, polyester thread has a good resistance to acids, but a relatively poor resistance to alkalis. Conversely, nylon has an excellent resistance to alkalis, but a relatively poor resistance to acids. Resistance to mould and mildew is important in outdoor applications such as camping equipment. This is due both to use of such products in unpredictable weather conditions, and storage in conditions with little airflow. In this respect, cellulosic fibres such as cotton are more susceptible to attack than synthetic polymers.

Abrasion is important, for example, in the context of safety belts for cars, industrial conveyor belts, buttonholes and hems. Cotton and silk are more susceptible to abrasion than either polyester or nylon. Yarn structure is also important, with continuous filaments being more resistant to abrasion than spun yarns. Resistance to temperature is important in the context of sewing at high speeds, where temperatures can reach 300°C. Breakage of synthetic sewing threads usually occurs in the range 240–260°C, whilst heat-set thermoplastics display stability up to 150°C. A subsequent reduction of 10% in length can occur if thermoplastic thread is heated above this temperature. Natural fibres appear to be stable at temperatures up to 400°C. There is some evidence to suggest that needle temperature is slightly lower when sewing with spun yarns than with continuous filaments, which may be due to the irregular surface texture of the spun yarn acting as a cooling vane. Threads can be made more resistant to temperature by lubrication, which decreases friction.

Extensibility is important for threads used to join fabrics which are themselves highly extensible. All sewing threads are extensible to some degree. The variables affecting extension-to-break include the substrate on which the stitching occurs, substrate thickness and the type of thread used. Thread extension-to-break also depends on such factors as yarn structure, twist and fibre type. Silk and a number of synthetic continuous filaments can extend up to 20% whereas a typical spun thread extends only to approximately 10% before breaking. Cotton threads, with low extensibility but high elastic recovery, increase in extensibility as a result of the stitching process, whereas synthetic continuous filament threads with high initial extensibility and low elastic recovery decrease in extensibility as a result of stitching. Increased stitching speed decreases the impact of the stitching process on extensibility.

The relationship between the strength of a thread and the corresponding strength of the seam is unclear. Uniformity of strength along the length of the thread is, however, important, and is achieved usually by combining and folding of the fibres/filaments. For stitch formation, it is the straight tenacity of the thread which is the important factor; however, for seam strength, the loop tenacity is the more important parameter. The relationship between these two factors is not constant for all fibres, and an expression of the relationship is given by the following:

$$\text{Loop efficiency} = (100) \times [(\text{loop} - \text{strength}/2)/\text{breaking load}]$$

The relationship between the straight tensile strength of the thread to the loop strength and seam strength is unclear and depends on both the bending of the thread to the required geometry, and on the stitch formation process. During loop formation, fibre–fibre friction and the bending of the thread itself concentrates the stresses in the bent part of the thread, thus reducing the load at which the fibre breaks. The loop strength is only approximately 80–90% of that of a single thread.

The stitch formation process is important as the speed at which the thread passes through the eye of the needle can reach 40–45 m/s, and at the rotary sewing hook, up to 2,000 m/s. The effects of resulting friction and stresses act repeatedly on the thread since, before stitch formation occurs, the thread passes many times through the fabric, the eye of the needle and the bobbin case. Such severe conditions can reduce the initial strength of the thread by up to 60%. The extent of thread abrasion during the formation of a stitch depends on the smoothness of the machine parts that contact the thread, the surface frictional properties of the thread itself including finishing treatment, the shape and type of needle, and the type of fabric in respect of density, hardness of the yarns, number of plies, and the number of passages through the eye of the needle before the thread is incorporated into the seam. The residual strength value, RSV, is expressed according to:

$$\text{RSV} = (100) \times [(\text{strength of thread after sewing})/(\text{strength of thread before sewing})]$$

3.9 Seaming quality problems

Various studies have discussed the key factors affecting seam sewability (McWaters and Clapp, 1994, Stylios, 1983; 1997; Stylios and Lloyd, 1989; Zunic-Lojen, 1998; Mallet and Du, 1999). Typical quality problems identified include seam slippage, seam damage, seam grinning, seam cracking and seam pucker. It is important to note that the practicalities involved on the production floor differ somewhat from a research-based approach.

McLaren Miller (1998) investigated lockstitch seam instability in the cross grain construction of woven fabrics. She comments upon the fact that previous research has suggested that the difference between subjective quality and objective quality is that the former is perceived while the latter can be quantified in part from the mechanical properties. It is suggested that by linking these two aspects, it is possible that irregularities arising in production may be better controlled. She describes a situation during a discussion with a production manager at a plant which supplies a UK retail chain; it was stated that:

> It is all very well having handling values for fabric, as numbers, to try to eliminate the pucker problem, but we have to work with what we're given and the turnover in fabric and design is high. We principally rely on the experience of the handlers and in-line assessment to eliminate the problems concerned.

Stylios and Sotomi (1996) have been at the forefront of establishing technologies for garment sewability and attempting to predict levels of seam pucker using laser and fuzzy logic techniques. They investigated the possibility of developing thinking sewing machines for intelligent garment manufacture. They mention the fact that there is reasonable progress in relating fabric properties to sewing machine settings and stitching quality. There are, however, areas that have not been numerically defined because of the complexity of the dynamic interactions between needle, fabric and machine parameters.

Ferreira (1994c) developed an on-line control system to optimise seam production during the sewing process. He correlated sewing thread tension, presser foot forces and needle penetration forces with seam quality to generate information that will allow for better knowledge of seam production. Theoretical models have also been developed to explain interactions between the needle and bobbin sewing threads (Ferriera *et al.*, 1994a, 1994b), which explain that seam balance is a function of the stitch and formation cycles.

Fabric feeding can be a major cause of seam deformation. Kennon and Hayes (2000) have investigated fabric feed timing on lockstitch sewing machines and conclude that by retarding the feed timing by 25°, the tension in the stitch formation has been reduced, therefore reducing seam pucker.

Zunic-Lojen and Gotih (2003) have analysed the needle bar kinematics with the thread take-up lever on a lockstitch sewing machine by using computer simulation. An image was drawn on the basis of modelling the kinematic simulation of a needle bar with the thread take-up lever and measurements of the thread tension forces on the sewing process.

McLoughlin and Hayes (2007) have developed a fabric sewability system which automatically analyses the results from Kawabata tests and used this information to successfully generate an automated textual report of the fabric properties. It also produced guidance as to the sewability of the material. The system was tested by a team of industrial experts from the apparel industry, the Fabric Sewability Panel. Experts from industry were invited to analyse a number of fabrics and render a judgement on their prospective sewability. The level of agreement amongst the experts was measured using Kendall's coefficient of concordance and significance testing techniques. Comparisons were then made between the judgement of the experts and the results from the fabric sewability system. The results confirmed that there was a correlation between the judgements given by the experts and the fabric sewability system. During this exercise similar discussions took place between the researchers and the experts. As one expert stated:

> We are not dealing with sheet metal here! Fabrics are flexible structures that have external factors introduced into them during make up. Sewing threads, machine settings and operator handling are all important factors that influence the performance of a fabric during sewing.

As has been noted, much of this state-of-the-art research, though important and necessary, does not seem to have an impact on the manufacturing shop floor. McLoughlin (2005) comments that:

> The gap between work practice methods and research methods could be seen to be large. Many companies do not have the resources to fund purchase of an objective measurement system. In fact, many do not know of the existence of such systems at all. It's apparent that there are major difficulties involved in joining fabrics together at the machine interface by the sewing process (pp. 99–100).

There are, however, a number of measures that may be taken in order to help alleviate this problem if not eliminate it completely. These were described as:

- collating historical machine settings data for each style and fabric sewn;
- establishing methods for dealing with seam pucker, understanding its causes and steps which may be taken to counter it;
- giving technicians and production staff greater understanding of the properties associated with a fabric, including knowledge of fibres, yarns, yarn twist, frictional properties, shear forces, extensibility and bending rigidity;
- extending the use of fabric objective measurement systems in fabric manufacturing companies in order to enable warnings of material

instability to be given prior to despatch at fabric apparel manufacturing companies.

Further research should be performed and a settings database may be created to determine optimum sewing conditions for each type of fabric sewn. The use of low cost instrumentation for machine optimisation should also be promoted; such equipment for measuring thread tensions and strain gauges currently exists and is inexpensive to purchase.

Seam quality problems can be time consuming, frustrating and costly to a manufacturer. There are many types of problems associated with the quality of seams on a product. Therefore, only the most common problems are considered in the following sections, starting with seam pucker. The elimination of such problems can be achieved by applying a commonsense approach of logical reasoning, team working and machine and production knowledge. There are many other factors that affect the sewing and stitching of textiles. Many of these factors have not been learned at college or university but have been acquired through many years of experience in industry. The following advice is offered as examples of good practice for determining potential seaming problems.

3.10 Seam pucker: causes and prevention

One of the most common problems associated with lightweight fabrics is seam pucker defined as an unequal, crinkling or a gathering of the seam. This phenomenon is common on lightweight materials such as shirting, blouses and microfibre fabrics. It also presents problems on fabrics for skirts, trousers and suits. There are four main causes of seam pucker:

- feed pucker
- tension pucker
- stitch density and fabric type
- inherent pucker.

3.10.1 Feed pucker

Feed pucker occurs when two plies of fabric to be joined are not fed uniformly through the sewing machine (Fig. 3.57). The bottom ply is usually fed more positively by the feed dogs while the bottom ply is only held and guided by the presser foot. The shortening of one of the fabric layers (usually the bottom one) creates a wavy appearance on one side and results in what is known as 'feed pucker'.

A great variety of feed mechanisms have been developed to try to improve feed pucker. The various feed systems have been described in detail above and it is also possible to tilt the feeder to provide the desired result on the

3.57 Example of feed pucker on surf trousers.

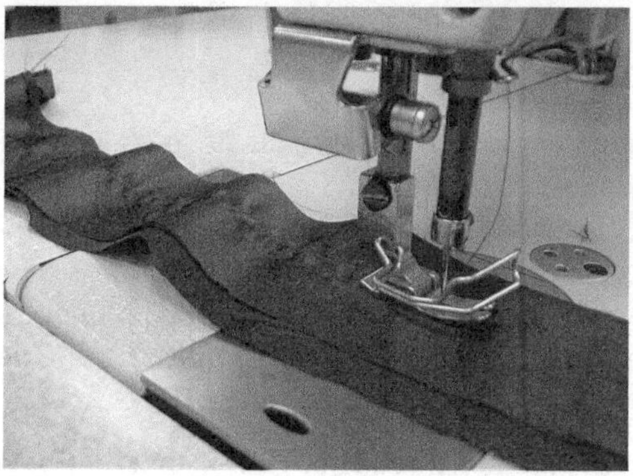

3.58 Example of tension pucker.

seam. This adjustment can only be undertaken by a skilled sewing machine engineer.

3.10.2 Tension pucker

The tension must be as slack as possible to produce a well-balanced stitch and also by using the smallest diameter needle possible and the correct needle-point type for the fabric (Fig. 3.58). These factors all contribute to a good quality seam and help to reduce tension pucker. The sewing threads must suit the seam position and the thread must be at a minimum diameter to produce a minimum of disruption within the yarns of the fabric whilst maintaining the strength of the seam. Other factors that can affect tension pucker are the extension properties of the sewing thread and the possibility of shrinkage due to moisture and heat. The yarn twist of the sewing thread and frictional properties both can have a significant effect on the regularity of stitch interloping and seam appearance.

3.10.3 Stitch density and material type

This type of pucker must also be considered when studying tension pucker. It is directly linked with the required thread tension and the length of thread required by the seam. Increasing the thread consumption in the seam increases the seam strength. In a lockstitch, for example, a 3% increase in stitch consumption can give an almost 60% increase in seam strength. A stitch is only complete after the fabric has moved past the needle. The fewer stitches per cm, the greater the distance the fabric must be moved for the next stitch insertion. Consequently, a greater force is required to present the correct thread length for a perfect stitch. This causes a higher thread tension in the seam which results in puckering that could have been avoided.

3.10.4 Inherent pucker

This type of pucker is the hardest to eliminate as it is caused by the displacement of the warp and weft yarns by the needle penetration and thread insertion into the fabric. If sewn in the warp direction, the warp threads will be displaced laterally causing an inevitable shortening of their length relative to adjacent yarns. The fabric structure becomes jammed, resulting in swelling and puckering of the seam. Other factors need to be addressed in order to eliminate inherent pucker. These include:

- the use of finer needles and finer threads is vital if inherent pucker is to be avoided;
- the use of fine holed needle plates is also essential for reducing seam pucker;
- the use of stitch type plays a very important part in reducing pucker;
- operator handling needs to be addressed;
- the sewing direction is important.

The cause of the particular problem highlighted in Fig. 3.57 is the fittings used on the machine, namely the feed dog and presser foot mechanism. With this type of garment, fine fittings must be used. A fine-toothed feeder and a foot would provide a smoother feeding process and reduce the amount of friction on the garment.

3.11 Thread breakage: causes and prevention

This factor is one of the most frustrating problems to a machinist (see Fig. 3.59). If the thread breaks during the sewing of a seam, the thread has to

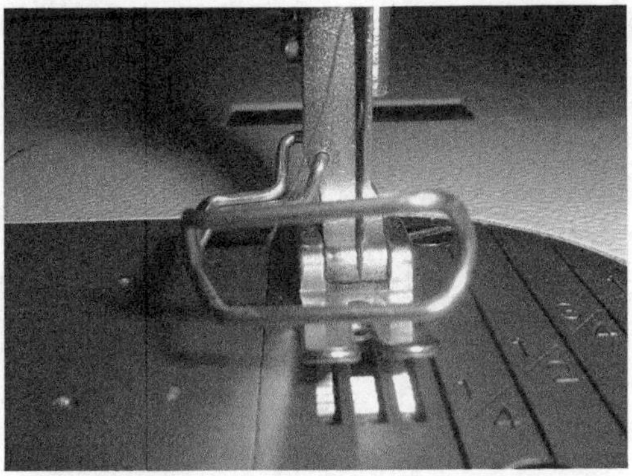

3.59 Example of sewing thread breakage.

be unpicked and the seam sewn again. This is because the joining of the stitch cannot be seen on the face of the garment as the aesthetics of the garment is compromised. Common types of breakage and possible methods of prevention are discussed below.

- Thread channels jagged: check and polish all thread guiding parts. Check for correct thread, run according to operating instructions.
- Faulty threading: check thread run according to operating instructions.
- Blunt, crooked or wrong needle type: insert new needle.
- Needle too high: adjust the needle height.
- Needle too low: insert needle as far as it will go; observe operating instructions.
- Wrong relationship between needle and thread: observe specifications and only use good quality thread.
- Needle threaded from the wrong side: always thread the needle from the long groove side.
- Needle hole nicked by needle: lightly trim the edges, polish the needle hole. If necessary fit new needle plate.
- Needle hole too small or excessively thick needle thread: fit throat plate with larger diameter hole or re-machine the needle hole. Use needle thread according to specifications. Observe relationship between needle and the yarn.
- Badly worn hook, sharp edges: fit new hook and adjust.
- Bobbin case tension spring screws too high, thread catches: tighten screws sufficiently. If this creates an excessive tension, slightly bend the check spring.
- Thread clearance between bobbin case support and hook inadequate: the clearance should be enough to allow the thickest thread gauge to pass through smoothly.

- Hook, bobbin case top and bottom sections nicked by needle: polish or use new parts.
- Too much loose thread as the needle enters/needle pierces loose threads: adjust take-up spring.
- Thread clearance insufficient between hook bottom and bobbin case bottom: increase clearance according to specifications.
- Excessive bobbin thread tension: adjust tension according to operating instructions.

3.12 Needle breakage

Needle breakage is less common but still has a significant impact on the condition of the machine and the sewability of the seam. Many factors can cause this problem. These include:

- wrong hook setting: adjust hook according to specifications
- the needle is bent and clears the hook point: fit new needle
- needle too small for the needle hole or material
- needle protection during its cycle
- wrong needle to thread ratio: observe specifications
- knotty or uneven thread: only use good quality thread
- needle breaks on entering into the material: feed wrong in relation to needle position
- bobbin case incorrectly fitted: press in bobbin case until it clicks into place
- throat plate incorrectly fitted: lightly tighten throat plate set screws in a diagonal sequence and then tighten firmly
- the material is pushed or pulled during sewing: check needle seats, guide more lightly
- feeder too high: move material along during its forward movement or back feeding: adjust feed dog height
- feed dog timing incorrect: adjust feed dog
- hook worn: fit new hook
- needle drops out during sewing: fit new needle screw in needle bar
- needle bar excessively worn: fit new needle bar or bush
- excessive thread tension, needle bends and is caught by the hook point: set correct tension to suit sewing material; see operating instructions.

3.13 Slipped/missed stitching: causes and prevention

Another common occurrence of seam instability and performance is slipped stitching (Fig. 3.60) which can significantly weaken the seam, particularly

3.60 Example of slipstitching.

with chain-stitch. The lockstitch is more resistant to the seam unravelling due to the nature of the stitch formation. However, it spoils the aesthetic look of the garment and reduces the strength of the overall seam. Factors that cause this problem are:

- needle wrong, bent or incorrectly fitted: fit new needle; for correct fitting refer to operating instructions
- faulty threading: observe thread run according to operation instructions
- faulty take up: adjust take-up spring
- wrong hook setting: set up properly
- needle thread ratio incorrect: use needle system
- hook point damaged: polish or insert new hook
- needle hole is too large and material is drawn in: use throat plate having a smaller needle hole; observe purpose of the machine
- needle too low: fit needle as specified in operation instructions
- needle too high: alter needle bar height
- poor needle quality: use good grade needles
- insufficient presser bar pressure, with thick materials the needle raises the material: adjust presser bar
- right-hand twisted thread: left-hand twisted thread should be used in most machines
- hook catches thread loop too early or too late: adjust loop lift according to instructions
- insufficient or excessive tension: adjust the tension to suit the material to be sewn, in accordance with operating instructions
- the thread 'twirls', irregular loop formation, thread too sharply twisted: only use good quality thread with the indicated gauge and twist
- thread unevenly thick and brittle: do not use thread, which has been

stored too long under dry conditions; only use good quality thread in the indicated gauge and thickness.
- needle too far from hook point: adjust clearance between needle and hook point.

3.14 Uneven seams: causes and prevention

Here, faults may be caused in the same manner as indicated under 'thread breakage' and 'slip stitches'. Some explanations are given below:

- looping of threads above or below the material: adjust needle and bobbin thread tension according to operating instructions
- poor and knotty thread: only use good quality thread; for gauge and twist, see instructions
- forward and reverse stitch of different length: observe adjustment instructions
- hook has run out of oil, guide groove rough, bobbin case on edge: observe oiling details in operating instructions; fit new hook and adjust
- faulty threading: observe thread run according to operating instructions
- tensioning discs dirty, fouled or sticky: clean components and, if necessary, re-polish; all parts should be able to move freely
- thread-guiding components rusty or rough: remove rust and re-polish
- thread take-up incorrect: adjust take-up spring; adjust resilience according to material thickness
- needle too high or too low: fit needle according to operating instructions or adjust needle bar height
- thread fails to pass smoothly over hook: re-polish all thread contact points; observe settings
- feed dog setting wrong: set to feed dog figures given in instructions
- fouled hook, hook prevented from rotating evenly: clean hook; observe settings
- machine sews in curve: check feed dog position; check contact of presser foot on feed dog
- bobbin irregularly wound, wrongly inserted or threaded: re-wind, insert and thread according to operating instructions
- machine fails to sew over seams and folds: incorrectly set presser foot pressure; adjust feed dog height; set presser foot pressure to suit material thickness according to operating instructions
- coarse feed dog teeth ruffle the material: use correct feed dog teeth as specified by the manufacturer – for thin, lightweight materials use fine-toothed feed dog; coarse and saw-tooth type feed dogs should be used only for such material for which they have been intended – observe the setting; feed dog follow-up movement

- irregularly wound bobbin thread: when winding up, make sure that the threads lie adjacent to one another.

3.15 Comparing welded and sewn seams

Several authors (Carr and Latham, 1994; Laing and Webster, 1998; Cooklin *et al.*, 2006) have passed comment on welded seams as an alternative to sewn seams, giving an overview of the various technologies employed. In addition, more methodological research has been conducted on welded seams and sewn seams separately. Shi and Little (2000) presented work which, as a result of investigating the possibility of using welded seams to incorporate optical fibres into garments as a data channel, gave experimental credence to the three important weld parameters of amplitude, pressure and weld time. Through a 2^3 factorial experimental design, the research indicated a positive increase in weld strength (of up to 15%) with an increase in pressure and, separately, weld time, until a point is reached when the strength begins to decrease due to polymer orientation or energy dissipation, respectively. It was found that weld strength increased by 80% when the amplitude was increased from 42–60 μm. This research suggests that seams welded at the optimum settings had equivalent strength to sewn seams.

Chung *et al.* (1999) presented the effect of sewn seams on the bending length of fabric samples. The research investigated the effect of seam allowance on the bending length of vertical seams and the effect of seam distance, from the free fabric end, on the bending length of horizontal seams. A rapid increase in the bending length of vertical seams was found with an increase in the seam allowance. As the seam allowance was increased, a gradual tailing off of the effect was experienced. This was attributed to the extra weight of the seam counteracting the increasing stiffness of the seam due to the profile change, relating to the second moment of area, when incorporating larger seam allowances. Conversely, the bending length of the sewn samples decreased as the distance of the horizontal seam from the free end of the sample increased. The work goes on to relate the aspects of bending rigidity, bending length and second moment of area in theoretical equations.

3.15.1 Experimental methodology

An experimental methodology was developed to investigate the load carrying (kg), extension (mm) and bending length (mm) of different seam constructions. Each of the seam types was constructed in the same windproof, waterproof, breathable non-porous polyester hydrophilic fabric with a mass per unit area of 106 g/m^2 and a thickness (measured under 2 gf/cm^2) of 0.28 mm). The seam types created were ultrasonically welded, sewn, and sewn and tape-sealed. Table 3.2 presents the basic machine settings used in the construction of these seams.

3.15.2 Bending length

All seams were constructed parallel to the warp yarns of the fabric (vertical seams; Chung *et al.*, 1999). Both the sewn and the sewn and tape-sealed samples were made into the ISO 4916 seam type 1.01.01 (Fig. 3.61) with an 8 mm seam allowance (the bottom ply being folded under from left to right when the tape was attached to seal the seam) (Fig. 3.62).

The welded seam was constructed as an ISO 4916 2.01.02 (Fig. 3.63) seam type with a narrow overlap of 4 mm which was entirely enclosed in the weld (this type of seam construction has been observed in many of the commercially available welded products).

Eight samples (warp 130 mm × weft 50 mm) of each seam type were cut from the central portion of a conditioned longer length of seam to negate any effects of acceleration or deceleration at the seam ends. The FAST 2 bending length meter was used to measure the length of sample required to achieve a bending angle of 41.5° (Fig. 3.64). Each of the samples was tested with the coating side upward and the sewn sample had the seam allowance folded over to match the sewn and tape-sealed sample.

Table 3.2 Basic machine settings

Setting	Welded	Sewn	Sewn and tape-sealed
Machine type	Pfaff 8304-082	Pfaff 1050	Pfaff 1050 + Pfaff 8304-020
Amplitude	55%		
Pressure	2 bar	40 N	40 N
Weld time	Speed 10		
Stitch length		2 mm	2 mm
Seam tape			19 mm supported film
Sewing speed		1750 spm	1750 spm
Sewing thread		8.4 Tex polyester	8.4 Tex polyester
Needle diameter		80 μm	80 μm

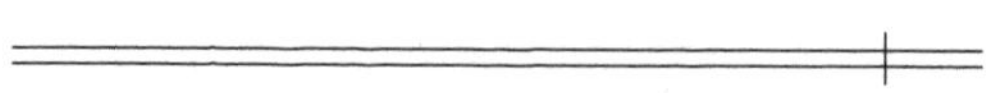

3.61 ISO 4916 seam type 1.01.01.

3.62 Sewn and tape-sealed seam profile.

3.63 ISO 4916 seam type 2.01.02.

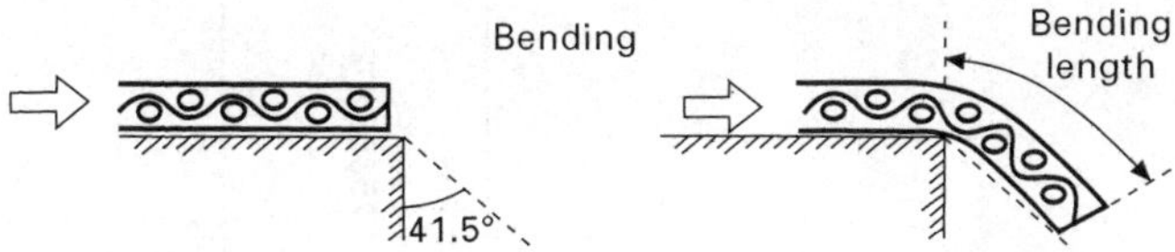

3.64 FAST 2.

3.15.3 Seam load and extension at break

48 samples (warp 200 mm × weft 50 mm) of each seam type were cut from the central portion of a conditioned longer length of seam to negate any effects of acceleration or deceleration at the seam ends. A Testametric constant rate load/elongation test rig was used with the jaws set at a gap of 115 mm. The sample was loaded with the seam central, transverse loading, and the seam allowance facing down and to the rear. With 'peak hold' on, the samples were 'strained' and the 'stop' button depressed at audible and visible signs of seam cracking which was beyond 'necking' or the rupture of only the first millimetres at either end. The load and extension at this point of seam failure were recorded.

3.15.4 Overall results of load carrying and extension tests

Initial ANOVA tests indicated a statistically significant difference (at the 1% level) between the three seam types with respect to their load carrying (Table 3.3) and extension to break (Table 3.4) characteristics but these results are distorted by the much lower variance exhibited by the sewn seams compared to the other two seam construction techniques. The ANOVA of the bending length at the 5% level revealed a statistically significant difference between the three seam construction techniques (Table 3.5).

3.15.5 Comparison of two seam types using the *t*-test

The sewn and tape-sealed seam was shown to carry a statistically significantly higher load than the welded seam at the 1% level (Table 3.6) and extend further to break (Table 3.7). This can be attributed to the double mechanism of the sewn seam and the tape acting to bear the load placed on the seam increasing the load carrying capacity and the gradual peeling of the tape from the main fabric adding to the seam's ability to extend further before failure. The sewn and tape-sealed seam had a longer bending length than the welded seam (at the 5% level) showing the added stiffness gained from the tape and the folded seam allowance contribute to a less compliant seam than a welded seam (Table 3.8).

A *t*-test (at the 1% level) between the welded and the sewn seam (Table 3.9) revealed a greater load carrying capacity in the welded seam (although

Table 3.3 ANOVA for load

Anova: Single factor

SUMMARY

Groups	*Count*	*Sum*	*Average*	*Variance*
Sewn & tape-sealed	48	1615.9	33.66458333	68.56871897
Welded	48	1241.7	25.86875	69.45836436
Sewn	48	648.2	13.50416667	**3.111471631**

ANOVA

Source of variation	*SS*	*df*	*MS*	***F***	*P-value*	***F crit***
Between groups	9921.605417	2	4960.802708	**105.4453769**	9.95849E-29	**4.758931027**
Within groups	6633.512083	141	47.04618499			
Total	16555.1175	143				

Table 3.4 ANOVA for extension

Anova: Single factor

SUMMARY

Groups	*Count*	*Sum*	*Average*	*Variance*
Sewn & tape-sealed	48	940.18	19.58708333	13.14493174
Welded	48	513.69	10.701875	13.69959003
Sewn	48	579.63	12.075625	**0.858833644**

ANOVA

Source of variation	*SS*	*df*	*MS*	***F***	*P-value*	***F crit***
Between groups	2196.09796	2	1098.04898	**118.9078684**	5.50519E-31	**4.758931027**
Within groups	1302.057704	141	9.234451803			
Total	3498.155664	143				

Table 3.5 ANOVA for bending length

Anova: Single factor

SUMMARY

Groups	*Count*	*Sum*	*Average*	*Variance*
Sewn & tape-sealed	8	291.5	36.4375	3.959821429
Welded	8	272.5	34.0625	2.245535714
Sewn	8	272	34	2.285714286

ANOVA

Source of variation	*SS*	*df*	*MS*	***F***	*P-value*	***F crit***
Between groups	30.89583333	2	15.44791667	**5.457939012**	0.012337615	**3.466794851**
Within groups	59.4375	21	2.830357143			
Total	90.33333333	23				

Table 3.6 t-Test sewn and tape-sealed versus welded: load

t-Test: Two-sample assuming equal variances

	Sewn and tape-sealed	*Welded*
Mean	**33.66458333**	**25.86875**
Variance	68.56871897	69.45836436
Observations	48	48
Pooled variance	69.01354167	
Hypothesised mean difference	0	
df	94	
t-Stat	**4.597280566**	
P (*T*<=*t*) one-tail	6.65775E-06	
t Critical one-tail	2.366668923	
P (*T*<=*t*) two-tail	1.33155E-05	
t Critical two-tail	**2.629149094**	

Table 3.7 t-Test sewn and tape-sealed versus welded: extension

t-Test: Two-sample assuming equal variances

	Sewn and tape-sealed	*Welded*
Mean	**19.58708333**	**10.701875**
Variance	13.14493174	13.69959003
Observations	48	48
Pooled variance	13.42226088	
Hypothesised mean difference	0	
df	94	
t Stat	**11.88120251**	
P (*T*<=*t*) one-tail	1.0086E-20	
t Critical one-tail	2.366668923	
P (*T*<=*t*) two-tail	2.01721E-20	
t Critical two-tail	**2.629149094**	

Table 3.8 t-Test sewn and tape-sealed versus welded: bending length

t-Test: Two-sample assuming equal variances

	Sewn and tape-sealed	*Welded*
Mean	**36.4375**	**34.0625**
Variance	3.959821429	2.245535714
Observations	8	8
Pooled variance	3.102678571	
Hypothesised mean difference	0	
df	14	
t Stat	**2.696653935**	
P (*T*<=*t*) one-tail	0.008685043	
t Critical one-tail	1.76130925	
P (*T*<=*t*) two-tail	0.017370087	
t Critical two-tail	**2.144788596**	

Table 3.9 *t*-Test welded versus sewn: load

t-Test: Two-sample assuming unequal variances

	Welded	*Sewn*
Mean	**25.86875**	**13.50416667**
Variance	69.45836436	3.111471631
Observations	48	48
Hypothesised mean difference	0	
df	51	
t Stat	**10.05592519**	
P (*T*<=*t*) one-tail	5.33085E-14	
t Critical one-tail	2.401720849	
P (*T*<=*t*) two-tail	1.06617E-13	
t Critical two-tail	**2.675733413**	

Table 3.10 Seam strength (source: Anon., 1981)

Fabric	Thread	s.p.cm.	Average seam strength (kg/cm)	Average welded strength (kg/cm)
Knitted Nylon 6	Koban Tkt. 120	4.7	4.40 material failed	0.64
Knitted Nylon 6	Koban Tkt. 120	4.7	4.80 material failed	1.07
Knitted Nylon 6	Koban Tkt. 120	4.7	6.14 thread and material failed	0.84
Knitted Nylon 6	Drima-T Tkt. 120	3.9	5.15 thread failed	1.59
Woven Nylon 6	Koban Tkt. 120	4.7	1.97 material failed	0.34
Woven Nylon 6	Koban Tkt. 120	4.7	4.83 material failed	1.13
Woven Nylon 6	Koban Tkt. 120	4.7	7.00 thread failed	1.27
Knitted Nylon 66	Koban Tkt. 120	6.3	4.92 material failed	1.20
Knitted Nylon 66	Koban Tkt. 120	6.3	3.02 material failed	0.66

the variance in these seams is much higher than in the sewn seams with comparative *Cv* values of 32.22% and 13.06%) compared to the sewn seams. This contradicts early work presented in the COATS technical manual (Anon., 1981) which proposed that excessive weld pressure could result in parallel orientation of the polymer chains leading to a weakening of the join due to reduced polymer elasticity. Welded seams were generally weaker than sewn seams (Table 3.10) and have poor peel bond strength. Welded seams may give waterproof seams (no stitching holes and no 'wicking' thread), but the fabric strength can be reduced by as much as 60% and the seam strength may be as little as 50% of a sewn seam (Table 3.10).

The strength of the welded ISO 4916 2.01.02 seam over the sewn ISO 4916 1.01.01 seam could be due not only to the weld strength but the increased surface area of fabric actually joined together and the negation of any peel effect which is a principal mechanism of seam failure. However, the nature of the sewn seam – with inter-yarn and sewing thread friction being the force holding the seam together – allows it to open/extend further before rupture than the welded seam as illustrated by the *t*-test (at the 5% level) in Table 3.11.

t-Tests at both the 1% and 5% (Table 3.12) level revealed no significant difference in the bending length between the welded and the sewn seams indicating that the main contributor to a stiffer, less compliant or malleable, seam is the sealing tape used to give a waterproof finish to sewn seams.

3.15.6 Conclusions

This small-scale study shows the welded seam to have a lower bending length than the sewn and tape-sealed seam giving some credence to the claims that welded seams in performance outdoor garments contribute to increased garment compliance and reduced resistance to bending. However, in both properties of transverse load carrying and extension (primary stress

Table 3.11 *t*-Test welded versus sewn: extension

t-Test: Two-sample assuming unequal variances

	Welded	*Sewn*
Mean	**10.701875**	**12.075625**
Variance	13.69959003	0.858833644
Observations	48	48
Hypothesised mean difference	0	
df	53	
t Stat	**−2.494429002**	
P (*T*<=*t*) one-tail	0.007884302	
t Critical one-tail	1.674115993	
P (*T*<=*t*) two-tail	0.015768603	
t Critical two-tail	**2.005745046**	

Table 3.12 *t*-Test welded versus sewn: bending

t-Test: Two-sample assuming equal variances

	Welded	*Sewn*
Mean	**34.0625**	**34**
Variance	2.2455357	2.2857143
Observations	8	8
Pooled variance	2.265625	
Hypothesised mean difference	0	
df	14	
t Stat	**0.0830455**	
P (*T*<=*t*) one-tail	0.4674955	
t Critical one-tail	1.7613092	
P (*T*<=*t*) two-tail	0.9349911	
t Critical two-tail	**2.1447886**	

Table 3.13 Summary of performance of the three seams

	Welded	Sewn and tape-sealed	Sewn
Load	Next best	Best	Worst
Extension	Worst	Best	Next best
Bending length	Best	Worst	Best

orientations leading to failure in seams) the welded seam underperformed compared to the sewn and tape-sealed seam. These findings go some way to giving more methodological support to anecdotal evidence that garments with welded seams are less durable than their counterparts with sewn and tape-sealed seams. The performance of the three seams is summarised in Table 3.13.

Further work in this area is to be conducted on larger numbers of seamed samples for the bending tests in addition to the completion of field trials currently being piloted.

3.16 Future trends

The sewing and joining of textiles is an infinite process that has evolved over many hundreds of years, and this method of production looks set to continue for the foreseeable future. Research into problems associated with identifying seam quality is useful in developing techniques and methods to combat production problems. This research can be disseminated in language that production people in manufacturing can understand and can provide useful information for production. This enables problems to be understood and particularly solved. The academic approach in the analysis of sewability problems and the practical approach to solving them will enable a greater understanding of how and why these problems occur and provide the expertise to solve them.

3.17 References and further reading

Anon. (1981), 'Coats: threads and seams', *Coats Technical Manual*, Coats, Uxbridge.

Carr, H. and Latham, B. (1994), *The Technology of Clothing Manufacture*, 2nd edn. Blackwell Scientific, Oxford.

Chung, S., Hu, J. and Lo, M. (1999), 'Effect of seam allowance of a plain seam on bending of woven fabrics', *Research Journal of Textiles and Apparel*, 3(1).

Cooklin, G., Hayes, S. and McLoughlin, J. (2006), *Introduction to Clothing Manufacture*, 2nd edn. Blackwell Publishing, Oxford.

Ferreira, F. B. N., Harlock, S. and Grosberg, P. (1994a), 'A study of thread tensions on a Lockstitch Sewing Machine (Part 1)', *International Journal of Clothing Science and Technology*, 6(1), 14–19.

Ferreira, F. B. N., Harlock, S. and Grosberg, P. (1994b), 'A study of thread tensions on a Lockstitch Sewing Machine (Part 2)', *International Journal of Clothing Science and Technology*, 6(5), pp. 26–29.

Ferreira, F. B. N., Harlock, S. C. and Grosberg, P. (1994c), 'A study of thread tensions on a lockstitch sewing machine (Part III): Further Stitch Formation Analysis', *International Journal of Clothing Science and Technology*, 6(5).

International Organisation for Standardisation (1991), Stitch types – Classification and Terminology. ISO, Geneva.

Kennon, W. R. and Hayes, S. G. (2000), 'The effects of feed retardation on lockstitch sewing', *Journal of the Textile Institute*, 91, Part 1, 509–522.

Laing, R. M. and Webster, J. (1998), Stitches and Seams. The Textile Institute, Manchester.

Mallet, E. and Du, R. (1999), 'Finite element analysis of sewing process', *International Journal of Clothing Science and Technology*, 11(1), 19–36.

McLaren Miller, J. (1998), 'An analysis of lockstitch seam instability in the cross–grain construction of woven fabrics', PhD Thesis, pp. 4, 15.

McLoughlin, J. (1998), 'The expanding role of the clothing machine engineer', *World Clothing Manufacturer*, 79, 7, p 37–41.

McLoughlin, J. (1999), 'Implementation of a zero breakdown strategy', *World Clothing Manufacturer*, 80, 1, p 12–16.

McLoughlin. J. (2000), 'Time to value the production worker', *World Clothing Manufacturer*, 81, 3, p 16–21.

McLoughlin, J. (2005), 'Development of an automated reporting method for the analysis of sewability measurement', MPhil, Thesis, University of Manchester, pp. 99–100.

McLoughlin, J. and Hayes, S. G. (2007), 'Automating objective fabric reporting', in Ariadurai, S.A. and Wimalaweera, W.A. (eds), *The Textile Institute 85th World Conference*, 1–3 May 2007, Colombo, pp. 568–582.

McWaters, S. D. and Clapp, T. G. (1994), 'Computer simulation of fabric deformation for the design of equipment', *International Journal of Clothing Science and Technology*, 6(5), 30–38.

Needles Eye (1996), 'The right adjustment needs the right gauge', *Needles Eye*, 390, 24–28.

Schmetz (2000), *Guide to Sewing Techniques*, 3rd edn. Schmetz GmbH, Herzogenrath, Germany.

Shi, W. and Little, T. (2000), 'Mechanisms of ultrasonic joining of textile materials', *International Journal of Clothing Science and Technology*, 12(5), 331–350.

Stylios, G. (1983), 'Seam pucker and structural jamming in woven textiles', *MSc Thesis*, University of Leeds.

Stylios, G. (1997), 'Automation of sewing machine settings in difficult-to-see fabrics using objective measurement technologies', *International Journal of Clothing Science and Technology*, 9(6), 7–9.

Stylios, G. and Lloyd, D. W. (1989), 'A technique for identification of seam pucker due to fabric structural jamming', *International Journal of Clothing Science and Technology*, 1(2), 25–27.

Stylios, G. and Sotomi, J. O. (1996), 'Thinking sewing machines for intelligent garment manufacture', *International Journal of Clothing Science and Technology*, 8(1–2) 44–55.

Union Special (1988), Stitch Formation Lockstitch Type 301, Technical Papers.

Zunic-Lojen, D. (1998), 'Simulation of sewing machine mechanisms using program package ADAMS', *International Journal of Clothing Science and Technology*, 10(3/4), 219–225.

Zunic-Lojen, D. and Gotih, K. (2003), 'Computer simulation of needle and take-up lever mechanism using the ADAMS software package', *Fibres and Textiles in Eastern Europe*, 11(4), 39–44.

4
Mechanisms of sewing machines

J. MCLOUGHLIN and A. MITCHELL, Manchester Metropolitan University, UK

DOI: 10.1533/9780857093967.1.123

Abstract: This chapter describes the evolution of the sewing machine from its basic origins into a hybrid of machine engineering capable of producing high-quality and high-precision clothing. A comprehensive review of all the different automated machines on the market is beyond the scope of this chapter; however, the following sections will outline the recent developments in machine design and development that has been devised towards the pursuit of automated sewing.

Key words: sewing machine, automation, automated sewing, stitching, machine engineering.

4.1 Introduction

The principles of the sewing machine and stitch formations have hardly changed since their development over 200 years ago. However, significant progress has been made to the technologies that enhance the machine performance. Major developments in machine equipment were created resulting in the Transfer Street System developed by Pfaff (Germany) in the 1960s. This system enabled the production of a shirt in a sequence of manufacturing operations. The technology used then would not be recognised today as the motors used were substantial. They were bulky and very heavy, slow running and prone to failing at inopportune moments. This system was also very expensive, putting it beyond the financial scope of even the biggest clothing manufacturing companies.

Since the mid 1980s many technological advances have been made to the sewing machinery which is used for the production of clothing. New construction materials and improved electronic systems have played an important role in these advances. Variable speed electronic motors, including machines such as bartackers and automatic seamers, have to sew over considerably different thicknesses of materials and seam constructions. If the machine had one sewing speed only, snapped threads and broken needles were a frequent occurrence when sewing over varying thicknesses. It is now possible to adjust the machine's sewing speed according to what has to be sewn, and this capability makes a positive contribution to the improvement of productivity.

Sewing machine manufacturers are now seriously taking into account the operators who will use their machines and are incorporating many of the following items into their designs:

- Machines now have large clearances in their main working areas, which facilitate the easy and comfortable manipulation of work during sewing. The positions and shapes of a machine's control elements are ergonomically designed so as to suit the natural movement range and convenience of the operator.
- In accordance with new safety regulations, machines have built-in safety devices, which protect the operator and the machine against operational errors.
- Specially designed stands are available for machines where the operator works in a standing instead of a seated position. The standing working position is a distinctive feature of quick-response types of production systems.
- Machines are painted in colours that are easy on the eye and which fit in with the dynamic colour schemes used in modern factories.

These are just a few of the innovations, which, along with many others, are gradually becoming standard features for all types of sewing machines.

This chapter describes the evolution of the sewing machine from its basic origins into a hybrid of machine engineering capable of producing high-quality and high-precision clothing. A comprehensive review of all the different automated machines on the market is beyond the scope of this chapter; however, the following sections will outline the recent developments in machine design and development which have been devised in the pursuit of automated sewing.

4.2 The evolution of the sewing machine

An explorer once followed the headwaters of the Yangtze Kiang River to a Tibetan village in the heart of Asia. So primitive was the place that he was confident that no other outsider had been there. He was congratulating himself when a familiar hum fell on his ear. Tracing the sound he found a native crouched on the floor of a mud hut using a Singer sewing machine. The ubiquitous Singer (Fig. 4.1) is the best known and was at one time the most widely used product in the world. These types of sewing machines evolved from the earliest machines that were developed in the 1750s, the earliest being made from wood. Barthélemy Thimonnier (France) built 80 wooden sewing machines, apparently all capable of producing efficient stitches, but his success was brief (Fig. 4.2). Tailors, worried about their livelihoods, raided his shop and completely demolished his premises and tools. He barely escaped with this life. It is important to point out that with

4.1 An early version of the Singer sewing machine.

4.2 An example of Barthélemy Thimonnier's machine.

advances in technology, significant job losses have followed. More information is given about this later in the chapter.

The first Singer sewing machine (Fig. 4.3) was built by Isaac M. Singer in accordance with his patent of 1851. It is a lockstitch machine of which the shuttle is propelled by a driver moved by a crank pin on the lower shaft beneath the table. The needle motion is obtained from a crank pin on the upper shaft and is straight and vertical. The cloth feed is provided by a finely grooved and serrated wheel which is worked intermittently by a band worked by a rocking lever from a cam on the underneath shaft. The machine was packed in a box, which, when opened, served as a stand and contained a treadle connected by a pitman rod with a handle on the balance wheel. The treadle was pivoted near the centre and was worked by heel-and-toe action. Singer did not realise that he could also patent the treadle, but when this was pointed out to him, it was too late, for it was then already in public use. Although this machine is very old, there are similarities between this type of machine and the modern machines used in production today.

4.3 Machine categorisation

Machines are normally characterised into three categories:

4.3 Example of one of the first industrial Singer sewing machines.

1. Integrated Sewing Units – ISU (basic sewing machines)
2. Mechanised (fixed long and short cycle)
3. Semi-automated including automated workstations and transfer lines.

Modern equipment used in production is capable of producing high sewing speeds. In lockstitch, speeds of 5,000–6,000 stitches per minute are not uncommon, and chain-stitch machines can produce up to 8,000–10,000 stitches per minute depending on the type of machine. Even though these machines are now generally classified as basic, it can be seen from the examples below that there is nothing basic about them.

4.4 Integrated stitching unit (ISU)

This machine is the basic workhorse of the clothing industry (Fig. 4.4) and has numerous applications due to the great variety of work aids and feed systems available for it. Some of the main features of this class of machine are:

- automatic foot lift
- automatic bartacking
- stitch counting (able to count a programmable number of stitches and then automatically stop)
- automatic under-bed thread trimming (eliminating the need to use scissors to cut the sewing threads)
- automatic fabric ply sensing.

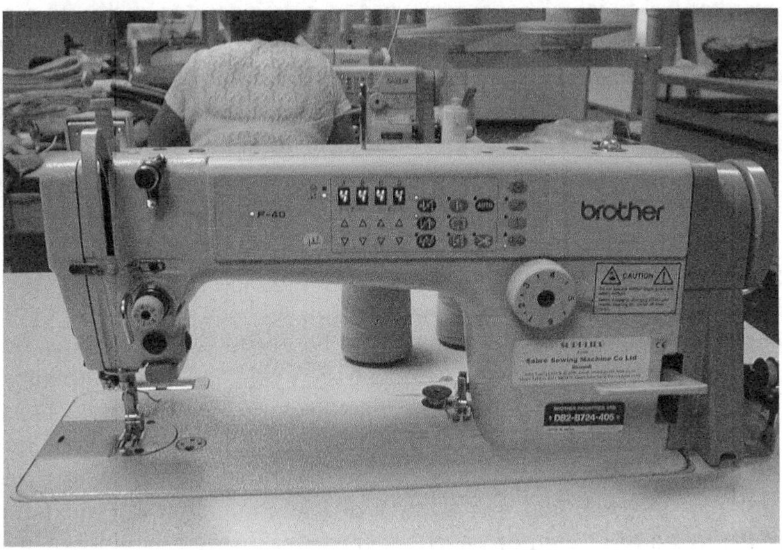

4.4 Integrated stitching unit (ISU) machine.

They are generally available in three versions for sewing light, medium or heavy weight materials.

With the addition of a programmable microprocessor control system, the machine can be programmed for repetitive operations such as sewing around labels or stitching simple decorative elements. Many versions of this machine have bobbins, which hold 50% more thread than the conventional bobbin. An electronic monitoring system offers an early warning that the bobbin needs changing. This ensures that the operative will complete a seam prior to the bobbin running out, thus avoiding unacceptable thread joins or costly re-working of seams.

Electronically controlled thread cutting, tacking, needle positioning and foot lifting systems have become standard features on these and many other machines. They have contributed to a reduction in operator fatigue by removing some of the repetitive strain actions such as lifting the foot manually by pressing a knee lifter or using scissors to cut the sewing threads, whilst other features facilitate accuracy and consistency in sewing methods.

Under-bed thread trimmers (UBT) are not a new idea; they have been used extensively on automatic cycle sewing machines for many years. Automatic cycle sewing machines, such as buttoners, buttonhole and bartack machines have been equipped with knives to sever the thread automatically at the end of each sewing cycle for over 50 years. What is relevant is that sewing machines other than automatic cycle sewing machines such as plain lockstitch, twin needle lockstitch, blind stitch and some chain-stitch machines are being designed and produced with options at extra initial cost to the purchaser, of added facilities, such as the automatic severing of threads whenever the operator has finished a seam or whenever the operator wishes to remove the work from under the presser foot. The design and construction of automatic cycle sewing machines (as distinct from automated sewing units) are still in many instances 100% mechanical, apart from the power supply which is, of course, electrical. The design of the feed mechanisms and the starting and stopping motions of automatic cycle sewing machines had to some extent a limiting effect on the maximum rpm to which the machine could operate successfully. Thus the inclusion in the cycle of a thread trimming mechanism posed very few problems. In recent times, developments in mechanical engineering have allowed speed control, and the ability to control speed has allowed under-thread trimmers to be added to all machine categories, including ISU categories as well as fixed cycle and semi automatics.

Most automatic cycle sewing machines operate at between 2,000 and 3,000 rpm, whereas plain lockstitch sewing machines far exceed this figure. Because the rpm is dependent on the design and manufacture, one of the problems to overcome was a quick deceleration from the maximum speed at the finish of a seam of up to 8,000 rpm to a speed which would allow an under-bed thread trimming mechanism to operate successfully and to allow the needle to be positioned accurately, first in the down position and then in the up position. The most suitable speed for the functions described will range between 180 and 220 rpm depending on the design and manufacture.

As mentioned previously, without doubt, the addition of UBT mechanisms to plain lockstitch sewing machines has been largely the result of developments to the electric and electronic motor. The development of the electronic circuit board of sewing machine motors has made it possible to have many other functions added to the normal start, drive and stop of the conventional electric sewing machine motor. Among the most notable of these are:

- needle positioning (down and up)
- automatic backtacking (beginning or end or both)
- pre-set number of stitches in the backtack
- automatic lifting of the presser foot after trimming (or in the middle of a seam)
- thread wiping (away from the presser foot)
- stitch counting.

When a plain lockstitch sewing machine is designed with the above facilities programmed to function in sequence, it is very often referred to as an integrated sewing unit (ISU).

This is not to say that an under-bed thread trimming attachment is wholly electric or electronic; the knife action, positioning and timing still remain mechanically controlled and operated, which is necessary to allow the thread catcher to operate at the correct time in the stitch forming cycle. The electric or electronic motor functions in under-bed thread machines can be summarised as follows:

- decelerate the sewing machine speed (rpm) to a speed consistent with the mechanical action of thread catcher mechanism
- automatically position the needle in the down position
- actuate the thread catcher mechanism
- automatically position the needle in the up position.

It is necessary at this stage to define the needle down position. On the earlier designs of the under-bed thread trimming machines, the needle down position was bottom dead centre (BDC). This posed a problem, as on occasion it was found that due to the needle being positioned at BDC and consequently the needle thread not reaching the loop forming stage, it was possible that the rotating hook failed to pick up the needle thread. This resulted in a slip stitch and as a result the thread catcher was unable to pick up the threads. Also, when the operator was required to turn a corner, the same would apply and invariably the machine would slip a stitch.

Later developments of under-bed thread trimming machines defined the needle down position as when the needle had reached BDC and then risen approx. 4 mm. This ensured that the rotating hook had secured the thread loop formed by the needle thread and so the possibility of a slipstitch was eliminated.

The next stage in the under-bed thread trimming cycle requires that the operator back heels the treadle. This action will automatically either mechanically or electronically (solenoid operated) actuate the thread catcher movement. The continuation of this movement will bring both the upper and lower threads to a position where they can be cut by the knife. At this stage the needle will rise to its highest position and the machine will stop.

Again we need to define the needle position. It is commonly referred to as the 'needle up' position, but this can prove misleading. In fact, the stopping position of the needle after the threads have been severed is when the needle has reached its highest position and has just begun its descent. This position is crucial for a successful thread trimming action. Why is it so crucial?

The synchronisation of the thread catcher and knife timing require that the thread catcher and the upper and lower threads are brought to a position where they will be cut by the knife at the precise moment when the take-up lever reaches top dead centre. Otherwise, if thread trimming occurs earlier, the upward travel of the take-up will draw thread from the needle, which at best will leave a short end of thread to be drawn completely out of the needle eye.

The length of the thread end at both the start and finish of a seam using a machine with a UBT attachment can be controlled within certain limitations, the main one being that the machine requires a definite length of thread to commence sewing, otherwise the stitching will not commence at the position intended. At the finish of a seam, the thread trimming occurs beneath the throat plate, and consequently, the thread end at the finish will be consistent with the distance between the under-bed thread trimming knife and the underside of the material.

Having established that an under-bed thread trimmer will leave ends of threads consistent with trouble-free sewing, this will vary with the design and construction of under-bed thread trimming mechanisms. The factors that control the thread end lengths, other than design, can be summarised as follows: elasticity of the thread being used, timing and positioning of the thread catcher, sharpness of the trimming knife, distance the knife is positioned from the path of the needle, tension release (degree of release and timing), auxiliary tension regulation, and thread wiper.

Elasticity of the thread being used

Awareness of thread elasticity is essential, specifically to the prevention of it relaxing and ravelling out of the needle eye at the moment of cutting.

This can be overcome by a certain degree of tension release applied to the thread, but unfortunately practice shows that the elasticity of the thread is not always constant, and therefore problems arise.

Thread catcher and knife timing

This has already been discussed; in short if the thread trimming occurs prior to the take-up lever reaching its highest position, too short an end could be the result.

Sharpness of the trimming knife

The degree of sharpness of the knife is very important; a dull knife will result in a variation of thread length end, and could even affect the cutting as a whole if the knife blade is very dull.

Distance knife is positioned from the needle

This can affect the length of the end of the thread to a minor degree. The majority of the designs of UBTs have the facility to allow the trimming knife to be adjusted in relation to the needle, but the positioning is to some extent governed by other factors such as the stitch-forming implements, and the finishing position of the needle height at the end of the trimming action.

Tension release (degree of release and timing)

The amount of tension release applied (irrespective of the normal tension release applied when raising the foot, which is common to all sewing machines) and the timing of the release during the thread cutting action is very important. Stated simply, the more the tension is under release as the thread catcher is making its forward travel, the less the thread is stretched, and consequently the longer the thread end.

Auxiliary tension setting

Later models of the plain lockstitch sewing machines equipped with UBTs were designed with an auxiliary tension, prior to the main tension, replacing the normal thread guide. This simplifies to some extent the regulation of the thread length end. The same principles apply as with the main tension release, but you can adjust the auxiliary tension to gain a more accurate and consistent length of starting thread by simple adjustment of the auxiliary tension-regulating nut.

Thread wiper

Some designs of sewing machines equipped with UBT attachments use a thread wiper, which is electronically operated. This is a mixed blessing. The principles of operation are as follows. As the thread is severed, the wiper takes the upper thread from the needle away from the area of the presser foot to avoid the possibility of the thread being trapped under the foot when starting to sew. In some instances, if the machine is less than 100% correctly adjusted with regard to the under-bed thread trimming attachment, it can result in the thread being drawn out of the needle eye on the first stitch. It should also be stated that whilst a thread wiper will assist in the removal of any excess starting thread being visible on the right side of the garment, this starting end of thread doesn't just disappear; rather, it is transferred to the underside of the garment and usually creates an untidy stitch on the underside of the garment for the first two or three stitches.

Most of the major sewing machine manufacturers supply a range of machines equipped with under-bed attachments and control panels attached to the sewing machine motor to carry out some of the handling operations which previously required the skill of the operator. This undoubtedly simplifies the training required for operators in the correct handling of the sewing machine. Thus handling time is reduced, productivity is increased and quality is improved, particularly on short operations and intricate small parts. Eliminating the need for scissors reduces handling time. Eliminating the need for the operator to position the needle in the down or the up position also increases productivity. Quality is improved by the programming of the automatic backtacking to the pre-set number of stitches required to secure the seam.

The design of under-bed thread trimming attachments and the control panels will vary with the individual sewing machine manufacturers, but the principles of the operation will remain the same.

4.5 Types of motors used in sewing machines

Several different types of motor exist today and include:

- manual friction clutch (Fig. 4.5, top)
- electromagnetic clutch motor (Fig. 4.5, bottom)
- direct current (DC) motor
- alternating current (AC) motor.

The first two can be described as continually turning motors; these are most commonly powered by three-phase electricity and have an output in the order of half a horsepower.

The manual friction clutch motor operates as follows. The sewing machine is connected to the pulley wheel of the clutch assembly via a V-belt. When

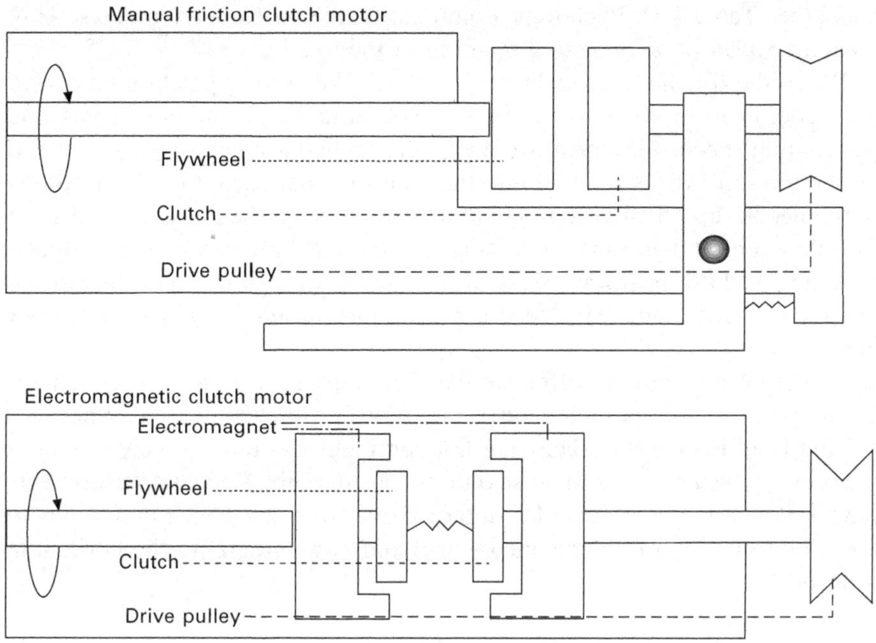

4.5 Manual friction (top) and electromagnetic (bottom) clutch motors.

the treadle is depressed, the clutch plate is pulled onto the continually revolving flywheel of the motor. The maximum speed of such a motor is around 2800 rpm. The speed of the machine could be altered to make the machine run faster or more slowly; by altering the ratio of drive pulley sizes, the machine can be made to run at speeds greater or lower than this. More recent developments as mentioned above have facilitated more control over electronic motors, which are explained in more detail in Section 4.10.

The electromagnetic clutch motor operates as follows. The operation is similar to that of the manual friction clutch motor in that a clutch plate is drawn onto a continually running flywheel. However, in this instance the clutch plate is drawn onto the flywheel by an electromagnet. The forward movement of the treadle governs the strength of the electromagnetic force attracting the clutch plate assembly. It was the use of this type of motor that allowed the development of electronic positioning motors. A positioning motor allows greater control over the action of the sewing machine. It means that each time the machine stops – or is required to do a certain task – the position of the top shaft (and hence all other mechanical elements) can be guaranteed. It is rare to find a lockstitch machine without one.

The needle positioning system operates as follows. The treadle is connected to a code switch, which generates a grey code binary output (a binary sequence where only one bit changes each time) on the motor control panel (see Table 4.1). Each logic combination relates to a step in the sewing machine speed or an actuating position of the treadle.

When the treadle is brought back through the resting position (show on grey code) the machine is running at 200 stitches per minute (spm), the positioning speed. An optical shaft encoder (called a synchroniser) mounted on the top shaft feeds a signal into the control panel each time the needle is in the needle-up position (take-up at top of stroke) or needle-down position (needle risen by 3–4 mm), indicating that the machine can now be stopped in either of these positions. This allows automatic trimming of threads or presser foot lift along with needle positioning to allow entry of the work piece.

Direct current motors offer similar functions in a more direct manner. There is no clutch assembly on a DC motor, which only runs when the sewing head is in action. They are not continuously running which reduces the power consumption and noise output considerably. Recent developments have seen the more compact DC motor move from sub-bench mountings to a position directly on the balance wheel and now integral to the horizontal arm itself.

4.6 Three-thread overlock with a microprocessor

Based on a regular overlock machine, this version is used for the assembly seaming of garments made up in lightweight knitted materials (Fig. 4.6). The microprocessor enables the accurate calibration of stitch tension without manually adjusting the spring tensioners, and permits automatic cutting of

Table 4.1 Example of the digital grey scale used on the ISU machine

Step	Logic code
FL + T	1100
FL	1000
0	0000
1/2	0100
1	0110
2	1110
3	1010
4	0010
5	0011
6	1011
7	1111
8	0111
9	0101
10	1101
11	1001
12	0001

4.6 Overlocking machine with a microprocessor.

the thread chain at the beginning and end of seams. A 'tractor-foot' can be fitted to this machine when sewing seams with varying thicknesses.

4.7 Mechanised sewing machines

These are machines, which still require the operator to place and control parts within the sewing area, but due to their characteristic fixed cycle design, they are able to automatically sew a pre-determined stitch line or pattern. Examples of these are:

- bartack machines
- buttonhole machines
- button sewing machines
- spot tack machines.

4.7.1 Lockstitch bartacker

This class of machine has a wide range of applications apart from regular bartacking operations (Fig. 4.7). It can easily be converted to sew an endless variety of stitch patterns in an area 6 cm where the pattern contains up to 72 stitches. Sewing speed is adjustable according to the thickness of the part being sewn, and the free cylinder arm makes handling easier when bartacking on cylindrical sections of garments such as belt loops, pocket corners and trouser flies.

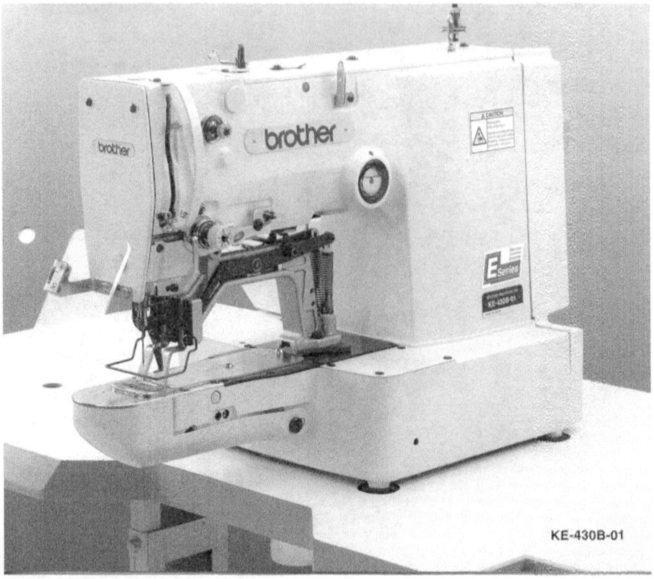

4.7 Lockstitch bartacker.

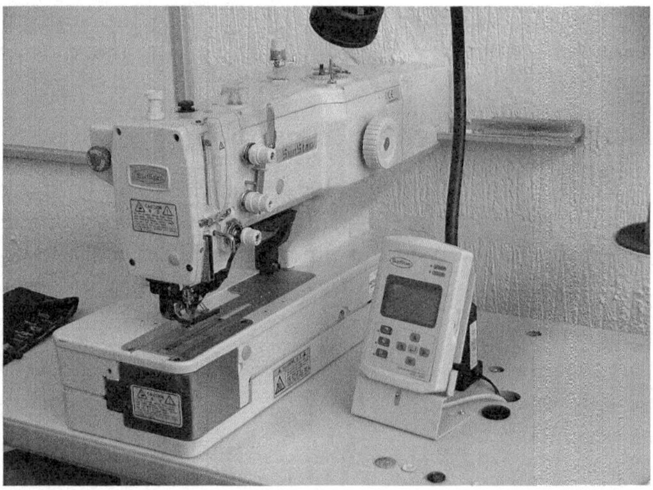

4.8 Lockstitch buttonhole machine.

4.7.2 Lockstitch buttonhole machine

The machine shown in Fig. 4.8 is for sewing buttonholes in lightweight materials, such as those used for blouses, dresses and shirts. In one continuous operation, the machine sews a rectangular shaped buttonhole by means of purl stitch, and both ends are secured by a bartack. The machine is operated by one pedal only, and the head can be mounted in the length or across the width of the work-top, depending on whether vertical or horizontal buttonholes are being sewn.

4.7.3 Blind spot tacker

These machines are designed for sewing components which are to be concealed within garments such as shoulder pads. They are available in different versions for blind stitching the hems and facings of garments made in materials ranging from sheer to heavy weight (Fig. 4.9). The machine forms a single-thread stitch and can sew up to 3,000 stitches per minute. Stitch penetration control is digitally displayed and most machines are fitted with thread trimmers and a stitch-condensing device, which prevents the end of the sewing from unraveling.

4.7.4 Single-thread button sewing machine

The machine shown in Fig. 4.10 has an electronically controlled system, which automatically feeds correctly positioned buttons directly from the button hopper on the right into the button clamp. It is easily adjusted for sewing two- or four-hole buttons, and changing the number of stitches, say from 10 to 20, just requires pressing a knob on the front of the machine. A special version of this machine has a needle penetration control whereby just two of the total stitches sewn are visible on the facing, whilst the others are sewn through the forepart and fusible only.

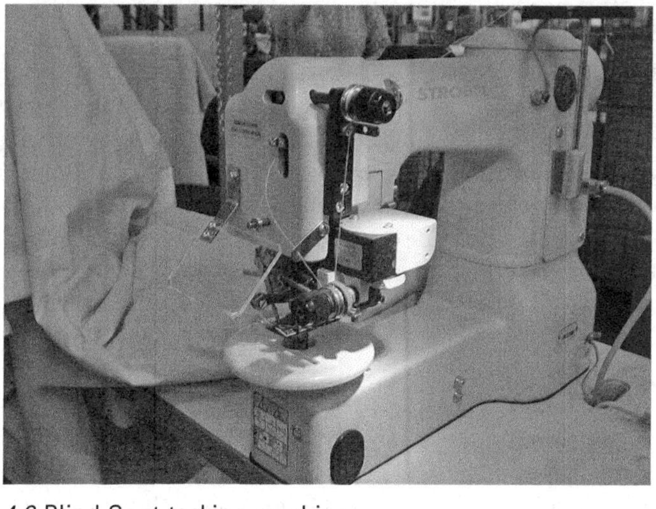

4.9 Blind Spot tacking machine.

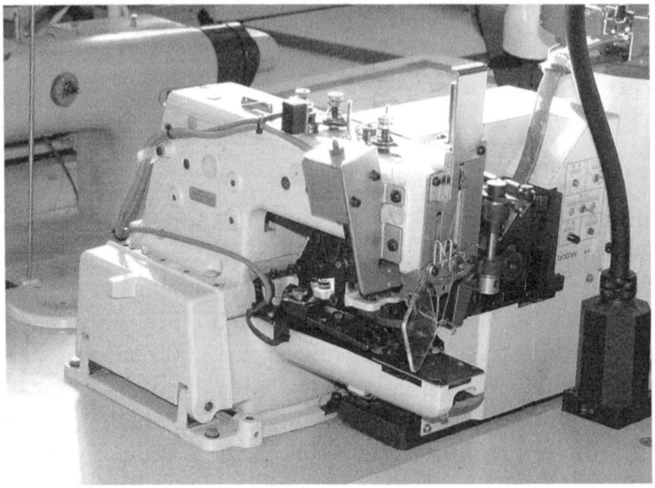

4.10 Single thread button sewing machine.

4.8 Semi-automatic machines, automated workstations and transfer lines

Semi-automatic machines are capable of doing one set of operations on the same garment. Examples of these would be:

- pocket setting machine (Fig. 4.11)
- profile stitching (Fig. 4.12 and 4.13)
- long seaming operations, e.g. on trousers, curtains, etc. (Figs 4.14 and 4.15)
- semi-automatic bartacking (Figs 4.16 and 4.17).

These machines employ state-of-the-art equipment enabling computerised technology to program, for example, the number of buttonholes on a garment. They are also used to produce consistent and accurate sewing on many types of products, for instance the precise sewing of pockets on a pair of jeans. Embroidery machines are good examples of stitching techniques. Each embroidered design has been developed to be accurate and defined, to the point where technology has advanced to a stage that allows the minutest detail of a motif or graphic to be produced.

4.8.1 Pocket setting machines

Pocket setting machines or pocket setters have been designed to sew various different pocket shapes (Fig. 4.11). The machine consists of pocket folding jigs that can automatically fold the pocket to shape. This equipment is very

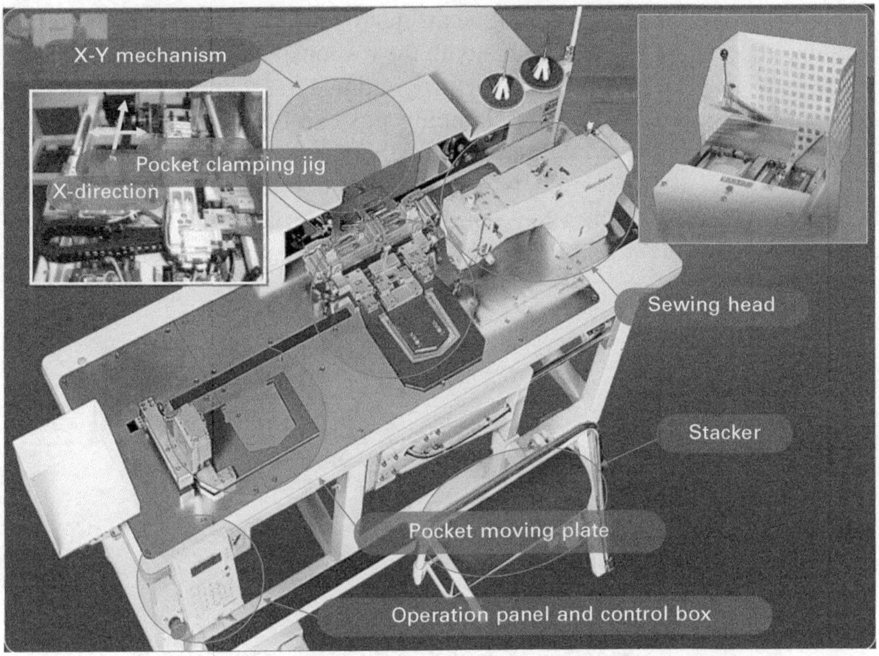

4.11 Pocket setting machine.

accurate in folding and sewing the pockets correctly, thus eliminating operator error and reducing operator fatigue.

The main limitation of this equipment relates to the fact that if the shape or dimensions of the pocket is re-styled at the design stage, costly new plates or jigs need to be purchased. Many manufacturers who use this type of equipment tend to compile an internal manual of available pocket shapes and dimensions which designers are required to adhere to when designing. Considerable attention will be given to the approval of purchasing new parts over using existing ones, as there is obviously a lead time attached to this type of acquisition which can specifically stifle fast fashion and quick response of new designs. An additional limitation relates to the reliability of the equipment, as production bottlenecks can easily result if such a machine breaks down and cannot be readily repaired. In this instance, it is obvious that trained engineers who are proficient in machine repair or skilled machinists who are similarly proficient need to be on hand should this fault occur.

4.8.2 Profile stitchers

These machines can achieve speeds in excess of 2,700 stitches per minute and can sew a variety of patterns for decorative stitching and for joining material parts such as sewing labels, airbags and other small product parts.

These machines offer substantial seam quality by preventing the material slipping underneath the foot, enhancing the responsiveness to multi-sections of the material. The machines can be programmed to produce many different types of stitching patterns. The clamping jigs can be changed for different garment parts to be sewn. Examples of these are given in Figs 4.12 and 4.13. More recent developments include pocket setters with double colour technology. Such machines have one sewing head, and two needles which alternate, giving the possibility to switch automatically from one colour to another.

4.12 Profile stitching for decorative stitching on a denim pocket.

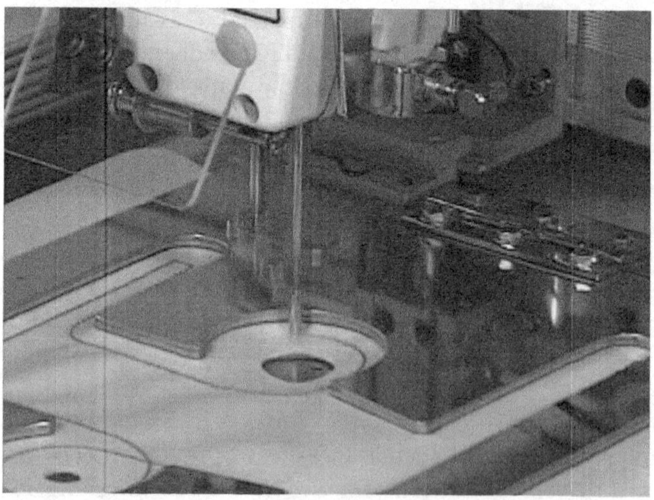

4.13 Profile stitching joining the seams on an air bag.

4.8.3 Long seaming operations

The workstations of these machines are engineered for specific operations on a garment. Examples of these machine types include serging on a pair of trousers, attaching waste bands and automatic sewing of side-seams and inner-seams. Examples of this equipment are given in Figs 4.14 and 4.15.

The automating operation of serging on a pair of trousers is an accurate method of edge stitching to prevent fraying of the fabric and neatening the edge of the material. An operator would be able to use more than one of these machines at the same time. The whole panel of the fabric is placed on the table and moved towards the sewing elements. Additional rollers or

4.14 Hemming of a pair of curtains.

4.15 Serging for a pair of trousers.

sometimes a conveyor are used to assist the continual feeding of the component past the sewing head or, in the case of trousers, can be further fed onto an automatic stacking device. Likewise infra-red photosensing devices enable the machine to operate, and pneumatic clamps hold the material in place to produce an accurate seam down the length of the fabric.

4.8.4 Semi-automatic bartacking

Figure 4.16 shows a mechanised bartacking machine. These machines have been around for many years. One of the main functions of these machines is the attaching of belt loops to waist bands. Belt loops would be folded by the operator and placed underneath the machine by hand. The development of microprocessor control and the advancement of pneumatic systems have enabled the operation to become more automated. The belt loops are continuously fed into the machine from a roll and are cut precisely to length. The belt loops are fed through to clamping fingers that form and fold the belt loops to shape. It can be seen that the operator only has to hold the garment in place. In this case, it is a pair of denim jeans. The belt loop is attached in one operation with the machine using a two sewing needle operation. An example of this is given in Fig. 4.16. Recent developments now allow programming of the bartack design as shown in Fig. 4.17.

4.8.5 Mechanised matching and automatic run stitching of straight and slanted breast welt pockets

These machines are ideal for processing chequered and striped breast welts (Fig. 4.18). However, their effectiveness is still highly dependent on accurate

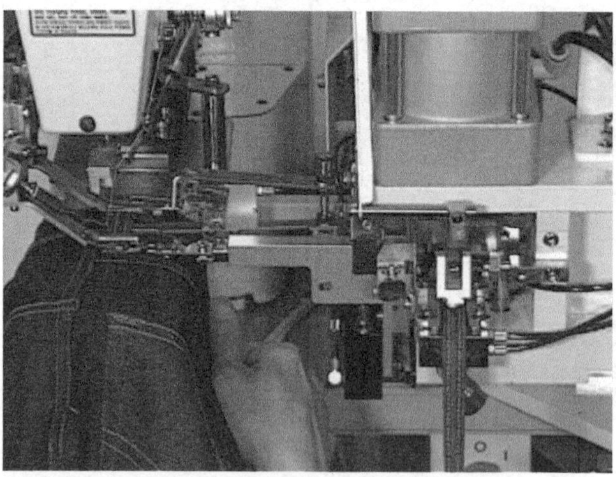

4.16 Semi-automatic belt looping machines.

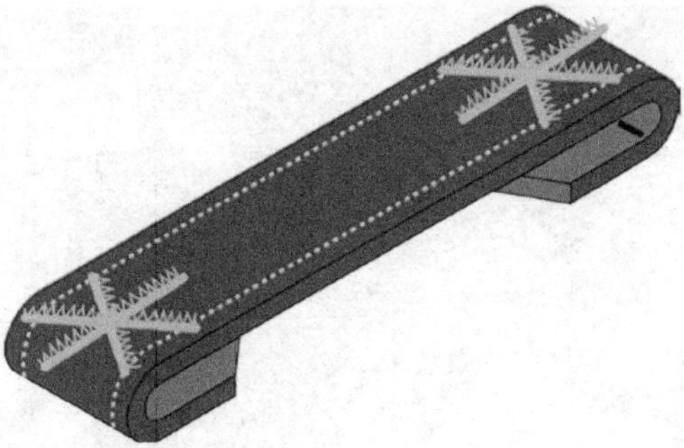

4.17 Programmable bartack design for belt loop attachment (source: Vi.Be.Mac. S.p.A. (2010) www.vibemac.com).

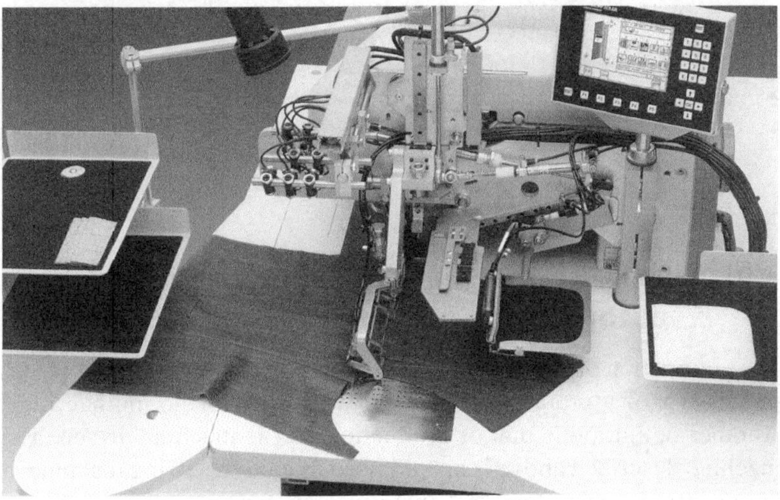

4.18 Run stitching of straight and slanted breast welt pockets.

cutting. Machine parameters can be called up in a coloured graphic display at the touch of a button. The user display can easily be adapted by the user interface depending on the design of the welt.

4.8.6 Run stitching of outside and inside of pockets of jackets and blazers

No mechanical adjustments to different pockets or materials are required as all machine parameters can be easily called up in the graphic display at the touch of a button (Fig. 4.19). The advantages of using this machine are:

- automatic feeding of pocket facing, pocket bag, flap or breast welt
- precise corner incision

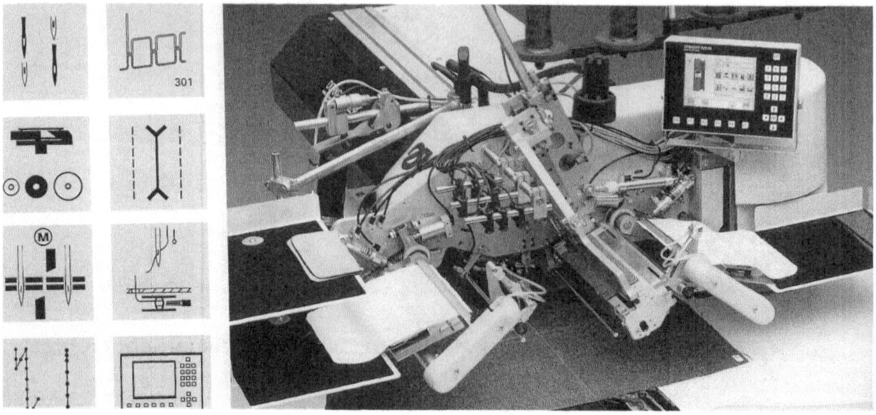

4.19 Automatic run stitching of jackets and blazers.

- automatic incision of piping ends
- a special folding system for piping strips, automatic adaptation to different material thicknesses
- programmable pocket sequences.

4.9 Advantages and limitations of machine automation

The automation of the sewing processes involves machines, by which machine processing of a product is achieved with little, or no, human intervention. Another definition is that of production that is machine driven. The type of machine described above is now classed as a basic sewing machine, but it still has electronic technology that enables sophisticated actions such as automatic presser foot lifting and automatic thread trimming to be performed. The evolution of sewing machine technology has advanced further to enable even greater advances in apparel production. However, although semi-automated machines currently exist, the function of fully automated sewing does not, because human interaction at some level is still required. The advantages of machine automation include:

- machine performing tasks reduces/eliminates operator fatigue
- a reduction in labour costs
- production becomes more streamlined and efficient
- reduced human intervention improves the quality of the product.

The disadvantages of machine automation include:

- the dexterity of fingertips in manipulating fabrics for sewing cannot, at this time, be managed by mechanical means

- the flexible structure of textiles makes it more difficult in handling fabrics when sewing
- textile materials tend to stick together and can be difficult to separate
- production flexibility is more difficult with regard to changing over to new styles/products
- availability of 'stand-by' manpower with appropriate skill in case of machine failure
- potential significant job losses.

4.10 Computer numerical control (CNC)

Many of the automated machines described above use some form of computer numerical control. In essence, CNC machines are controlled by computer generated data which is initially developed using a specialised program and then transferred to a machine using a direct cable link or removable storage medium. The required sequence of movements by the motor is converted into X and Y coordinates. CNC machines do not possess a conventional feed system for fabric movement; instead, the feed of the material is dependent on the movement of the clamp that holds the material in place.

Mechanised machines also have a system of movement of the clamp to feed the material. However, the main difference is that the mechanised machines are not programmed using computerised technology. The motors used for feeding the material are servocontrolled, enabling the fabric to be moved in very small sequences. These small degrees of freedom are the motions that enable stitches to be formed. The computer program continues to deliver small impulses to the X and Y coordinates to the servomotors acting on the needle head or the feeding system. Another example of computer numerical control is on embroidery machines (Figs 4.20 and 4.21).

4.11 Achieving fully automated apparel manufacture through the application of robotics

Successful application of robotics seems to be a long way off in the clothing industry. As one quality and assurance manager described a long time ago, 'we are not dealing with sheet metal here'. 'Textile materials are flexible structures that are susceptible to external factors such as moisture, static and heat'. Therefore, the ability to control these materials using robotic methods, particularly in the production of apparel, seems a long way off. There are, however, some applications where the use of fully automated systems are more able to be implemented, for example in the production of products such as towels, where there are fewer component parts. More automated processes can also be implemented for domestic products, such as curtains and bedding. The attaching of labels has been used as an example of using

4.20 Multi-head embroidery machine.

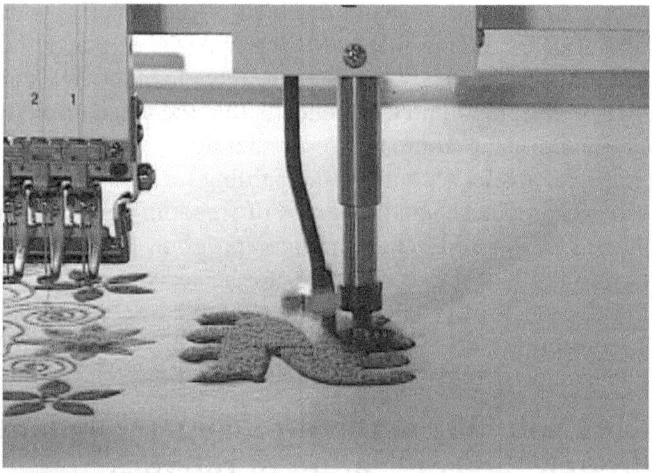

4.21 Multi-head embroidery machine motif stitching.

robotic arms. This production manufacturing method is still not completely satisfactory, however, as more than one label can be picked up at the same time due to friction and static from the threads on the label.

4.12 Conclusion

Ever since the 1960s when Pfaff embarked on the journey to automate a system which would carry out all of the operations on the front panel of a shirt style, many others have followed but been likewise unsuccessful.

Most other attempts since then occurred between the 1980s and early 1990s at a time when cheap imports were having devastating effects on apparel manufacturing employment in the West. During this time, significant efforts to improve competitiveness through automation were made by The Textile & Clothing Technology Corporation (TC^2) who returned to Pfaff's Transfer Line (Tyler, 2008), this time hoping to develop robotics to facilitate the movement of fabric. Other major attempts to mechanise sewing construction included research conducted by the Japanese Technology Research Association (TRAASS) whose work was described by Ono *et al.* (1992) and Marlowe (1992) as highly complex and doubtful in terms of economic viability.

To date, there have been no commercially viable solutions capable of simulating the human hand, which is still the only viable 'limp fabric handling device' which can perform the tasks associated with material handling for sewing construction. Therefore achieving full automation of apparel construction is still unobtainable. Some of these tasks include the following, but greater detail is provided by Tyler (2008):

1. Separation and pick up.
2. Orientation of pieces in the correct direction and plane.
3. Aligning or superimposing two or more plies or components.
4. Application or insertion of a third or fourth component such as a zip.
5. Insertion of ease or fullness.
6. Sewing along or manipulating the cloth to enable irregular shaped profile edges to be married together.
7. Accurate stacking, transfer and re-positioning to allow subsequent operations to be performed.

It is unlikely that investment into this research area will ever match that available in the 1960s and 1980s, as the change in market conditions renders research and development into this quest economically unviable. Furthermore, this quest has been regarded as a 'fait accompli' in respect of the general belief that 'the equipment technology route to competitiveness does not deliver a competitive advantage' (Tyler, 2008).

For now, the tangible option appears to suggest that machine designers and manufacturers should continue to invest their efforts to further automate aspects of sewing machinery which is able to:

- improve a factory's ability to respond quickly to changes in demand
- further progress quality
- improve throughput time
- be affordable.

4.13 Sources of further information and advice

www.brother.com
www.durkop-adler.com
www.sunstar.co.kr

4.14 References and further reading

Cooklin, G., Hayes, S. and McLoughlin, J. (2006), *Introduction to Clothing Manufacture*, 2nd edn. Blackwell Publishing, Oxford.

Marlowe, P (1992), 'Programmable sewing progress: but where to?', *Bobbin,* 33(5), 22–42.

McLoughlin, J. (1998), 'The expanding role of the clothing machine engineer', *World Clothing Manufacturer*, 79(7), 37–41.

McLoughlin, J. (1999), 'Implementation of a zero breakdown strategy', *World Clothing Manufacturer*, 80(1), 12–16.

McLoughlin, J. (2000a), 'Steps to improving quality at the needle point', *Industrial Focus, Mauritius*.

McLoughlin, J. (2000b), 'Time to value the production worker', *World Clothing Manufacturer*, 81(3), 16–21.

Ono, E., Ichijo, H. and Aisaka, N. (1992), 'Flexible robotic hand for handling fabric pieces in garment manufacture', *International Journal of Clothing Science and Technology*, 4(5), 16–23.

Tyler, D. (ed.), (2008), *Carr & Latham's Technology of Clothing Manufacture*, 4th edn. Blackwell Publishing, Oxford.

5
Problems relating to sewing

M. CARVALHO, H. CARVALHO and L. F. SILVA,
University of Minho, Portugal

DOI: 10.1533/9780857093967.1.149

Abstract: This chapter will mainly be concerned with the analysis of seam failure through rupture or slippage. The mechanisms of seam failure under stress or due to wear-out will be highlighted. Test methods for seam performance will be presented briefly. Some common sewing defects that can be the cause of seam failure, or simply result in minor faults, will be examined. Finally, the chapter presents future trends in technological tools for defect avoidance and detection.

Key words: seam failure, seam slippage, seam defects, stitch formation, automatic sewing monitoring.

5.1 Introduction

This chapter deals mainly with the most significant problem in joining textile materials with seams: seam failure through rupture or slippage. Seam failure can be produced by mechanical stress on the seam, or due to seam wear-out from abrasion. It can occur in the direction of the seam or transverse to it. Several mechanisms of seam failure are possible and will be described. The most influential factors on seam strength and failure will be highlighted. The chapter will, furthermore, describe some common defects in sewing that can be the cause of seam failure or that can result in minor faults mainly from an aesthetic viewpoint. These include defects produced by needle penetration, seam pucker and seam undulation, seam grinning, ply-shift, folds, stitch distortions, inadequate tensions, improper stitch balance, irregular stitch length, and skipped stitches. New methods for automatic detection of some of these defects will be reviewed.

The most severe problem that can occur in a sewn joint between textiles is the rupture of the seam. This is true for all types of seams, but especially for joining seams, in which a major functional fault is produced. According to Amann Group (n.d.a), the quality criteria of seams can be summarised as shown in Table 5.1. Of these criteria, the first four describe the functional behaviour of the seam. This chapter will deal mainly with the first three of them, the fourth being very specific and thus outside the scope of this chapter.

Table 5.1 Seam properties (adapted from Amann Group, n.d.a)

Seam stability
Seam breaking strength
Seam slippage
Seam elasticity
Abrasion resistance
Individual product-related criteria
Heat resistance
Weather-proofness
Chemical resistance
Electrical conductivity
Watertightness
Economic aspects – seam profitability
Care properties (harmony of care properties)
Seam feeling (touch, softness)

5.2 Seam elasticity

When referring to seam elasticity, normally the behaviour of the seam when stressed in the direction of the stitch line is meant. Elasticity is a very important property of the seam, particularly when sewing stretch fabrics, in which the seams may be stretched along with the fabric. To achieve high elasticity, normally stitches of classes 400, 500 and 600 are preferred. (The exception is the 304 zig-zag lockstitch that assures excellent extensibility through its particular stitch path.) Thread tensions are set as low as possible to provide a loose stitch which can be stretched up to a point by deforming its geometry before stretching the thread itself. A loose stitch will, however, result in higher seam grinning (Carr & Latham, 1994). If higher extensibility is needed, the choice of thread is decisive. Textured polyamide threads, for instance, are very often used for sewing knitwear due to its elasticity, given by both the raw material and the thread structure. Thread manufacturers offer other alternatives for highly elastic sewing threads (Amann Group, n.d.b).

5.3 Seam failure

5.3.1 Seam failure under load due to rupture of the sewing thread

The first type of seam failure to consider is the rupture of the sewing thread upon stress in the direction of the seam line, when the extensibility of the stitch and thread are exceeded. Seam rupture can be divided in two types of events: rupture of the seam itself, i.e., of the sewing threads, or of the fabric at the seam line (Burtonwood & Chamberlain, 1966). In some situations, the thread will not break, but when the seam is stretched, the thread runs back at the seam end and/or beginning. This happens when tensions are low, the surface frictional properties of fabric and/or thread allow the slip of the thread through the fabric, and when the seams are not conveniently fixed at beginning and end by a cross-seam or by a backtack. The backtack operation is easily achieved in lockstitch by reversing sewing direction, but it is technically more complex in other types of stitches.

5.3.2 Seam failure under load perpendicular to the stitch line

The next important situation for seam performance analysis is when seams are put under load transversely to the stitch line. The first observation that can be made in most of the seams when stressed is seam *grinning*. Seam grinning can be described as 'the separation of the interface line between two sewn fabrics' (Uçar, 2002). It is worse with rigid than with extensible fabrics, increases with the extensibility of the sewing thread and decreases at higher stitch densities. It is normally also much more visible in loose stitch constructions, such as 400, 500 and 600 class stitches, and can be reduced by tensioning the threads (although this will reduce elasticity). Seam grinning is easily noticed and sensed as a sewing defect. It is a reversible effect, although it increases under cyclic loading (Gurarda & Meric, 2010).

If load is increased beyond the value that can be absorbed by the seam's structure without damage, two effects, or a combination of both, take place: seam rupture and seam slippage. Seam slippage is a phenomenon that occurs in woven fabrics, defined as the displacement of yarns of one yarn system over the other by a seam, when transverse stress is applied to the seam (Yildirim, 2010). The slippage produces an opening in the fabric. Figure 5.1 shows examples of seam grinning and seam slippage.

Seam rupture and seam slippage are quantified using tensile testing machines. In these, samples of seams are gripped by two jaws. One of the jaws is fixed, whilst the other moves away at a pre-determined speed, thus pulling the sample apart (Fig. 5.2). The tensile testing machine measures force and extension during the test and this data is represented graphically as shown in Fig. 5.3.

There are several normalised test methods for seam slippage and rupture testing:

- Determination of maximum force to seam rupture using the strip method (ISO Standards, 1999a): The strip method uses samples of two plies of fabric sewn at the edge, with 100 mm width and about 350 mm of length. The sewn assembly is cut on the corners such that the seam line maintains 100 mm width, but the sample ends are only 50 mm wide (see Figs 5.4a and b). The samples are then pulled apart at a predetermined

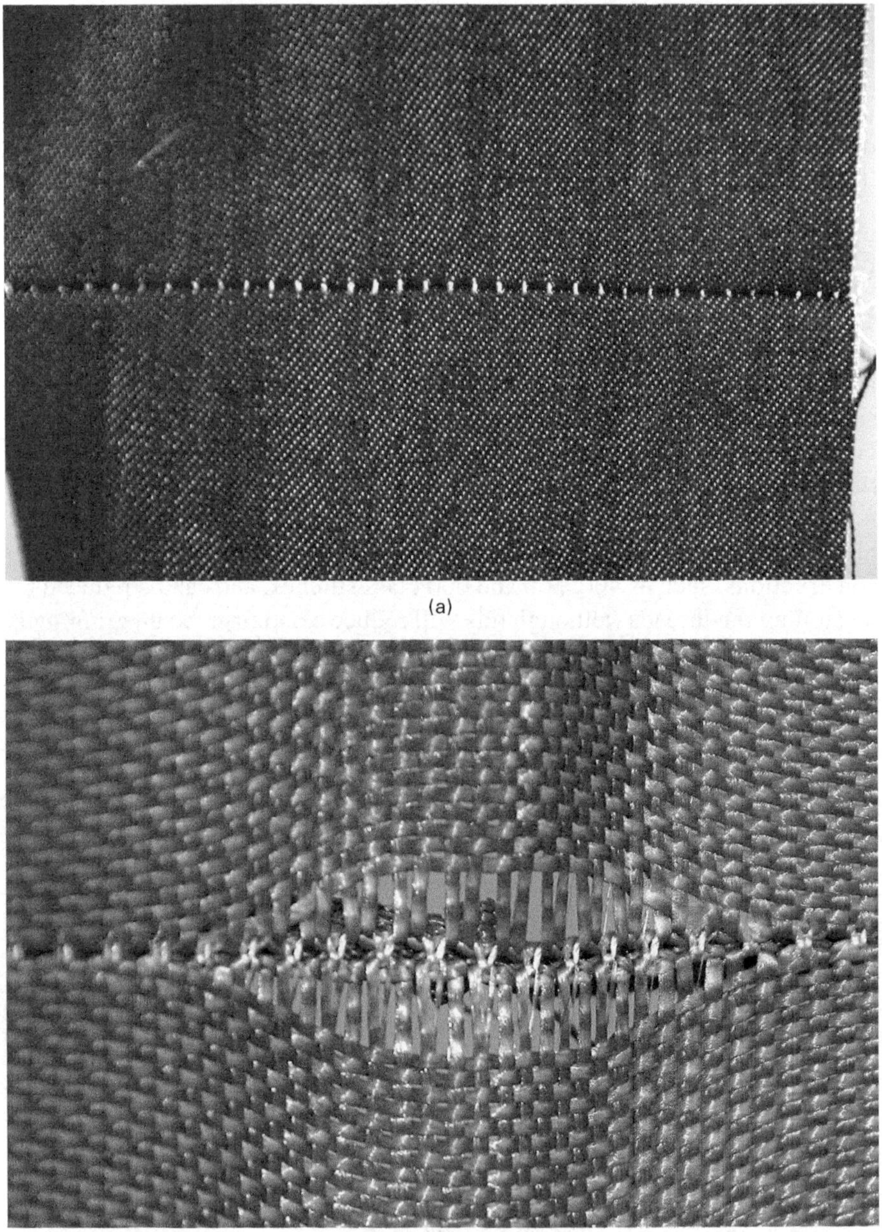

(a)

(b)

5.1 (a) Seam grinning; (b) seam slippage.

constant rate of extension, and the test is completed when the assembly ruptures due to thread breakage, fabric breakage or a combination of both. The maximum force and the stress–strain curve are recorded.

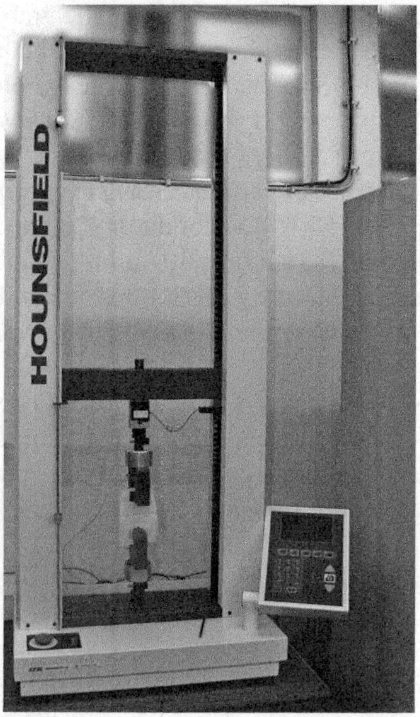

5.2 Tensile testing machine.

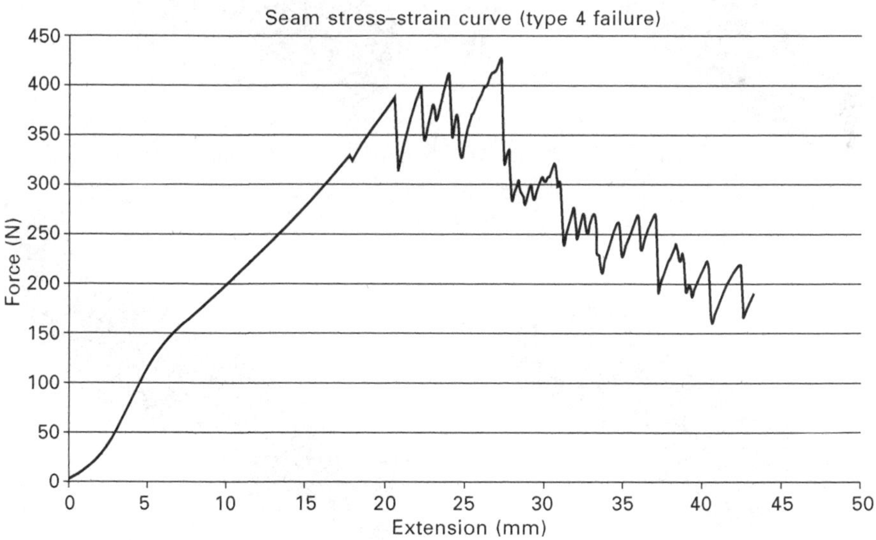

5.3 Stress–strain curve. In this particular case, the curve describes a type 4 failure in which, as explained later, the seam withstands the load, and the yarns in the fabric are progressively torn by the seam.

- Determination of maximum force to seam rupture using the grab method (ISO Standards, 1999b): The grab method is a variation of the previous one. In this case, the jaw does not hold the sample along its whole width. Samples are prepared with a width of 100 mm, but the jaws will only grab them over a width of 25 mm (see Figs 5.4c and d).
- Determination of the slippage resistance of yarns at a seam in woven fabrics using the fixed seam opening method (ISO Standards, 2004a): In this method, a fabric sample is sewn with stitch type 301 with predetermined machine settings, and cut in a way that provides an unsewn and a sewn sample of the same dimensions. The sewn sample is cut open on the sewn end assuring a seam margin of 12 mm. Both of the samples are stretched at a constant rate of extension up to a load of 200 N. Seam slipping is defined as the difference between the extension of the sewn and the unsewn samples, compensated by the residual displacement value at 5 N. Five samples each are tested in total. The average forces

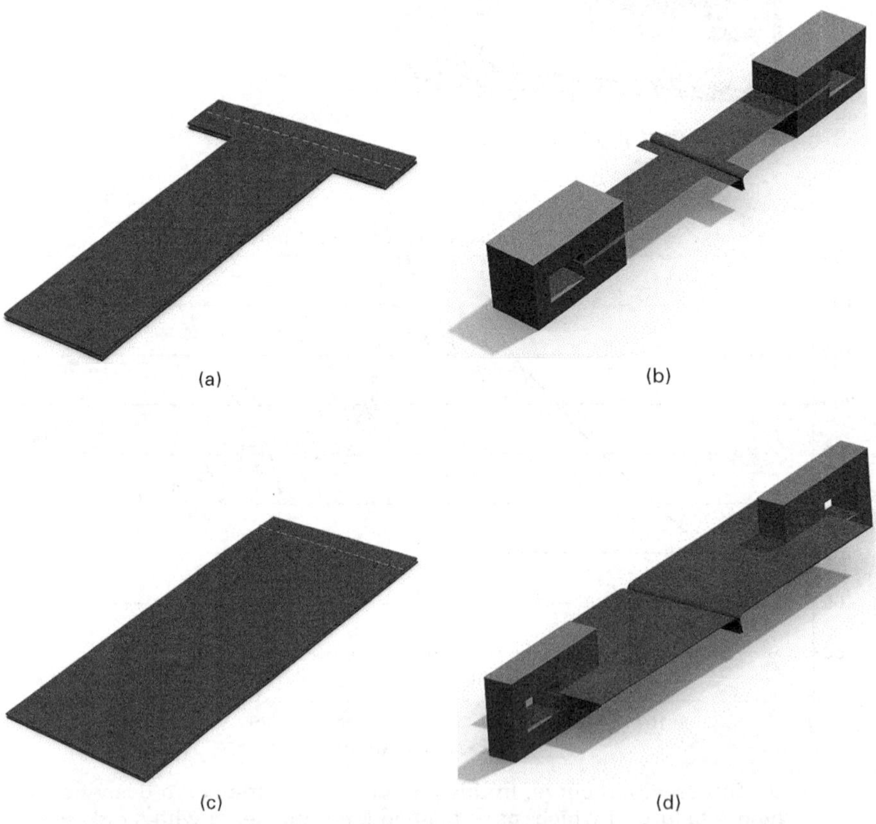

5.4 (a and b) Strip method versus (c and d) grab method.

measured at certain seam openings (2, 3, 4, 5 and 6 mm) are the result of the test.

- Determination of the slippage resistance of yarns at a seam in woven fabrics using the fixed load method (ISO Standards, 2004b): The fixed load method uses only sewn samples, prepared under the same conditions as with the previous method. The samples are stretched up to a predetermined load, depending on the type of fabric that is being tested (fabric for garments, home textiles or furniture). When this load value is reached, the jaws are drawn back very quickly until there is a force of 5 N applied to the sample. The slippage is measured as the longest dimension of the hole in the load direction.
- Determination of the slippage resistance of yarns at a seam in woven fabrics using the needle clamp method (ISO Standards, 2007): The objective of this method is to avoid the influence of seam preparation and sewing thread variation: the method does not use a seam. Instead, the samples are stretched with a special needle-clamping device that is devised with needles on one side and the corresponding holes on the other, in which the sample is fixed. The construction of the needle clamping device depends on the kind of material being tested. In apparel fabrics, for instance, 17 needles per 5 cm are used, whilst for furnishing and upholstery fabrics, seven needles in a width of 5 cm are used. Samples are 50 mm wide. The result of the test is the difference in extension between the stretching of a sample with a normal and with a needle clamping device, at a force value of 250 N, compensated by the residual displacement value at 5 N.

Depending on the balance between seam strength and fabric strength, determined by the fabric properties, thread properties, seam edge dimension, stitch density and other factors, slip, rupture or a combination of both occur. Gribaa *et al.* (2006) classified the behaviour of the seam under transverse stress into four types, reproduced in Fig. 5.5.

- Type 1: Breaking of the sewing thread.
- Type 2: Intact seam, with yarns of the fabric pulled out at the edge of the sample.
- Type 3: Rupture of the sewing thread, leaving holes at the stitch line due to yarn slippage.
- Type 4: Damage of the fabric along the stitch line (with seam intact).

5.3.3 Factors affecting seam strength under load

Seam rupture tests are made with a simple superimposed seam, the worst-case situation for the behaviour of a seam under stress. In practice, seam construction can significantly improve breaking strength. An obvious example is given in Fig. 5.6.

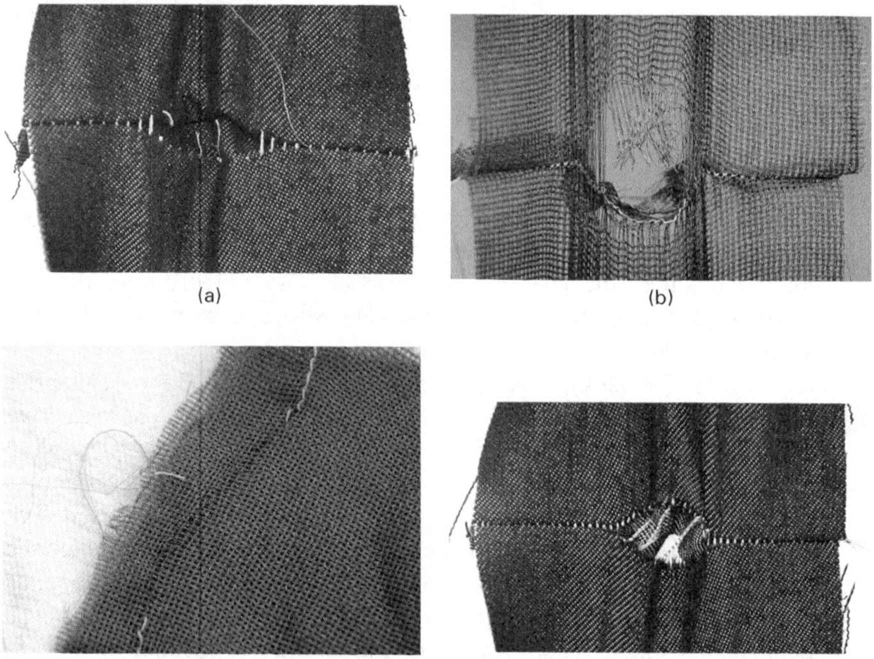

(a) (b) (c) (d)

5.5 (a) Breakage of the sewing thread with negligible damage to the fabric; (b) fabric yarns pulled out of the sample edge; (c) breakage of the sewing thread with damage to the fabric at the seam line; (d) intact seam, fabric rupture.

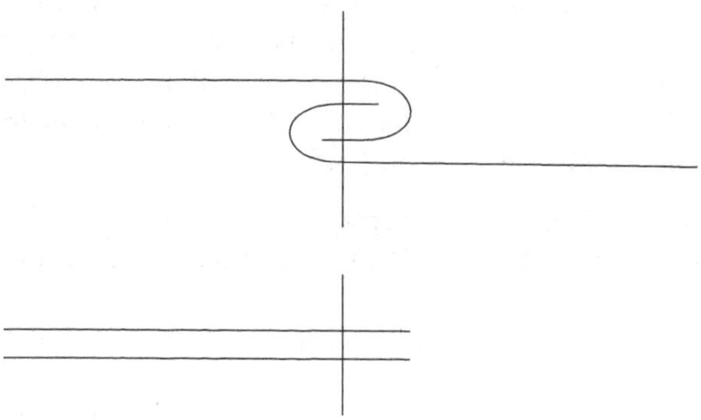

5.6 Two different seam constructions with different strengths.

Under the same conditions, and assuming there is no slippage or rupture of the fabric, the strength of a seam is proportional to the breaking strength of the thread (more precisely, to its knot strength after sewing), to stitch density and to the number of stitch rows (Burtonwood & Chamberlain, 1966). Thread breaking strength depends on the type of fiber used (Gribaa *et al.*,

2006), the thread's structure, and the thread linear density (higher breaking strength for coarser yarns; Mukhopadhyay *et al.*, 2004). Although seam strength increases with increasing stitch density, there is an upper limit for this. An excessive stitch density can result in an increased abrasion of the threads and fabric damage during the seam, actually reducing seam strength. Gribaa *et al.* (2006) found no difference in seam breaking strength between lockstitch (stitch type 301) and double chain-stitch (stitch type 401), but observed much more extension of the latter, as expected. Burtonwood and Chamberlain (1966) had found higher breaking strength for double chain-stitch.

Threads suffer abrasion and structural changes during the sewing process. They do not exhibit the same breaking strength before and after they have been used in the seam. Crow and Chamberlain (1969) confirmed some increase in degradation of linear tenacity of threads for increasing stitch density, smaller needles, increasing sewing speed, increasing thread tension and increasing number of plies being sewn. The most influential factor, though, was found to be the relation of the size of the machine's throat plate hole and the machine needle. Although straight thread tenacity was reduced during sewing, Crow and Chamberlain (1969) observed that minimum knot strength did not necessarily suffer during sewing.

Sundaresan *et al.* (1998) have found reductions in thread linear tenacity of 14 to roughly 31% after sewing. Thread degradation along the seam takes place at all the thread guides, during contact with the needle (especially in the needle eye), during the contact with the fabric and even when interacting with the remaining threads in the stitch. It depends on the fabric properties, such as frictional properties of the yarns, tightness factor of the fabric, as well as on the thread's properties, such as material, structure and linear density (Sundaresan *et al.*, 1997).

The reduction of breaking strength in threads during high-speed sewing has been found to be dependent on fabric and thread properties, but has also been related to thread consumption during the seam (Sundaresan *et al.*, 1998). It is known that the sewing thread is (over-)fed and drawn back again during the stitch formation process, i.e., much more thread is provided to stitch formation than is actually consumed by the stitch. This cyclic process causes the thread to travel back and forth in the thread paths. In fact, each section of the thread passes a point in a thread path many tens of times. The lower the thread consumption, the higher will be the number of cycles – and thus the abrasion – that the thread will go through. Unbalanced stitches tend to increase consumption of one thread whilst reducing consumption of the other(s). This situation may result in an increased stress on the thread(s) with reduced consumption. Thread breakage strength also suffers reduction after laundering. Mukhopadhyay *et al.* (2004) found reduction in the order of 20% breaking strength and increase of strain at rupture.

5.3.4 Seam failure due to wear-out

The issue of laundering and daily use is an important aspect of seam performance. The abrasion of the threads during daily use is one of the causes of seam failure. The rupture of the thread is particularly serious in chain-stitch (type 101) or double chain-stitch (type 401), where the stitch unravels easily from the end to the beginning at a broken stitch. Multi-needle class 400, class 500 and 600 stitches can also be unravelled, although not as easily. In this respect, lockstitch presents a clear advantage. Besides not being possible to unravel it, lockstitch is also much less voluminous and less exposed to abrasion, with thread interlacings being concealed within the fabric. On the other hand, lockstitch machines generally work with higher thread tensions, requiring more resistant thread. Thread wear-out also depends strongly on the thread materials and structures. Polyamide multifilament thread has been found to be by far the most resistant type of thread from this point of view. In general, multifilament threads present advantages over staple threads, whilst polyamide and polyester materials are more resistant than cotton (Amann Group, n.d.b).

5.3.5 Seam failure due to slippage

Seam slippage is the other phenomenon that may occur when seams are put under load. As previously described, when the seam itself is strong (resistant thread, high stitch density) the fabric may fail first, either by slippage (behaviour type 2 or 3 as classified by Gribaa *et al.*, 2006), or by rupture (type 4 behaviour). Assuming that the seam does not fail, the occurrence of seam slippage depends both on the seams as well as on the fabric properties. Seam slippage decreases with increasing stitch density, since there is a better distribution of the load. It depends also on the sewing thread's material and structure.

The fabric properties, namely the frictional properties of the yarns and the fabric density, cover factor and weave type also play an important role. The frictional properties of the yarns are influenced by the yarn's twist, size, type of fibre and density. The width of the fabric edge (the seam margin) is also relevant: the fabric has more tendency to slippage if the margin is narrow (Gribaa *et al.*, 2006; Yildirim, 2010; Uçar, 2002). Many researchers have reported these factors as being the most influential to the propensity of a fabric to seam slippage. Miguel *et al.* (2005) related slippage to the fabric's material and type of finishing (which influence the frictional interaction between the yarns in the structure).

The interaction between yarns in the fabric is also influenced by the structure of the fabric yarns and the weave type. Filament yarns have been found to cause higher slippage due to their more slippery surface, and weave type has an important influence on slippage (Kalaoglu & Meric, 2005). Regarding weave type, the latter work also describes the influence of the direction of stress in combination with the directional properties of the fabric structure. The analysis of seam slippage is of interest not only in static but also in dynamic conditions. Yildirim (2010) has found that seam slippage is generally worse under dynamic than under static loads, and the importance of the causal factors varies depending on the type of load.

5.3.6 Weakening of seams due to fabric damage induced by stitching conditions

Finally, one has to consider the possible deterioration of the fabric itself during the sewing process, which can contribute to the type 4 failure. During sewing, the fabric suffers potential weakening by (a) bursting of weave threads due to the collision between the needle tip and the fabric yarns, (b) friction between the fabric yarns and the sewing thread, (c) friction between the fabric yarns and the needle, and (d) thermal aggression of the needle on the fabric yarns. Blackwood and Chamberlain (1970) have found that the damage produced on a fabric does not change significantly with thread in the needle, compared to a situation in which the fabric is stitched without thread.

Regarding the interaction with the needle, this is a subject that has been studied by several researchers since the 1960s. Hurt and Tyler (1975) established that finishing processes are able to modify the frictional properties of fabrics, decreasing the needle penetration forces required. These frictional properties have a major influence on seam damage. Needle size is the main variable affecting damage (Hurt and Tyler, 1976).

Leeming and Munden (1978) confirmed this assertion and experienced that the fabrics exhibiting high penetration force values were generally those which exhibited sewing damage using standard sewing tests. Their L&M sewability tester (US Patent 3979951, 1976) simulates a sewing machine by penetrating a fabric sample with an unthreaded needle at 100 penetrations per minute. Force is measured and the device counts the number of times the value is higher than a preset threshold.

A set-up closer to the industrial condition was developed by Rocha (1996), in which a piezoelectric sensor was inserted into the needle bar of an industrial overedge sewing machine. This measurement system would later be further developed by Carvalho (2010) and allows the measurement of needle penetration and withdrawal force at industrial speed. Rocha was able to relate material and machine variables to the needle penetration forces. Among other parameters, bending rigidity and drape factor were found to be related to needle penetration forces. Needle size, sewing speed, fabric finishing and needle tip state could be related to needle penetration force;

forces increase with increasing needle size and sewing speed, and obviously with worn-out and damaged needle tips Carvalho *et al.* (2009). Gurarda and Meric (2005) found that both pre-setting temperature and the finishing process have significant effects on the seam performance, needle penetration force and elastane fibre damage during the sewing of cotton/elastane woven fabrics.

Needle heating is another problem during high-speed sewing, particularly in the sewing of synthetic materials. Besides causing damage to the fabrics, they also produce the deterioration of the thread, a factor for a reduction in seam strength, as explained previously. In extreme cases, hot needles may even result in thread breakage. This problem is once again particularly critical with synthetic threads. Liasi *et al.* (1999) measured needle temperatures with infrared radiometry and determined temperatures above 180ºC in some conditions. They found that the thread is weakened as much as 40% at 140ºC. Sewing speed, the type of material being sewn, and thread tension were found to be the main factors contributing to needle heating. Needle temperature can be reduced by reducing sewing speed or thread tension, by using lubricants or air cooling.

5.4 Seam problems related to material feeding

In line with many of the aspects mentioned above, the problems that cause seam defects and failure can also be addressed to the technology used in the production of seams. Although some research and improvements have been carried out in this field, the sewing process has remained almost unchanged over the years, staying largely dependent on the skills of the operator to set up sewing parameters and to handle the fabrics being sewn. Slight changes in the machine settings can have a large influence on the quality of the obtained seams and of the end-products, as well as sewing operation time.

The sewing machine feeding system, for instance, is responsible for the occurrence of several defects if not properly adjusted. Although home and industrial sewing machines, particularly the latter, use a wide range of feeding system types, a sewing machine feeding system is basically composed of a presser foot, a throat plate and a feed dog (see Fig. 5.7). The feed dog has an approximately elliptical movement and, soon after the withdrawal of the needle from the fabric plies, it rises above the throat plate to engage the fabrics against the underside of the presser foot. The hit of the feed dog on the presser foot, with the fabrics in between, generates a lack of fabric control, especially at high sewing speeds, producing different sewing defects.

Several studies have been conducted and a better scientific understanding, different findings and achievements have already been accomplished; some of these studies will be presented in Section 5.7. The consequences and causes of the sewing feeding system operation have already been the

5.7 Components of a typical drop feed system in an industrial Zoje lockstitch sewing machine.

subject of study by several researchers. The analysis of the performance of typical feeding systems, in industrial lockstitch, chain-stitch and overedge (or 'overlock') machines, enabled the determination of correlations between the feeding process itself and the occurrence of several defects, as well as with the objective measures of the fabrics, to predict its behaviour during the sewing operation. Dorkin and Chamberlain (1961) identified some of the causes that are related to seam pucker and its occurrence, and, five main causes for seam pucker, as determined by these researchers, will be outlined in Section 5.6.

Eileen Johnson (1973) analysed the feeding system of a lockstitch sewing machine, to describe its operation and the influence of high-speed operation on its performance. Although the properties of the sewing threads and fabrics (and its finishing) may lead to the occurrence of seam defects, Johnson calculated and measured the phenomenon of presser foot 'bouncing', testing two lockstitch sewing machines, in which contact losses between the presser foot and feed dog occurred without any fabric. Stitching on paper, and without threads, Johnson also concluded that:

- the stitch density did not remain constant over a range of speeds and presser foot pressures, partly due to the presser foot 'bounce', which results in the loss of fabric control;
- the stitch density decreased with increased speed;
- the stitch density was always greater at maximum presser foot pressure than at minimum pressure settings of the presser foot;
- the difference between obtained and nominal stitch densities was as high as 35–40%.

These variations of stitch densities can result in the production of irregular and uneven seams, which can influence and vary seam characteristics and strength. Poor material feeding can even result in skipped stitches. Another problem is ply-shift, originating from misaligned seams: two fabric plies, perfectly aligned at the beginning and end before sewing, appear misaligned at the end of the seam after sewing. This happens because the feed dog has better contact with one ply than with the other: there is a little slip between the plies that results in ply-shift.

Although there are many variations of feeding systems and designs of presser feet that can solve some of the above-mentioned problems (such as compound feed, unison feed, drop feed with variable top feed, differential bottom feed, differential bottom feed with variable top feed, and others like the puller feed system or top and bottom belt feed), presser foot 'bouncing' is a problem in most of them. For this reason, this problem has been analysed by many researchers, including Frank and Mo (1974), Matthews and Little (1988), Clapp *et al.* (1992), Araújo *et al.* (1992), Rocha *et al.* (1992a,b, 1996a,b), Bühler & Hennrich (1993, 1994), Chmielowiec and Lloyd (1995), Alagha *et al.* (1996), Silva (2002) and Carvalho (2010).

To avoid this adverse presser-foot dynamic, sewing test-rigs have been designed and built, and some innovations have been proposed and tested. Barrett (1992), Barrett and Clapp (1995), Stylios and Sotomi (1995) and Silva *et al.* (2010) presented alternative actuating solutions for the presser foot, as well as monitoring systems and control strategies for fabric feeding, based on the use of electromagnetic and pneumatic linear actuators, combining fuzzy logic schemes with neural control.

5.5 Problems in stitch formation

5.5.1 Stitch balance

Stitch balance is crucial for a quality seam. The experience of the operator or technician during machine set-up for each type of sewing conditions will determine a quality seam. A balanced stitch is the result of the ideal relationship between the thread pre-tensions and the material and thread properties, according to the objectives of the desired seam type. For example, on an overedge sewing machine performing the 504 stitch type, the balanced stitch, or adjusted stitch, is obtained when the looper threads cross at the centre of the edge of the material being sewn, and the needle thread holds the seam efficiently conferring the necessary resistance and elasticity of this stitch type (Fig. 5.8). Therefore, the seam quality is directly related to the pre-tension values defined during machine set-up. A good indicator for correct stitch balance and tension is the resulting thread consumption values for each thread. Theoretical thread consumption values can be computed from the geometrical parameters of the seam, i.e, material thickness, stitch length

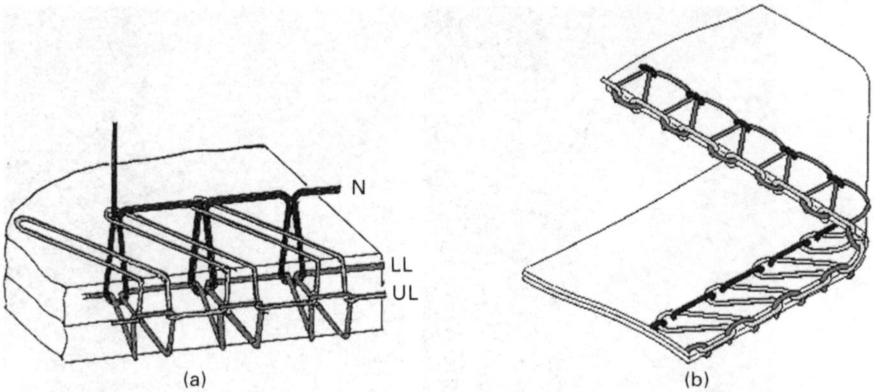

5.8 Correctly balanced overedge stitch type 504: (a) schematic representation with UL: upper looper thread, LL: lower looper thread, N: needle thread; (b) realistic representation.

and width (Amann Group, n.d.b). Incorrect adjustment results in deviation from the predicted consumption values.

When the needle thread loop is too loose, the result is excessive seam grinning, as previously described. If any of the interlacing spots of the threads do not match the positions shown in Fig. 5.8, there is an imbalance between the thread tensions. It can be said that threads pull one another. If one thread is excessively tensioned, it will pull the other threads towards it. This can be easily observed in the seam and corrected by appropriate adjustment (in the example given, by loosening the thread's tension). A correctly adjusted stitch will have the interlacings on the proper spots. The overall tension will be chosen according to the most important functional objective: loose, if elasticity is desired, tight if seam grinning is to be avoided (Carvalho and Ferreira 2006).

5.5.2 Stitch distortions

A spontaneous imbalance in the stitch is shown in Fig. 5.9. In this case, the thread tension adjustment is not the cause of the imbalance, because the stitch shows balance before and after the defect. Many factors may have caused this defect: thread irregularity, improper handling by the operator, sharp speed variations, etc.

5.5.3 Skipped stitches

Skipped stitches are one of the most serious sewing defects, because they are visible in the final product and represent a weak point in the seam. They

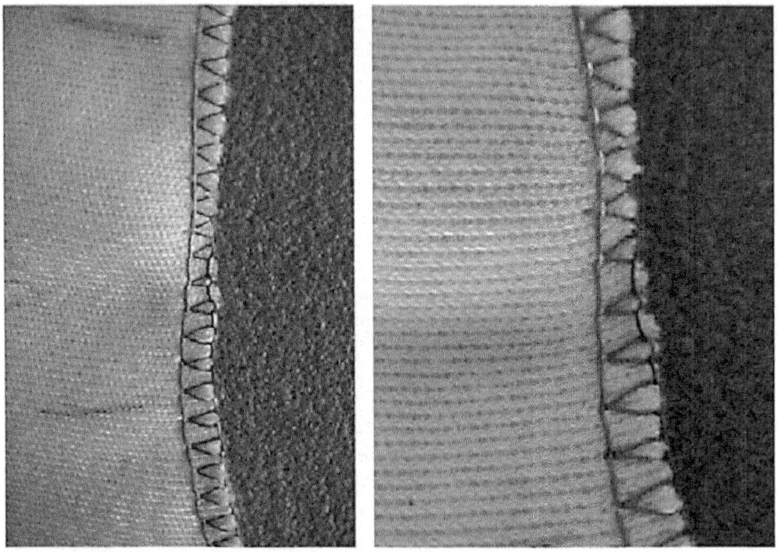

5.9 Distorted stitch.

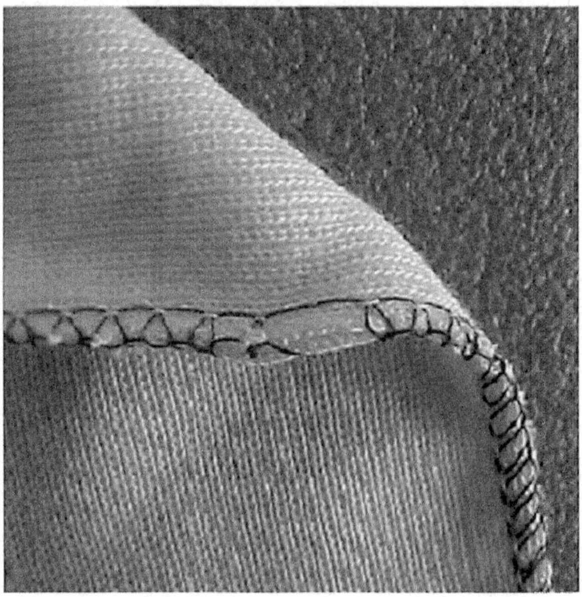

5.10 Skipped stitch.

are the result of the failure in the interlacing of the sewing threads (Fig. 5.10). They occur in different situations and for different reasons, namely (Carvalho, 2003):

- sewing threads with inadequate twist amount and/or direction, not allowing the loop formation needed for the interlacement of the sewing threads (in general sewing machines require threads with 'Z' twist);
- inadequate combination of needle diameter and thread linear mass;
- sewing thread wrongly guided in the machine;
- damaged needle or needle incorrectly positioned;
- needle and loopers at inadequate distance or damaged.

5.5.4 Automatic detection of some of these defects

It has been shown that defect situations regarding stitch balance, stitch distortions and skipped stitches can be detected automatically using instrumented sewing machines. Simple solutions for skipped stitch detection are already being proposed commercially by the sewing machine manufacturer Pfaff. Using an overedge sewing machine equipped with sensors to monitor thread tension, thread consumption and pressure foot displacement, Carvalho *et al.* (2006) and Carvalho (2010) have shown the detection of several defect situations, namely the occurrence of skipped stitches and distorted stitches, besides the assessment of correct stitch balance.

Analysing the peaks of tension in the sewing threads involved in stitch formation, it is quite straightforward to detect a skipped stitch. Figure 5.11 shows an example of this. Thread tension on the lower looper thread was measured during a seam on an interlock knitted fabric. The machine had been set up with a very loose needle thread tension. On the thread tension waveform, it is possible to identify a sharp drop in amplitude in two of the three lower looper tension peaks normally observed. The tensions on the upper looper thread reflect a similar effect, as expected, since the interlacing failed between these two threads.

It has also been shown that the detection margin can be enhanced by observing tension peak ratios instead of observing the measured values directly. Figure 5.12 represents the ratio between the peaks of tension in the needle thread and the lower looper thread. The skipped stitch is clearly depicted in the graph.

Peak ratios do not need to be computed between values of tension on different threads. In Fig. 5.11, the computation of ratio P3LL/P1LL would yield a similar result. With peak ratios, it has been shown that stitch distortions such as the one shown in Fig. 5.9 are also clearly detected. In general, they are excellent indicators of the regularity of stitches. Further work has to be done on determining which peak ratios are more appropriate for detecting the different types of skipped stitches and stitch distortions.

The assessment of correct stitch balance and overall stitch tensioning can be performed by the measurement of thread consumption. Figure 5.13 shows a situation in which thread consumptions for a perfectly tensioned and

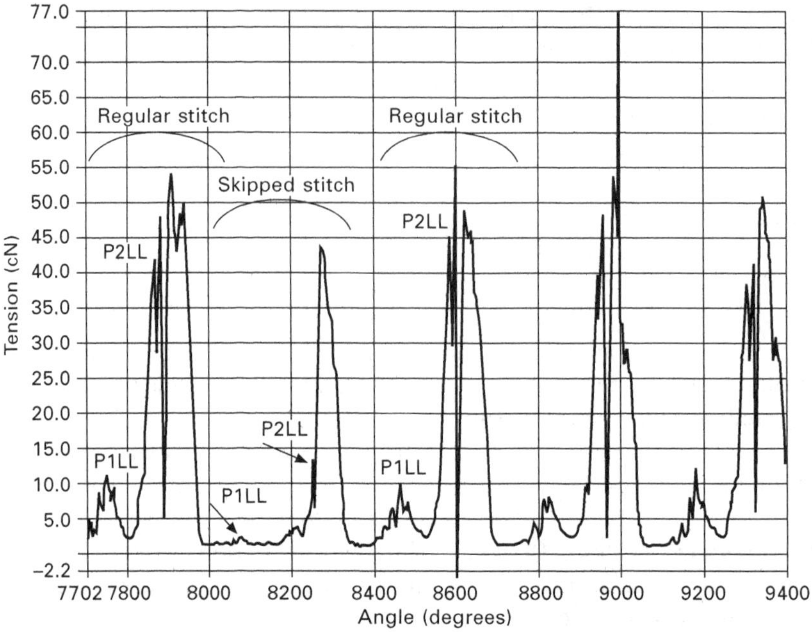

5.11 Skipped stitch detection during monitoring of lower looper thread tension. (P1LL: lower looper tension peak 1, P2LL: lower looper tension peak 2.) The first two peaks suffer a sharp decrease. The third peak, P3LL, not labelled in the graph, suffers a less significant decrease.

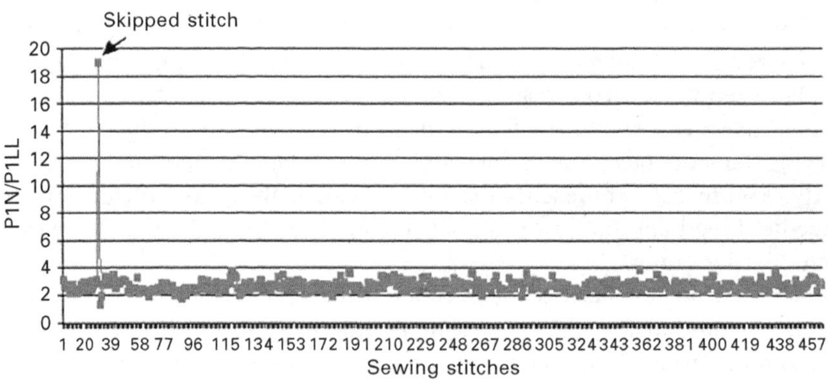

5.12 Ratio of peak 1 of needle thread tension by peak 1 of lower looper thread tension.

balanced seam are compared with the consumptions after unbalancing the seam. The seam was unbalanced by loosening thread tension on the upper looper thread and tightening the lower looper thread.

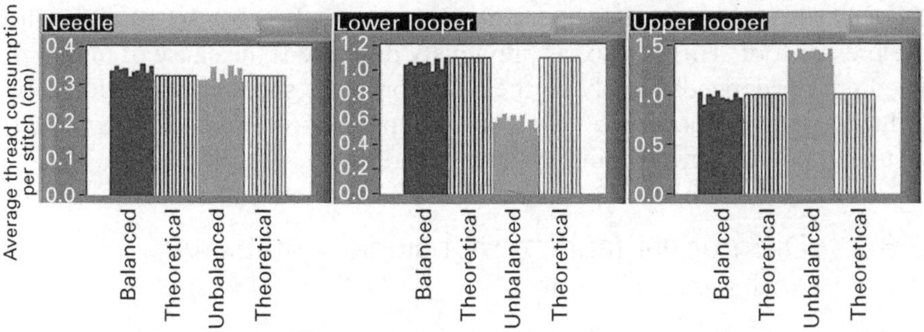

5.13 Average thread consumption per stitch (cm) in 10 seams of about 40 stitches each. The transparent bars at the right side are the theoretical value of thread consumption calculated for each seam.

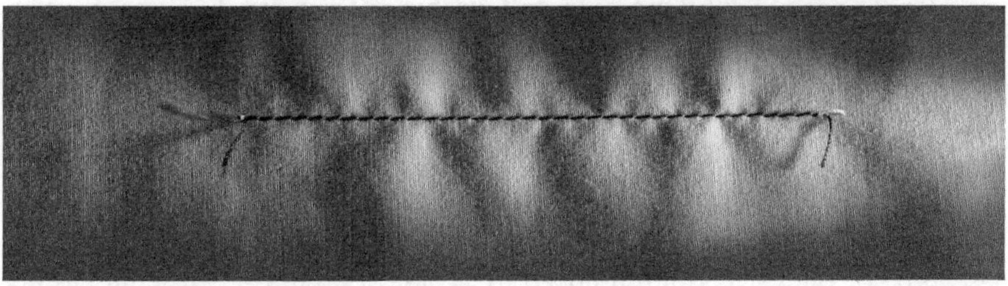

5.14 Seam pucker.

5.6 Seam pucker and other surface distortions

Seam pucker and other modifications of the seam's surface causing the seam to appear rippled, undulated or, in general, not laying flat, are normally a purely aesthetic defect of a seam, although on stretch fabrics they may be related to the elasticity of a seam. Figure 5.14 shows an example of seam pucker.

The term 'seam pucker' is generally used in association with the deformation of woven fabrics. The mechanics of deformation in woven fabrics are generally very different from those occurring with knitted fabrics. Even so, some authors use the term 'pucker' for seam deformations in all types of situations. Dorkin and Chamberlain worked on this subject as early as 1961 and described five main causes of seam pucker.

5.6.1 Differential fabric stretch/'feed pucker'

This effect, also known as 'feed pucker' (Amann Group, n.d.c), takes place when the fabric plies are not all fed by the material feeding system in the same condition, causing different stretch to the different plies. Feed pucker appears very often when sewing two plies on a drop-feed system, where there is a tendency for the upper ply to get stretched by the friction with the presser foot. This is also (as previously described) the cause of ply-shift. Feed pucker depends mostly on the fabric properties and can be reduced by reducing presser foot force, using specific presser feet (e.g. Teflon coated) or using a compound feeding system.

5.6.2 Differential fabric dimensional instability and seam pucker

In this case the fabrics being sewn have different dimensional changes after the seam, causing the seam to pucker. This can happen right after sewing, when the fabrics are recovering from extensions suffered by the feeding system, but the recovery may also take longer. The dimensional change can also be produced by pressing and washing. To avoid this type of seam pucker, the dimensional behaviour of the fabrics being sewn should be studied carefully.

5.6.3 Sewing thread dimensional change/'tension pucker'

This type of pucker is often referred to as 'tension pucker' (Amann Group, n.d.c). Sewing threads suffer high tensions during sewing, thus being stretched in the process. After the operation is completed, they tend to recover from this extension. If the thread's size reduction after sewing is higher than that of the fabric, it will contract the fabric and thus result in pucker. (In the opposite situation, the thread may be loose and the seam may lose strength and exhibit excessive grinning.) As previously stated, the process may be immediate or take longer time spans. A similar situation with the same result is the shrinkage of the sewing threads after pressing or washing. To reduce this problem, less thread tension should be used, or the feeding system of the machine may be set up to stretch the fabric slightly (using negative differential feed or tilting the feed dogs appropriately), in order to compensate for the thread's shrinkage. This can also be achieved by sewing the fabric in a direction of higher elasticity, causing it to stretch a little.

5.6.4 Structural jamming/'displacement pucker'

Also known as 'displacement pucker' (Amann Group, n.d.c), in this case the modification of the fabric due to the presence of sewing thread in its structure is the cause of seam pucker. Densely woven fabrics have very little space to accommodate the sewing thread. The insertion of the thread causes the structure to distort. The problem is worse when stitch density and thread thickness increase; therefore, thin threads and needles and lower stitch densities should be used. Sewing at an angle (15° or more) to the weft/warp yarns also helps, because threads find more space in a diagonal direction to accommodate between fabric yarns.

5.6.5 Pattern mismatch and seam pucker

In this final case, the machine operator tries to compensate variations in length of the pieces being sewn, by stretching one and easing in the other, so as to achieve a match on the seam ends. In this situation, it is clear that the seam will be prone to a surface distortion, and the solution is also clear: solve the cutting inaccuracy that causes the mismatch in length.

5.6.6 Analysing seam pucker

The extent of seam pucker has traditionally been classified by comparing the seams with a set of photographic standards defined by the American Association of Textile Chemists and Colorists. Under this classification, seam pucker is quantified in a scale from 5 = no seam pucker to 1 = very intense seam pucker. Researchers have long been developing more accurate and complete methods for seam pucker objective evaluation. Stylios (2005) describes the PLMS: Pucker Laser Measurement System, which combines laser sensing and artificial vision to sense the surface of a seam and quantify parameters of the seam pucker such as the amplitude and frequency of the 'waves' produced by seam pucker. Park and Kang (1999a) combine this sensing method with spectral analysis of the measured waves and used neuro-fuzzy algorithms to classify seam pucker and relate the measured parameters with sewing conditions (Park and Kang, 1999b).

5.7 Future trends

Many of the problems described in this chapter result from sewing defects – systematic or sporadic. These are produced by unfavourable sewing conditions due to bad properties of fabrics, threads, due to wear-out of sewing machine components or needles, due to incorrect setting of machines or bad choice of components such as feed dogs, presser feet, needles, etc. Most of these defects go unnoticed for whole batches, because sewing machines are traditionally blind regarding their process variables. Sewing technicians set the machines based on their experience, in a trial-and-error process, without any quantitative feedback of the machine during its operation. Sewing defects can only be detected by visual inspection or by quality control *after* production. The best way to avoid these problems at an early stage is still an aware and sharp-visioned seamstress. But her work in this respect should be eased, and much more control should lie in the hands of production managers and technicians, as well as much more autonomy in the machines. By this, the monitoring and control of the sewing process itself, and thus the quality and functionality of the end-product, is meant – not material handling itself, as that is an entirely different subject.

Stylios (1996) defines the principles of complete intelligent manufacturing systems, which are composed of four tools:

- A fabric measurement system to determine the set of properties completely defining each material.
- A sewability/tailorability prediction system which will predict the interaction between the machines and the material, helping to set up the machine and/or engineer the material for optimum sewability/tailorability.
- Intelligent sewing machines which can set up themselves and dynamically control their mechanisms as a function of the material properties.
- Self-learning systems attached to the machines or centrally integrated into the whole system, to adapt the control models dynamically.

Fortunately, extensive work has already been accomplished by many researchers that has enabled the development of industrial applications pointing the way to intelligent textile and garment manufacture. Even commercial products have been presented and are successfully being marketed. Pfaff's Doku-Seam System (Pfaff Industrial, 2009a) is a sewing system for monitoring and documentation of seams, being able to measure thread force, stitch length setting and the backtack lever of the machine. It was developed to assemble car seats, assuring complete monitoring of the seams supposed to tear when side airbags are deployed. A backtack during this seam, an incorrect stitch length setting, excessive tension, skipped stitches or other factors may cause malfunction of this critical piece of safety equipment. Pfaff also offers the SRP system (speed responsive presser foot), a system that can adjust force on the presser foot according to sewing speed, and reduce feeding pucker (Pfaff Industrial, 2009b).

A comprehensive and integrating approach has been followed by several academic researchers, pursuing both offline sewability testing/process planning tools, as well as the 'intelligent and self-learning' sewing machines and control systems. Some illustrative examples include neuro-fuzzy control systems able to set up the machines automatically according to material properties (Stylios & Sotomi, 1995); the control of the machine's presser foot to adapt itself to the material being sewn, by use of sensors and adaptive control systems (Silva *et al.*, 2010); methods for optimum thread tension control, thread tension balancing and stitch formation defect monitoring (Carvalho, 2006); integrated software for testing and development of machine control/monitoring algorithms and material testing tools (Carvalho, 2010). It has been possible to demonstrate skipped stitch and stitch deformation detection, thread balance and tension prediction and monitoring, presser foot bouncing detection, needle penetration and withdrawal force measurement, among other functionalities.

The adoption of these tools in industrial practice will certainly be an incremental process. Some of them require costly equipment, demanding careful cost–benefit analysis, others have still to be further developed and tested, but it is certain that most of them will become mainstream in the future. This is especially plausible considering a new need in the textile industry: the need to assure reliability of technical textiles, guaranteeing both functionality as well as the safety of users of this whole new breed of products.

5.8 Sources of further information and advice

Besides the extensive reading available in textbooks and scientific publications, good advice on sewing problems, within extensive information on all aspects of sewing, can often be found on the websites of sewing machine, needle and thread manufacturers. Here is a selection of some interesting websites:

- Pfaff-Industrial, http://www.pfaff-industrial.com (look in FAQs)
- Pegasus, http://www.pegasus.co.jp/
- Duerkopp-Adler, http://www.duerkopp-adler.com
- American & Efird thread supplier, http://www.amefird.com/
- Amann sewing threads, http://www.amann.com/
- Schmetz needles, http://www.schmetz.com/
- Groz-Beckert needles, http://www.groz-beckert.com

5.9 References

Alagha, MJ, Amirbayat, J & Porat, I 1996, 'A study of positive needle thread feed during chainstitch sewing', *Journal of the Textile Institute*, Vol. 87 Part 1, No. 2, pp. 389–395.

Amann Group, n.d.a, 'No quality product without quality seams', in *Focus – Basic knowledge, Vol. 2: Sewing and Embroidering*, retrieved 22 September 2010 from http://www.amann.com/en/focus-basic-knowledge.html

Amann Group, n.d.b, 'Determining your sewing thread requirements', retrieved 4 October 2010 from http://www.amann.com/en/download/industrial-sewing-threads.html

Amann Group, n.d.c, 'Prevention of seam pucker', retrieved 4 October, 2010 from http://www.amann.com/en/download/industrial-sewing-threads.html

Araújo, M, Little, TJ, Rocha, AM, Vass, D & Ferreira, FN 1992, 'Sewing dynamics: Towards intelligent sewing machines', NATO ASI on Advancements and Applications of Mechatronics Design in Textile Engineering, Side, Turkey, 5–16 April.

Barrett, GR 1992, 'Sewing dynamics: The Autodamp, Technology for the elimination of presser foot bounce', North Carolina State University Leaflet, 7/92.

Barrett, GR & Clapp, TG 1995, 'Coprime factorization design of a novel Maglev presser foot controller', *Mechatronics Journal*, Vol. 5, No. 2/3, pp. 279–294.

Blackwood, WJ & Chamberlain, NH 1970, 'The strength of seams in knitted fabric', Technical Report No. 22, Clothing Institute, Leeds.

Bühler, G & Hennrich, L 1993, 'Control "On-Line" de la Formación de las Costuras', *Punto Técnica y Moda*, Vol. 11, No. 4, pp. 239–242 and 244.

Bühler, G & Hennrich, L 1994, 'Control "On-Line" de la Formación de las Costuras', *Punto Técnica y Moda*, Vol. 12, No. 1, pp. 42–46 and 51.

Burtonwood, B & Chamberlain, NH 1966, 'The strength of seams in woven fabrics (Part 1)', Technological Report No. 17, Clothing Institute, Leeds.

Carr, H & Latham, B 1994, *The Technology of Clothing Manufacture*, 2nd edn, Wiley-Blackwell, Oxford.

Carvalho, H 2010, *Optimisation and Control of Processes in Apparel Manufacturing: New tools for efficiency and quality assurance in industrial sewing technology*, VDM Verlag Dr. Müller, Saarbruecken.

Carvalho, H, Rocha, AM & Monteiro, JL 2009, 'Measurement and analysis of needle penetration forces in industrial high-speed sewing machine', *Journal of the Textile Institute*, Vol. 100, No. 4, pp. 319–329.

Carvalho, MAF 2003, 'Estudo das Relações entre os Parâmetros de Controlo, Propriedades dos Materiais e Condições de Regulação numa Máquina de Costura Corta-e-Cose' [Study of the Relationship between Control Parameters, Material Properties and Set-up Conditions in an Overlock Sewing Machine], PhD diss., University of Minho, Portugal.

Carvalho, MAF & Ferreira, FBN 2006, 'High performance in sewing – guaranteeing seam quality through control of sewing dynamics', in *Textiles for Sustainable Development*, Nova Science Publishers, New York, pp. 161–171.

Carvalho, MAF, Ferreira, FBN & Ferreira, AM 2006, 'The effect of the sewing threads in seam quality and machine set-up – objective evaluation'. *Proceedings of the International Fiber Conference 2006 – Extreme and Aesthetic Textiles*, 30 May–3 June, Seoul National University, Seoul, Korea.

Chmielowiec, R & Lloyd, DW 1995, 'The measurement of dynamic effects in commercial sewing machines', *Proceedings of the 3rd Asian Textile Conference*, Vol. II, pp. 814–828.

Clapp, TG, Little, TJ, Thiel, TM & Vass, DJ 1992, 'Sewing dynamics: objective measurement of fabric/machine interaction', *International Journal of Clothing Science and Technology*, Vol. 4, No. 2/3, pp. 45–53.

Crow, RM & Chamberlain, NH 1969, 'The performance of sewing threads in industrial sewing machines', Technological Report No. 21, Clothing Institute, Leeds.

Dorkin, C & Chamberlain, N 1961, 'Seam pucker, its cause and prevention', Technological Report No. 10, Clething Institute, Leeds.

Frank, TFP & Mo, ESM 1974, 'A simple test for assessing presser foot bounce', *Clothing Research Journal*, Vol. 2, No. 2, pp. 81–84.

Gribaa, S, Amar, SB & Dogui, A 2006, 'Influence of sewing parameters upon the tensile behavior of textile assembly', *International Journal of Clothing Science and Technology*, Vol. 18, No. 4, pp. 235–246.

Gurarda, A & Meric, B 2005, 'Sewing needle penetration forces and elastane fiber damage during the sewing of cotton/elastane woven fabrics', *Textile Research Journal*, Vol. 75, No. 8, pp. 628–633.

Gurarda, A & Meric, B 2010, 'Slippage and grinning behaviour of lockstitch seams in elastic fabrics under cyclic loading conditions', *Tekstil Ve Konfeksiyon*,Vol. 20, pp. 65–69.

Hurt, FN & Tyler, DJ 1975, 'Seam damage in the sewing of knitted fabrics. II – Material variables'. HATRA Research Report, No. 36.

Hurt, FN & Tyler, DJ 1976, 'Seam damage in the sewing of knitted fabrics. III – The mechanism of damage'. HATRA Research Report, No. 39.

ISO standards 1999a, Seam tensile properties of fabrics and made-up textile articles – Part 1: Determination of maximum force to seam rupture using the strip method, ISO 13935-1:1999.

ISO standards 1999b, Seam tensile properties of fabrics and made-up textile articles – Part 2: Determination of maximum force to seam rupture using the grab method, ISO 13935-2:1999.

ISO standards 2004a, Determination of the slippage resistance of yarns at a seam in woven fabrics – Part 1: Fixed seam opening method, ISO 13936-1:2004.

ISO standards 2004b, Determination of the slippage resistance of yarns at a seam in woven fabrics – Part 2: Fixed load method, ISO 13936-2:2004.

ISO standards 2007, Determination of the slippage resistance of yarns at a seam in woven fabrics – Part 3: Needle clamp method, ISO 13936-3:2007.

Johnson, EM, 1973, 'Some factors affecting the performance of high speed sewing machines', *Clothing Research Journal*, Vol. 1, No. 1, pp. 3–35.

Kalaoglu, F & Meric, B 2005, 'Investigation of the performance of linings', *International Journal of Clothing Science and Technology*, Vol. 17, No. 3/4, pp. 171–178.

Leeming, CA & Munden, DL 1976, Testing fabric sewing properties, US Patent 3979951.

Leeming, CA & Munden, DL 1978, 'Investigations into the factors affecting the penetration force of a sewing needle in a knitted fabric and its relationship with fabric sewability', *Clothing Research Journal*, Vol. 6, No. 3, pp. 91–118.

Liasi, E, Du, R, Simon, D, Bujas-Dimitrejevic, J & Liburdi, F 1999, 'An experimental study of needle heating in sewing heavy materials using infrared radiometry', *International Journal of Clothing Science and Technology*, Vol. 11, No. 5, pp. 300–314.

Matthews, BA & Little, TJ, 1988, 'Sewing dynamics, Part I: Measuring sewing machine forces at high speeds', *Textile Research Journal*, Vol. 58, pp. 383–391.

Miguel, RAL, Lucas, JM, Carvalho, ML & Manich, AM 2005, 'Fabric design considering the optimisation of seam slippage', *International Journal of Clothing Science and Technology*, Vol. 17, No. 3/4, pp. 225–231.

Mukhopadhyay, A, Sikka, M & Karmakar AK 2004, 'Impact of laundering on the seam tensile properties of suiting fabric', *International Journal of Clothing Science and Technology*, Vol. 16, No. 4, pp. 394–403.

Park, CK & Kang, TJ 1999a, 'Objective evaluation of seam pucker using artificial intelligence. Part II: Method of evaluating seam pucker', *Textile Research Journal*, Vol. 69, No. 11, pp. 835–845.

Park, CK & Kang, TJ 1999b, 'Objective evaluation of seam pucker using artificial intelligence. Part III: Using the objective evaluation method to analyze the effects of sewing parameters on seam pucker', *Textile Research Journal*, Vol. 69, No. 12; pp. 919–924.

Pfaff Industrial, 'Pfaff Docu-Seam System', 2009a, retrieved 7 October 2010 from http://www.pfaff-industrial.com/pfaff/en/news/reports/PFAFF%20Docu-seam-system/view?searchterm=doku

Pfaff Industrial, 2009b, 'What are the causes of seam pucker', retrieved 7 October 2010 from http://www.pfaff-industrial.com/pfaff/en/service/faqs/generalsewing/seampucker/view?searchterm=Speed responsive

Rocha, AM 1996, 'Contribuição para o Controlo Automático dos Parâmetros de Costura: Estudo da Dinâmica da Penetração da Agulha e da Alimentação do Tecido' [Contribution to automatic control of sewing parameters: Study of needle penetration and fabric feeding dynamics], PhD diss., University of Minho, Portugal.

Rocha, AM, Sousa, EJ, Lima, M & Araújo, M 1992a, 'Estudo da Dinâmica do Processo de Costura numa Máquina de Corta-e-Cose', *Revista de Robótica e Automatização*, No. 9, pp. 12–17.

Rocha, AM, Lima, M, Sousa, E & Araújo, M 1992b, 'Evaluation of sewing performance and control of sewing operation', Presentation at the 2nd International Clothing Conference, University of Bradford, UK, June.

Rocha, AM, Lima, M, Ferreira, FN & Araújo, M 1996a, 'Developments in automatic control of sewing parameters', *Textile Research Journal*, Vol. 66, No. 4, pp. 251–256.

Rocha, AM, Ferreira, FN, Araújo, M, Monteiro, J, Couto, C & Lima, M 1996b, 'Mechatronics in apparel: control, management and innovation on the sewing process', *Proceedings of the Mechatronics'96 Conference*, Vol. II, pp. 109–114, Universidade do Minho, September.

Silva, LF 2002, 'Estudo de Mecanismos Alternativos de Controlo do Sistema de Alimentação de Máquinas de Costura Industriais', PhD thesis, School of Engineering, University of Minho, Guimarães, Portugal.

Silva, LF, Carvalho, H, Soares, F & Guhr, F 2010, 'Adaptive control of an electromagnetically actuated presser-foot for industrial sewing', 15th IEEE International Conference on Emerging Technologies and Factory Automation, Faculty of Engineering of the University of the Basque Country, Bilbao, Spain, 13–16 September.

Stylios, G 1996, 'Principles of intelligent textile and garment manufacturing systems', *Assembly Automation*, Vol. 16, No. 3, pp. 40–44.

Stylios, GK 2005, 'New measurement technologies for textiles and clothing', *International Journal of Clothing Science and Technology*, Vol. 17, No. 3/4, pp. 135–149.

Stylios, G & Sotomi, JO 1995, 'A neuro-fuzzy control system for intelligent overlock sewing machines', *International Journal of Clothing Science and Technology*, Vol. 7, No. 2/3, pp. 49–55.

Sundaresan, G, Hari PK & Salhotra, KR 1997, 'Strength reduction of sewing threads during high speed sewing in an industrial lockstitch machine: Part I – mechanism of thread strength reduction', *International Journal of Clothing Science and Technology*, Vol. 9, No. 5, pp. 334–345.

Sundaresan, G, Hari PK & Salhotra, KR 1998, 'Strength reduction in sewing threads during high speed sewing in industrial lockstitch machine: Part II: Effect of thread and fabric properties', *International Journal of Clothing Science and Technology*, Vol. 10, No. 1, pp. 64–79.

Uçar, N 2002, 'Grinning of ISO 514 stitched seams on knitted fabrics under the effects of repeated extension and recovery', *Textile Research Journal*, Vol. 72, pp. 944–948.

Yildirim, K 2010, 'Predicting seam opening behavior of woven seat fabrics', *Textile Research Journal*, Vol. 80, pp. 472–480.

6
The quality and performance of sewn seams

A. MUKHOPADHYAY and V. K. MIDHA,
National Institute of Technology Jalandhar, India

DOI: 10.1533/9780857093967.1.175

Abstract: Study of the structural properties and behaviour of seams is very important for the performance of integrated assembly of textile products, and a range of non-textile products. The performance parameters of seams include strength, elasticity, durability, security and appearance. However, priority given to any of the characteristics will vary, depending on the end-use of the seamed product. Basic understanding about the seam performance parameters and factors affecting them is discussed in this chapter. The chapter also discusses strategies to improve seam performance through judicious selection of material, setting of operating parameters at correct levels and also by improving the design of the sewing machine.

Key words: degradation of seam, seam construction, seam performance.

6.1 Introduction

The primary function of a seam is to provide uniform stress transfer from one piece of material to another, thus preserving the overall integrity of the assembly. Although developments in fabrication processes such as fully-fashioning, three-dimensional knitting and thermoplastic moulding have resulted in the production of three-dimensional textile products, seaming remains fundamental to the development of textile and a range of non-textile products. Ultrasonic welding, fusing and glueing have limited use in joining textile materials due to their low strength, low extensibility and relatively high cost.

Study of the structural properties and behaviour of seams is very important for the performance of integrated assembly. The performance parameters of seams include strength, elasticity, durability, security and appearance. However, priority given to any of the characteristics will vary, depending on the end-use of the seamed product. For example, seam appearance will be accorded greater priority for a dress shirt constructed from woven fabric. On the other hand, appearance is seldom very important for seams that are primarily functional, such as seams on seat belts or parachutes, although some consideration is usually given to appearance if a change in appearance is likely to be associated with deteriorating performance (Carr and Latham,

1994; Dorkin and Chamberlain, 1961; Mukhopadhyay *et al.*, 2004, 2005; Ukponmwan *et al.*, 2000). In the subsequent sections factors affecting seam performance parameters and ways to improve seam performance under varied situations are discussed.

6.2 Seam strength

Seam strength is the force required to break the stitching thread at the line of stitching. Relatively inextensible fabrics and stable warp knit structures are tested for transverse seam strengths, whereas extensible structures like knitted, elastic and highly resilient fabrics are tested for longitudinal seam strengths (Hu, 2004). Seams must have suitable strength and elasticity to withstand loads during wearing and prevent seam breakage or visible damage to the fabric in the seam region. The knowledge of strength characteristics is necessary not only under quasi static strain but also under impact strain, especially for the technical products which have to resist special and often very hard conditions of application.

The amount of stresses to which the seams are subjected depends on the garment application and the position of the seam in the garment. For severe applications, such as military combat forces, very high seam strengths are required. Crow and Dewar (1986) recommended minimum seam strength of 14,900 N/m for garments intended for severe use. However, if fabric strength is lower than seam strength, fabric failure before seam failure makes the assembly unrepairable. Therefore it is advisable to select the right fabric which can hold the required seams. The capacity of a material to carry a seam is referred to as seam efficiency. This is the ratio of seamed fabric strength and unseamed fabric strength expressed as a percentage (Burtonwood and Chamberlain, 1966). Frederick (1952) recommends 80% seam efficiency for United States military cotton fabrics. Seam efficiencies of 60–80% are common and easily obtainable, whereas obtaining efficiencies of 80–90% is more difficult. It is important to note that seam efficiency for different materials can be compared only when the type of seam failure is similar. In an earlier study, it was found that with the decrease in thread linear density, seam efficiency of the lighter denim fabrics increases and that of heavier fabrics decreases (Behera *et al.*, 1997). In another study, Gurarda (2008) observed that at the same thread densities, twill fabric samples possess higher seam efficiency values than plain woven fabrics.

6.2.1 Impact of stresses and seam failure

The forces acting on seams in clothing are a function of body movements, garment design, and seam characteristics. Garments made from stretchable fabrics are more comfortable than those made from non-stretchable fabrics due to their extension with body movements (Kirk and Ibrahim, 1966). The seams in such garments should also provide corresponding stretch to avoid garment distortion, wearer discomfort and seam failure.

In order to estimate the seam performance for clothing in different applications, it is important to understand the stances causing maximum stresses, the location of these stresses in the garment and their levels during the stances. Even putting on and taking off the garment cause considerable stress in the clothing (Nestler and Schlegel, 1978). Normal wearing may produce lower levels of stress than applications where active performance is required, such as military combat forces, mountaineering, and industrial workplaces. Military armed forces, in particular, require to move, live, survive and fight in hostile environments, which often demand garments of high serviceability. Most stress-causing stances include (Kirk and Ibrahim, 1966): bending the elbow fully, forcibly bending the knee, crossing arms in the front, bending over for maximum spinal curvature and sitting with the knees up.

The level of stresses during these actions may lead to the failure of the garment. In a military combat garment, maximum stresses occur in the shirt (1,650 N/m) and the coverall top (3,430 N/m), while crossing the arms in front of the body. Back seams of the trousers and coveralls are also subjected to stresses of the order of 2,410 N/m and 2,900 N/m respectively, during squatting (sitting with knees and hips fully bent and legs spread apart) (Crow and Dewar, 1986). However, certain seams were subjected to very high stresses even in normal circumstances, for instance the crotch seam of pantyhose and jeans.

In general, a seam may fail as a result of longitudinal or transverse loading; transverse seam strength is a prime requirement in the joining of woven fabrics by sewn seams (Tsui *et al.*, 1984). However, in real-life situations, a seam may be loaded in both directions resulting in stresses applied multiaxially or biaxially. Rupture of the assembly is preceded by contraction or slippage of the seam and the failure of seam may result from (Chmielowiec, 1987a; Burtonwood and Chamberlain, 1966):

1. fabric yarn rupture,
2. sewing thread rupture,
3. fabric yarn slippage, and
4. a combination of two or more reasons.

Most commonly, seam failure may be either due to the failure of sewing threads or rupture/damage of fabric (Fig. 6.1). Gribaa *et al.* (2006) categorized transverse seam failures into four types.

- Type 1: Contracting of seam followed by breaking of the sewing thread without damaging the fabric.

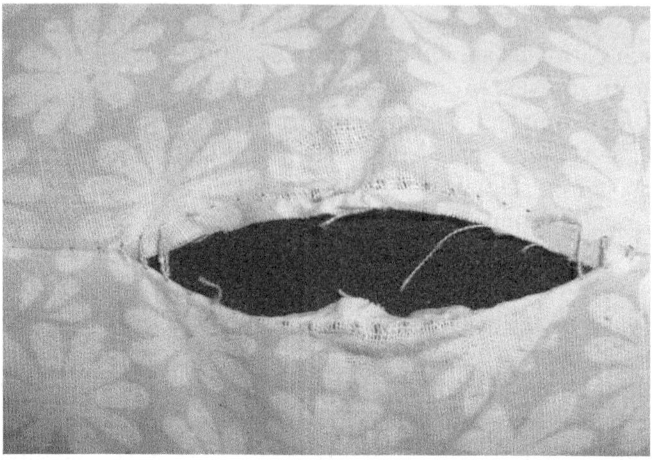

(a)

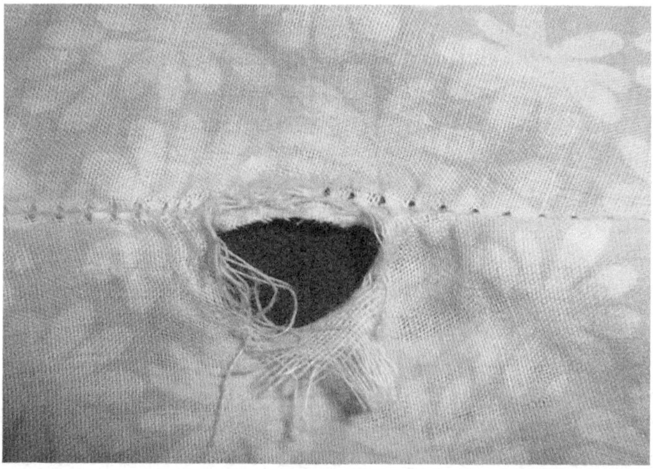

(b)

6.1 Types of seam failure: (a) failure of sewing threads (b) rupture of fabric.

- Type 2: Seam slippage occurs and the seam remains intact but yarns in the fabric pull out of the seam from the edge.
- Type 3: Sewing thread gets ruptured, leaving holes in the fabric due to yarn slippage.
- Type 4: Fabric gets damaged along the stitch line.

The type of failure occurring in an assembly depends on factors such as fabric strength, thread strength, stitch density and seam edge level. In practice, the majority of seams that fail on commercial garments show type 1 breakage. In situations where the seams fail showing type 2 breakage, a strong thread is combined with a comparatively weak fabric. However, when two very strong fabrics are joined, as in some workwear garments, a stronger sewing thread should be used, although failure may occur by rupture of the sewing thread. Some sewn articles are broken within the seam allowance, when compressive load is applied by sewing threads. For woven fabrics, this type of rupture is caused by unsuitable seam allowance length, and is introduced into fabric yarn slippage corresponding with type 3. Non-woven fabric or polymer film is broken by expansion of stitch holes at seam allowance. In addition, for non-woven fabrics, rupture often happens at intra-fabric elements except the seam part.

In the case of longitudinal seam failure, the first break in sewing thread is probably most important, since once the continuity of sewing thread is destroyed, further stitch unravelling can occur. The extent of unravelling is likely to depend on the frictional characteristics of the sewing thread and fabric yarns, and the position of the thread interlacing in relation to the fabric structure.

6.2.2 Factors affecting seam strength

Seam strength is affected by sewing thread strength, type of seam, stitch type, stitch density, machine parameters, and weight and construction of the fabric being sewn. Common sites of seam failure in garments examined after normal wearing and cleaning are the front torso, the sleeve, the seat (crotch seam) and zippers. The predominant cause of seam failure was identified as thread breakage. Selection of suitable sewing thread for a fabric is important for better seam performance and appearance. A stronger sewing thread normally results in a stronger seam provided the failure takes place at the sewing threads.

The important thread parameters include:

- fibre type
- thread construction (staple spun, core spun, textured or multifilament)
- thread size
- thread finish (soft, mercerized, glaced or bonded).

Use of stronger fibres with higher loop strength contributes to higher seam strength. For example, a 100% spun polyester thread provides greater seam strength than a 100% cotton thread of the same size. Synthetic fibres like polyester and nylon are much more resistant to abrasion and chemical degradation (such as bleach) than cellulosic fibres. Cellulosic fibres on the other hand, have greater heat resistance. Threads made of high performance or functional fibres can be used for specific applications to improve the performance of garments. In general, continuous filament threads retain more strength than staple fibre threads after exposure to high temperatures. As regards the impact of fibre material, PPTA (poly p-phenylene terephthalamide) sewing threads (Kevlar by Dupont and Twaron by Teijin) resist high temperatures and retain strength after exposure to high temperatures. Thus, they can provide a greater margin of safety in applications where such strength properties are critical, such as open flame resistant bedding (Hall and Knoff, 2008). In another application, polypropylene sewing threads can be advantageously used in soft containers which can be shipped under different weather conditions (Sakhnik and Rodionov, 2005). Application of moisture on polypropylene thread increases its strength by eight per cent on average and decreases the elongation due to the plasticizing effect of the moisture.

Among multifilament, core spun and staple spun polyester threads of same ticket number, multifilament threads produce the strongest seams, whereas the weakest seams are produced using staple spun threads. It is to be noted that usually core spun thread is used where higher seam strength is demanded. Between the two-ply and three-ply threads, three-ply threads show higher seam strength value than two-ply threads. In the situation where strength of sewing thread improves with the increase in twist in plied structure, corresponding seam strength will also improve (provided the breakage of seam is due to the tensile failure of sewing threads). Further, for needle threads, a high level of twist helps impart a circular cross section to the thread and lower area of contact between the thread and the needle, leading to lower thread strength loss during sewing. Higher residual strength of needle thread provides higher seam strength. However, the study by Meric and Durmaz (2005) showed that the amount of twist in sewing thread does not have any significant influence on seam strength.

Given a specific fibre type and thread construction, seam strength depends largely on the thread linear density. Seam strength and seam efficiency of fabric stitched with coarser thread is higher than the fabric stitched with finer thread due to higher breaking load of the coarser thread than the finer thread, as shown in Fig. 6.2 (Mukhopadyay *et al.*, 2004; Mukhopadhyay, 2008). However, it is very important that the thread size is compatible with the fabric weight, otherwise the seam performance may deteriorate rather than improve. A smaller thread size will embed itself in the seam, making it less prone to surface abrasion as compared to coarser threads used on heavy fabrics, which remain on the surface and therefore are prone to more abrasion. Mercerized threads have lower tensile strength but are more resistant to abrasion than soft cotton threads of the same fibre type and size. Glace or bond finished thread generally has superior abrasion resistance to soft finished threads. Thread of higher abrasion resistance leads to a stronger and more durable seam.

As regards impact of fabric characteristics on seam strength, the following parameters are important:

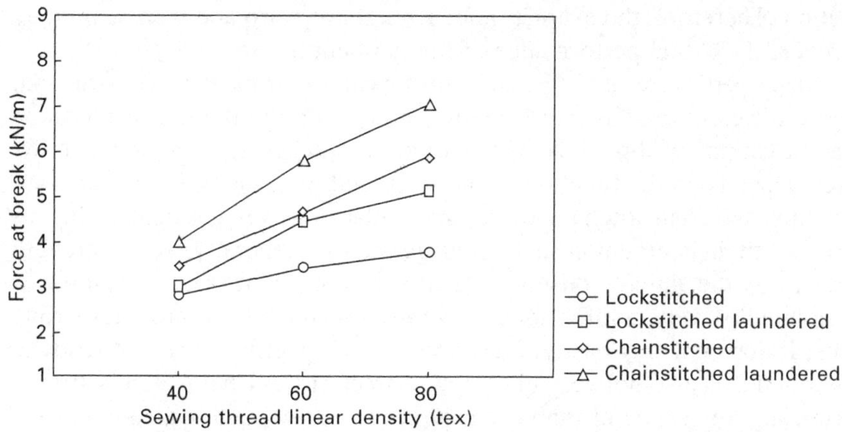

6.2 Effect of sewing thread linear density on force at break. (Reprinted from A. Mukhopadhyay, *Journal of Engineered Fibers and Fabrics*, 3(1), 2008. Copyright is held by JEFF and its Co-Sponsors; AATCC, INDA, TAPPI and The Fiber Society, P.O. Box 1288, Cary, NC 27512, USA.)

- fabric weight and thickness
- fibre content
- fabric construction
- pattern placement and seam direction
- propensity of the yarns in the seam to shift or pull out of the seam.

Earlier studies showed that stronger fabrics with compatible sewing threads in the seam produce stronger seams (Behera *et al.*, 1997; Sajn, 2007). Seam strength also increases with the increase in fabric weight due to greater fabric strength and higher resistance during yarn slippage (Behera *et al.*, 1997). Fabric weave is also known to have a significant effect on seam strength in both the warp and weft directions. The plain weave fabrics exhibit highest seam strength in both directions, followed by twill fabric. Since plain weave has the highest number of interlacements per unit area, it does not permit easy movement of yarns when a seam is subjected to loading. On the contrary, twill weave has longer floats allowing more free space for yarn movement (Cheng and Poon, 2002). Domingues *et al.* (1997) reported that seam direction (warp and weft) also has a significant effect on the seam strength. In an experiment on plain and twill weave fabrics, it was observed that the seam strength is higher for seams along the warp direction than those along the weft direction.

It is important to note that many seams in garments are constructed with the position of the seam in relation to fabric structure not identical for each of the component parts or, alternatively, not parallel to the warp or the weft direction. Therefore, the evaluation of strength of warp and weft seam may not reveal the actual performance of the garment in use.

Seam properties are also affected by the type of stitch and seam construction, because different threads interlace differently with the fabric and consume different lengths of thread. A wide variety of stitches are available and the choice depends on the function of the seam and its location in the garment. Lockstitch and chainstitch are the most popular in woven garments, whereas overlock stitch is common in knitted garments. Generally, seam strength increases as the thread consumption in the stitch increases. Lockstitch is flat, non bulky, very comfortable, consumes less thread and provides strong seams; it does not fray easily, does not unravel or grin easily, but tends to pucker and it is also not very elastic in characteristic. Chainstitch is strong, bulkier, slightly less comfortable, consumes more sewing thread, durable and elastic; seams may grin but are less likely to pucker. It is to be noted that chainstitch constructions are least prone to shearing or cutting of needle and looper threads when interlooped together in the seam as compared to needle and bobbin threads in lockstitchs. An overlock seam provides considerably larger longitudinal extensibility, particularly when the seam is opened flat and pressed. Threads used in 301 lockstitch seams are more susceptible to shearing each other than 401 chainstitch and 504 overedge seams because of the way the threads are interlocked together rather than interlooped (Rao, 2006).

Chmielowiec (1987b) investigated the effects of seam type and sewing thread on seam efficiency in workwear garments that are subjected to heavy laundering. In the context of denim-based garments, particularly trousers and jackets, sewing thread is also used for decorative purposes along the seamline. The thread possesses a linear projection on the surface of the garments and is subjected to more abrasion than the garment (Behera *et al.*, 1997). Scott (1987) analysed the effect of new stonewashing methods on trims and sewing threads in denim garments. The abrasion, wear and damage caused by the conventional aqueous stone-washing process were compared with that caused by dry stone-washing. Less damage was found with dry stone-washing than with aqueous stone-washing.

Many seam constructions are more resistant to both stress and abrasion than other constructions. Among felled seam, five-thread overlock stitch, combination of five-thread overlock stitch with single needle lockstitch, and combination of five-thread overlock stitch with two needle lockstitch, felled seams show the highest seam strength and combination of overlock with two needle lockstitch show the second highest seam strength. These seams are recommended for areas on the garment that are subjected to higher tensions. Overlock stitch shows the lowest seam strength and the combination of overlock with single needle lockstitch shows the second lowest seam strength (Yeşilpinar and Bahar, 2007). Since quantity of thread consumed is more in lap seams, the seams are stronger than safety stitched seams and superimposed seams. Different thread combinations also influence seam performance. In lap seam, when needle thread is changed to a weaker thread, seam strength is reduced by 12.8%, whereas a change in the bottom thread has no significant influence on lap seam strength. This is because in chainstitch, the bottom thread suffers less frictional damage during loop formation compared to needle thread. Loss in needle thread strength during sewing leads to decrease in seam strength (Chmielowiec, 1987a). Observation also showed that among polyester-cotton core spun threads and spun polyester threads, lap seam strength remains unchanged.

There are also few studies related to other geometric parameters in a specific seam type such as stitching in multiple rows and length of overlap. It is important to note that the number of rows of stitching in a particular seam type affects seam breaking strength, as stresses applied to the seams may be carried by several stitching lines. What proportion of the total applied load is borne by each depends on the position of the stitching rows in relation to each other and to the load and stitch tension (Burtonwood and Chamberlain, 1966). In a study on the strength of lapped and superimposed joins in four stitch-joined parachute webbings, the effect of length of overlap (between 80 and 400 mm) was found to be insignificant (Webb, 1976).

As regards impact of stitch density, the greater the number of stitches per inch in a seam, the greater is the seam strength, as shown in Fig. 6.3

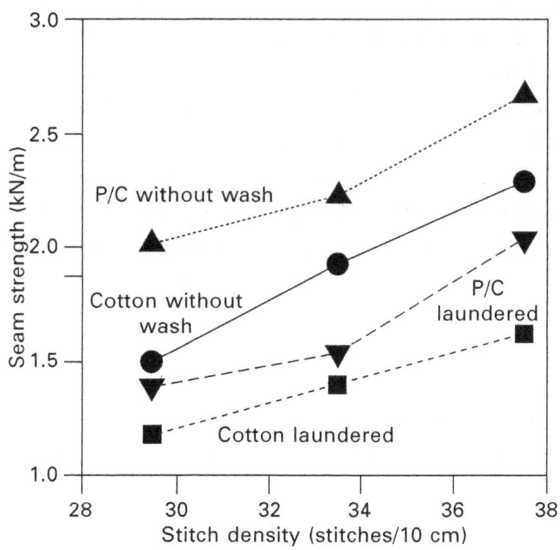

6.3 Effect of stitch density on seam strength. (Reprinted from A. Mukhopadhyay, M. Sikka and A.K. Karmakar (2004), 'Impact of laundering on the seam tensile properties of suiting fabrics', *Int. J. Clothing Sci. & Tech.*, 16, 394–403; an Emerald Publication.)

(Mukhopadhyay *et al.*, 2004). A higher number of stitches per unit length provides more thread into the seam, leading to higher seam strength. Moreover, at higher stitch density, fabric feed is reduced and therefore the fabric is securely held and controlled during stitch insertion. In a fabric assembly with lower stitch density, lower adherence between the fabric and the seam causes type 1 failure in the fabric assembly. Contraction of the seam on application of tensile forces leads to appearance of holes in the fabric, which progressively enlarge until its rupture. Higher stitch densities in the seam result in strong adherence between the fabric and the seam, thereby the stitch line piles up the weft yarns of the fabric and warp yarns get slipped from the edges. Yarn slippage from the edge causes tearing of the fabric leading to type 2 or type 3 failures (Gribaa *et al.*, 2006).

However, higher stitch densities beyond a certain limit can cause the needle to puncture the fabric a greater number of times. The fabric tensile strength is reduced due to cutting of the yarns and damaging of the structure. In lockstitch sewing, the needle thread strength is also significantly reduced during sewing because it is subjected to more loading cycles through the machine before getting incorporated into the seam (Midha *et al.*, 2009a, 2009b, 2010a, 2010b). Therefore, at very high stitch densities, both the fabric and the needle thread get weakened, leading to decrease in seam strength (Chmielowiec, 1987a; Brain, 1970; Aditya and Bose, 1983). Wang *et al.* (2001) showed that the extent to which the impact of stitch density on the strength, extensibility, and stress withstands retention is different for three stitch types commonly used for knitted fabrics. When stitch density is increased, the tensile strength at break for 301 lockstitch is reduced; whereas the value is increased gradually for 504 three thread overlock. In case of a 514 four thread overlock seam, initial increase in strength is followed by gradual decrease with the increase in stitch density. From the above findings, it can be inferred that for every stitch and seam type, the optimum level of stitch density is different, which is also inherently linked with substrate, sewing machine parameters and sewing thread characteristics. Excessive stitches per inch can also contribute to seam puckering and reduce the speed through the machine resulting in loss of production.

It is possible to obtain stronger seams through appropriate stitch balance by monitoring the ratio of needle to bobbin thread consumption. The interlacing of the two threads should be positioned approximately in the middle of the substrate (Fig. 6.4a) by adjusting needle thread and bobbin thread tensions. Improper tensions can cause various problems, such as puckered seams and thread breakage, besides making the seam unbalanced (Fig. 6.4b) (Amann, 2007). Apart from the above, seam strength and overall seam performance can be improved through eliminating/minimizing various seam defects such as skipped stitches, seam grinning, puckered seam, needle cutting on knits, stitch run-off, etc.

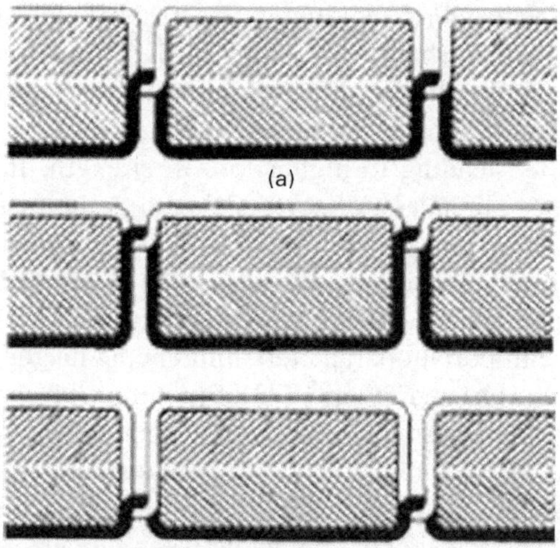

6.4 (a) Thread balance for a lockstitch; (b) needle thread tension too high or bobbin thread tension too low (top), needle thread tension too low or bobbin thread tension too high (bottom).

There are numerous machine parameters that can influence sewing thread characteristics and ultimately the seam properties. During lockstitch sewing, the compression force of the tension discs, force of tension spring and the action of the take-up lever cause tensile loading of the thread. The thread is also subjected to inertia forces due to rapid acceleration in both forward and backward directions, in addition to friction at various guides. Further, during the formation of the stitch, it passes through the needle and fabric assembly and is exposed to high tensile forces, frictional contact with the needle and fabric assembly, thermal loading due to heating of the needle, and bending through small radius of curvature (Mori and Niwa, 1994). In the process of stitch formation, it also interacts with the bobbin thread and is subjected to bending, abrasion and tensile deformation before getting incorporated into the seam. The needle passing at high speed through the substrate generates a lot of heat and the temperature of the needle reaches 300–350°C (Dorkin and Chamberlain, 1963; Hurt and Tyler, 1971, 1972a,b; Sondheim, 1953; Worthington, 1975; Li *et al.*, 2001). Such a high temperature causes harmful heating of the sewing thread and the fabric, especially synthetic materials.

Due to the above-mentioned factors during the sewing process, a substantial loss in thread strength (up to 30%) occurs (Midha *et al.*, 2009a, 2009b; Sundaresan *et al.*, 1997). Therefore the thread is weaker than expected after getting incorporated into the seam, and this influences seam strength. In a study on the effect of fabric type, stitch density and type of sewing thread on seam strength, it is observed that seam strength of coarser yarn fabric is greater than that of finer yarn fabric, although coarser yarn fabric has lower tensile strength than finer yarn fabric (Midha *et al.*, 2011). This is because loss in needle thread strength is about 40% more while sewing finer yarn fabric than coarser yarn fabric. Higher thread density in finer yarn fabric results in more abrasion of the needle thread during its repeated passage through the fabric assembly, leading to higher loss in strength. It may be added that in the above case, breakage of sewing thread is the main cause of seam failure.

It is to be noted that in general sewing speed, needle size, needle point shape, needle surface finish, material sewn, thread tension, fabric tightness and fabric finishing are the important parameters influencing needle heating (Howard and Parsons, 1968; Hurt and Tyler, 1973). Stitching with continuous filament threads results in higher needle temperature than with staple spun threads. This is due to a higher area of contact between the continuous filament thread and the needle surface. Sewing threads of bulked or textured yarns have both a reduced surface area of contact between thread and needle and facilitate air movement through the needle eye. Staple sewing threads also transport more cooling air to the needle, allowing sewing speed to be increased by up to ten per cent, relative to that with comparable continuous filament threads, without adverse needle temperature effects (Dorkin and Chamberlain, 1963). The problem of needle heating can be reduced by the addition of lubricant, introduction/increase of air flow, reduction of the sewing speed, shortening of the sewing stroke, using pilot needles, improving needle finish, and improving the needle design (Hersh and Grady, 1969).

In a study on the effect of needle thread strain and load on presser foot on seam strength, it was found that as the load on the presser foot increases, seam strength increases (Krasteva and Petrov, 2008). This is due to better gripping of the fabric assembly, reducing variations in stitch density, which allows uniform stress transfer during tensile loading. As the strain on the needle thread increases, the seam strength increases up to an optimum level and then decreases.

6.2.3 Prediction of seam strength

To judge the performance of the fabric assembly, it is important to predict the seam strength. It would then be possible to determine the required thread strength and stitch density to match the required seam strength. In the recent past, there has been increasing interest in predicting the seam strength of fabric for a given thread strength. The most commonly used equation uses sewing thread strength for predicting the seam strength:

$$\text{Estimated seam strength (lbs)} = \text{Stitches per inch} \times \text{Thread strength (lbs)} \times C$$

where C is a factor based on normal loop strengths of threads. For lockstitch, $C = 1.5$ and for chain stitch, $C = 1.7$ (American and Efird, 2007).

Some researchers used average loop strengths to predict the seam strength. Burtonwood and Chamberlain (1966) argued that the seam strength depends on the strength of the weakest stitch in the seam and therefore introduced the concept of minimum loop strength. They predicted the seam strength by using minimum knot strength. However, Brain (1970) observed the knot strength to be about 30% weaker than loop strengths and therefore used thread tensile strength, average loop strength, average minimum loop strength and corrected minimum loop strength to predict the seam strength of woven fabrics. Close examination of a seam loaded to near breakage showed that the thread loops are not composed of parallel lengths of thread as in the loop strength test. Compression of yarns in the stitch causes the widening of each thread stitch loop, making an angle of 65° at the loop point. The effective loop strength was therefore corrected by a factor $\cos(\theta/2)$, to get the corrected minimum loop strength. The use of corrected minimum loop strength improved seam strength prediction compared to the tensile or average loop strength. In an attempt to predict the seam strength of notched webbings for parachute assemblies using Taguchi's design of experiment and artificial neural networks, it was established that prediction by ANN is better than Taguchi's method (Zeydan, 2008).

Recently, seam strength prediction has been made using the residual thread strength (Midha *et al.*, 2011). A closer estimate of the seam strength is observed by considering the thread strength loss during the sewing process. Seam strength is envisaged as the multiplication of three parameters, namely, loop strength at 50 mm gauge length, stitch density and a factor indicating loss in thread strength:

$$\text{Seam strength (N)} = 2 \times \text{Stitches per inch} \times \text{Loop strength} \times C$$

where C is a factor for loss in thread strength and the typical value is 0.84, when polyester–polyester core spun is used as bobbin thread and 0.9 when polyester–cotton core spun thread is used as bobbin thread. It may be added that in all the above cases, prediction of seam strength is based on type 1 seam failure.

6.3 Seam extensibility and recovery

To accommodate body movement, 25–30% extension is needed for normal comfortable clothing and 40–60% for sportswear (Schmid and Mecheels, 1981). Extensibility greater than about 60% is not normally required except in fabrics designed for special applications like pantyhose. The extensibility of a seam should match the extensibility of the fabric, otherwise seam thread breakage (known as seam cracking) may result. Woven fabric structures (unless elastomeric yarns are used) limit fabric extension to less than 10% at low loads, whereas weft-knitted structures extend by 80–100%. Thus for a woven fabric, all seam structures are more extensible than the substrate, while for knitted fabrics the reverse is true.

Both seam extensibility and seam extension-to-break are dependent on a large number of variables such as elasticity of sewing thread, stitch density, stitch type, substrate thickness and the balance ratio. Seam strain at fracture increases with increase in thread linear density (Fig. 6.5) and is higher for chainstitch seams compared to lockstitch seams (Mukhopadhyay, 2008). In a more detailed study, it is observed that with the increase in stitch density, the extensibility at break of 301 lockstitch increases gradually for crosswise fabric samples, whereas it decreases gradually for lengthwise samples. It is also observed that breaking elongation of 504 three thread overlock stitch increases gradually for both sewing directions, whereas in the case of 514 four thread overlock stitch, breaking elongation increases initially and then decreases (Wang *et al.*, 2001).

Blackwood and Chamberlain (1970) compared the extension properties of ISO 301, 401 and 504 seams on knitted fabrics. ISO 504 seams observed maximum extensibility, whereas ISO 301 and 401 seams provided 80% extensibility. Extensibility of seam is influenced by initial modulus of the thread and the fabric. Higher initial modulus of the sewing thread or fabric usually reflects lower extensibility of the seam. Higher stitch density provides better gripping of the fabric assembly and therefore the seam initial modulus increases as shown in Fig. 6.6. Seam elongation is also influenced by the needle thread strain and the load on the presser foot. With the increase in the load on the presser foot, seam elongation increases. As the strain on needle thread increases, seam elongation also behaves similarly to seam strength, i.e. increases up to an optimum level and then decreases.

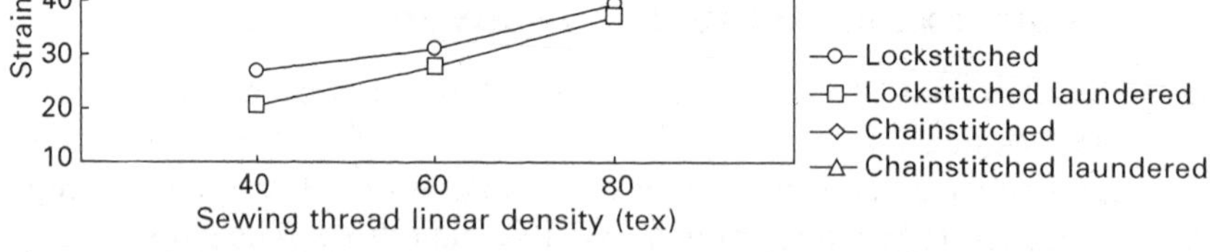

6.5 Effect of thread linear density on seam strain at break. (Reprinted from A. Mukhopadhyay, *Journal of Engineered Fibers and Fabrics*, 3(1), 2008. Copyright is held by JEFF and its Co-Sponsors; AATCC, INDA, TAPPI and The Fiber Society, P.O. Box 1288, Cary, NC 27512, USA.)

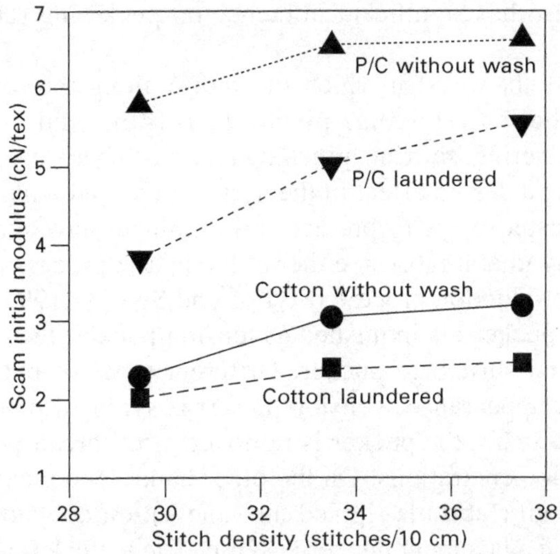

6.6 Effect of stitch density on the seam initial modulus. (Reprinted from A. Mukhopadhyay, M. Sikka and A.K. Karmakar (2004), 'Impact of laundering on the seam tensile properties of suiting fabrics', *Int. J. Clothing Sci. & Tech.*, 16, 394–403; an Emerald Publication.)

6.4 Seam puckering

The Oxford Dictionary defines seam pucker as 'a ridge, wrinkle or corrugation of the material or a number of small wrinkles running across and into one another, which appear when sewing together two pieces of cloth'. Among many quality parameters of sewing garments, puckering of a seam is one of the major obstacles to quality. It downgrades the appearance of the final product sewn from lightweight fabric. AATCC 88B provides a set of photographic replicas for the subjective evaluation of seam puckering (Shiloh, 1971). It was first identified during the 1960s, and the problem was solved with extensive use of permanent-press woven materials for outerwear (e.g. polyester/cotton shirting materials). During the late 1980s and early 1990s, seam pucker re-emerged as a problem as the use of lighter suiting materials and microfibre-based materials increased.

Seam puckering is observed when seam contractive forces exceed the fabric buckling resistance inside a stitch. The problem of seam puckering is more common with woven fabrics than knits; and it is prominent on tightly woven fabrics. Puckering may be apparent after sewing or it may develop later on during use (McIntyre and Daniels, 1995). During the sewing process, it is influenced by the properties of sewing threads and fabrics, seam construction, processes of needle penetration, stitch formation, sewing thread tension, fabric feeding and compatibility of the sewing thread, needle and the material being sewn (Fan and Leeuwner, 1998). The shrinkage of threads after sewing also has significant influence on puckering (Dobilaite and Juciene, 2006).

In general, there are six ways in which the fabric along the seam gets distorted: fusion, gathering, shirring, pinching, ruffling and cockling. Distortion by way of gathering, shirring, pinching and ruffling are deliberately obtained for introducing a design effect in the garment (Xi and Ajiki, 2002). Researchers have indicated ways to predict and eliminate unwanted seam pucker in lightweight synthetic fabrics, either at the textile processing stage or later in the apparel production process (Stylios and Sotomi, 1993). There are four main types of pucker distinguished as tension pucker, feed pucker, displacement pucker and shrinkage pucker. Different types of pucker can be identified from their appearance. Tension pucker is symmetric pucker to the left and right of the seam; the pucker is removed after thread is clipped between adjacent needle penetrations. On the other hand, feed puckering is asymmetrical to the seam (fabric is rippled and uneven on one side of the seam only) (Fig. 6.7). Displacement pucker is symmetric at the left and right of the seam; the pucker remains visible after the thread is clipped between adjacent needle penetrations. Shrinkage puckering appears similar to tension pucker and occurs due to differential shrinkage of fabric to that of sewing threads. In this section, causes of different types of pucker and their remedies are discussed.

Tension pucker

Tension pucker occurs when the over-stretched sewing thread shrinks to its original length and gathers up the fabric along the line of the seam. However,

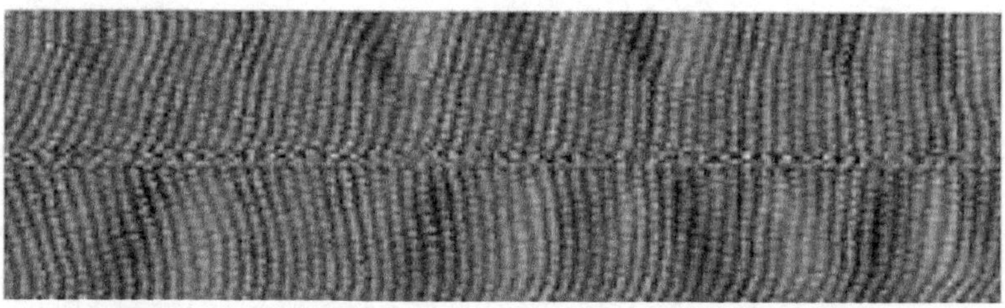

(a)

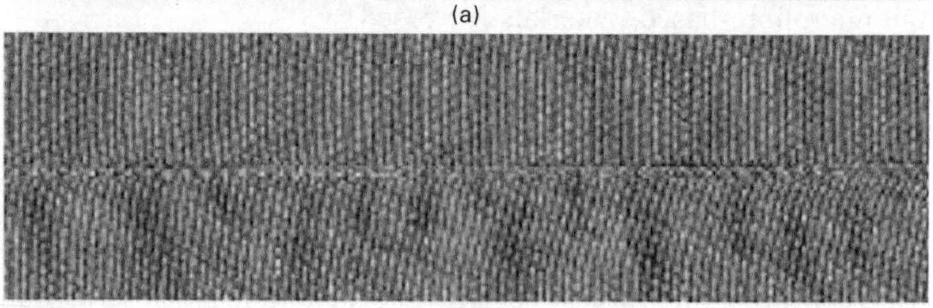

(b)

6.7 Appearance of different types of seam pucker: (a) tension pucker, (b) feed pucker (Amann & Söhne GmbH & Co. KG, n.d.).

sometimes the sizing or resins in the fabric initially maintain a flat seam but later on or after laundering, the seam appears to be puckered. The main causes of tension pucker are:

- excessive thread tensions
- incorrect thread balance
- tightly wound bobbin thread and
- excessive stitch lengths.

Sewing thread possessing lowest reversible strain is better for avoidance of seam pucker in lightweight fabrics. It is observed that after relaxation of tension with time, core spun polyester sewing threads tend to shrink more, the reversible strain of which is highest compared to staple spun threads. Further, as the number of layers in the sewing assembly increases, the seam pucker decreases due to increase in thickness and rigidity of sewn material. After washing and drying, the highest pucker is observed in garments sewn with cotton sewing threads, due to yarn swelling (Dobilaite and Juciene, 2006).

The following steps help in avoiding tension pucker.

- Tensions of the needle and bobbin threads should be set at the lowest levels. However, a certain amount of tension is necessary for correct stitch formation, and if the tension is too low the seams may not be tight enough. Smaller thread size requires lower thread tensions compared to larger thread size.
- Threads possessing low elongation or high initial modulus minimize the extent of stretching during sewing which reduces the tension pucker.
- If structural jamming in the fabric is not a problem, larger size needles or ball eye needles help to open up a larger hole in the fabric and the stitch can be set with the lightest tension.
- A perfect balance between the needle and bobbin thread is essential to avoid tension pucker. In lockstitch sewing, the needle and bobbin threads should interact in the middle of the assembly, which is indicated by equal consumption of the needle and bobbin threads. In chainstitch sewing, the thread tensions need to be correctly balanced in such a way that the needle thread forms pointed loops on the underside of the fabric, through which the looper thread can easily be pulled when unpicking the seam. An optimum balance is indicated by a thread consumption of approximately two-fifths for the needle thread and three-fifths for the looper thread.
- If the bobbin thread is wound too tightly, there is a risk of stretching it along its entire length, and it will get stitched into the seam in the stretched condition, leading to seam puckering after relaxation.

However, when sewing extremely fine and delicate fabrics, a small amount of puckering is unavoidable.

Feed pucker

Differential feeding of the two fabrics during sewing leads to feed pucker in the seam. The common drop feed transport mechanism only applies force on the underside of the two layers of fabrics being stitched. Transport of the top layer takes place only by the friction between the two layers. There is a risk of uneven feeding of the two fabrics, especially when using smooth fabrics. In a study by Bertoldi and Munden (1974), it was observed that for a plain-woven lightweight wool fabric, the upper ply of the fabric moves forward by about two per cent less than the lower ply. This effect is increased further, if the pressure of the presser foot is high.

Feed pucker can be avoided by following steps:

- Pressure of the presser foot should be set to the minimum level at which the feed rate of the two fabric layers is uniform.
- The lift of the feed dog is to be adjusted so that the teeth protrude only slightly above the stitch plate. Use of fine-toothed feed dogs can reduce feed pucker.
- Manual guidance of fabric under tension, both in front and behind the needle can reduce feed pucker.
- Use of floating presser foot (where only the sewing foot moves instead of the entire presser foot system) allows much lower pressures and helps to avoid uneven feed of fabric layers. This system also allows higher sewing speeds, due to reduced inertia of the moving parts.

Displacement pucker

Displacement pucker is observed in cases where, subsequent to the displacement of fabric yarns from their position during sewing, yarns are unable to return to their original positions after positioning of sewing thread. Figure 6.8 shows schematic representations of displacement pucker. The extent of displacement depends on the density of the fabric, stitch density, the needle and the sewing thread size. Displacement pucker increases with the increase in stitch density, because more yarns are displaced per unit length of the seam. When coarse threads enter heavyweight fabrics during sewing, high seam puckering is observed. Decreasing the diameter of the sewing thread and linear stitch density reduces the possibility of jamming and puckering (Schwartz, 1984). It has been observed that displacement pucker is lower in chainstitch than lockstitch, because the loop of thread on the underside of the fabric causes less displacement of the warp and weft yarns in chainstitch. Fabric bending, fabric shear and the relative magnitudes of the bending stiffness of both fabric and sewing thread affect the displacement pucker. A compressible fabric allows threads to recover without causing puckering but in a structurally jammed fabric, seam puckering is likely

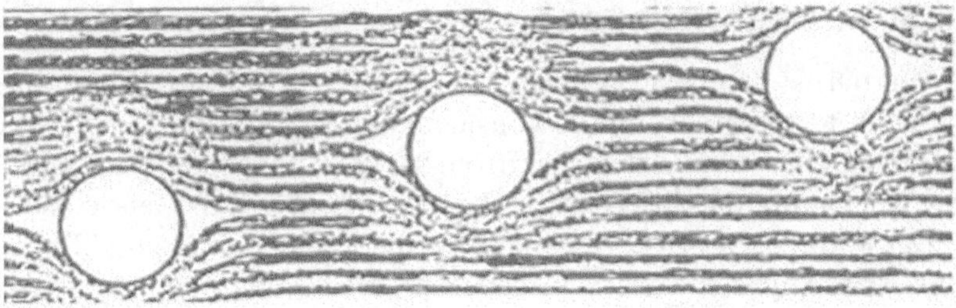

6.8 Schematic representations of displacement pucker.

to occur (Stylios and Lloyd, 1990) due to higher inter-yarn resistance in fabric.

The following steps help in avoiding the displacement pucker.

- Use the finest possible needle for a given thread size.
- Displacement of the warp and weft threads is initially caused by the needle rather than by the sewing thread. A damaged (blunt) needle will cause more disturbance of the fabric along the line of the seam. Regular checking of the condition of the tip of the needle is therefore strongly recommended.
- Care should also be taken to select a needle with the correctly shaped tip for the fabric to be sewn.
- Using fabrics with lower cover factor can reduce displacement pucker.
- Stitching the fabrics at an angle to the warp or weft (5–10°) can also reduce pucker.

Shrinkage puckering

Shrinkage puckering occurs when one of the components sewn into a seam shrinks at a different rate from other components. Typical components include the shell fabric, interlining, zipper tapes, stay tapes and the thread. The dimensions of fibres, yarns and fabrics (and garments made from them) alter with changes in ambient temperature and humidity. A difference in the moisture content of components in a seam before and after seaming may lead to differential dimensional changes and therefore cause seam puckering. All the components of fabric assembly should have minimum shrinkage to produce the flattest pucker-free seam. Even two per cent shrinkage in the fabric is enough to cause seam puckering (Behera *et al.*, 1997). Lightweight shirting fabric, typically polyester/cotton blends with a smooth texture, should be stitched with threads which have less than one per cent shrinkage to avoid seam puckering. Heat setting of finished products may remove puckering permanently.

6.5 Seam slippage

A partial or complete loss of seam integrity manifested by yarn slippage parallel to the stitch line is considered as seam slippage. It is caused by pulling of the yarns in the fabric from the seam under strain and is determined by calculating the difference in the extensibility of the fabric and the seam extensibility (ISO 13935). BS 3320 provides another method for measuring the seam slippage: the seam gape observed when a standard seam is put under a fixed load. Various observations related to seam slippage are listed below (Morris and Brain, 1975; Galuszynski, 1985; Miguel *et al.*, 2005; Behera *et al.*, 1997; Bhalerao and Jain, 1996).

- Resistance to slippage increases with the increase in yarn-to-yarn friction, contact angles between threads (fabric geometry), number of threads in fabric being pierced during sewing, stitch density, sewing thread diameter or flexural rigidity.
- Increasing sewing tension (i.e., tightening the stitch) generally has no effect on minimizing seam slippage. Complex seams such as flat felled seams (ISO 2.04.05) and oversewn plain seams (ISO 1.01.05) can overcome the problem of seam slippage in some fabrics.
- Open fabric constructions, i.e., a low number of interlacing of warp and weft yarns per unit area permits movement of yarns over each other when a seam is subjected to a load leading to seam slippage.
- Fabrics constructed from smooth warp or weft yarns or application of lubricants or softeners to the fabric as part of the finishing process make the fabrics susceptible to seam slippage.
- Seam slippage is dependent on direction of seam. In general, occurrence of seam slippage is greater for weft direction seams than for warp direction seams. However, in the case where either warp or weft yarns possess a low level of crimp, seams sewn along the relevant direction are prone to seam slippage.
- Incompatible fabric–thread combinations, such as sewing low weight fabrics with coarse threads, or low extensible fabrics sewn with high extensible core spun threads, cause more seam slippage. Fabrics sewn with cotton threads generally give comparatively lower seam slippage due to minimum difference of extensibility between the fabric and the threads.
- Seam slippage decreases after washing the garments due to relaxation which in turn results in stronger gripping of threads during tensile loading of seams.
- Seam slippage is also influenced by the location of the seam. In the side joints, seam slippage is comparatively lower in garments stitched from plain fabrics than twill.

6.6 Drape and bending

Investigation of the impact of a seam on fabric drape performance helps in understanding, evaluation and assurance of the appearance of the final garment. Drape describes the ability of a textile material to orient itself into folds in more than one plane under its own weight. This unique characteristic offers a sense of fullness and a graceful appearance, which distinguish textiles from other materials such as paper or polymer films. In practice, drape determines apparel design and influences fabric quality for specific end uses. The drape and bending behaviour of fabric is affected by the seaming process of combining two or more pieces of fabric, because seamed parts of the fabric bend comparatively less than the unseamed parts. The drape response of knitted fabrics with overlock seams is different from both the lockstitched woven fabrics and seamless knits investigated previously (Gaucher and King, 1983; Hu and Chung, 1998; Chung *et al.*, 1999) due to differences in both fabric and stitch structure (Ucar *et al.*, 2004). Garment drapeability is also intrinsically linked with seam allowance, seam positions and seam directions in addition to the base fabric characteristics.

The seams can be of four types with reference to their bending axis.

1. horizontal seams
2. vertical seams
3. radial seams
4. circular seams.

When the seam is parallel to the bending axis of the fabric (hanging edge of the cantilever), it is known as a horizontal seam, while that which is perpendicular to the bending axis is a vertical seam. Radial seam is sewn across a circular specimen through its centre, while a circular seam is sewn around a circular specimen with a radius from the specimen's centre, as shown in Fig. 6.9.

Based on the earlier research findings (Hu, 2004; Hu *et al.*, 1997; Hu and Chung, 1998, 2000; Ucar *et al.*, 2004), the following conclusions can be made regarding woven fabric:

1. Fabrics with horizontal or vertical seams drape in two dimensions, whereas radial and circular seams drape in three dimensions. Therefore, drape properties of horizontal and vertical seams can be evaluated using bending length, whereas for radial and circular seams, drape coefficient (ratio of the projected area of the draped specimen and its undraped area, after deduction of the area of the supporting disc) and drape profiles are used.
2. Horizontal seams have little effect on the bending rigidity of the fabric, whereas bending rigidity increases significantly with vertical seams on the fabric. Since horizontal seam can be parallel to any points from

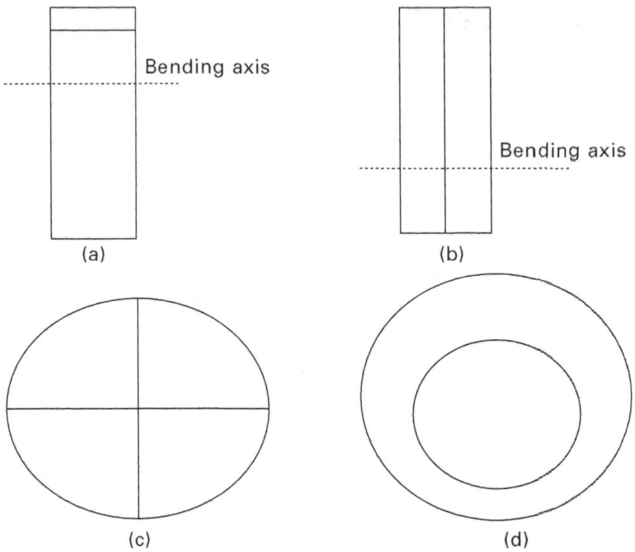

6.9 Types of seam: (a) horizontal; (b) vertical; (c) radial; (d) circular.

the free end to the hanging edge of the fabric cantilever, the bending length varies depending on the different position of seam and seam allowance. Bending length increases as the horizontal seam is moved away from the free tip end. Addition of horizontal seam and the increase in seam allowance cause a decrease in bending length and an increase in drapeability.

3. In vertical seams bending rigidity and/or bending hysteresis is affected by seam allowance, number of fabric layers, sewing thread crimp and thread tension during sewing. The bending length/bending rigidity increases with the increase in seam allowance but as seam allowance increases further, there is a small increase in bending length or rigidity (Hu and Chung, 2000). Bending rigidity increases with the increase in number of fabric layers (Shishoo *et al.*, 1971). Bending rigidity of seamed multi-layered fabric was about four to ten times that of a single layer. Asymmetric bending moments are obtained on each side of the fabric, if the thread tension is not balanced.
4. For radial seams, the important parameters influencing the drape coefficient are the number of radial seams and seam allowance. The drape coefficient increases with increase in number of radial seams, especially in the case of heavyweight fabrics, but it is slightly affected by increased seam allowance. Drape profile is significantly affected by the addition of radial seams. The more seams added to a fabric, the more stable the drape profile, i.e., the fabrics have a more regular node arrangement, especially in heavyweight fabrics. The drape profile of

lightweight fabrics cannot be altered significantly by introduction of radial seams.

5. For circular seams, the drape coefficient (%) is affected by the seam position. The drape coefficient is highest when a seam is located just off the edge of the pedestal. If the seam is moved outward from the pedestal, the drape coefficient (%) drops quickly and is lowest when the seam is located near the edge of the specimen. The drape profile of the fabric with a circular seam is entirely different from the drape profile of a fabric with radial seams. Nodes do not stay at any specific positions as in the case of radial seams.

In the case of knitted fabric, when one seam is added to the fabric in the wale direction, the drape coefficient is slightly higher compared to one seam added in the course direction. This is due to the loop structure of the fabric, where the rigidity in the course direction of a seamless fabric is higher than the rigidity in the wale direction. An increase in fabric density causes an increase in drape coefficient and a decrease in nodes due to increased bending and shear rigidity of the fabric. It is important to note that nodes on the seamed parts of samples are wider than nodes on the seamless parts of samples. Further, the fabric density has more influence on drape coefficient values, and seam number has greater influence on node values (Ucar *et al.*, 2004).

6.7 Seam grinning/gaping

Seam grinning or gaping occurs when a seam is subjected to a transverse load and is defined as 'unwanted opening of the seam, which reveals the sewing thread spanning the gap' (McIntyre and Daniels, 1995). Figure 6.10 shows a typical seam grinning in fabric. Seam grinning may also be described as the separation of the interface line between two sewn fabrics. It is one of the most important seam defects when considering the appearance, aesthetics and usage of both garments and upholstered furniture (Ucar, 2002).

Gaping can also contribute to functional failure of the seam and hence garment. Based on the earlier studies, the following conclusions can be made:

1. Chainstitch is more prone to gaping compared to lockstitch, especially when the linear stitch density is low (Blackwood and Chamberlain, 1970).
2. Low tension in the needle thread leads to grinning, when two ends of cloth are pulled outwards. The amount of twist, number of plies in the thread and lubrication ratio do not have any significant influence on seam opening. Among multifilament thread, core spun and staple spun

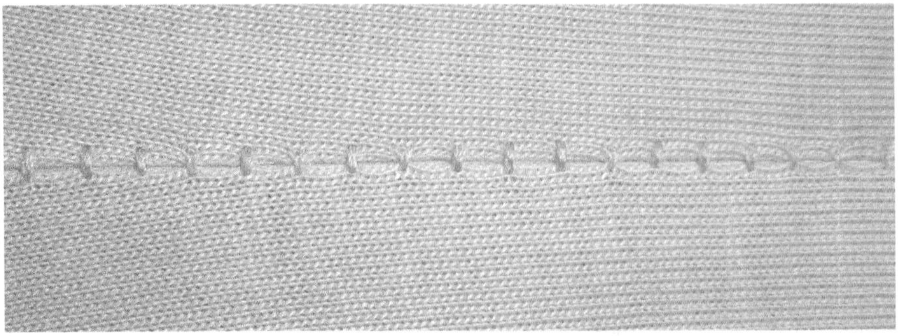

6.10 Seam grinning.

threads, it is observed that core spun threads show maximum resistance to seam opening (Meric and Durmaz, 2005).

3. Shimazaki and Lloyd (1990) showed that the effect of the loading number on seam grinning is very small, although the effect of thread properties and stitch density is noticeable.
4. The behaviour of an overlock stitched seam in knitted fabric will inevitably be different from a lockstitch seam, as explained by Shimazaki and Lloyd (1990), due to the differences in seam geometry. The seam grinning mechanism for woven fabrics will also be different from that for knitted fabrics due to fabric structural properties, i.e., seam slippage that contributes to seam grinning in woven fabrics will not occur in knitted fabrics, and knits are more extensible (less rigid) than woven fabrics.
5. Seam grinning increases with decreased stitch density and increased thread extensibility (Ucar, 2002). The said parameter decreases with higher values of fabric extensibility and fabric possessing a lower level of shear parameters (shear rigidity, hysteresis of shear force at 5° and tensile resilience). Since knitted fabrics are usually more extensible and less rigid than woven fabrics, seam grinning occurs less frequently in knitted fabrics than woven fabrics under same sewing and deformation conditions (Ucar, 2002).

6.8 Barrier properties of seams

Overall product performance/barrier behaviour in protective clothing is also determined by the characteristics of the seam. Depending on the application, the seam might require barrier against particulate, liquid and gas/vapour, including chemical resistance or biological barrier characteristics. Conventional sewing techniques of interlacing or interloping of threads with the help of needles may leave holes in the fabric. Liquid penetration through these holes may result in the failure of functional garments like raincoats, gloves, lifejackets, inflatable boats, inflatable arm bands, wet-weather clothing, chemical protective clothing, footwear and hospital staff uniforms.

Barrier to liquid penetration into the fabric can be developed by suitable construction or applying a finish. It is to be noted that puckering can also adversely affect the barrier properties of fabric. In a study on seam characteristics of breathable waterproof fabrics with various finishing methods, it was observed that puckering in the said fabric is mainly due to inherent (differential shrinkage of sewing thread and fabric) and feed pucker (Jeong and An, 2003). Pucker grade of laminating type is generally much higher than that of the coating type. It is important to note that the wavelength and amplitude of pucker are more important than the number of waves in breathable waterproof fabrics.

In the above case, sewing threads made from polyester fibre with a water-repellent finish reduce the wicking of water. Further, seamed fabrics are sealed by waterproof sealing tape to prevent water from penetrating through the holes caused by stitching. However, ultrasonic welding or fusing is preferred for medical garments where completely closed seams are required. These will resist penetration of any extraneous agent, e.g. blood-borne pathogens. It may be noted that the Occupational Safety and Health Administration (OSHA), USA, requires that medical garments have completely closed seams (Devine, 1994). Materials resistant to chemical penetration and permeation (e.g., pesticide solution) at the joints were the focus of earlier studies by several researchers (Berardinelli and Cottingham, 1986; Dimit *et al.*, 1992; Lloyd, 1986). Pesticide penetration in the thigh and scrotum area was found to be highest in these studies. However, there is lack of systematic study on the above aspect.

6.9 Flame retardancy of seams

Advice is given on seam construction and the selection of nylon and polyester-fibre sewing threads for garments required to meet flammability regulations, particularly the US children's sleepwear standard DOC FF3-71 (J. & P. Coats Ltd., 1980). To achieve a flare-free seam, it is necessary to use a polyester–fibre or nylon thread in a garment made of thermoplastic fibres, or a cotton thread in a cotton fabric of low flame retardance. Seams consisting of sewing threads made from 100% polyester fibre or 100% nylon will melt but will not be responsible for conveying a flame up a seam unless treated or coated with highly flammable additives. From the flammability viewpoint, blended-fibre threads, e.g. polyester-fibre/cotton core spun threads, are quite flammable and should not be used where the flammability of the end-product is the major consideration.

Jakes *et al.* (1975) observed that, with flame-retardant thermoplastic fabrics (polyester–fibre and polyester–fibre/acetate), the char length of cotton and core spun thread seams increases proportionately with increasing MR (ratio of thread mass to fabric mass), whereas polyester fibre thread seams are insensitive to MR. For a flame-retardant cotton fabric, the seam char length increases with MR for certain seam types with all threads tested. The flammability of sewing threads must be considered in the context of the end-product utilizing them and not of the threads themselves. In the manufacture of flame-retardant garments, needle coolants and thread lubricants should not be used, since they increase the flammability of the seam (Taylor, 1990).

6.10 Degradation/damage of seams

During wear, garments are subjected to a combination of tensile, bending, abrasion and shear for a repeated number of times and the resultant forces could be high depending on the applications. Abrasion resistance of sewing threads is an important consideration in stitching safety belts for cars, industrial conveyor belts, buttonholes on outerwear, and hems of trouser cuffs. The impact of the above, often coupled with the effect of other factors like moisture, temperature, sweat, light and chemicals used in washing or dry cleaning, leads to the degradation/damage of the seam. There are many other factors which affect seam performance depending on applications. Damage usually occurs in the seam area rather than in the middle of the cloth. In the following sections degradation/damage due to different factors are discussed.

6.10.1 Damage due to repeat loading

Most seams on garments, upholstery as well as technical textiles are subjected to repeated loading during daily use such as walking, sitting, squatting down and seat belts. However, in some applications loading may be only under smaller loads. This repeated loading causes several seam defects, such as slippage for woven fabrics, seam grinning and seam failure. The damage to seams under repeated loading depends on several factors such as the stitch geometry, the strength and extension characteristics of the sewing thread, the mechanical properties of the fabric, etc.

Fabric type and extension limit influence the load required to extend a lockstitch seam during repeated longitudinal extension/recovery cycles at low loads. Fabric type and extension limit are important in determining the load at maximum initial extension and the load change for 0–100 cycles, while fabric type is the only variable that influences load change for 100–5,000 cycles. Fabric type and thread type are the most important variables affecting the breaking load and breaking extension of seams under cyclic stress (Webster *et al.*, 1998).

6.10.2 Photodegradation

Exposure of stitched seams to ultraviolet radiation in products like curtains, tents or outdoor furniture, may result in the loss in breaking energy, extension and breaking load of sewing thread in addition to discolouration, disintegration or 'rotting' and a gradual deterioration in the fabric's mechanical characteristics (Wall and Frank, 1971). All these will affect seam performance. Polyester and nylon core spun sewing threads with non-cotton covering are suitable for these applications. A thread made of acrylic, synthetic resins and plasticizers constructed from continuous filaments, plied and bonded, has been developed by Robinson Thread Co. Inc. for products which need to resist ultraviolet radiation.

6.10.3 Biodeterioration

Any undesirable change in the properties of textiles (including the fabric and sewing thread) caused by enzymes, fungi, bacteria or other organisms is termed as biodeterioration of textiles (Hamlyn, 1983). Textile materials used in contact with soil, those used outdoors, such as tents, awnings, umbrellas, and those used in moist conditions, such as shower linings, bath mats and towels, are at higher risk because favourable conditions are provided for growth of such organisms (Hearle *et al.*, 1989). Cellulosic fibres are more susceptible to attack by micro-organisms compared to synthetic fibres. Certain finishes or coatings applied on the fibres may result in biodeterioration of even synthetic fibres (Higginbotham, 1976).

6.10.4 Mechanical/chemical treatment

Mechanical and chemical treatments during laundering, dry-cleaning, bleaching and brushing also affect the properties of seams (Slater, 1991). Fabric damage due to needle action during the sewing process becomes more apparent following cleaning. Agitation during the laundering may result in the abrasion of seams with other fabrics, agitator or with the walls of the container. In general, resistance to abrasion is affected by:

- mechanical properties of the fibres, fibre type, fibre fineness and its cross-sectional shape; nylon and polyester threads are more resistant to abrasion than cotton or silk threads,
- yarn structure – continuous filament threads are more resistant to abrasion than staple spun threads,
- the type of finishes used,
- stitch and seam construction – a seam in which the sewing thread remains on the surface of the fabric, is likely to be abraded prematurely. Hearle *et al.* (1989) observed preferential severe abrasion on the raised surfaces

of the seams on 100% cotton twill woven overalls after six years' use in a dirty industrial environment.

Mukhopadhyay *et al.* (2004) observed that, in general, number of washings or laundering does not show any significant influence on seam strength. However, a small reduction in seam strength has been observed after laundering in some cases, and the reduction is more when sewing is carried out with coarse threads compared to fine threads. It is justified on the basis of greater projected surface area of coarse threads which is susceptible to greater damage during laundering compared to finer threads. Although seam strength marginally decreases after laundering, seam elongation shows a marginal increase after laundering. Seam initial modulus decreases after laundering (Fig. 6.6), which is due to a substantial fall in initial modulus of the fabric. In agreement with the above, in a later study by Sajn (2007), it was found that ten washings do not significantly influence the breaking force of the seam.

The chemical action of soaps, detergents or bleaching agents used during the cleaning process may be detrimental to the seam or sewing thread. In various situations, sewing threads are required to be resistant to acids and/or alkalis. Nylon threads have relatively poor resistance to acids but excellent resistance to alkalis, whereas polyester threads have good resistance to acids but lower resistance to alkalis.

6.10.5 Adverse environmental conditions

Seams (fabric and sewing threads) must possess adequate fastness to light, hot pressing, rubbing, severe washing, water, perspiration, sea water and chlorinated water, depending on the applications to which they are to be put. For instance, in the case of workwear fabric, the dye must not change shade or stain other materials under severe laundering, nor must the dye sublimate if the garment is processed at high temperature; on the other hand, in swimwear, the thread must be fast to salt and chlorinated water.

In adverse environmental conditions seam failure may be site-dependent. In the case of close-fitting garments, underarm seams may be affected by perspiration, as sweat is often in a pH range between 6.8 and 4. In the case of motorcyclists' clothing following a crash, Woods (1996) identified various prominent areas of seam failure on protective garments such as seams on the armhole/shoulder/underarm, seams on the lower back/buttock, seams on the lateral aspects of the hip/thigh, and seams on the lateral aspects of the elbows and forearm.

6.11 Sources of further information and advice

The readers may consult the following book for some details on the subject:

Laing R M and Webster (1998), *Stitches and seams*, Manchester, The Textile Institute.

6.12 References

Aditya R N and Bose S K (1983), 'Seam strength of jute bags', *The Indian Text. J.*, 93(7), 79–86.

Amann and Sohne GmbH & Co. (2007), 'Sewing of ultra lightweight fabrics', *Information for the Sewing Industry*, 1–7.

Amann & Söhne GmbH & Co. KG (n.d.), Prevention of seam pucker, Bönnigheim. Available from: www.amann.com/download/industrie-naehgarne.html (Accessed 28 February 2010).

American and Effird (2007), Selecting the right SPI. Available from: http://www.amefird.com/wp-content/uploads/2010/01/Selecting-the-right-SPI-2-5-10.pdf (Accessed 26 March 2010).

Behera B K, Chand S, Singh T G and Rathee P (1997), 'Sewability of denim', *Int. J. Clothing Sci. & Tech.*, 9, 128–140.

Berardinelli S P and Cottingham L (1986), 'Evaluation of chemical protective garment seams and closures for resistance to liquid penetration', in Barker R G and Coletta G C, *Performance of protective clothing*, ASTM STP 900, Philadelphia, PA, American Society for Testing and Materials, 263–275.

Bertoldi A M and Munden D L (1974), 'The effects of sewing machine variables on fabric pucker', *Clothing Res. J.*, 2, 68–80.

Bhalerao S and Jain Y S (1996), 'Seam slippage of woven fabrics', *Man-made Textiles in India*, 39(1), 23–25.

Blackwood W J and Chamberlain N H (1970), 'The strength of seams in knitted fabrics', Clothing Institute Technological Report No. 22, London, The Clothing Institute.

Brain D H (1970), 'The prediction of strengths of lockstitch seams in woven fabrics', *J. Text. Inst.*, 61, 493–505.

Burtonwood B and Chamberlain N H (1966), 'The strength of seams in woven fabrics', Clothing Institute Technological Report No. 17, London, The Clothing Institute.

Carr H and Latham B (1994), *The Technology of Clothing Manufacture*, Oxford, Blackwell Science.

Cheng K P S and Poon K P W (2002), 'Seam properties of woven fabrics', *Textile Asia*, 33(3), 30–34.

Chmielowiec R (1987a), 'The effect of seam type and sewing thread on seam efficiency', *Japan Text. News*, 140–143.

Chmielowiec R (1987b), 'Seam strength factors', *Textile Asia*, 18(3), 94–97.

Chung S, Hu J and Lo M (1999), 'Effect of seam allowance of a plain seam on bending of woven fabrics', *Res. J. Text. Apparel*, 3, 54–64.

Crow R M and Dewar M M (1986), 'Stresses in clothing as related to seam strength', *Text. Res. J.*, 56, 467–473.

Devine J (1994), 'Understanding ultrasonics', *Bobbin*, 35, 74–85.

Dimit C A, Easter E P and De Jonge J O (1992), 'The effect of seams and closures on pesticide penetration', in McBriarty J P and Henry N W, *Performance of protective clothing*, ASTM STP 1133, Philadelphia, American Society for Testing and Materials, 418–426.

Dobilaite V and Juciene M (2006), 'The influence of mechanical properties of sewing threads on seam pucker', *Int. J. Clothing Sci. & Tech.*, 18, 335–345.

Domingues J P, Manich A M, Sauri R M and Barella A (1997), 'Assembling textile structures: wear simulation', *Int. J. Clothing Sci. & Tech.*, 9, 75–87.

Dorkin C M C and Chamberlain N H (1961), 'Seam pucker – its causes and prevention', Clothing Institute Technical Report No. 10, London, The Clothing Institute.

Dorkin C M C and Chamberlain N H (1963), 'The facts about needle heating', Clothing Institute Technological Report, No. 13, London, The Clothing Institute.

Fan J and Leeuwner W (1998), 'The performance of sewing threads with respect to seam appearance', *J. Text. Inst.*, 89, 142–154.

Frederick E B (1952), 'Development of a sewability test for cotton fabrics', *Text. Res. J.*, 22, 687–692.

Galuszynski S (1985), 'Some aspects of the mechanism of seam slippage in woven fabrics', *J. Text. Inst.*, 76, 425–433.

Gaucher M L and King M W (1983), 'Predicting the drape coefficient of knitted fabrics', *Text. Res. J.*, 53, 297–303.

Gribaa S, Amar S B and Dogui A (2006), 'Influence of sewing parameters upon the tensile behaviour of textile assembly', *Int. J. Clothing Sci. & Tech.*, 18, 235–246.

Gurarda A (2008), 'Investigation of the seam performance of PET/Nylon-elastane woven fabrics', *Text. Res. J.*, 78, 21–27.

Hall W R and Knoff W F (2008), 'Heat aged tensile strength retention of poly (p-phenylene terepthalamide) sewing thread', *J. Engineered Fibers & Fabrics*, 3(4), 15–22.

Hamlyn P F (1983), 'Microbiological deterioration in textiles', *Text. Mag.*, 12, 73–76.

Hearle J W S, Lomas B, Cooke W O and Duerdon I J (1989), *Fibre failure and wear of materials: an atlas of fracture, fatigue and durability*, Chichester, Ellis Horwood.

Hersh S and Grady P (1969), 'Needle heating during high speed sewing', *Text. Res. J.*, 39, 101–102.

Higginbotham R S (1976), 'Deterioration of textiles in use', *Text. Mag.*, 5(2), 40–47.

Howard G M and Parsons D (1968), 'Sewing needle temperature, Part I: Theoretical analysis and experimental techniques', *Text. Res. J.*, 38, 606–614.

Hu J (2004), *Structure and mechanics of woven fabrics*, Cambridge, Woodhead Publishing.

Hu J and Chung S (1998), 'Drape behavior of woven fabrics with seams', *Text. Res. J.*, 68, 913–919.

Hu J and Chung S (2000), 'Bending behavior of woven fabrics with vertical seams', *Text. Res. J.*, 70, 148–153.

Hu J, Chung S and Lo M (1997), 'Effect of seams on fabric drape', *Int. J. Clothing Sci. & Tech.*, 9, 220–227.

Hurt F N and Tyler D J (1971), 'An investigation of needle heating and associated problems in machine sewing – I', HATRA Research Report, No. 19.

Hurt F N and Tyler D J (1972a), 'An investigation of needle heating and associated problems in machine sewing – II', HATRA Research Report, No. 20.

Hurt F N and Tyler D J (1972b), 'An investigation of needle heating and associated problems in machine sewing – III', HATRA Research Report, No. 21.

Hurt F N and Tyler, D J (1973), 'Influence of fabric finishing conditions on sewing needle temperature', *Clothing Research Journal*, 1, 47–52.

J. & P. Coats Ltd. (1980), 'Flame resistant threads', *Manufacturing Clothier*, 61(3), 55–57.

Jakes K A, Smith B F and Spivak S M (1975), 'The influence of thread mass and fabric mass on seam flammability', *Text. Res. J.*, 45, 830–834.

Jeong W Y and An S K (2003), 'Seam characteristics of breathable waterproof fabrics with various finishing methods', *Fibers and Polymers*, 4, 71–76.

Kirk W and Ibrahim S M (1966), 'Fundamental relationship of fabric extensibility to anthropometric requirements and garment performance', *Text. Res. J.*, 36, 37–47.

Krasteva D G and Petrov H (2008), 'Investigation on the seam's quality by sewing of light fabrics', *Int. J. Clothing Sci. & Tech.*, 20, 57–64.

Li Q, Liasi E, Simon D and Du R (2001), 'A study on the needle heating in heavy industrial sewing: Part II: Finite element analysis and experiment verification', *Int. J. Clothing Sci. & Tech.*, 13, 351–367.

Lloyd G A (1986), 'Efficiency of protective clothing for pesticide spraying', in Barker R G and Coletta G C, *Performance of protective clothing*, ASTM STP 900, Philadelphia, American Society for Testing and Materials, 121–135.

McIntyre J E and Daniels P N (1995), *Textile terms and definitions*, Manchester, Textile Institute.

Meric B and Durmaz A (2005), 'Effect of thread structure and lubrication ratio on seam properties', *Indian J. Fibre & Text. Res.*, 30, 273–277.

Midha V K, Mukhopadhyay A, Chattopadhyay R and Kothari V K (2009a), 'Studies on the changes in tensile properties of sewing thread at different sewing stages', *Text. Res. J.*, 79, 1155–1167.

Midha V K, Kothari V K, Chattopadhyay R and Mukhopadhyay A (2009b), 'Effect of high speed sewing on the tensile properties of sewing thread at different stages of sewing', *Int. J. Clothing Sci. and Tech.*, 21, 217–238.

Midha V K, Chattopadhyay R, Kothari V K and Mukhopadhyay A (2010a), 'Effect of process and machine parameters on changes in tensile properties of threads during high speed industrial sewing', *Text. Res. J.*, 80(6), 491–507.

Midha V K, Kothari V K, Chattopadhyay R and Mukhopadhyay A (2010b), 'Effect of workwear fabric characteristics on the changes in tensile properties of sewing threads after sewing', *J. Eng. Fibers Fabrics*, 5 (1), 31–38.

Midha V K, Mukhopadhyay A and Kaur R (2011), 'An approach to seam strength prediction using residual thread strength', *Res. J. Tex. & Apparel*, 15(3), 75–85.

Miguel R A L M, Lucas J M, Carvalho M L C and Manich A M (2005), 'Fabric design considering the optimisation of seam slippage', *Int. J. Clothing Sci. & Tech.*, 17, 225–231.

Mori M and Niwa M (1994), 'Investigation of the performance of sewing thread', *Int. J. Clothing Sci. & Tech*, 6, 20–27.

Morris P A and Brain D H (1975), 'Seam slippage', *Clothing Res. J.*, 3, 135–144.

Mukhopadhyay A (2008), 'Relative performance of lockstitch and chainstitch at the seat seam of military trouser', *J. Engineered Fibers & Fabrics*, 3, 21–24.

Mukhopadhyay A, Sikka M and Karmakar A K (2004), 'Impact of laundering on the seam tensile properties of suiting fabrics', *Int. J. Clothing Sci. & Tech.*, 16, 394–403.

Mukhopadhyay A, Ghosh S & Kaur R (2005), 'Seam performance of differently processed military khaki fabrics', *Melliand Int.*, 11, 311–313.

Nestler R and Schlegel W (1978), 'Determining the stresses to which wearing apparel is subjected', *Textiletech*, 28, 113–117.

Rao J V (2006), *Stitches*, Northern India Textile Research Association, Ghaziabad, India.

Sajn D (2007), 'Influence of the material on the seam quality', *Tekstilec*, 50, 77–92.
Sakhnik M V and Rodionov V A (2005), 'Effect of moisture on the breaking characteristics of polypropylene twisted threads', *Fibre Chemistry*, 37, 48–50.
Schmid I U and Mecheels D (1981), 'Forces acting on textiles and clothing seams as functions of body movements and cut of garment', *Melliand Textilber.*, 10, E314–E320.
Schwartz P (1984), 'Effect of jamming on seam pucker in plain woven fabrics', *Text. Res. J.*, 62, 176–180.
Scott K (1987), 'Denim's trials and tribulations', *Manufacturing Clothier*, 68, 33–37.
Shiloh M (1971), 'The evaluation of seam puckering', *J. Text. Inst.*, 62, 176–180.
Shimazaki K and Lloyd D W (1990), 'Opening behavior of lockstitch seams in woven fabrics under cyclic loading conditions', *Text. Res. J.*, 60, 654–662.
Shishoo R L, Klevmar P H, Cednas M and Olofsson B (1971), 'Multilayer textile structures: relationship between the properties of a textile composite and its components', *Text. Res. J.*, 41, 669–679.
Slater K (1991), 'Textile degradation', *Text. Progress*, 21, 1–150.
Sondheim W S (1953), 'Causes of seaming damage: needle heating', *J. Text. Inst.*, 44, T580–T585.
Stylios G and Lloyd D W (1990), 'Prediction of seam pucker in garment by measuring fabric mechanical properties and geometric relationships', *Int. J. Clothing Sci. & Tech.*, 2, 6–15.
Stylios G and Sotomi J O (1993), 'Investigation of seam pucker in lightweight synthetic fabrics as an aesthetic property, Part I: A cognitive model for the measurement of seam pucker', *J. Text. Inst.*, 84, 593–600.
Sundaresan G, Hari P K and Salhotra K R (1997), 'Strength reduction in sewing threads during high speed sewing in an industrial lockstitch machine: Part I – Mechanism of thread strength reduction', *Int. J. Clothing Sci. & Tech.*, 9, 334–345.
Taylor M A (1990), *Technology of Textile Properties*, 3rd edn, Forbes Publications, London.
Tsui W C, Burtonwood B, Burnip M S and Estakhrian H V A (1984), 'Aspects of seam strength prediction: Part I', *J. Text. Inst.*, 75, 432–445.
Ucar N (2002), 'Grinning of ISO 514 stitched seams on knitted fabrics under the effects of repeated extension and recovery', *Text. Res. J.*, 72, 944–948.
Ucar N, Kalaoclu F, Bahtiyar D and Blac O E (2004), 'Investigating the drape behavior of seamed knit fabrics with image analysis', *Text. Res. J.*, 74, 166–171.
Ukponmwan J O, Mukhopadhyay A and Chatterjee K N (2000), 'Sewing threads', *Textile Progress*, 30, 1–94.
Wall M J and Frank C G (1971), 'A study of the spectral distribution of sun-sky and xenon-arc radiation in relation to the degradation of some textile yarns, Part I: Yarn degradation', *Text. Res. J.*, 41, 32–38.
Wang L, Chan L K and Hu X (2001), 'Influence of stitch density to stitches properties of knitted products', *Res. J. Text. Apparel*, 5, 46–53.
Webb W (1976), 'Some observations on the behaviour of superimposed and lap sewn joints', *Clothing Res. J.*, 4, 74–85.
Webster J, Laing R M and Niven B E (1998), 'Effects of repeated extension and recovery on selected physical properties of ISO-301 stitched seams: Part 1: Load at maximum extension and at break', *Text. Res. J.*, 68, 854–864.
Woods R I (1996), 'Specification of motorcyclists protective clothing designed to reduce road surface impact injuries', in Johnson J S and Mansdorf S Z, *Performance of protective clothing*, ASTM STP 1237, Philadelphia, PA, American Society for Testing and Materials, 3–22.

Worthington A P (1975), 'Effect of sewing machine needles in clothing manufacture – literature survey 1950–1974', Manchester, Shirley Institute.

Xi X and Ajiki I (2002), 'Application of a special sewing thread to gathering', *Res. J. Text. Apparel*, 6, 82–90.

Yeşilpinar S and Bahar S (2007), 'The effect of sewing and washing processes on the seam strength of denim trousers', *AATCC Review*, 7, 27–31.

Zeydan M (2008), 'Modelling the woven fabric strength using artificial neural network and Taguchi methodologies', *Int. J. Clothing Sci. & Tech.*, 20, 104–118.

7
Intelligent sewing systems for garment automation and robotics

G. K. STYLIOS, Heriot-Watt University, UK

DOI: 10.1533/9780857093967.1.208

Abstract: Sewing machine interactions at different speeds have been used to construct qualitative rules mapping fabric properties to optimum sewing machine settings for intelligent sewing machines. The inference procedures of fuzzy logic have been implemented in a neural network to allow for optimisation of output membership functions and, subsequently, self-learning. The technique is successfully applied to develop intelligent sewing machines and further implemented in textile and garment manufacturing. An intelligent manufacturing environment has been put forward in which fabric properties predict the sewability of any fabric, determine the minimum change of fabric properties required, and control in real time the stitching of a garment by using the feedback closed loop of the Neuro-Fuzzy model. The system has been successfully tried in an industrial setting. Optimum settings were achieved under static and dynamic machine conditions, including for the properties of difficult fabrics and compensation for mishandling by the operator over the speed range of the sewing machine.

Key words: intelligent sewing machines, Neuro-Fuzzy, garment manufacture, sewing, fabric properties, sewability, fabric measurement, fabric mechanics, seam pucker, sewing damage, feedback, closed loop, industrial trials.

7.1 Introduction

Despite the globalisation of markets and the high labour costs, apparel products are still labour intensive and are produced using traditional methods and machinery. A very important factor that inhibits the true transformation of fashion apparel and retailing is the making-up or joining process, which remains predominantly manual, complex and costly. Consequently, the significant interaction between machinery, fabric and skill has to be considered in finding generic solutions for full sewing automation.

7.2 Developments in the automation of sewing

Over the last 30 years considerable research and development has been carried out focusing on this problem: BRITE/EURAM in Europe,[1] TRASS in Japan[2] and Drapers in the USA.[3] A number of automated machines (e.g., robotic sewing arm[4]) have been invented as a result of this work, but the very much needed universal flexible solution for the manipulation and especially the joining of the fabric has yet to be realised. All these attempts, including the recent Leapfrog project (http://www.leapfrog-eu.org), do not consider the fabric effectively as a non-linear, non-homogeneous, diverse engineering material that has its own mechanics, which define its behaviour under stresses and strains usually small, and being capable of taking up large curvatures, in cutting, handling and sewing. Consequently every attempt can only work for a limited, at best, group of fabrics.

The mechanics of fabric have been much researched in the last 30 years and the efforts of Grosberg,[5] Kawabata and Niwa,[6] Postle[7] and Stylios[8] have to be highlighted; and in the context of clothing the work of Stylios, and Kawabata and Niwa are key. However, the interaction of the material with sewing in the so-called fabric/machine interface still needs to be established, and this is a fundamental stumbling block for further advances, such as in the area of mass customisation.[9]

The concept of intelligent garment manufacture[10] emphasises the importance of fabric properties and their interaction with the whole manufacturing process, and especially with sewing. In order to achieve this, computational intelligence and engineering is applied to research, to develop intelligent textile and apparel manufacturing systems, and to introduce desired flexibility, leading the way for effective sewing garment robotics.

7.2.1 Defining fabrics as engineering materials

The use of fabric measurement technologies with machinery parameters can offer powerful feedback on-line models that can numerically control and optimise processing parameters prior to and during garment manufacture.

Arguably, there has been reasonable progress in relating fabric properties to stitching quality, but there are still areas that have not been numerically defined because of the complexity of the dynamic interactions between fabric–sewing machine parameters.[11] The influence of low stresses and strains on a fabric alters its physical or mechanical properties and hence its behaviour which is non-linear. The effect of frictional dynamics of fabrics is something that academe and industry have had to deal with over many years. Owing to incomplete mathematical understanding of these non-linear interactions, the behaviour of each material cannot be effectively modelled using conventional techniques.

The latest research has enabled the modelling of fabrics as non-linear discrete parametric engineering materials by modification of Dahl's dynamic frictional model as in Eq. [7.1]:

$$\frac{dFr}{dx} = \quad 1 - sign \; \frac{dx}{dt} \; \frac{Fr}{fr} \qquad [7.1]$$

where σ is the slope coefficient, Fr is the dynamic friction force, x is the displacement and fr is the Coulomb friction force. A complete fabric map of stress–strain relationships can hence be established from fabric shear, bending, tensile and compression curves using low stress precise measurement systems such as KESF or FabLAB-A (formerly FAMOUS[12]) and shown in Fig. 7.1.

Figure 7.2 shows the dynamic modelling of tensile and shear of a woven plain fabric and Fig. 7.3 modelling of fabric pure bending hysteresis from buckling. These non-linear models have been devised from FabLab-A data curves, which is a completely automatic fabric measurement system working along the principles of measurement of the KESF, but with improvements in shear which is measured under constant load and bending which is extracted from buckling. FabLab-A is an all-in-one instrument, covering an area of less than 1.0 m^2, and it is based on advanced microelectronic and sensor technology.

Seam quality itself can have many definitions which are still subjective and, even if measured, have to retain the analogous subjective understanding: unbalanced stitch, seam pucker, seam slippage, seam damage. Systems such as lasers or sensors on machines may be used on-or off-line to measure these faults, such as in the case of seam pucker,[13] or during sewing accounting even for the occurrence of discrete problems, such as, for example, pulling the fabric and hence introducing undue high thread tensions. Effective solutions to all these problems have to be provided in the design of a new control model, in which control algorithms need to be robust enough to function successfully.

7.1 FabLab-A fabric automated measurement.

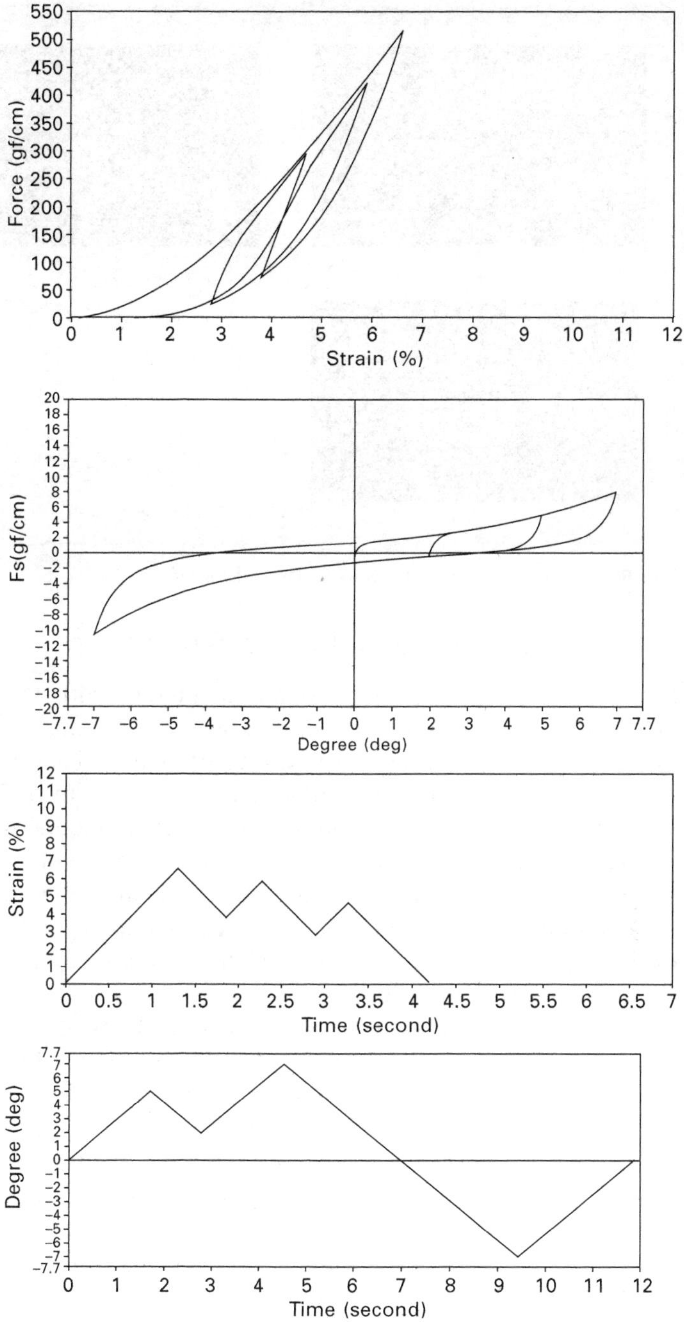

7.2 Time-effect dynamic simulations of fabric tensile and shear behaviour, by FabLab-A.

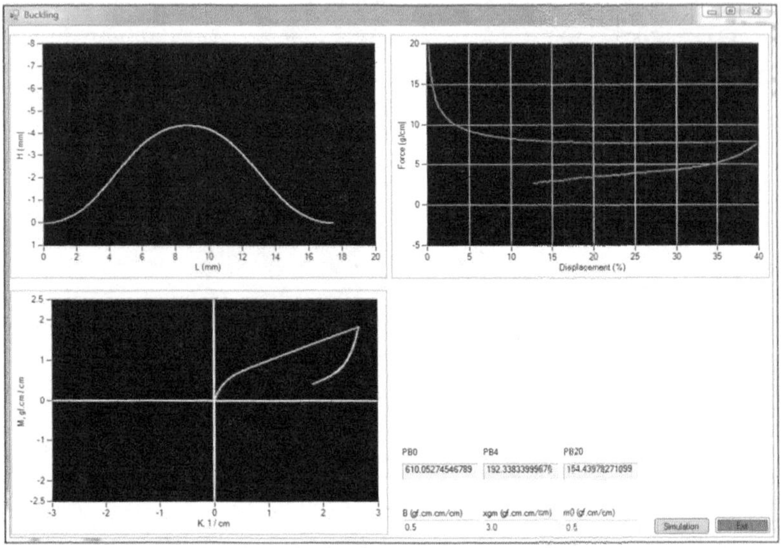

7.3 Fabric bending from buckling, by FabLab-A.

7.2.2 On-line control model for machine sewing

The following properties have been found to interact with the sewing process: fabric tensile, shear, bending, thickness and compression, and yarn and thread tensile and friction. These properties have been measured automatically using automated measurement systems such as the FabLab-A. For a predetermined fabric type, a central rotatable experimental design of Box and Hunter[14] for three variables has been used to determine the effect of foot force (gf), sewing speed (rpm) and thread tension. Combinations of thread tension, foot force and speed have then been used in experiments and, for each of the experimental combinations of the testing methodology, the stitching quality has been assessed.

Analysis of the results has established a theoretical sewing model and the appropriate rules for a sewing machine control system based on Neuro-Fuzzy. A sewability index, a kind of sewing fingerprint, for any given fabric is pre-determined prior to input into the Neuro-Fuzzy system. This pre-processing has been performed by an artificial neuron and a weight on the connection between the neuron and each fabric property, trained using gradient descent methods (in error backpropagation), to predict fabric sewability accurately from the relevant physico-mechanical properties of any fabric. This establishes a fabric sewability system named the Sewability Prediction System (SPS), used as input to the intelligent sewing machines as we will see later. Fuzzification of inputs has been carried out at the input layer while defuzzification is performed at the output layer, as shown in Fig. 7.4.

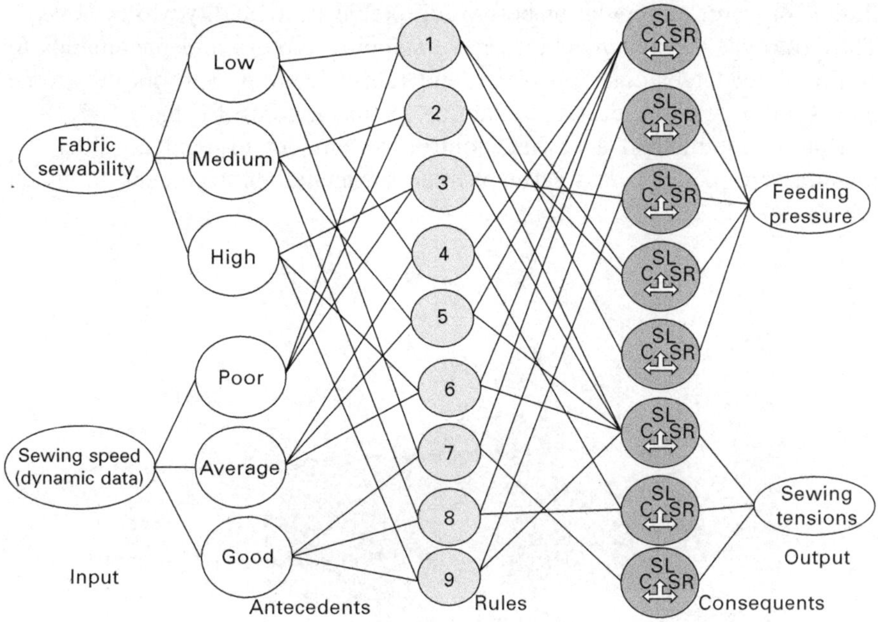

7.4 The sewing control system.

Triangular membership functions may be fully described by three parameters: namely, the centre, left and right spreads, to allow for asymmetry. Automatic tuning can be achieved by replacing the standard sigmoid function $f(z) = (1 + e - z) - 1$ employed in forward propagation of neural networks by another function, which incorporates these parameters in a network which implements the Neuro-Fuzzy system. For linear functions:

$$\mathrm{m}(x) = (1 - |(x - c)/s|)$$

where c is the centre of the membership function, s is s_L for left spread or s_R for right spread, and x is a value of the output variable.

Furthermore, the learning capabilities of the neural network has been used to optimise the output membership functions of the Neuro-Fuzzy system and implemented as an adaptive fuzzy system for closed loop feedback control as we will see later.

If all the input membership functions are left unchanged, automatic tuning can be performed on output membership functions. It is expected that the membership functions representing low optimum outputs be the most severely affected by the tuning, in terms of magnitude of change and the direction of tuning.

The Neuro-Fuzzy control system has been applied on a number of instrumented intelligent sewing machines and has successfully been tried towards eliminating the learning curve introduced by a new fabric, with limited operator manipulation before acceptable seam quality was achieved.* The prototype sewing machines have stepping motors and pneumatics to control thread tensions, foot force and fabric feeding. Sensors determine their state of tension, pressure enabling feedback control (Figs 7.5–7.7).

Having established a sewing control system, let us next consider the manufacture of fabric and garment in an automated environment, in which

(a)

(b)

7.5 Intelligent sewing machines: (a) overlock and (b) lockstitch.

**Automation of Sewing Machine Settings in Difficult to Sew Fabrics Using Objective Measurement Technology* in collaboration with Bellow Machine Co., Claremont Garments Plc and Burberrys of London from 1993 to 1995, carried out at Bradford University.

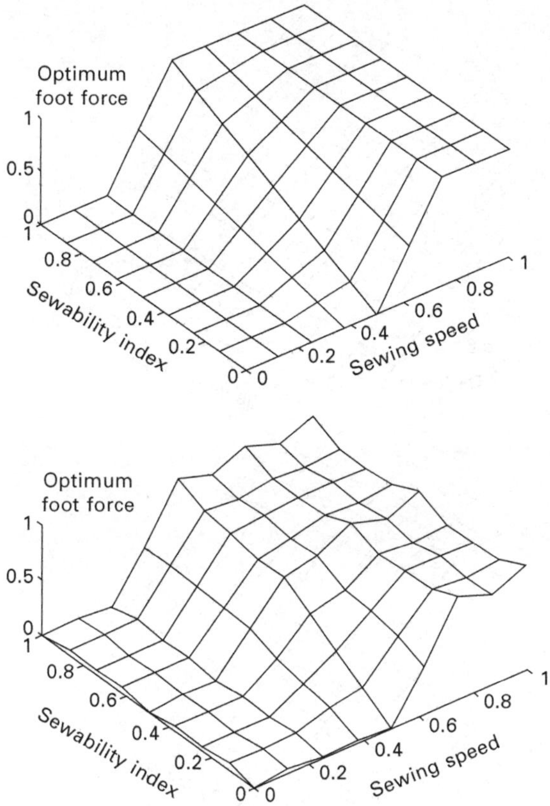

7.6 The control surface of the presser foot force (simulated (top) and achieved (bottom)).

fabric production upstream and garment manufacture downstream are being fused together. In this approach fabric mechanics are central because the behaviour of the fabric is defined and its characteristics can be precisely engineered through weaving or knitting, and especially through finishing (upstream). Or knowing the behaviour of the fabric, the sewing manufacturing process can be monitored, controlled and optimised dynamically and in real time by the sewing control system (downstream). These processes have the connotation of intelligent manufacturing as will become more apparent later. The chosen route depends on the fabric, its properties and the predicted seam severity.

7.2.3 Intelligent garment automation

Figure 7.8 shows a schematic explanation of the intelligent sewing environment (ISE). It consists of three basic systems interconnected with each other on-line: the sewability prediction system (SPS), the intelligent sewing system

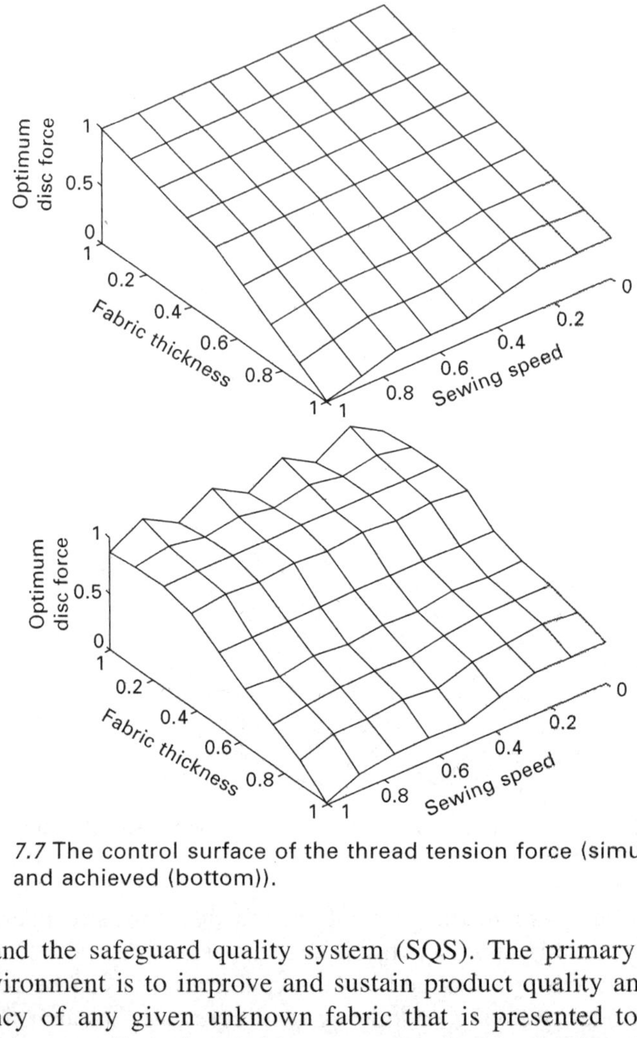

7.7 The control surface of the thread tension force (simulated (top) and achieved (bottom)).

(ISS) and the safeguard quality system (SQS). The primary objective of the environment is to improve and sustain product quality and production efficiency of any given unknown fabric that is presented to the ISE, by eliminating manufacturing problems associated with stitching of the material, while at the same time minimising skill requirements of the operator. This approach takes the command of the process from the operator (deskill) to the fabric mechanical properties through the control system, and hence it directly deskills garment sewing rendering garment robotics a realistic proposition. Let us now consider the sub-systems that constitute the ISE and how they are integrated.

Automatic measurement of fabric mechanical properties

The FabLab-A, Fig. 7.1, is a robotic fabric measurement system, designed and built with modern sensors, motors, slides and electronic control based

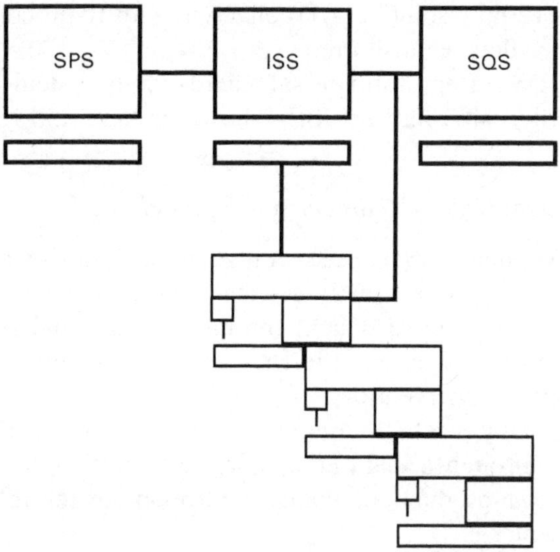

7.8 A schematic drawing of the intelligent sewing environment (ISE) concept.

on CANopen, which can measure tensile, buckling, shear, compression and surface properties without any operator intervention. The system can work as a stand-alone measurement instrument or it can be integrated within the manufacturing process as in this case. The measured fabric properties become inputs to the SPS described earlier. KESF data can also be used, but the operation of the instrument is not automatic and cannot be on-line.

On-line automatic measurement and monitoring of seam appearance

The seam laser system (SLS), which monitors and measures the deformation along the stitch line, ensures stitching quality during fabric sewing. The measurement model is based on wave formation, and measurement is made at each side along the seam line of the fabric during sewing and on the simulation of eye and brain in determining seam pucker in garments.[15–17]

On-line automatic measurement and monitoring of sewing damage

The sewing damage system[18] (SDS) which is incorporated in an instrumented sewing machine monitors and measures the penetration forces of the sewing needle as a measure of damage during the sewing process of a given material. The sewing plate of a single needle overlock sewing machine has been strain gauged with a semi-conductor two-element force sensor (220 gauge factor). The shaft encoder of the machine has been used for determining positions and stitch counting. A 12-bit resolution A/D converter, with 10 ms conversion time, has been used for data acquisition.

Both the SLS and SDS systems form the safeguard quality system (SQS) of the intelligent sewability environment (ISE), shown as 'test' in Fig. 7.9.

Intelligent sewing system (ISS), intelligent sewing machines

The intelligent sewing system (ISS) consists of instrumented sewing machines shown in Fig. 7.5 able to automatically set their static, and monitor and change their settings (dynamic feedback) through sensors and actuators, to ensure optimisation of production efficiency and quality with minimum operator/mechanic skill or intervention.[18]

These intelligent sewing machines are integrated by means of modelling the desired functions through a main computer, but can also be operated as stand-alone systems in production, quality control and/or fabric/garment design and development.

7.3 Operational principles of the intelligent sewability environment (ISE)

To understand the working principles of this system, let us consider two paradigms from two distinctly different groups of troublesome textile materials:

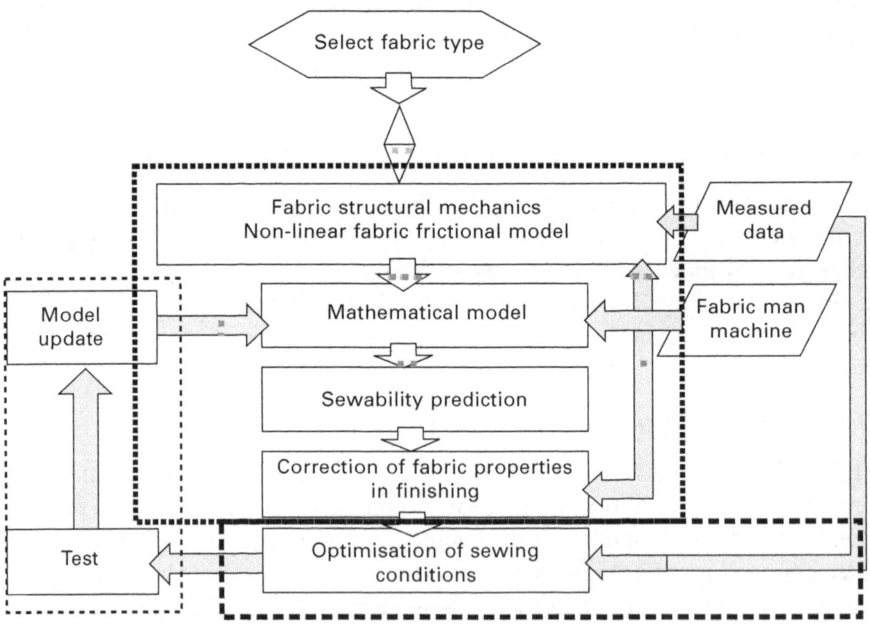

7.9 Complete ISE system schema.

lightweight woven and highly dense knitted fabrics. These materials have severe problems during manufacture due to their tendency to deform during stitching; seam pucker, or damage of their structure by cutting their constituent yarns; and sewing damage, respectively. The knitted fabrics consist of single knit as well as double knit groups, adding to the complication, since they behave differently. In industry, these problems require experience, skill, subjective assessment, and in many cases an ad-hoc trial-and-error approach. Let us see how the ISE combats these two problems. Figure 7.9 shows a schematic diagram of the whole system.

7.3.1 Fabric measurement and sewability prediction

Woven fabrics are cut and tested for weight, thickness and flexural rigidity, using on-line automatic measurement. Seam pucker severity is predicted by the SPS described earlier. Knitted fabrics are cut and tested for weight, thickness and tensile properties as well as breaking extension of their constituent yarns using on-line automatic measurement. Sewing damage is predicted by the fabric sewability system described earlier.

7.3.2 Correction of fabric properties

When a fabric is predicted to be unacceptable in terms of seam pucker, for example, the issue is how to correct some of the fabric properties so that the fabric has less tendency to pucker or become free from pucker. In order to preserve most of the characteristics of the original fabric, the correction in its properties should be minimised. Fabric finishing in severe fabrics may play an important role in such cases. Irrespective of any improvement of fabric property, the system then continues downstream into the actual sewing process, where it optimises automatically the settings of the first in the line of sewing machines.

7.3.3 Intelligent sewing system

In this case, sewing machine speed, sewing thread tension and presser foot pressure are changed automatically in the intelligent sewing system. A strip of material under these conditions is then stitched and the safeguard SLS system ('test' in Fig. 7.9) automatically assesses the quality of the seam. If the quality of the seam is acceptable, the system allows the rest of the sewing machines in the production line to use the same settings. If not, it repeats the same task in a loop until it reaches the desired seam quality. The values of the quality of the sewn fabric are then read back into a self-learning model.

7.3.4 Self-learning feature

The self-learning feature of the model is achieved by adaptive fuzzy optimisation of the output membership functions of the Neuro-Fuzzy system for closed loop feedback control. This feature is desirable in a scientific sense because it simulates the behaviour of the human brain in its ability to learn by experience, and in a practical sense since it enables the system to continuously become more accurate in predicting a problem.

After the self-learning has been performed, the dynamic optimisation of the sewing machine settings is then taken over by the Neuro-Fuzzy model, which ensures optimum stitching quality at any operational sewing speed.[11,19] This control system, shown in Fig. 7.4, receives two inputs, one from the processed fabric properties, and the other from machine speed. The system then controls the optimum settings of feed dog pressure and sewing thread tensions of the sewing machine achieved by the sensors, stepping motors and pneumatic pressure.

The simulated control surfaces and the actual surfaces followed by the Neuro-Fuzzy control system have been tested. There is good agreement between the actual disc force and simulated control surfaces, as shown in Figs 7.6, and 7.7, developed mainly from expert experience (operator/mechanic) and the experimental fine tuning from fabrics. The role played by thread tension after achieving seam balance in knitted fabrics is not critical and is practically insignificant when compared to the role of thread tension in woven fabrics, especially on lockstitch machines. However, the thickness of knitted and woven fabrics has a significant effect on seam balance, although more so in knitted fabrics.

7.3.5 ISE System implementation

The ISE has been tested with unknown woven and knitted fabrics without any mis-prediction. Problematic fabrics such as microfibre fabrics and shingosen have been tried and have shown reduction or, in most cases, elimination of seam pucker and/or sewing damage problems. A large pilot commercial trial of this system was performed at the Burberrys of London making-up unit in Castleford, UK, on a classical raincoat, and has shown that this technology could equally produce the required flat seam appearance in the garment without the use of any interlining.

7.4 Conclusions

Although this chapter could not cover in detail all aspects of intelligent garment manufacture, it has shown that the fabric/machine/human interface can be connected with true garment automation and ultimately sewing garment robotics. This chapter reports the methodology, the model base and the mechatronic engineering systems that have been developed and integrated in the so-called integrated sewability environment to combat problems associated with the manufacture, handling and joining of diverse limp materials. The principles of sewability prediction of fabric faults and optimisation of fabric properties which is analogous to 'fabric genetic engineering' have been established. Although safeguard quality systems (test) are desirable for ensuring quality of sewn products in their own right, it has been shown how they are used for model updating, mimicking human-like self-learning. The next generation of intelligent sewing machines have also been shown within the apparel manufacturing context. The synergism of neural and fuzzy logic has been employed to model the dynamic control of sewing machinery for complex interactions with fabrics, so that it becomes possible to optimise sewing machinery settings automatically, even dynamically, for any manufacturing task. The deliverables of this work should be used as catalysts for further research and development in garment robotics, which will make mass customisation a reality.

7.5 References

1. BRITE-EURAM Programme (1991), *Synopses of Current Projects 1990–1991*, Commission of the European Communities, Brussels.
2. Iguchi, K. (1990), 'Automated sewing system', *Journal of Textile News*, pp. 90–96.
3. Tyler, D. (1989), 'The development phase of the textile/clothing technology corporation apparel automation project', *International Journal of Clothing Science and Technology*, 1(2), 11–16.
4. Kochan, A. (1996), 'Moll: sewing to perfection', *Assembly Automation*, 16(1), 36–39.
5. Grosberg, P. (1966), 'The mechanical properties of woven fabrics. Part II: The bending of woven fabrics', *Textile Res. J.*, 36(3), 205–221.
6. Kawabata, S. and Niwa, M. (1991), 'Objective measurement of fabric mechanical property and quality', *Int. J. Cloth. Sci. Technol.*, 3(1), 7–18.
7. Postle R. (1991), 'Fabric objective measurement technology: present status and future potential', in Stylios, G. (ed.), *Textile Objective Measurement and Automation in Garment Manufacture*. Ellis Horwood, Chichester, pp. 27–45.
8. Stylios, G. and Sotomi, J.O. (1996), 'Thinking sewing machines for intelligent garment manufacture', *International Journal of Clothing Science and Technology*, 8(1–2), 44–55.
9. Pine, B.J. (1993), *Mass Customisation: The New Frontier in Business Competition*, Harvard Business Press, Cambridge, MA.
10. Stylios, G., Fan, J., Sotomi, O.J. and Deacon, R. (1992), 'A new concept in garment manufacture: the sewability integrated environment incorporating automatic objective measurement systems', *International Journal of Clothing Science and Technology*, 4(4), 45–48.
11. Stylios, G., Sotomi, O.J., Zhu, R., Fan, J., Xu, Y.M. and Deacon, R. (1994), 'A

sewability integrated environment (SIE) for intelligent garment manufacture', *IEE, Factory 2000 – Advanced Factory Automation,* Conference Publication Number 398, 543–551.

12. Stylios, G.K. (2005), 'New measurement technologies for textiles and clothing', *International Journal of Clothing Science and Technology*, 17(3–4), 135–149.
13. Stylios, G.K. (1998), 'Method and Apparatus for Assessment of Seam Pucker and other Surface Irregularities', British Patent No. 9819614.0.
14. Box, G.E.P. and Hunter, J.S. (1957), 'Multi-factor experimental designs for exploring response surfaces', *Annals of Mathematical Statistics*, 28(1), 195–241.
15. Stylios, G. and Sotomi, J.O., 'Investigation of seam pucker in lightweight synthetic fabrics as an aesthetic property. Part I: A cognitive model for the measurement of seam pucker', *J. Textile Inst.*, 84(4), 593–600.
16. Stylios, G. and Sotomi, J.O. (1993), 'Investigation of seam pucker in lightweight synthetic fabrics as an aesthetic property. Part II: Model implementation using computer "vision"', *J. Textile Inst.*, 84(4), 601–610.
17. Stylios, G. and Sotomi, J.O. (1993), 'A new instrument for routine objective assessment of seam deformations in limp materials', in Blackshaw, D.M.S., Hope, A.D. and Smith, G.T. (eds), *Laser Metrology and Machine Performance*, Lamdamap 93, Computational Mechanics Publications, Southampton, pp. 233–238.
18. Stylios, G and Sotomi, O.J. (1992), 'Automatic settings of optimum sewing machine conditions for the manufacture of high quality garments made of lightweight synthetic fibre fabric', *Proceedings of the 21st Textile Research Symposium on Basic Properties of Fibres and Fibre Assemblies, Performance and Design of New Fibrous Materials*, 7–9 August, Japan.
19. Stylios, G. and Sotomi, O.J. (1994), 'A neural-fuzzy control system for intelligent sewing machines'. *IEE Intelligent Systems Engineering,* Conference Publication Number 395, 241–246.

Part II

Adhesive bonding of textiles

8

Adhesive bonding of textiles: principles, types of adhesive and methods of use

E. M. PETRIE, Independent Consultant, USA

DOI: 10.1533/9780857093967.2.225

Abstract: This chapter reviews the fundamental principles of all types of adhesive bonding for textiles. Adhesives and adhesion are of fundamental importance to the textile and clothing industries. They are necessary in nearly every type of textile-based product. This chapter discusses the various theories supporting adhesion, the common types of adhesive used with textiles, and important bonding processes that are employed in the textile industry. It concludes with a review of recent trends in textile adhesives.

Key words: adhesives, adhesion, waterborne, solvent, hot melt, powder, film, polymer, dispersion, coating, surface preparation, plasma, curing, drying.

8.1 Introduction to adhesives in the textile industry

Adhesives and adhesion are of fundamental importance to the textile and clothing industries. They are necessary in nearly every type of textile-based product. The adhesive bonding of textiles is one of the oldest of joining technologies, and adhesives continue to play an increasing role in the manufacturing and design of textile-based products. New adhesive bonding technology meets the continuing challenges of this important industry. The production of textiles is a global industry, and the end-uses for textile-based materials continually expand. Figure 8.1 illustrates a 2003 world overview of textile bonding adhesives in volume (weight) and value (US dollars).

Industries that are dependent on textile fibers include apparel, transportation (e.g., automobile and aerospace), filtration, and geotextiles among others. Adhesion plays an important role in all coating, laminating, bonding, and fusing processes that are used in the manufacture of modern textile products. The requirements of the adhesive and adhesive bond are as varied as the products manufactured with textiles. Demands for higher performance and for adhesion to newly developed fibers continue to increase due to the sophistication of products in which textiles are used.

Adhesion to textile fibers and fabrics may be either temporary or permanent. Temporary adhesion is generally required at some point in the yarn or manufacturing process in order to move the product through to

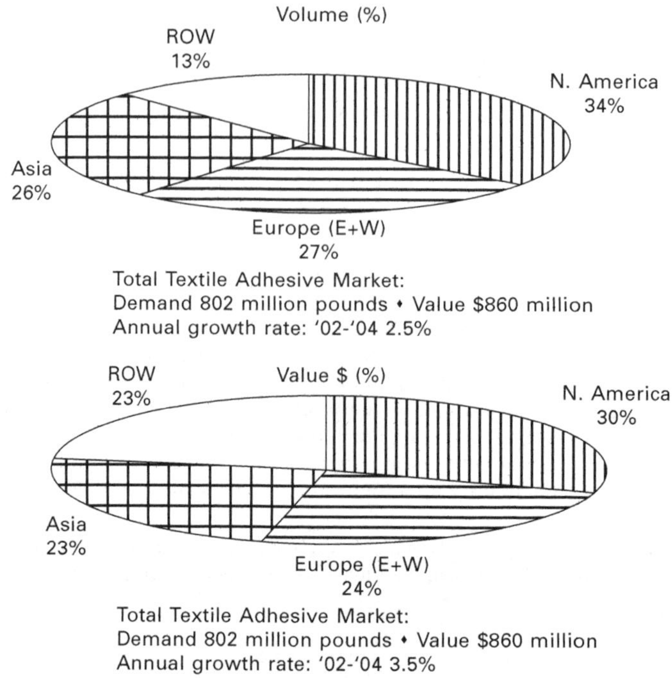

8.1 World overview of textile adhesive usage, 2003 (from Nick, 2003).

subsequent parts of the process. Permanent adhesives are more demanding, as the adhesive must last the life of the final product. In most cases, if the adhesive fails, the product also fails. Generally, products that are produced by bonding textiles can be categorized into four segments: (1) woven, (2) non-woven, (3) laminated products, and (4) others (knits, etc.). Several of the textile-based products that depend on adhesion for their commercial success are noted in Table 8.1.

8.1.1 Definition of adhesive and adhesion

An *adhesive* is a substance capable of holding at least two surfaces together in a manner that is appropriate for the application. Modern adhesives are manufactured by formulating or compounding a base resinous material with fillers, pigments, stabilizers, plasticizers, and other additives to yield a product with the desired end-use characteristics and acceptable cost. The method involved can range from simple mixing to sophisticated copolymerization processes.

Adhesives function primarily by the property of *adhesion*. Adhesion is the attraction of two different substances resulting from intermolecular forces between the substances. The substances in this case are the adhesive

Table 8.1 Textile materials that depend on adhesives and/or adhesion

Woven	Non-woven	Laminates	Other
• Chemical protective clothing • Inflatables • Furniture upholstery and curtains • Automotive interiors, tire cords, airbags • Apparel and accessories • Footwear • Hose • Belting	• Diapers • Incontinence pads • Feminine hygiene • Uniforms • Filter media • Apparel linings • Fashion accessory linings	• Signage • Membranes • Tents and awnings • Tarpaulins • Supported sheet roofing • Tank bladders • Linings • Carpet backing • Industrial blankets	• Bandages and surgical dressings • Composites • Sail technology • Aerospace • Geotextiles • Specialty industrial tapes • Bullet-proof protection • Helmets

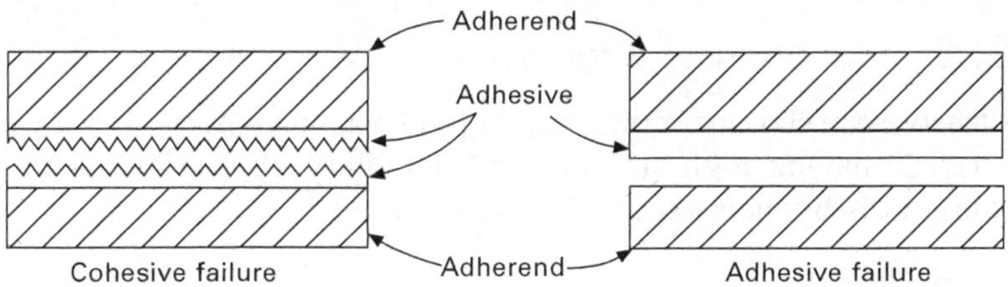

8.2 Examples of cohesive and adhesive failure.

material and the substrate or adherend. These adhesion forces develop when the substances are brought into contact, and work is required to separate them. Adhesion is generally caused by molecular interactions between the substrate and the adhesive and not necessarily by chemical bonds. Chemical bonds, however, do occur in a few instances.

Adhesion is distinctly different from *cohesion*, which involves the intermolecular attractive forces within a single substance. Cohesion relates to the forces within the molecule itself or those that connect molecules within a substance. To better understand the difference between adhesion and cohesion, consider the failed joints illustrated in Fig. 8.2. Bonded joints will fail either adhesively or cohesively or by some combination of the two. Adhesive failure is an interfacial bond failure between the adhesive and the adherend. Cohesion is the inner strength of a material (i.e., determined by molecules in the bulk), and this is manifested mainly by interconnecting forces between the molecules.

Cohesive failures could exist within either the adhesive material or the substrate itself. Cohesive failure of the adhesive material occurs when stress allows a layer of adhesive to remain on both substrates (i.e., the attachment of the adhesive to the substrate is stronger than the internal strength of the adhesive itself, and the adhesive fails within its bulk). When the adherend fails before the adhesive and the joint area remains intact, it is known as a cohesive failure of the adherend.

Cohesive failure could also occur in a weak boundary layer that may exist between the adhesive and the substrate. This may often appear as an adhesive failure, but it is a cohesive failure within the weak boundary layer. Common weak boundary layers are contaminants (e.g., oil, grease, or dirt), oxide layers, moisture, and migrating chemicals. Many adhesive specialists feel that weak boundary layer failure is the most common type of unforeseen failure. It is important to understand whether the failure mode is adhesive or cohesive. If cohesive failure occurs, the adhesion has reached a maximum, and all stress is transferred to the bulk adhesive or to the substrate. To further increase the strength of a joint, one must manipulate material parameters that are not the same as those contributing to adhesion.

8.1.2 Function of adhesives

Adhesives are chosen for their holding and bonding power. They are generally materials having high strength and durability. A major classification of adhesives is by their expected service life:

- permanent
- non-permanent/temporary.

'Permanent adhesive' is a term generally used to define an adhesive whose strength is critical to the success of the assembly. This term is usually reserved for adhesives with high shear strength and good environmental resistance. The joined textile is generally required to have similar strength at the point of joining as the substrate material. Permanent adhesives are generally made from formulations where the main component is a thermoset polymeric resin such as epoxy or polyurethane or a thermoplastic resin with sufficient heat and moisture resistance to resist the service environments.

'Non-permanent' or 'temporary' adhesives are adhesives with much lower strength and permanence, yet sufficient for the application. They are generally used to hold layers of fabric together firmly until they can be joined together more securely at a later stage. Examples of these adhesives are pressure-sensitive films, hot melts, and elastomeric adhesives. These adhesives are generally made from formulations where the main component is an elastic thermoplastic resin. Naturally occurring adhesives (e.g., starch and dextrin) are also normally considered to be temporary adhesives because of their susceptibility to moisture and other common service environments.

Permanent adhesives are normally rated in terms of their shear strength or the holding power of an overlap joint divided by the bond area. Temporary adhesives are often rated in terms of their peel strength or resistance to edge separation divided by the bond width. The load bearing capabilities of adhesives are provided in Table 8.2, and Fig. 8.3 illustrates the major forces to which adhesive bonds are subjected.

Whether permanent or temporary, adhesive strength is dependent not only on the adhesive but also on the substrate, joint design, rate of loading, and environmental service factors. The effect of several environments common to textile applications on thermosetting and thermoplastic adhesives is provided in Table 8.3.

Also, the strength of specific adhesives varies significantly due to its formulation, even for adhesives within the same chemical family. Thus,

Table 8.2 Load-bearing capabilities of selected adhesives

Adhesive type	Shear strength (MPa)	Peel strength (N/m)
Pressure-sensitive	0.005–0.02	300–600
Rubber-based	0.3–7	1000–7,000
Thermoplastic emulsion	10–14	–
Hot melt	1–15	1,000–5,000
Polyurethane	6–17	2,000–10,000
Acrylic	6–20	900–6,000
Epoxy	14–50	700–18,000
Phenolic	14–35	700–9,000
Polyimide	13–17	350–1,760

Notes: To convert from MPa to pounds per square inch, multiply by 145.
To convert from N/m to pounds per inch of width, divide by 175.

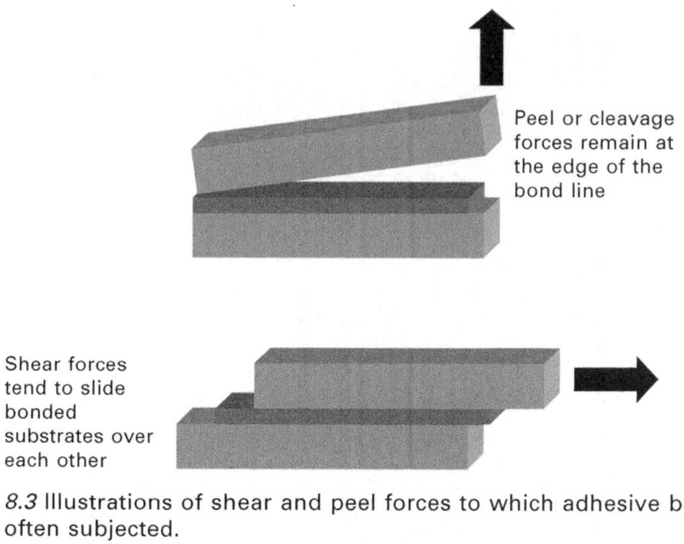

8.3 Illustrations of shear and peel forces to which adhesive bonds are often subjected.

Table 8.3 Relative resistance of adhesives to common textile chemical environments

Adhesives	Hot water	Acid	Alkali	Oil, grease	Alcohols	Ketones	Esters	Aromatics	Chlorinated solvents
Thermosetting adhesives									
Cyanoacrylate	6	6	6	3	5	5	5	4	4
Epoxy + polyamine	2	2	2	2	1	6	6	1	–
Epoxy + polyamide	6	3	6	2	1	6	6	3	–
Silicone (RTV)	2	3	3	2	3	3	3	3	3
Polyester + isocyanate	3	3	2	2	3	2	2	6	2
Thermoplastic adhesives									
Acrylic	3	2	2	2	2	3	2	3	4
Polyurethane	3	3	3	2	2	5	5	–	5
Polyvinyl acetal	–	6	3	2	3	3	6	3	6
Polyvinyl alcohol	6	5	5	2	3	1	1	1	1
Polyamide	6	6	2	2	6	2	2	2	6
Cellulose acetate	6	1	2	–	4	6	6	6	6
Vinyl vinylidene	3	–	–	2	2	2	3	–	–

there can be thousands of adhesive formulations classified as 'polyurethane' or 'acrylic' – all with unique characteristics to meet specific application requirements.

The properties of the adhesive joint are very much dependent on the substrate and the way that the adhesive and substrate are processed to create a joint. The adhesive–adherend interactions that affect the adhesive bond are reviewed in Section 8.2.2. It would be ideal if a single adhesive was universally applicable, cost very little, experienced long-term performance characteristics, and provided no safety, health, or environmental problems. The reality is that this will never be the case, and the selection of an adhesive will depend on many parameters, many of which are conflicting and require a compromise.

Although the primary function of an adhesive is to hold substrates together, adhesives are also useful for providing secondary functions. For example, an adhesive may also be used as a sealant, coating, vibration damper, insulator, or gap filler. Multiple functions may be required in the same application. Many product designers feel that multifunctionality is one of the most valuable characteristics of adhesive bonding.

Because adhesives are viscoelastic materials, they can act as vibration dampers to reduce the noise encountered in some assemblies. In the automotive and office furniture industries, for example, adhesives are used in combination with textiles and structural partitions to provide noise reduction.

Adhesives can also sometimes be used to perform sealing functions, offering a barrier to the passage of fluids and gases. For example, rainwear may be constructed of six or more layers of fabric, and the outermost layer can be made water repellant by chemical treatment and using a water impermeable adhesive to bond it to the interior layer.

Another property of adhesives that is often advantageous is their ability to function as electrical and thermal insulators in a joint. The degree of insulation can be varied with different adhesive formulations and fillers. Adhesives can even be made electrically and thermally conductive with silver and boron nitride fillers, respectively.

In addition to their primary and secondary functions, adhesives may have to provide properties that are highly dependent on the industry in which they are used. In the textile industry these properties could include:

- textile feel or hand
- permeability against water vapor and air
- barrier properties against liquid water
- washability at elevated temperatures and resistance to dry cleaning chemicals
- sterilization (medical textiles)
- tear and abrasion resistance

- thermal, chemical, and weather stability
- flame resistance.

8.1.3 Competitive methods of joining

Numerous joining methods are available for textile materials, garments, and other finished products. The joining processes can be broadly divided into mechanical fastening and bonding. Bonding can further be classified into adhesive bonding, solvent bonding, and thermal welding. There are several mechanical bonding and thermal welding processes as shown in Fig. 8.4.

Arguably, solvent bonding and thermal welding can be considered to be a subset of adhesive bonding. With these processes the substrate itself (in molten or solubilized form) acts as the adhesive. Once the substrate is liquefied, the principles of adhesion come into play to determine the strength and durability of the bond. Furthermore, the thermal methods noted in Fig. 8.4 can also be used to activate a hot melt adhesive rather than the textile substrate. In this chapter, the focus will be on adhesive bonding, where a separate component (i.e., adhesive) is used to join two or more substrates together without melting or dissolving the existing textile or coating.

In evaluating the appropriate joining method for a particular application, a number of factors must be considered. Usually, the decision of which fastening method to use involves several trade-offs. An analysis of requirements can be useful in identifying potential fastening methods. When this is performed, the possibility of using adhesives over other methods becomes apparent.

Some fastening methods can quickly be eliminated from consideration, such as when attaching textiles to non-polymeric materials that cannot be heated to form a melt or liquefied by contact with solvent. Adhesive bonding

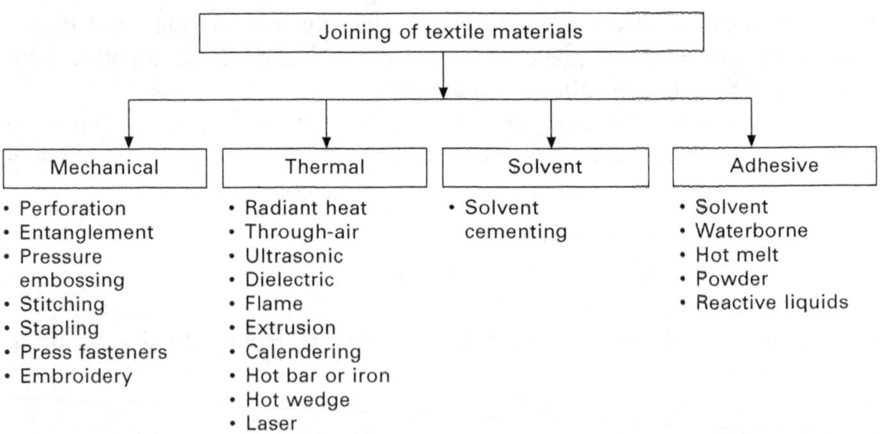

8.4 Methods of combining textiles.

is also generally appropriate when there are large areas to join, when fatigue strength is a concern, when adhesives can provide increased manufacturing productivity, or when continuous, non-perforated seams are required. Adhesives are also usually the first choice in applications where they can perform a secondary function such as providing water impermeability, electrical or thermal conductivity, abrasion resistance, and so forth.

8.1.4 Advantages and disadvantages of adhesive bonding

Adhesive bonding presents several distinct advantages and disadvantages when compared to other methods of joining. These pros and cons are summarized in Table 8.4.

Adhesives are often used to join technical textiles requiring high seam strength (e.g., moving belts, parachutes, packaging). Stresses are transmitted more effectively by adhesives than by other joining methods. Plastics, elastomers, and certain metals can be reliably joined to textiles only with adhesives. With some organic fibers, heat welding is usually at too high a temperature, and mechanical fastening destroys the lightness and aesthetics of the final product. Certain examples of less obvious applications where adhesive bonding is a practical method of assembly are shown in Table 8.5.

Table 8.4 Advantages and disadvantages of adhesive bonding

Advantages	Disadvantages
• Provides large stress bearing area • Provides excellent toughness and abrasion resistance • Damps noise • Can be used to provide a moisture-resistant barrier • Can be made to be thermally and/or electrically conductive • Joins all shapes and thicknesses • Joins any combination of similar or dissimilar materials • Provides smooth contours • Provides continuous, non-perforated seams • Heat, if required, is too low to affect heat-sensitive substrates • Provides attractive strength-to-weight ratio • Often less expensive and faster than other methods of joining	• Surface must be carefully cleaned or chemically treated (if a low surface energy textile) • Long cure times may be needed and fixturing could be necessary • Limitation on upper continuous operating temperature (generally 175–200°C) • Heat and pressure may be required for cure • Rigid process control usually necessary • Inspection of finished joint is difficult • Useful life depends on the service environment • Environmental, health, and safety considerations are necessary • Special training may be required; selection of the optimal adhesive and bonding process may be difficult and relates only to the intended application

Table 8.5 Examples of applications where bonding is a practical method of assembly

Optimum application areas for adhesive bonding	Examples
Dissimilar materials	Combinations of textiles with metals, rubbers, plastics, foamed materials, fabrics, wood, ceramics, glass, etc.
Heat-sensitive materials	Thermoplastics, magnetic materials, electronics
Laminated structures	Sandwich construction; sheet laminates, core laminates, office partitions, auto and aircraft interior parts
Structural applications	Load-bearing structures, geotextiles, civil engineering industries
Bonded inserts	Plug inserts, studs, rivets, tubes, frame construction
Sealed seams	Rainwear, geotextiles, container seams
Fragile components	Substrates cannot withstand applied pressures, thermal expansion changes, etc., from other joining methods
Components of particular dimensions	Where bonding areas are large or there is a need for shape conformity between bonded parts
Temporary fastening	Where the intention is to dismantle the bond later, the use of various labels, surgical and pressure-sensitive tapes, adhesives for positioning and locating parts prior to assembly by other means
Thin or aerodynamic shapes	Laminates, geotextiles, coverings, clothing and sportswear

In many applications adhesive bonding is the only logical choice. Table 8.6 describes several common applications where adhesive bonding provides a significant technical and/or commercial advantage. This table also indicates the specialized requirements for each application. It is apparent from the diversity of applications and requirements that there is no single adhesive system that is appropriate and that the selection of the proper adhesive for an application can be a time-consuming but critical step in the design process.

An excellent recent example of a product that could not be considered without the use of adhesives is perhaps the most notable performance enhancing invention in sports, the Speedo LZR Racer swimsuit. Many considered it to be the winner at the Beijing Olympics in 2008 for it seemed that within a very short time every swimmer was wearing the space-age looking suits. Since the suit was introduced in February of 2008, more than 50 world records were broken by athletes wearing it.

Adhesives allow manufacturers of these swimsuits to create new, ultra-lightweight swimwear that allows the swimmers to glide through the water. In addition to adhesives, modern joining technology such as ultrasonic

Table 8.6 Common adhesive applications in the textile industry (adapted from Fries, 1983)

Application	Examples	Common substrates	Adhesive requirements
Apparel, accessories	Laminations in clothing and fashion accessories such as handbags and belts	• Fabric-to-fabric • Fabric-to-foam • Fabric-to-metallized film	• Wash and/or dry cleaning resistance • Soft hand • Good elongation
Upholstery	Home, office, and automotive interior parts	• Woven fabric-to-non-woven fabric • Fabric-to-foam	• Resistance to dry cleaning solvents and/or steam cleaning • Elongation
Home furnishings	Table covers, draperies, shower curtains	• Fabric-to-fabric • Fabric-to-film • Fabric-to-foam	• Variable substrate types (e.g., PVC film) • Wash or dry cleaning resistance
Tarpaulins, covers, industrial blankets	Chemical treated fabric or textile for industrial or civil engineering applications	• Fabric-to-film • Film-to-fabric-to-film	• High shear strength • Weatherability • Heat, cold, chemical resistance
Pressure-sensitive applications	Shoe construction tapes, wall-board sealing tapes	• Fabric • Fiberglass	• High peel strength • Immediate tack or early strength • Resistant to the end-use environment

welding is used to make flat, drag-resistant seams. In the sport of swimming, one of the main challenges is to reduce drag. About 90% of the swimmer's propulsive energy is used to just overcome the drag force caused by water. One competitive swimmer claimed that swimming in these new suits is 'like swimming downhill'.

8.2 Reasons for the success and failure of adhesives

There is, unfortunately, no universal theory of adhesion on which to accurately model the interactions that take place between the adhesive and the adherend. Instead, several theories exist that provide a means by which one can rationalize certain observations. They are generally useful in understanding why adhesives stick and why, at times, they fail.

Adhesion theories also allow us to make predictions and even obtain a semi-quantitative realization of joint strength. By being familiar with these theories, one can develop a knowledge base and an awareness of how adhesives and sealants work in practical situations.

8.2.1 Theories of adhesion

The actual mechanism of adhesive attachment has not yet been explicitly defined. Several theories attempt to describe the phenomenon of adhesion. The most common theories of adhesion are based on:

1. adsorption and wetting,
2. diffusion,
3. electrostatic interactions,
4. mechanical,
5. chemical bonding, and
6. weak boundary layers.

No single theory explains adhesion in a general, comprehensive way. Some theories are more applicable for certain substrates and applications; other theories are more appropriate for different circumstances. Each theory has been subjected to much study, question, and controversy.

Adsorption theory

The adsorption theory states that adhesion results from molecular contact between two materials and the surface forces that develop. A bond develops from the adsorption of adhesive molecules on the substrate and the resulting attractive forces, usually designated as secondary or van der Waals forces. For these forces to develop, the respective surfaces must not be separated more than several angstroms in distance. Therefore, the adhesive must make intimate, molecular contact with the substrate surface.

The process of establishing continuous contact between an adhesive and the adherend is known as 'wetting.' Figure 8.5 illustrates good and poor wetting of an adhesive spreading over a surface. Good wetting results when the adhesive flows into the valleys and crevices on the substrate surface; poor wetting results when the adhesive bridges over the surface perturbations. Obtaining intimate contact of the adhesive with the surface essentially ensures that interfacial flaws are minimized or eliminated. At a minimum, poor wetting causes less actual area of contact between the adhesive and adherend, and stress risers at the small air pockets along the interface. This results in lower overall joint strength.

Wetting can be determined by the angle a drop of water or adhesive makes when placed on the substrate. Complete, spontaneous wetting occurs when the contact angle is zero, or the material spreads uniformly over a substrate to form a thin coating (Fig. 8.6, left). Wetting is favored when the substrate

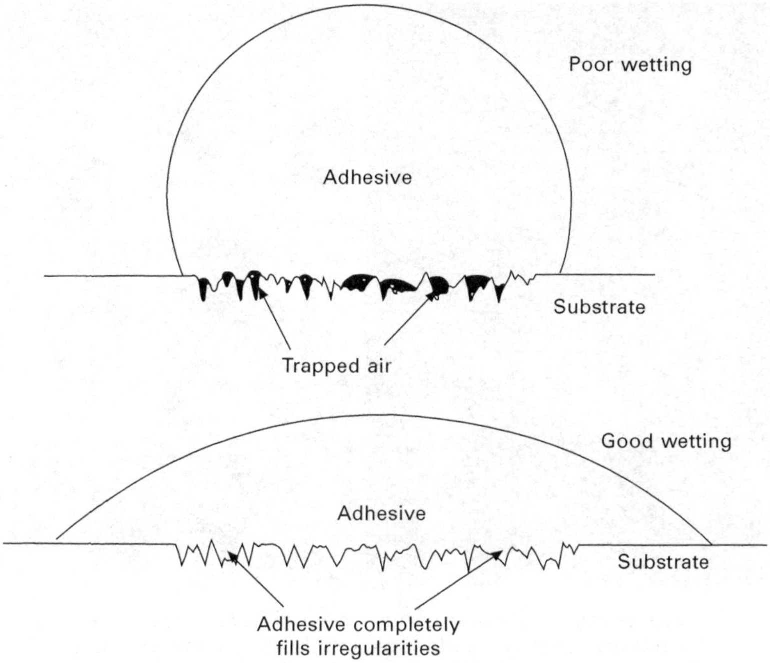

8.5 An illustration of good and poor wetting by an adhesive spreading over a surface.

surface energy is high, and the surface tension of the wetting liquid is low. Low energy adhesive polymers, therefore, easily wet high energy substrates such as metals or glass. Conversely, polymeric fibers and other polymeric substrates, such as fluorocarbons and polyolefins, have low surface energies (Fig. 8.6, right) and will not be readily wet by water or conventional adhesives unless they have a lower surface energy.

The fact that good wetting requires the adhesive to have a lower surface tension than the substrate explains why organic adhesives, such as epoxies, have excellent adhesion to metals, but offer weak adhesion on many untreated polymeric substrates, such as polyethylene, polypropylene, and the fluorocarbons.

Some important concepts develop out of these principles. You would expect, for example, that polyethylenes, if used as adhesives, would provide excellent adhesion to a variety of surfaces including both polymers and metals. In fact, they do provide excellent adhesion. However, commercial polyethylene generally has many lower molecular weight constituents that create a weak boundary layer, thus preventing practical adhesion. Polyethylene makes an excellent base for hot melt adhesives once the weak low molecular weight constituents are removed.

To optimize joint strength, it is often necessary to reduce the surface tension

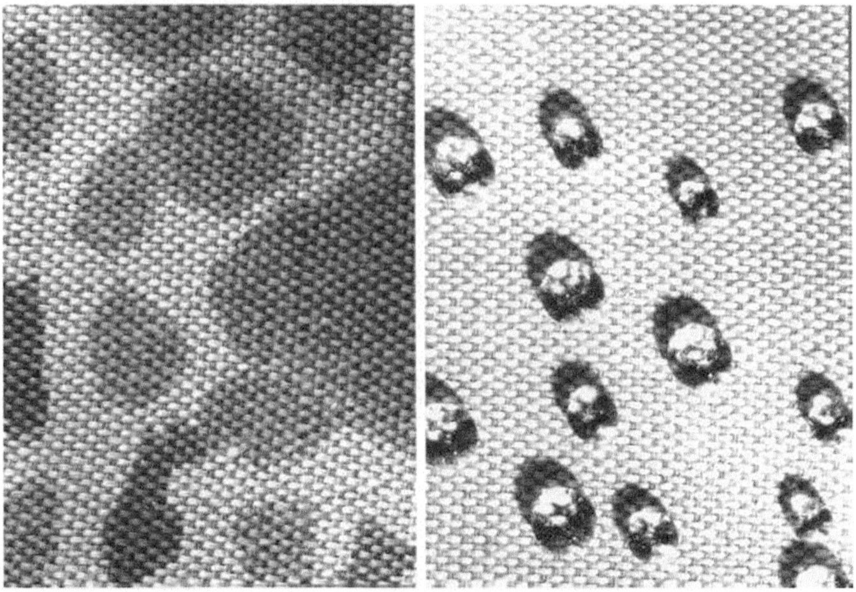

8.6 Example of water contact on a high energy textile fabric (left) and on a low energy textile fabric (right) (courtesy of Kruss, USA).

Table 8.7 Main categories of substrate surface treatments used to alter the surface energy (Court, 2004)

Mechanical	Chemical	Energetic
• Grit blast • Abrasion • Cryogenic blasting • Vapor honing • Peel ply (composites)	• Solvent clean • Detergent wash • Etch – either acid or alkali • Anodizing – electrochemical • Primer • Vapor or ultrasonic cleaning	• Plasma (atmospheric and low pressure) • Corona discharge • Flame • Eximer laser

of the adhesive or increase the surface tension of the substrate. Reduction in surface tension of the adhesive is generally achieved by substituting or adding a lower surface energy base polymer or by adding a 'wetting agent' or 'adhesion promoter'.

To increase the surface energy of the substrate, a surface treatment generally must be employed. Prebond surface treatments are employed to (1) remove weak boundary layers, (2) increase the surface energy of the substrate, or (3) provide a consistent surface for bonding. There are three main categories of surface treatments: mechanical, chemical, and energetic. Within each of these categories there are several processes as shown in Table 8.7. Of these, the chemical or energetic surface treatments are most appropriate for textiles.

Once intimate contact is achieved between the adhesive and substrate through wetting, it is believed that permanent adhesion results primarily through forces of molecular attraction called van der Waals forces (Petrie, 2007). However, other types of forces can also prevail due to the close proximity of the adhesive and substrate, and these forces provide the basis of the other common theories of adhesion.

Mechanical theory

Mechanical interlocking can be used as a method of adhering textiles by flow or penetration of fibers from one substrate to another to provide mechanical attachment as well as additional strength. Similarly, the adhesive can lock onto the micro-roughness on the surface of a substrate to create a mechanical attachment.

At one time, adhesion was thought to occur only by the adhesive flowing and filling pores, holes, crevices, and micro-voids on the substrate. When the adhesive then hardens, the substrates are held together mechanically. The surface of a solid material is never truly smooth but consists of a maze of peaks and valleys. According to the mechanical theory of adhesion, in order to function properly the adhesive must penetrate the cavities on the surface, displace the trapped air at the interface, and mechanically lock on to the substrate.

One way that surface roughness aids in adhesion is by increasing the total contact area between the adhesive and the adherend. If interfacial or intermolecular attraction is the basis for adhesion, increasing the actual area of contact will increase the total energy of surface interaction by a proportional amount. Thus, the mechanical interfacing theory generally teaches that roughening of surfaces is beneficial because it (1) gives 'teeth' to the substrate and (2) increases the total effective area over which the forces of adhesion can develop. Surface roughness generally aids in adhesive bonding by providing a mechanical interlocking effect. However, if the adhesive does not wet the substrate well, as shown in the top part of Fig. 8.5, roughening can only increase the amount of air entrapped at the interface. In these cases, roughening is not suggested and a chemical surface pre-treating process may be necessary.

Electrostatic theory

The electrostatic theory of adhesion is generally not considered as frequently as the other theories. However, there are certain applications where it is appropriate, and this theory has become popular recently due to the knowledge that electrostatic adhesion plays a critical role in many biological functions, self-cling films, and particle adhesion.

The electrostatic theory states that forces in the form of an electrical double layer are produced at the adhesive–adherend interface. These forces are primarily dispersion forces and forces arising from the interaction of permanent dipoles. Such interactions are only operational on a very short range (less than 0.5 nm), and their effectiveness decreases with the sixth or seventh power of the distance of separation.

This theory gathers support from the fact that electrical discharges have been noticed when an adhesive is peeled from a substrate. It was not until recently that electrostatic forces were confirmed without actually destroying the bond. However, it is believed that for conventional adhesive bonds, the electrostatic contribution to the total work of adhesion is quite small in comparison to van der Waals forces (Possart, 1988).

Diffusion theory

The fundamental concept of the diffusion theory is that adhesion arises through the interdiffusion of molecules from one material to another across the interface. The diffusion theory is primarily applicable when both the adhesive and adherend are polymeric, having compatible long-chain molecules that are capable of movement. The phenomenon is a two-stage process. Wetting is followed by interdiffusion of chain segments across the interface to establish an entangled network of molecules (Fig. 8.7).

For diffusion to occur, the adhesive and the adherend must be chemically compatible in terms of diffusion and miscibility. This means that the materials must be mutually soluble, and the diffusing molecules must be mobile. The mobility requirement generally necessitates that the materials be at a sufficiently high temperature to form a melt or dissolve in a solvent solution. Dissimilar polymers are usually incompatible because of their long chain nature and their low diffusion coefficients.

Situations in which the adherend and adhesive are soluble in one another are relatively uncommon. Therefore, the diffusion theory of adhesion can be

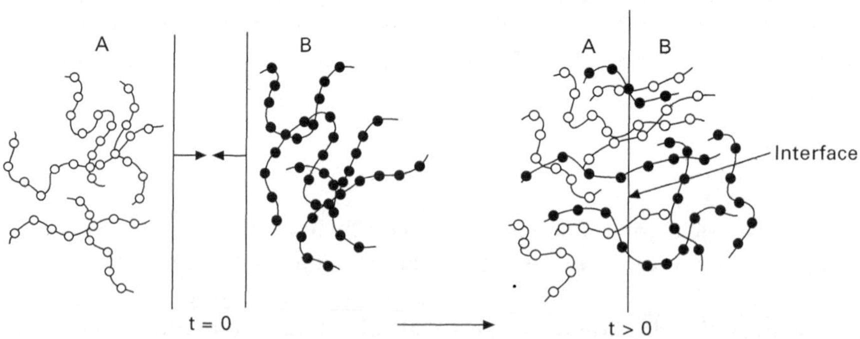

8.7 Interdiffusion of polymer molecules across an interface.

applied in only a limited number of cases. The theory is applicable primarily to the solvent or heat welding of thermoplastic substrates.

The diffusion theory also explains why certain monomers that are used in adhesives and primers provide high bond strength on untreated low energy surfaces such as polypropylene. The adhesive or primer molecules diffuse readily into the substrates and create molecular interlocking. These products offer a method of obtaining high bond strength to low energy substrates without a surface treating process.

Chemical bonding theory

In certain applications the formation of covalent chemical bonds occurs across the interface. Chemical bonding requires that there be mutually reactive chemical groups tightly bound on the substrate and in the adhesive. These strong (60–700 kJ/mol) and durable bonds are generally the result of close contact or adsorption of the adhesive on the surface followed by a chemical reaction. Thus, the term 'chemisorption' is often used to describe this mechanism.

Adhesives containing reactive functional groups, such as hydroxyl or carbonyl, tend to adhere more tenaciously to substrates containing similar groups. Hydroxyl bonding is one of the reasons why, for example, polyurethane base polymers are often used in adhesive formulations.

Weak boundary layer theory

According to the weak boundary layer theory, as first described by Bikerman (1961), when bond failure seems to be at the interface, usually a cohesive rupture of a weak boundary layer is the real event. The failure occurs cohesively within the weakest component of the joint, similar to the weakest link in a chain (Fig. 8.8). This theory largely suggests that true interfacial failure (i.e., adhesion failure) seldom occurs. Failure may occur so near to the interface that it is apparently at the interface, but in most cases it is ductile plastic deformation or cohesive failure of a weak boundary layer. Weak boundary layers can originate from the adhesive, the adherend, the environment, or a combination of any of the three. Examples of common weak boundary layers include moisture, oils from the factory or human touch, migration of plasticizers out of the substrate, and antistatic agents and other applied surface coatings.

Weak boundary layers occur at the interface between adhesive and substrate. When failure occurs, it is the weak boundary layer that fails, although failure may seem to occur at the adhesive–adherend interface.

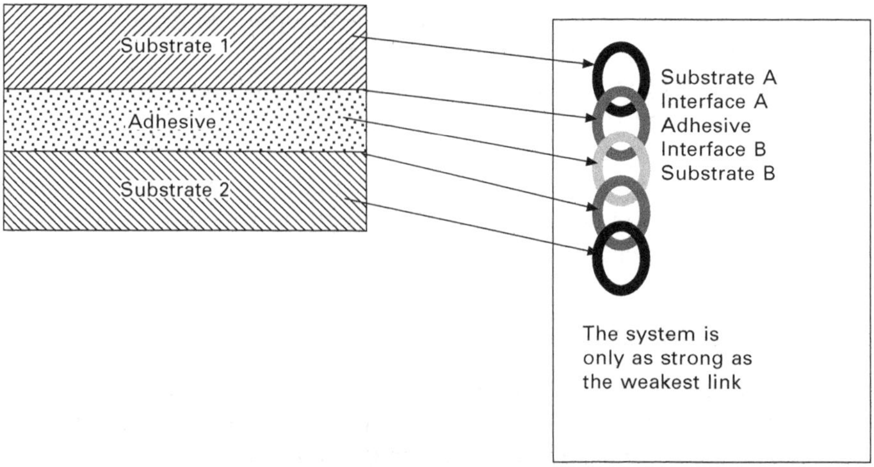

(a) (b)

8.8 In an ideal joint the substrate should be the weakest link. The adhesive joint (a) can be divided into at least six regions that are similar to the links (b) in a chain.

8.2.2 Adhesive–adherend interactions

The mechanism for adhesion in textile manufacture may be either physical (mechanical interlocking, diffusion, electrostatic) or chemical (intermolecular attractions and covalent bonding). There are generally three problems that interfere with satisfactory adhesion:

- contaminants or weak boundary layers
- required flexibility
- dimension stability of the fiber–adhesive interface relative to the environment.

There is a wide range of natural and manufactured textile fibers available, but satisfactory adhesion depends on the nature of the fiber surface, the presence of natural or added impurities, and the effects of physical and chemical treatments prior to bonding. Holme (1999), provides an excellent review of the subject of adhesion to textile fibers, addresses some of the problem areas relative to adhesion, and provides information regarding chemical pretreatments to enhance the adhesion to fibers.

Textile and all other types of bonding require clean surfaces that are free from dirt, grease, and other contaminants. In textile manufacture the presence of contaminants or 'weak boundary layers' can be particularly troublesome. For example, applied fiber lubricants are purposefully used to avoid build-up of undesirable friction or static electrification during textile manufacture. Additionally, processing treatments such as bleaching, dying, printing and chemical finishing can affect the nature of the fiber surface and the durability of the adhesive bond.

Often modern fibers are low surface energy materials that inhibit the adhesive or coating from spreading on and wetting the fiber substrate. To resolve these problems, prebond surface treatments ranging from simple washing to chemical or physical (e.g., flame, corona) treatment may be necessary to remove weak boundary layers and/or to provide a surface that is more wettable (Suo and Ooij, 2002).

A second major problem in adhesive bonding of textiles is that nearly all textile products are designed to be flexible. As a result, the bond must be deformable and have a relatively low level of stress concentration. Ideally, the adhesive or coating must deform and recover with the textile if satisfactory adhesion is to be obtained.

A third area of concern is the effect of the manufacturing and service environments on the fiber itself. Many fibers (e.g., cotton, wool, nylon) are hydrophilic and absorb water from the atmosphere or on immersion in water. As a result, they can swell or shrink in diameter depending on the environment. This dimensional change can increase the stress on the adhesive bond. Many fibers are also thermoplastic and, thus, are heat sensitive. These fibers may shrink in length or even melt on exposure to thermal conditions. For apparel, the potential environmental exposures can be significantly diverse. These include washing, tumble or line drying, ironing, and pressing treatments or dry cleaning.

Modern fabrics such as those based on microfibers with sophisticated surface finishes (e.g., 'peach skin' effects) are especially sensitive to both temperature and pressure. Extreme care has to be taken to prevent irreversible damage while insuring a satisfactory and even bond strength. This is a complex field of textile technology, which has not yet received sufficient attention. One approach, however, has been the development of reactive liquids or hot melt with low application temperatures, sometimes called 'warm melts'. These adhesives are applied at a temperature that is below the heat sensitivity of the fiber.

8.3 Classification of adhesives used in textile applications

Adhesives may be classified by many methods, and there can be several levels and sub-levels to these classification schemes. Perhaps, the broadest classification scheme is to categorize an adhesive as being manufactured from materials that are either synthetic or naturally occurring. Synthetic adhesives are manufactured from man-made materials such as polymers. Natural adhesives are manufactured from naturally occurring materials such as animal or agricultural by-products.

Most adhesives are made from high organic polymers, although there are several adhesive systems of inorganic origin. The oldest high polymers used for adhesives were of natural origin. Often naturally occurring adhesives are thought to be inferior to synthetic polymers because of their lower strength and limited freedom in processing. However, in many applications, such as bonding of paper or packaging, where the emphasis may be on the adhesive being biodegradable or repulpable, naturally occurring adhesives find a strong market. Modern epoxies, urethanes, acrylics, and other adhesive systems that are used in demanding, long-lasting applications are generally made from synthetic polymers.

8.3.1 Classification of adhesives

The classification of adhesives into synthetic and naturally occurring categories is usually far too broad for many practical purposes. The industry has settled on several common methods of classifying adhesives that satisfy most purposes. These classifications are by:

- mechanism of reaction,
- chemical composition (as indicated by the base polymer in the formulation), and
- end-use application.

See Table 8.8.

Mechanism of reaction

Some adhesives solidify simply by losing solvent while others harden as a result of heat activation or chemical reaction. Yet others solidify by cooling from a melt. Some systems may require several mechanisms to harden. For example, thermosetting solvent-based adhesives require the solvent to

Table 8.8 Classification of textile adhesives

Technology	Base polymer	End-use application
• Water-based • Solvent-based • Non-volatile liquids that cure by: – heat – chemical components – radiation • Hot melt (unreactive and reactive)	• Acrylic • Styrene block copolymers • Natural rubber • Polyamides • Polyesters • Polyurethane • Vinyls (PVAc, EVA, PVC)	• Laminating adhesives • Pressure-sensitive adhesives • Contact adhesives • Paste dot adhesives • Other

first evaporate and then chemical cross-linking to occur before a bond is achieved.

The most common textile adhesives are available as solutions in solvent or water, as dispersions in water, or as solids that melt under the application of heat. However, 100% reactive liquids such as moisture-cured urethanes are also used.

The mechanism of reaction is an important determinant in the adhesive selection process. For example, if the processing requirements dictate that only a room temperature curing adhesive system can be used, then temperature resistance may be sacrificed. There are very few room temperature curing adhesives that exhibit good elevated temperature resistance because of the lack of extensive crosslinking and the nature of the polymers formed with room temperature curing formulations.

The advantages and disadvantages of these various technology types are summarized in Table 8.9. New approaches, particularly 100% reactive, liquid systems designed for cure by exposure to ambient moisture, UV or electron beam radiation have recently been introduced.

Adhesives can be further classified by form and reaction mechanism. This hierarchy of adhesive types can be summarized as follows:

1. Adhesives that harden by chemical reaction:
   a. Two-part systems
   b. Single-part, cured via catalyst or hardener
   c. Moisture-curing adhesives
   d. Radiation (light, UV, electron beam, etc.) curing adhesives
   e. Adhesives catalyzed by the substrate
   f. Adhesives in solid form (tape, film, powder, etc.)
2. Adhesives that harden by solvent or water loss:
   a. Contact adhesives
   b. Pressure-sensitive adhesives
   c. Reactivatable adhesives
   d. Resinous adhesives
3. Adhesives that harden by cooling from the melt:
   a. Hot melt thermoplastic adhesives
   b. Hot melt applied pressure-sensitive adhesives
   c. Powder and film reactive adhesives

Most textile adhesive formulations are available as a liquid, but some compositions are also provided in film form, either as unsupported, reinforced, or coated on both sides of a carrier web. These are reactivated by heat or by solvent wetting. Film adhesives offer convenience, since no coating is needed by the user, but they have a higher materials cost. Because film adhesives are available as continuous stock, they are ideal for laminating fabrics.

Table 8.9 Advantages and disadvantages of various types of textile adhesives

Type	Advantages	Disadvantages
Solvent-based	• Oldest technology • Widest range of formulations • Faster drying than waterborne • High bond strength and durability	• Regulatory restrictions • Waste disposal • Safety (flammable), toxicity • High cost due primarily to petrochemical solvent
Waterborne	• Inexpensive to moderate cost • Generally not regulated • Fewer health and safety problems • Easy clean-up and storage • Increasingly wide range of formulations available	• High energy required to evaporate water • Corrosion of equipment may be an issue • Slow drying process • Generally low solids content (low thickness build) • Wetting of surfaces sometimes difficult • Poorer heat and moisture resistance than solvent-based types
Thermoplastic hot melts	• Simplicity • No curing mechanism needed • Low energy demand • Low footprint • Instant bond in many cases • Easy storage • No waste	• Limited heat and chemical resistance • Specialized application equipment required • Heat may damage substrates • Short open time and loss of tack on cooling
Reactive liquids	• No volatiles emission • Low energy demand • Maximum heat and chemical resistance • Maximum strength and durability	• Expensive • May require long cure times • Heat may be necessary to initiate or accelerate cure • Multiple component systems must be carefully metered and mixed • Storage life or working life (once mixed) could be too short

Base polymer

A variety of different chemical base polymers are available, each with their own properties, such as suitability for specific substrates and durability to water, solvents, heat, etc. The chemical type of base polymer employed in an adhesive formula generally determines the properties. For example, polyurethanes are flexible and have good adhesion, but some may discolor. Polyvinyl acetate emulsion adhesives are relatively inexpensive, but they are brittle and have limited water resistance.

Base polymers are classified as being either thermoplastic or thermosetting. Thermoplastic polymers are polymers that melt when heated and then resolidify when cooled. Thermoset polymers are those that do not melt when heated, but at sufficiently high temperatures they exhibit creep under load and even decomposition.

With thermoplastic polymers, only the secondary intermolecular bonds hold the individual polymer chains together. There is no chemical bonding between molecules. Thus, as the temperature or stress on the polymer is raised, a point will be reached where the forces holding the chains together will be reduced, and they are then free to slide past one another.

Thermoset polymers consist of a continuous network of polymer chains that are chemically crosslinked. The crosslinks are formed by reaction of one polymer molecule with another or with a second polymer. These reactions can occur either at room or elevated temperatures, and there are catalysts that can accelerate the reaction mechanism. The connecting chains of the thermoset polymer are held together with the same primary covalent bonds as are the atoms in the main chain. Thus, the molecules are interlocked and do not easily slip by one another. With sufficient heat or stress the covalent bond breaks and the polymer fails suddenly or degrades.

The base polymers that are more generally used in textile adhesives and their properties are summarized in Table 8.10. Most common textile adhesives can be either thermoplastic or thermosetting depending on the formulation, whether a crosslinker is included, and the processing conditions.

Textile adhesives, like other adhesives, are generally highly compounded formulations, and different base polymers may be blended to produce specific properties. For example, a hot melt adhesive may consist of a blend of copolyesters and modified polyethylene.

Modern-day adhesives are often fairly complex formulations of components that perform specialty functions. Very few polymers are used without the addition of some modifying substance such as plasticizer, tackifier, or inert filler. The selection of the actual ingredients will depend on the end properties required, the application and processing requirements, and the overall cost target of the adhesive. The various components that constitute an adhesive formulation and their primary functions are described below:

- base resin or binder (provide the main structural and chemical characteristics)
- catalysts or hardeners (provide chemical curing)
- catalysts (accelerate curing)
- inhibitors, and retarders (decelerate curing)
- solvents (provide low viscosity for easier coating and application)
- thickeners and thixotropic agents (provide increased viscosity and with thixotropes even non-sag properties)

Table 8.10 Properties of textile adhesives formulated from various base polymers

Base polymer	Types available	Properties
Acrylics	Solutions and aqueous emulsions	• Very wide adhesion range • Excellent resistance to discoloration, light, and oxidation • Curing types are available that have wash and dry cleaning resistance
Vinyl acetate	Hot melt and aqueous forms	• Good adhesion range • Flexibility • Limited heat and chemical resistance • Relatively low cost
Rubber-based (synthetic and natural)	Solvent solutions; aqueous dispersions (latex), hot melt (block copolymers)	• Properties vary with type • Some heat cure formulations are available • Good heat and chemical resistance • Some types are prone to oxidation
Polyamide	Hot melts	• Generally restricted to fusible interlining applications
Polyester	Hot melts, solvent solutions	• Chemically related to the polyester fiber • Good water and solvent resistance • Reactive types have good heat and chemical resistance • Good adhesion especially to polyester fabrics and films
Polyurethane	One and two component liquid (100% solids, solvent-based, waterborne), reactive hot melts	• High adhesive strength to many materials • Good elongation, toughness, and abrasion resistance • Good water and solvent resistance • Reactive types have good heat resistance • Some types yellow on exposure to UV
Polyvinyl chloride	Solvent solutions, 100% solids plastisols	• Limited adhesion • Good water resistance

- diluents (like solvents, they are used to lower viscosity, but diluents do not evaporate from the formulation)
- extenders (reduce cost)
- fillers (increase density, provide adjustment to physical properties)
- carriers and reinforcement (carriers provide a means of getting the adhesive to the joint; reinforcements provide improvement to the cohesive strength)
- plasticizers (provide flexibility)
- flexibilizers and tougheners (provide elongation and toughness)
- tackifiers (provide immediate self-adhesion or tack)

- film formers (used so that the formulation will more easily form a film once applied and dried)
- antioxidants, antihydrolysis and antifungal agents, and stabilizers (provide improved thermal stability and environmental resistance)
- soaps, surfactants and wetting agents (used to achieve better wetting or better dispersions of fillers in the adhesive).

End-use application

Adhesives may also be classified according to their end-use. Thus, metal adhesives, wood adhesives, and vinyl adhesives refer to the substrates to which they will bond; and acid-resistant adhesives, heat-resistant adhesives, and weatherable adhesives indicate the environments for which each is best suited.

Adhesives are also often classified by the method in which they are applied. Depending on viscosity, liquid adhesives can be considered to be sprayable, brushable, or trowelable. Heavily bodied adhesive pastes and mastics are considered to be extrudable; they are applied by syringe, caulking gun, or pneumatic pumping equipment.

8.3.2 Selection of the adhesive for an application

To achieve optimum performance when bonding two materials, you must carefully plan every stage of the bonding process. The selection of an adhesive is a critical factor that will influence each step. The adhesive selection will be dependent primarily on:

- the type and nature of substrates to be bonded,
- the methods of curing that are available and practical, and
- the expected environments and stresses that the joint will see in service.

The adhesive selection process is difficult because many factors must be considered, and there is no universal adhesive that will fulfill every application. It is usually necessary to compromise when selecting a practical adhesive system. Some properties and characteristics that are desired of the adhesive will be more important than others, and a thoughtful prioritization of these criteria will be necessary in selecting an adhesive. One must first go about finding an adhesive that will satisfy the high priority requirements of the application. The lower priority 'requirements' may then need to be compromised to find the best fit.

One needs to optimize the entire bonding process and not merely a single part of the process. Consideration needs to be given at the same time to the substrates, joint design, surface pretreatment, quality control, application and curing methods, and other sub-processes (Fig. 8.9). Seldom is everything except the adhesive fixed. Very often, nothing is rigidly established, and it is up to the design engineer to not only choose the adhesive but also to decide on all of the peripheral processes. Alternative substrates, processes, etc., should always be considered. For example, a slight change in the application methods could open the door for consideration of an entirely new family of adhesives.

The criteria for the proper selection of an adhesive for a particular application can be a bit overwhelming unless it is approached with a disciplined process. Table 8.11 is a list of considerations that may be examined in order to make the appropriate adhesive choice for the application. The factors listed are necessary when considering an adhesive. Each one of these could be developed more fully, and there are several that probably could be added to the list. The goal is to have a methodology and not necessarily an inflexible, universal process.

The number of available adhesives to choose from can be very great. However, with a knowledge of the functional product requirements and possible production capabilities, one can start the process of selecting the adhesive system.

The various adhesive families overlap somewhat with regard to characteristics. Several adhesive families can provide similar properties. There is also a significant variation available within certain adhesive families. Specifying a 'polyurethane' adhesive, for example, is not sufficient. Virtually thousands of formulations are available within this family with different characteristics regarding end properties, application requirements, curing processes, and so forth.

Choose the appropriate adhesive in cooperation with an adhesive supplier. Once the general product requirements, joint design, and service requirements

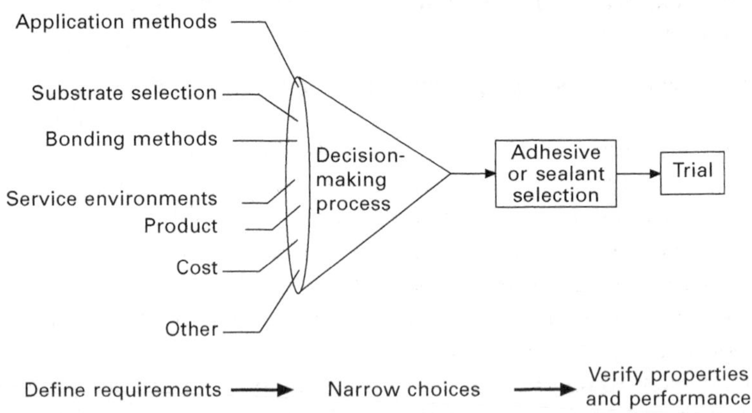

8.9 Adhesive selection requires a number of early considerations.

Table 8.11 Considerations when selecting an adhesive

Consideration	Importance
Properties of the substrate	The adhesive must wet the substrate surface. Thus, the adhesive must have a lower surface energy than that of the substrate. This can be accomplished by either modifying the adhesive or subjecting the substrate to a surface treatment.
Joint design and load	Most adhesives are better when the joint is designed so that they are under shear. Peel and cleavage stresses are notoriously harsh on adhesives. Toughness and flexibility are required when peel, impact, or thermal shock forces will be present.
Service environments	Certain polymeric systems withstand certain environments better than others. However, multiple environments (e.g., humidity and elevated temperatures) may provide unexpected problems. All adhesives should be tested in the service environment.
Physical properties of the adhesive	Often the elastic modulus will give an indication of the adhesive's value in certain applications. Tough, flexible adhesives are better in peel, impact, thermal shock, and when substrates of differing thermal expansion must be bonded. Brittle, rigid adhesives generally have better heat and chemical resistance, tensile strength, and creep resistance.
Application methods	The availability of production resources must be considered. Such parameters will determine the best form (film, paste, mastic, powder, liquid, etc.) required from the adhesive.
Assembly needs and rate of cure	Many adhesives will require fixtures or jigs until the adhesive cures to a 'handling strength'. The cost of such resources will often determine how much set or cure time can be acceptable. Some adhesives (e.g., cyanoacrylates, hot melts, pressure-sensitive adhesives) provide fast production.
Vendor selection	Reliability, price, willingness to meet demand and offer technical assistance, and training are all important characteristics in selecting a vendor. It is likely that several vendors are available to supply the needed adhesive.
Testing	Standard tests are suitable for comparing the relative characteristics of different adhesives, but they do little in determining the performance or longevity of the joint in actual service. Prototype testing under selected environmental conditions is generally required to insure an efficient product.

are known, adhesive systems with good durability potential can be selected. The desired form of the adhesive and method of application can then be chosen based on the formulation possibilities available within the class of adhesives chosen and on the availability of production equipment and scheduling requirements.

The tendency to over-specify the adhesive requirements must be avoided. Requirements for higher strength or greater heat resistance than is actually necessary for a specific application will exclude from consideration many formulations that may be equally optimal for the job, at lower cost, and with simpler production capabilities.

8.4 Bonding processes

8.4.1 Application of the adhesive

The method used to apply the adhesive or sealant can have as much to do with the success of the joining operation as the kind of material applied. The selection of an application or dispensing method depends primarily on the form of the adhesive: liquid, paste, powder, or film. Table 8.12 describes the advantages and limitations that are realized in using each of these four basic forms. Other factors influencing the application method are the size and shape of parts to be bonded, the areas where the adhesive is to be applied, and the production volume and rate.

The actual specification for such processes are dependent on the adhesive material that will be used, on the part design and its required end-use properties, and on the overall economics and production speeds required in the operation. The physical and chemical properties of the adhesive formulation will have a significant effect on the application methods that can be used and on the choice of equipment. Some of these effects are shown in Table 8.13.

Table 8.12 Characteristics of various adhesive application methods (Carroll, 1965)

Application method	Viscosity	Operator skill	Production rate	Equipment cost	Coating uniformity	Material loss
Liquid:						
• Manual brush or roller	Low to medium	Low	Low	Low	Poor	Low
• Roll coating, reverse, gravure	Low	Moderate	High	High	Good	Low
• Spray: manual automatic, airless, external mix	Low to high	Moderate to high	Moderate to high	Moderate to high	Good	Low to high
• Curtain coating	Low	Moderate	High	High	Good to excellent	Low
Bulk:						
• Paste and mastic	High	Low	Low to moderate	Low	Fair	Low to high
Solid:						
• Powder	–	Moderate	Low	High	Poor to fair	Low
• Dry film	–	Moderate to high	Low to high	Low to high	Excellent	Lowest

Table 8.13 How adhesive properties affect the equipment and process used for their application (Killick, 1983)

Property	Characteristics
Viscosity	High viscosity often requires low speed handling with complex high pressure systems. Mixing difficulties may arise when an extreme differential exists in the viscosities and the components. Pressure drops become critical. Pump priming becomes difficult.
Tack	Tack will sometimes have an effect similar to high viscosity. It makes materials difficult to move, usually hinders cleaning, and makes clean material cutoff in dispensing operations difficult.
Thixotropy	Usually thixotropic materials are easy to pump because the viscosity drops considerably when the material is moved or agitated. However, on occasion, pump pressure or shear action of a mixer will temporarily destroy the thixotropic characteristics of the material.
Presence of fillers	Many types of filler are abrasive and cause wear of piston cups, packing seals, and precision parts. The finer the filler particles, the less their effect on equipment will be. Some fillers settle so radically that they tend to pack in dead areas of valves, fittings, and seals, thereby making the use of dispersing equipment impractical. Compressible fillers, such as micro-balloons, make accurate metering extremely difficult.
Corrosiveness	The presence of acids, amines, and other corrosive elements in material systems requires the use of stainless steel parts. These are costly and are difficult to machine.
Presence of solvents	The action of solvents in material systems presents problems related to the elastomeric seals. There is no perfect elastomer in its resistance to solvents. Even fluorocarbon seals are not necessarily a suitable seal because of their cold flow tendencies and poor resistance to abrasion.
Temperature characteristics	Cooling of components may be necessary in order to extend the work life of the mixed system. Heating of individual components may be required to reduce viscosity, to improve control of the curing process, or to improve the flow characteristics of the mixed product.
Component ratio	In general, the wider the ratio of catalyst to base material, the more critical is the required accuracy of proportioning. Close ratio materials (e.g., 1:1 mix ratios) can tolerate mixing errors as high as ± 10%, but wide ratio materials (e.g., ratios greater than 20:1) may require ± 1%. This often entails more costly and complex equipment.
Material stability	Lack of stability (e.g., evaporation of solvent, reaction with moisture, dispersion instability) can create serious equipment problems and change the mixing proportion or ratio.
Pot life or working life	In general, long pot life materials do not present equipment problems. It is desirable to use materials with as long a pot life as possible consistent with the production requirements. Short pot life materials require the use of mixing equipment with small volume chambers, rapid ejection means, and a means for

Table 8.13 Continued

Property	Characteristics
	flushing the system. Certain materials have a tendency to cure progressively in thin sections on the surfaces of a mixer.
Hazards	Many components may be toxic and many are irritating to the skin. Other problems arise in connection with fire hazards or with the possibility of explosions. Generally, equipment suppliers have the means of minimizing such hazards.

Table 8.14 Main types of spray equipment

Type	Characteristics
Air spray	Utilizes air pressure to spray a fine mist of atomized adhesive on the substrate. Generally unsuitable for small parts. Maintenance is high due to nozzle clogging and overspray. Several passes may be necessary for thick coating build-up.
Hot spray	Adhesive is first heated permitting the use of high viscosity materials. Heavier coatings are possible than with air spray. Drying is accelerated due to evaporation of solvent in the spray.
Airless spray	Uses hydraulic pressure rather than air pressure. Saves energy and overspray. Can be applied both hot and cold.

Liquids

Liquids are the most common form of adhesive, and they can be applied by a variety of methods. Liquids have an advantage in that they are relatively easy to transfer, meter, and mix. They also tend to wet the substrate easily and provide uniform bond-line thickness. However, they have the disadvantages of sometimes being messy, requiring clean-up, and producing a relatively high degree of waste.

Spray, dipping, and mechanical-roll coaters are generally used on large production runs of textile parts. Spraying usually is evaluated against other methods on the basis of cost. Spraying applies the liquid resin quickly, reduces the required drying time through better evaporation of solvent, and is capable of reaching areas that are inaccessible to other manual application tools.

Spray bonding is used for applications which require high loft or bulk. Several spraying methods can be used to apply adhesives, including conventional air spray, hydraulic cold airless spray, hot spray, and hot airless spray. The main types of spray equipment are described in Table 8.14.

Spray methods can be used on both small and large production runs. The liquid to be sprayed is generally in solvent solution or water dispersion. Sizable amounts of product may be lost from overspray. Two-component adhesives are usually mixed prior to placing in the spray-gun reservoir. Application systems are available, however, that meter and mix the adhesive within the spray-gun barrel. This is ideal for fast-reacting systems, but guns must be thoroughly cleaned to avoid build-up of polymerized product.

When large flat surfaces and webs of materials are to be coated, mechanical-roller methods are commonly used to apply a uniform layer of adhesive via a continuous roll. Such systems are used with adhesives that have a long working life and low viscosity. Table 8.15 summarizes the roll coaters that are commonly used and several of these are illustrated in Fig. 8.10.

Roll coating via gravure rolls is used for print bonding. Print bonding is used for applications that require only a part of the area of the web to be bonded. Printing patterns are designed to enhance strength, softness, hand, absorbency, and drape.

In addition to roll coating methods, saturation bonding is used with liquid adhesives. Saturation bonding is used in conjunction with processes

Table 8.15 Common types of coaters

Type	Characteristics
Kiss	No back-up roll; coater is in light contact with substrate; contact is controlled by web tension. Best with adhesives having good flow characteristics and relatively slow drying rates.
Pressure or squeeze	Also called nip coaters; can provide greater penetration into open web structures. Often used to apply coatings to both sides simultaneously.
Reverse roll	Popular for web precision coating because of better control over thickness.
Two roll reverse	For additional control over adhesive coating; can handle wet film thicknesses from 1–20 mils.
Dip roll	Simplest of the basic types of coaters; used mostly with low viscosity adhesives; works well where impregnation or saturation is required.
Engraved roll	Provides accurate control over coating uniformity; best for extremely low adhesive weights.
Air blade	Similar to the kiss coater in that the adhesive is applied where the web touches it; possible to apply coating uniformly at high speeds; limited use with solvent-based adhesives.
Bead	Two rollers set at specific clearance; a rod between the bottom roll and the web makes a bead of adhesive that coats the web passing above it.
Knife spreader	Usually used to apply high viscosity adhesives to flexible webs; web is drawn between a precision knife edge and a rubber roller or flat bed plate.
Flow or curtain	Uses a head with a slot in the bottom through which the adhesive is poured or lowered onto the parts or surfaces as they pass beneath the head by a conveyor; provides materials savings and ability to coat irregular surfaces uniformly.

which require rapid and heavy adhesive addition. It is used, for example, on fabric applications which require strength, stiffness, and maximum fiber encapsulation. This is achieved by totally immersing the web in an adhesive bath and removing excess adhesive by vacuum or roll pressure.

A process called 'foam bonding' is used where it is necessary to apply adhesive at a low water content and high solids concentration. Air is mechanically agitated into the waterborne adhesive to cause foaming. Foam-

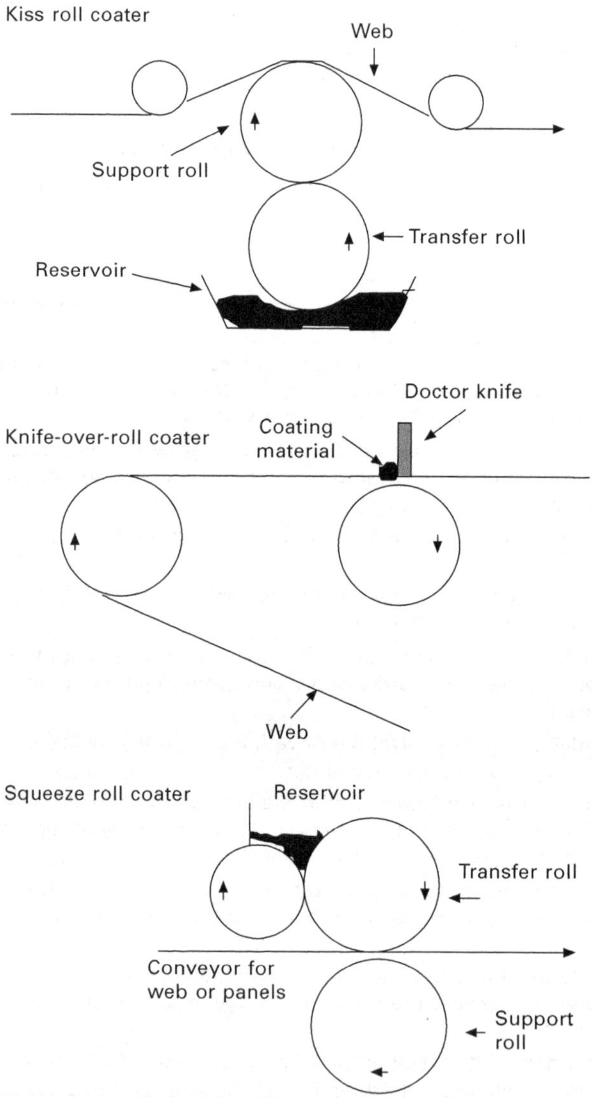

8.10 Methods of machine application of adhesive coatings (from Martin, 1970).

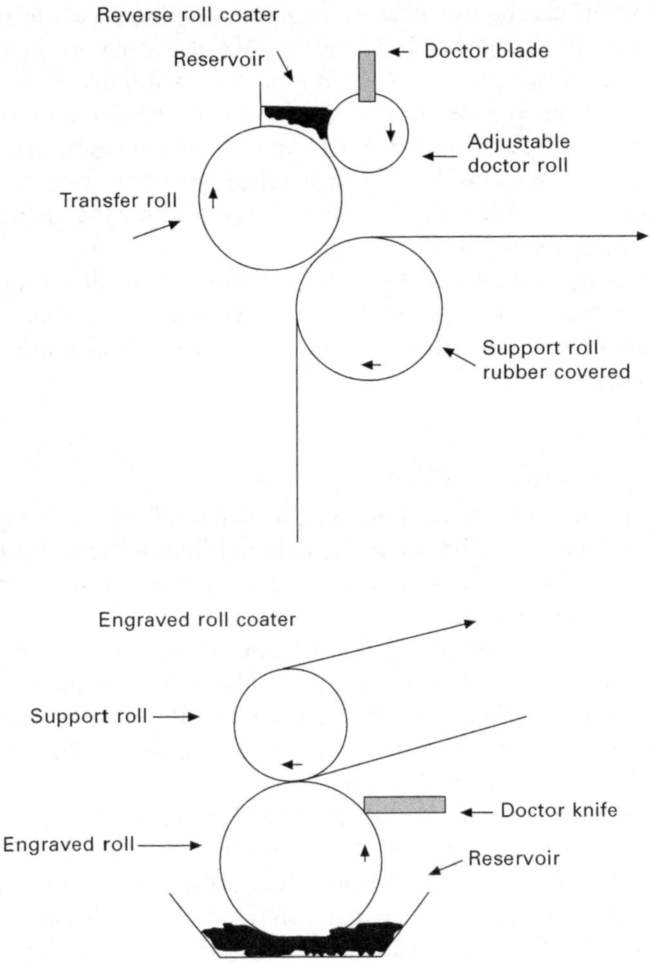

8.10 Continued

bonded products require less energy in drying, since less water is used. With foam bonding, the softness of the resulting product can be more easily controlled than with other bonding processes. The major disadvantage with the process is in controlling the foaming.

High viscosity liquids, pastes, and mastics

Bulk adhesives such as pastes or mastics are the simplest and most reproducible adhesives to apply. They produce heavy coatings that fill voids, bridge gaps, or seal joints. However, the process is limited to high viscosity materials. They can be in the form of high viscosity extrudable liquid or trowelable mastic.

These systems can be troweled on or extruded through a caulking gun. Little operator skill is required. Since the thixotropic nature of the paste prevents it from flowing excessively, application is usually clean, and not much waste is generated. With trowel application, the depth of notch and the spaces between them help regulate the amount of adhesive applied. Shallow, rounded and closely spaced notches are often used with lower viscosity adhesives to allow the lines of adhesive to flow together and form a continuous, unbroken film.

High viscosity liquids, pastes, and mastics are ideal adhesives for application by robots. Robotic application is commonly used in industries such as automotive and consumer products to increase quality and to reduce labor and material costs.

Powders

There are a number of applications for powdered adhesives. They can be applied to all sorts of textile products, although non-wovens are the most popular substrate. Often the powder is either used as a hot melt adhesive in solid form or in the form of a paste dot.

A thermoplastic adhesive powder can be applied onto a web by gravimetric feed and then bonded with heat and pressure. The process reduces waste and maintenance costs associated with liquid adhesives. It is most often used with non-woven articles. The major disadvantage is controlling the distribution of the particles.

For the bonding of non-woven fabric to itself or to other materials, usually a paste-like thermoplastic adhesive dispersion in water is used. These powders are generally dispersed in water and can be conveniently applied by a variety of processes in a predetermined pattern and density. The base polymer is generally a low VOC polymer such as a copolyamide or copolyester.

Once applied, the paste dots can be laminated with all types of textile materials under the influence of heat and pressure. The paste dot system is most effective when working with lightweight substrates such as non-woven interlinings, which have a low tolerance to high temperatures. These end-uses include:

- fusible interlinings for woven, knitted, and non-woven fabrics
- fusible dot coatings on breathable membranes (e.g., Goretex, Sympatex, Helsapor)
- print bonding on non-wovens
- anti-slip properties for automotive fabrics, tablecloth, etc.

B-staged (partially cured) thermosetting or thermoplastic powder adhesives can also be used as a hot melt type adhesive. They can be electrostatically applied like powder coatings or they can be applied gravimetrically by being dispersed onto a hot substrate surface. Heat and pressure must be applied for full curing and considerable care taken to obtain uniform thickness over large areas. If coating uniformity is poor, large variations of joint strength may result with powder adhesives. The main advantages of using powder adhesives are their cleanliness and minimal waste. Spills and overspray may be collected and reused if not contaminated.

Films

Adhesives are commonly available in film form (Fig. 8.11). Most of these tend to be thermoplastic, but reactive thermosetting films are also commercially used in structural applications. Adhesives applied in the form of dry films offer a clean, hazard-free operation with minimum waste and excellent control of film thickness. However, the method is generally limited to parts with flat surfaces or simple curves. Optimum bond strength requires curing under heat and pressure. Film material cost is high in comparison to liquids, but waste or material loss is the lowest of any application method.

Film adhesives are made in both unsupported and supported types. The carrier for supported films is generally fibrous fabric or mat. Film adhesives are supplied as heat-activated thermosetting or thermoplastic hot melt, pressure-sensitive, or solvent-activated forms. The type of dry film adhesive used will usually determine the technique by which it is applied.

- Pressure-sensitive adhesives available in the form of single coated or double coated films are tacky and bond instantly when hand pressure is applied.
- Thermoplastic adhesive films are normally positioned between the mating surfaces, and heat and pressure are applied to make the bonded joint; this joint can be processed very quickly.

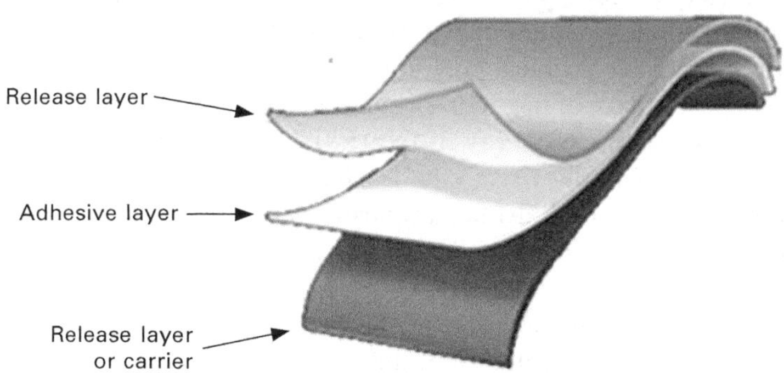

8.11 Adhesive hot melt films are commonly used to bond large pieces or to provide a method of bonding continuous textile substrate (roll or hot nip).

- Thermosetting adhesives in film form are applied between the two mating surfaces and the assembly is subjected to heat and pressure for a predetermined period of time.
- Solvent reactivatable adhesives in film form are reactivated by wiping with a cloth that has been dampened with solvent usually into a pressure-sensitive state.

Hot melt adhesives

Thermoplastic hot melt adhesives are generally used in non-structural applications where convenience and production speed are important. The main advantages of this process are that it is fast and free of solvent.

Hot melt equipment consists of hot spray devices (air and airless), wheel and extrusion dispensers, and heated handguns to handle preforms of various shapes. Hot melt materials and equipment are usually integrated into high volume production lines, and the equipment is very compatible with robotic application. Hot melt adhesive can also be applied to continuously laminated products.

Hot melt adhesives are available in a variety of solid forms. Usually they are supplied as billets and applied to the substrate through a hot melt glue gun. The hot melt application system is equipped with a temperature-regulated glue reservoir and insulated hosing to the gun. Once out of the gun, the adhesive solidifies quickly. A high degree of operator skill is necessary as the hot melt gels quickly. Hot melt adhesives that are stored in molten baths may require a blanket of nitrogen above the adhesive to eliminate the possibility of oxidation.

8.4.2 Surface preparation processes

A wide variety of fiber, yarn, and fabric parameters can affect the quality of adhesion. In addition, aqueous processing treatments carried out to finish the fabric or garment (e.g., bleaching, dyeing, printing, and chemical finishing) can modify the nature of the fiber surface and affect adhesion. Such treatments can modify the surface energy of the textile and interfere with the wetting and adhesion of an adhesive to the textile fiber (Fig. 8.12).

The removal of natural fiber impurities and sizes may be necessary to achieve optimal adhesion. In order to improve adhesion to both thermosetting and thermoplastic resin, fibers are usually subjected to controlled surface treatments. The main purpose of surface modification is to modify the chemical and physical structure of the surface to make it more amenable to adhesives but without influencing the fiber's bulk mechanical properties.

Thermoplastic materials often have a lower surface energy than thermosetting materials. Thus, physical or chemical modification of the

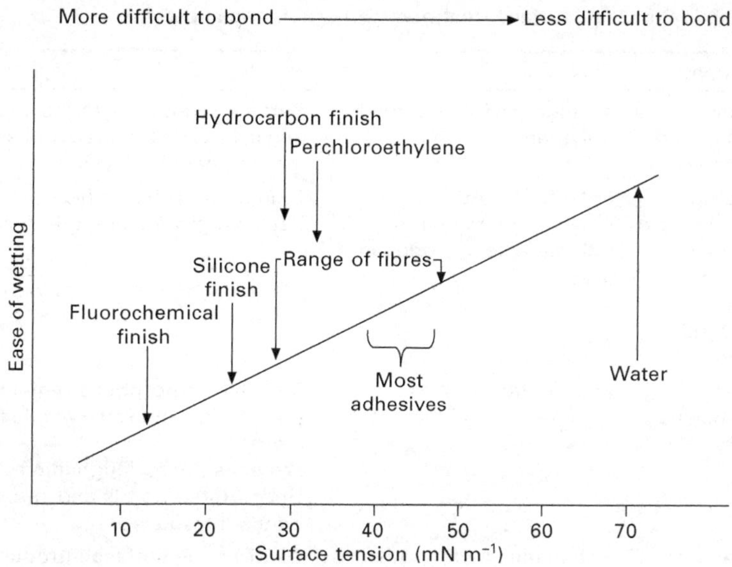

8.12 Effect of surface treatment on wetting (from Holme, 1999).

surface is necessary to achieve acceptable bonding. This is especially true of the crystalline thermoplastics such as polyolefins, linear polyesters, and fluoropolymers. Methods used to increase the surface energy and improve adhesion include the common surface treatments shown in Table 8.16. As with metal substrates, the effects of plastic surface treatments decrease with time, so it is important to carry out the priming or bonding as soon as possible after surface preparation.

The methods of surface modification of materials can be divided into four categories: mechanical, chemical, flame, and plasma. With textile fibers, chemical and plasma treatments are most widely used since the mechanical and flame treatment process will often degrade thin fibers.

Chemical pre-bond treating methods

Chemical methods have been used widely for treating textiles prior to bonding. The major disadvantage of chemical treating properties is that many of these processes use flammable or hazardous materials. Chemical pre-bonding treatments of textile fibers generally result in either:

1. controlled oxidation to remove the weak surface layer from the fiber or roughen the fiber surface, or
2. the production of carbonyl, carboxyl, or hydroxyl groups on the fiber surface which can then chemically bond to functional groups within the adhesive.

Table 8.16 Common surface treatments for polymeric substrates

Treatment	Plastic	Result
Solvent cleaning and abrasion	All high surface energy polymers	Surface roughness forms a greater effective bonding area and sites for mechanical interlocking
Oxidizing acids and acid mixtures: • Chromic • Nitric • Sulfuric • Formic	Polyolefins, ABS, polycarbonate, nylon, polyphenylene oxide, and acetal	Oxidation of the surface Reactive groups are introduced
Sodium naphthalene solution	Fluorinated plastics	Dissolves amorphous regions on the surface and removes fluorine atoms Provides micro-roughening Unsaturated bonds and carbonyl bonds introduced
Flame	Polyolefins, nylon, other low surface energy plastics	Oxidizes the surface introducing polar groups
Corona or electrical discharge	Polyolefins, polyethylene terephthalate, PVC, polystyrene, cellulose, fluorocarbons	Oxidation and introduction of active groups
Plasma	Nearly all low surface energy polymers including silicone rubber and other low surface energy elastomers	Crosslinking of the surface Surface oxidations with the formation of polar groups Grafting of active chemical species to the surface

Controlled oxidation has primarily been used on carbon fibers for use as reinforcements in composites. Chemical modification of the fiber surface to create reactive groups has more generally been used for improving the bond strength of organic synthetic fibers.

Plasma treating processes

Gas plasma treating processes have been developed for surface treatment of many polymeric materials. It is a dry, less hazardous process that is becoming a very common method of treating many low energy materials, including polyolefins, polytetrafluoroethylene, poly(ethylene terephthalate), and nylon. Relative bond strength improvements of ten to several hundred times are possible depending on the substrate and gas plasma. Low pressure plasma treatment has become a very popular treatment for small- to medium-sized parts that can fit into a vacuum container and for production volumes that are amenable to a batch type process.

Although providing excellent surface adhesion on many hard-to-bond and long shelf life plastics, conventional plasma treatment also has shortcomings. It requires a low pressure (partial vacuum) and, thus, the surface must be processed in a vacuum chamber. This generally necessitates a batch process. Continuous webs, films, etc., are difficult to process with plasma.

A newer process, atmospheric plasma treatment, has been developed that negates many of the disadvantages of low pressure plasma treatment.[1,2] With atmospheric plasma treatment the plasma is sustained at atmospheric pressure. It can be configured to work with a variety of feeding systems including continuous web. With atmospheric plasma, a high concentration of atoms and radicals are delivered to a surface to clean, remove, modify, or deposit materials of choice. The reaction chemistry depends on the gases fed to the plasma. These gases will depend on the substrate, and they can include oxygen, hydrogen, nitrogen, and carbon tetrafluoride. Atmospheric plasma has been used effectively on films, papers, foams, non-wovens, wovens, fibers, metals, and powders. Atmospheric plasma treatment equipment has been developed for any extrusion, lamination, or production line width. Operational speeds vary by substrate and can be up to 1,000 ft/min.

Adhesion promoters or coupling agents

Adhesion promoters or coupling agents are liquids that form a very thin (usually monomolecular) layer between the textile fiber and the adhesive. They form a chemical bridge between the fiber and the adhesive. These bonds are considered to be stronger than the internal molecular bonds within the adhesive. These new bonds also provide an interface region that is more resistant to chemical attack from the environment.

Specific coupling agents have been developed for bonding glass fibers to polyester resin, carbon fibers to epoxy resins, talc particles to nylon, and reinforcing tire cord to rubber tire compounds. In these applications, adhesion promoters not only improve the bond strength and permanence of the interface but also increase the physical properties of the resulting bulk material. Bulk properties, such as tensile strength and modulus, can be significantly improved. Virtually all glass fibers used in glass fiber reinforced composites are silane treated, and the resistance to deterioration by moist environments is greatly enhanced in this way.

Usually, these materials are grouped by their chemical composition, such as silane, titanate, and chlorinated polyethylene. The selection of a primer or adhesion promoter is normally less of a problem than selecting the adhesive

[1]Atomoflo, Surfix Technologies, LLC.
[2]Plasma3, Enercon Industries Corporation

since the adhesive supplier can usually offer a complete package or make recommendations.

Other adhesion-promoting surface treatments

Several other surface treatments have been reported to enhance adhesion of low energy polymers. These have generally used ultraviolet (UV) or ultrasonic energy sources. Recently, UV-laser irradiation has been used to modify fiber surfaces such as polyaramids (Kevlar), ultra-high strength polyethylene fibers, and others. UV radiation results in the incorporation of oxygen on the fiber surface. This makes the fiber more polar and wettable. The surface roughness has also been noticed to increase.

An Austrian research organization, at the Institute for Environmental Physics, Johannes Kepler University, Linz, Austria, has developed a new method in which laser radiation is used to modify fluorocarbon surfaces so that adhesives and coatings will bond strongly. The effect is achieved by irradiating the surface of fluorocarbons with UV light at extremely low wavelength (172 nm). The plastic substrate is modified only in those places where it will be bonded or coated. This is an advantage over other surface preparation processes where the entire part must be treated. The unmodified and bulk plastic retains its major properties. This is considered to be of special advantage in medical engineering, where the application of UV-irradiated fluorocarbons significantly facilitates the use of artificial blood vessels, cardiac valves, grafts, and artificial skin. The new technology ensures tissue adhesion and compatibility of the surface. The method is also being explored in the textile and the microelectronics industries, where surface modification of predetermined areas is desired.

8.5 Trends in adhesive types

Textile bonding is a rapidly growing and ever-changing field. Many of the adhesives and processes that are well accepted today were unknown a decade ago. Similar dynamic change is certain to occur in the next decade. The evolution of adhesive systems has been primarily due to regulatory and economic pressures as well as technical requirements. These forces have tended to focus end-users away from solvent-containing adhesives to waterborne, hot melt, and moisture reactive (100% solids) liquid systems.

8.5.1 Solvent-based systems

Solvent-based systems may be considered the root base for any trend line in adhesive development. They represent the oldest technology and the widest range of chemistries. Solvent-based adhesives have been valued through the years for providing fast drying conditions and good adhesive properties especially durability. The major disadvantages of solvent-borne adhesives, and the issues triggering their replacement, are regulatory restrictions on solvent emissions, waste disposal, and factory vapor levels. Many solvent systems are also flammable and require health and safety management processes. Another disadvantage of solvent-based adhesives is their cost. The solvents are only in the formulation to reduce viscosity and allow good coating characteristics. Once the adhesive is applied, the evaporated solvent must either be recovered or treated. Solvents are generally of petrochemical origin and as a result have a high price and price volatility associated with crude oil.

Table 8.17 summarizes the characteristics of the major solvent-based adhesives used in the textile industry. It is a major goal to reduce dramatically the use of solvent-based adhesives to only those applications where there is no adequate substitute.

8.5.2 Waterborne systems

Waterborne textile adhesives consist of a dispersion of very small particles of polymer in an aqueous medium. Waterborne adhesives have become increasingly popular for regulatory compliance and economic reasons. Solvent-based adhesives are generally environmentally unfriendly and are the subject of regulatory restrictions. In addition, many solvent-based adhesives are flammable, and their fumes can be harmful if inhaled. Solvents are also more expensive than water due primarily to the petrochemical nature of the solvent.

High molecular weight products can be made and dispersed in water to provide much higher solids content than solvent-borne adhesives. Waterborne adhesives can have a solids content up to 65% by weight, whereas solvent systems are only 20–30%. This provides the waterborne systems with high coating build-up for a relatively low viscosity system. It also reduces shipment costs in that more adhesive binder can be shipped per unit volume with waterborne systems.

The need for additional energy and/or time to dry waterborne coatings is one of the major drawbacks of switching from a solvent-based to a waterborne adhesive. As a result, significant effort has been devoted to determine the mechanism of drying and possible optimization strategies.

Other serious possible disadvantages of waterborne adhesive coatings include the following:

- swelling of water-sensitive substrates such as fabrics
- corrosion of ovens, drying towers, etc., due to water vapor
- poorer performance once cured, especially in moist environments, due

Table 8.17 Characteristics of common solvent-based adhesives used in the textile industry

Adhesive	Special characteristics	Usual adherends	Price range
Cellulose acetate, cellulose acetate butyrate	Water-clear, more heat resistant but less water resistant than cellulose nitrate. Cellulose acetate butyrate has better heat and water resistance than cellulose acetate and is compatible with a wider range of plasticizers.	Leather, fabrics, paper, plastics, wood, glass	Low
Cellulose nitrate	Tough, develops strength rapidly, water resistant. Bonds to many surfaces. Discolors in sunlight. Dried adhesive is flammable.	Cloth, glass, metal, plastics	Low
Vinyl vinylidene	Tough, strong, transparent and colorless. Resistant to hydrocarbon solvents, greases, and oils.	Particularly useful with textiles, also porous materials and plastics	Medium
Acrylic	Good low temperature properties, poor heat resistance, excellent resistance to UV. Clear and colorless.	Textiles, paper, metallic foils, plastics	Medium
Polyurethane	Excellent cohesive and adhesive properties. Good toughness and abrasion resistance.	Fabric, elastomers, metal, plastic	Medium to high
Styrene block copolymers	Similar to natural rubber, but better aging properties. Useful at both high and low temperatures (–40°C to 70°C). Blend of both high peel strength and high cohesive properties.	Textiles, disposables, foils, plastic films, foam, elastomers	Low
Neoprene	Superior to other rubber adhesives in most respects. Fast strength development. High temperature and weathering resistance. Resistant to UV, mild acids, and oils.	Fabric, leather, elastomers, metal, building materials	Medium to high

to the nature of the base polymer and the additives employed in the waterborne formulation

- freezing and destabilization of stored product due to exposure to low temperatures.

Modern waterborne adhesive formulations have provided significant strides in overcoming the last two objectives. However, in certain applications, such as shoe manufacturing, waterborne adhesive systems are still not on a par with solvent-based systems from a performance standpoint. Durability in moist environments is especially a concern because most waterborne adhesive formulations are formulated with moisture-sensitive dispersants, surfactants, and emulsion aids.

There are many types of waterborne laminating adhesives with widely differing applications and performance properties. Common waterborne textile adhesives include natural rubber latex and polymers, copolymers of acrylic esters and acids, vinyl and vinylidene chloride, vinyl acetate, and chloroprene. Other, less common, adhesives include starches, dextrin, and other polymers of natural origin. The characteristics of common waterborne adhesives used by the textile industry are given in Table 8.18.

Although waterborne adhesives generally have poorer moisture and thermal resistance than their solvent-based counterparts, the introduction of

Table 8.18 Characteristics of common waterborne adhesives used in the textile industry

Adhesive	Special characteristics	Usual adherends	Price range
Natural rubber latex	Excellent tack, good strength. Surface can be tack-free and yet bond to similarly coated surface. Poor aging and UV resistance. Good moisture resistance.	Felt, fabric, paper, metal, elastomers	Medium
Polyvinyl acetate	Versatile in terms of formulation and applications, but rather rigid bonds. Odorless, good resistance to oil, grease, acid. Fair weather resistance.	Paper, wood, fabric and most porous substrates	Low
Vinyls	Odorless and fungus resistant. Excellent resistance to grease and oils. Good bond strength.	Paper, wood, fabric and most porous substrates	Medium
Acrylic	Good low temperature properties, poor heat resistance, excellent resistance to UV. Clear and colorless.	Textiles, paper, metallic foils, plastics	Medium
Polyurethane dispersions	Excellent cohesive and adhesive properties. Good toughness and abrasion resistance.	Fabric, elastomers, metal, plastic	Medium to high
Neoprene latex	Superior to other rubber adhesives in most respects. Fast strength development. High temperature and weathering resistance. Resistant to UV, mild acids, and oils.	Fabric, leather, elastomers, metal, building materials	Medium to high

crosslinkers into the formulation has enabled waterborne adhesive to meet many of the performance criteria required and narrowed the performance gap between solvent-based and waterborne adhesives. The addition of crosslinkers, however, will reduce the working life of the adhesive, and this may be a barrier to its use.

Acrylic emulsions offer a low cost adhesive with moderate performance properties. They are tremendously versatile due to the large number of different monomers and resins available. Acrylic waterborne laminating adhesives can provide bonds ranging from flexible and tough to hard and rigid depending on the formulation. Acrylic adhesives inherently have very good UV and oxidative stability and are generally preferred for outdoor applications.

Although acrylic emulsions can be used as laminating adhesives directly, they are often formulated for specific applications. Tackifiers, for example, are added to provide improved peel strength especially to low energy surfaces as shown in Fig. 8.13. For bonding applications, the acrylics can also be formulated with varying degrees of tack and flexibility. Generally the polymers with lower glass transition temperatures are used to manufacture flexible laminates. Examples include bonding printed vinyl film to fabric for wallpaper and vinyl to foam for furniture covering.

Typically, waterborne polyurethane adhesives also have high performance properties. Waterborne polyurethane adhesives have been developed for standard laminating equipment and fast line speeds. They generally provide excellent adhesion to a wide range of flexible substrates and, when crosslinked, have strength greater than the substrates to which they are attached. The primary disadvantage of polyurethane adhesives is their relatively high cost.

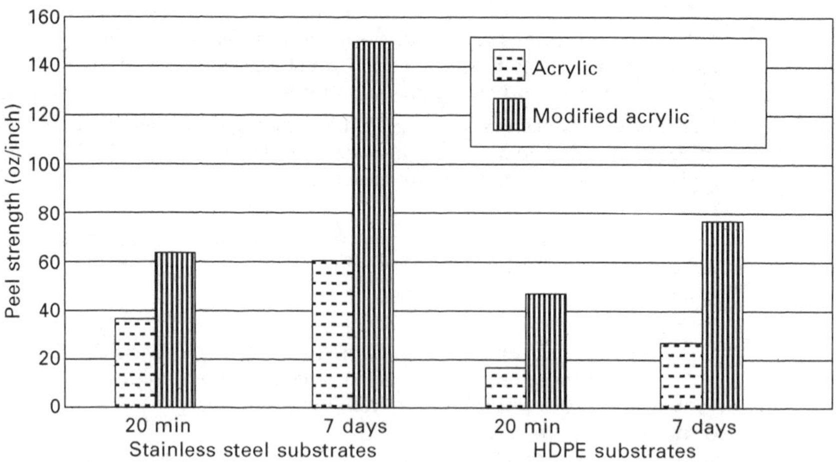

8.13 Relative peel strength of an acrylic latex adhesive that is unmodified and modified with tackifier.

Polyurethane dispersions formed mainly of polyester polyols and aliphatic isocyanates are preferable for textile adhesives because of their higher degree of adhesion and resistance to UV light. By varying the product's molecular weight, one can produce a wide range of end properties. Low molecular weight polyurethane dispersions yield adhesive with high hot tack and very fast crystallization. The crystallization rate is an important parameter for thermoplastic polyurethane adhesives as a higher degree of crystallization provides a faster strength development rate.

Since waterborne polyurethanes are primarily linear polymers, their performance can be greatly improved by the addition of crosslinking agents. A variety of crosslinkers are available. These systems differ significantly in reactivity, pot life, and the effect on final properties. Although crosslinkable waterborne acrylic and polyurethane adhesives are often dried at temperatures up to 95°C, the majority of the crosslinking reaction takes place in the finished product at ambient temperatures. There is generally sufficient integrity after the laminating process so that slitting or die cutting can be accomplished immediately after lamination even though the crosslinkers have not fully reacted.

Another common waterborne adhesive is a vinyl acetate–ethylene (VAE) copolymer. VAEs offer significant improvement over the older polyvinyl acetate adhesives. Incorporation of ethylene into the vinyl acetate backbone decreases the glass transition temperature and provides superior flexibility and adhesion.

A more recent development has been vinyl acetate acrylic (VAA) copolymers. VAA adhesives have excellent strength, setting speed, compatibility with common formulating raw materials, and emulsion stability (some are even freeze/thaw stable). VAA copolymers can be made to be soft and pressure-sensitive or hard and tack-free. VAAs generally offer lower glass transition temperature than VAEs, which results in improved flexibility and better low temperature properties. The incorporation of carboxyl functionality provides reactive groups for crosslinking.

8.5.3 Hot melt adhesives

Hot melt adhesives are gaining in popularity due primarily to their lower processing costs and regulatory requirements. They also provide the fastest method of bonding, and hot melt equipment occupies a much smaller factor footprint than other methods of bonding. However, they must be carefully chosen. For good durability, their softening and melting point must be well above their service temperature. As with other adhesives, there are many different types of hot melt base polymers and hot melt formulations available, each with advantages and disadvantages.

In addition, there are many forms of hot melt adhesives (pellets, blocks, ropes, sticks) and application patterns (dots, spiral spray, random spray, continuous lines, etc.). The form of the adhesive is important. For example, hot melt films are ideal for lamination of wide webs and discontinuous coating is preferred for applications requiring soft hand. The form of the hot melt adhesive is also an important cost consideration. Hot melt adhesive powders are available in most chemical types and in varying particle size ranges. There are a number of processes by which hot melt adhesive powders can be applied. A common method in the textile industry is the paste dot process.

Chemical types of hot melt adhesive include polyolefins, polyamide, polyester, and polyurethane. These are described in Table 8.19. Their serviceability must be balanced with ease of application. There are copolymer variants of each type, allowing a wide range of application and performance properties. Although nonreactive hot melt adhesives have been available for

Table 8.19 Properties of hot melt textile adhesives

Adhesive	Special characteristics	Usual adherends	Price range
Low density polyethylene (LDPE)	Low cost, inert	Used for fusible interlinings, automotive carpets	Low
High density polyethylene (HDPE)	Low cost, inert, higher temperature resistance than LDPE	Used in carpets and for fusible interlining, shirt collars and cuffs	Low
Ethylene vinyl acetate (EVA)	Good tack and adhesion, relatively low cost	Used in footwear and to bond leather and paper, good flexibility	Low
Polyamide and copolymers	Wide range of properties and melting points, better heat resistance than polyethylene and EVA but more expensive, some reactive types available	Used in garments and applications where solvent resistance is required (e.g., dry cleaning)	Medium to high
Polyester and copolymers	Wide range of properties, temperature resistance, good durability, high cost, some reactive types available	Automotive interior trim, specialty applications where heat and chemical resistance is important	High
Polyurethane	Good adhesion to many substrates, good durability, good flexibility and toughness, expensive, reactive types are available	Used in automotive interior trim, laminates for protective clothing, shoe manufacture, reactive hot melt polyurethane used in structural applications	High

many decades and are widely used in many applications, they have certain performance limitations, such as poor heat resistance, water or solvent permeation, and creep. These limitations generally prevent their use in many critical or structural bonding applications.

Hot melt adhesives are used for bonding fusible interlinings, in seam and hem sealing, fabric lamination, and a variety of other textile applications. The melting point of hot melt adhesives must be above the minimum temperature that the application will see in service. This could provide a problem in that many textile substrates are heat sensitive, and conventional hot melt adhesives must be applied at high temperatures. To avoid this problem, formulators have developed hot melts with lower application temperature, sometimes known as 'warm melts'. The most popular adhesive in this category is polyurethane thermoplastic and reactive polyurethane adhesives described below.

8.5.4 Moisture cured polyurethanes (reactive liquid and hot melts)

As noted above, many solvent-based and waterborne adhesive systems are capable of crosslinking to provide improved heat and chemical resistance. These most notably are acrylics and polyurethanes where a crosslinking agent is added to the adhesive immediately before application. In order to reduce the cost and potential problems associated with metering and mixing, reactive one-component adhesives have been developed. In the textile industry, this technology centers on polyurethane compositions that can be crosslinked via reacting with the moisture in the air (relative humidity).

Moisture-cured polyurethanes allow strong adhesive bonds with relatively little energy expense since they do not require heating. They are available as liquids or as solid hot melts. As liquids they can be applied by spraying or gravure roller coating. Polyurethanes form durable and flexible adhesive bonds with good drape and abrasion resistance. For this reason they are used in breathable garments, protective clothing, and car seat covers. However, these adhesives are more expensive compared to the others, and control of moisture exposure during storage and application is critical. These adhesives have been available for many years, but it is only in the last few years that significant improvements in open time and initial tack have been achieved to increase their usefulness.

More recently, reactive hot melt adhesives have become available. They have many of the desirable processing characteristics of conventional hot melts, such as no solvents present, no mixing requirements, and immediate green strength. Their unique feature, however, is that once applied they subsequently react to become a crosslinked, thermosetting polymer with enhanced physical properties. The most common reactive hot melt system is a polyurethane based upon moisture-curing technology. These are sometimes known as HMPUR (hot melt polyurethane rubber) or 'warm-melt' adhesives (to emphasize that these systems are applied at lower temperatures than conventional, non-reactive hot melts).

By being capable of curing to a thermosetting material, reactive hot melt adhesives eliminate or greatly reduce many of the performance problems associated with conventional hot melt adhesives. In addition, reactive hot melt adhesives have the ability to bond many different substrates, have lower application temperatures, and have longer service lifetimes. The advantages of these systems, however, do not come without some cost – generally a high material price and more difficult application processes. Table 8.20 summarizes the advantages and disadvantages in using reactive hot melt adhesives.

8.5.5 Driving factors for development

In addition to the move toward water-based adhesive products and away from solvent types, several trends have been noticeable in the textile industry over the past decade. These include:

Table 8.20 Advantages and disadvantages of hot melt polyurethane rubber (HMPUR) adhesives

Advantages	Disadvantages
➢ Lower application temperatures than conventional hot melts allow for reduced energy costs and lower safety concerns. ➢ Low or no VOCs. ➢ No mixing required (single component). ➢ Immediate green strength and fast development of cohesive strength (crosslinking). ➢ When compared to conventional hot melts: • improved high temperature resistance • enhanced moisture and solvent resistance • improved adhesion to the substrate • improved creep resistance • tough and flexible. ➢ Significant freedom in formulation ability. ➢ Tensile strengths ranging from 2,000 psi (1,400 N/cm^2) to 3,500 psi (2,400 N/cm^2). ➢ Elongation at break from 290 to 750%.	➢ Higher material cost. ➢ Full cure takes several days. ➢ Open time may be short (seconds) depending on the substrate. ➢ Moisture sensitivity could cause material to 'set' in the applicator necessitating purging and cleaning process. ➢ Special application equipment may be needed depending on the form and formulation of the product – to maintain moisture-proof environment until application.

- the use of adhesives (nonblocking hot melts and reactive 'B' staged adhesives) for off-line combining,
- heat sealed multilayer laminates,
- flame retardant adhesive systems (especially for clothing, aircraft, and household furnishings),
- high strength contact cements (for shoe manufacture and installing liners to outerwear).

The most significant volume trends, however, are the increasing use of hot melt adhesives and moisture-curing polyurethanes as replacements for either solvent- or water-based systems.

However, higher standards of performance in modern applications are pushing the boundaries of adhesives and sealants. This, of course, requires the development of newer and improved application and performance properties. But perhaps the leading driver is for 'performance on demand'. Significant demands have been placed on systems to be fast, easy, and low cost. This has led to several new generations of adhesives and sealants. Faster processing and efficiency are driving innovation in many businesses, and the adhesive industry is no exception. The electromagnetic spectrum can be looked upon as a roadmap for much of this activity, because many areas of the spectrum are used to initiate and accelerate adhesive cure.

Light-curing adhesives based on cyanoacrylate and acrylic polymers have become prime candidates for fast adhesive bonding processes given their ability to cure within milliseconds. Light-curing adhesives contain photoinitiators which absorb light to start their deterioration, which also starts the curing reaction. The free radicals that are produced during the photoinitiators' deterioration are very reactive and, therefore, provide an extremely fast speed of cure. Depending on the intensity and range of the light source, full adhesive cure can easily be achieved within seconds as long as the optimal light wavelength is present for the specific photoinitiator used.

Just as the raw materials that go into adhesives and sealants are changing, so are the materials that these products must bond. In the textile industry, new fibers have been developed that have increased the need for higher bond strength, and often these fibers are difficult to bond to because of their low surface energy or other properties. Examples of these are ultra-high molecular weight polyethylene, aramid (Kevlar), and carbon fibers. Flame retardants in protective clothing also could interfere with adhesion. These newer substrates provide significant challenges to the adhesive formulator and end-user. They have necessitated the development of advanced formulations and prebond surface treating processes.

8.6 References

Bikerman, J.J. (1961), *The Science of Adhesive Joints*, Academic Press, New York.

Carroll, K.W. (1965), 'How to apply adhesives', *Production Engineering*, Nov. 22.

Court, R. (2004), 'What treatments exist for altering the surface energy of a material?', TWI World Center for Materials Joining Technology, www.twi.co.uk, TWI Ltd., Cambridge.

Fourche, G. (1995), 'An overview of the basic aspects of polymer adhesion: Part I: Fundamentals', *Polymer Engineering and Science*, 35, 957–967.

Fries, J.A. (1983), 'Adhesive for fabric lamination', *Journal of Coated Textiles*, 12, 174–186.

Holme, I. (1999), 'Adhesion to textile fibres and fabrics', *International Journal of Adhesion & Adhesives*, 19, 455–463.

Killick, B.R. (1983), 'Dispensing high viscosity adhesives and sealants', in, G.L. Schneberger (ed.), *Adhesives in Manufacturing*, Marcel Dekker, New York.

Martin, R.A. (1970), 'How to apply adhesives', *Materials Engineering*, July.

Nick, D.P. (2003), 'Adhesive use in textile bonding and laminating', presentation at IFAI Expo 2003.

Petrie, E.M. (2007), 'Theories of adhesion', in *Handbook of Adhesives and Sealants*, 2nd edn, McGraw-Hill, New York.

Possart, W. (1988), 'Experimental and theoretical description of the electrostatic component of adhesion at polymer/metal contacts', *International Journal of Adhesives and Adhesion*, 8(2), 77–83.

Suo, S. and Ooij, W.J. (2002), 'Surface modification of textile fibers for improvement of adhesion to polymeric matrices: a review', *Journal of Adhesion Science and Technology*, 16(13), 1715–1735.

9
Adhesive bonding of textiles: applications

E. STAMMEN and K. DILGER, Technical University
Braunschweig, Germany

DOI: 10.1533/9780857093967.2.275

Abstract: This chapter provides an overview of adhesive bonding processes and products and their various applications in the textiles industry. Surfaces that can be bonded show broad variation in their chemistry and structure; consequently, adhesives used for bonding processes must be designed and selected according to their specific application. Furthermore, there is an increasing demand for adhesive joints to fulfil multiple additional functions. This chapter shows the versatility of adhesive bonding in everyday textile products and discusses state-of-the-art possibilities and future trends and requirements.

Key words: textile structure and surface, adhesives (viscous, film, powder), application processes, properties of adhesive in textiles, applied test specifications.

9.1 Introduction: textiles and adhesive joining

A wide range of natural and man-made synthetic fibres is available on the global market. These fibres are employed in a number of fields, including apparel and home textiles, automotive applications, geotextiles, technical textiles, sanitary products and many more.

Adhesive bonding is a good alternative for joining textile fabrics which cannot be satisfactorily joined by other techniques. The integration of an adhesive bonding step in two-dimensional production lines is easy and can reduce costs. Adhesive bonding is therefore growing in importance in all applications, both in inline and batch processes. This has led to a demand for a flexible bonding technology with the possibility of incorporating additional functions into the textile product.

9.1.1 Materials

Textile fabrics are made from fibres from different sources, natural or chemical. The chemical base of the surface to be bonded is therefore subject to significant variation, and covers the whole range of possible physical and chemical interactions (Kastner, 2008). The nature of the fabrication and textile finishing processes means that every surface to be adhesively bonded, whether synthetic or natural, must be closely examined. The chemical base, resulting surface energy and macroscopic structure all influence adhesive bonding applications.

Natural fibres are those of vegetable (e.g., cotton, cellulose), animal (e.g., wool, silk) or mineral (e.g., glass, carbon) origin. Vegetable and animal fibres vary in thickness, quality and homogeneity, which may cause problems in adhesive bonding, where a defined surface is needed. Synthetic fibres are split into two groups: natural polymer fibres and synthetic polymer fibres. The first group is cellulose derivates, made by so-called acetate or viscose processes (e.g., Lyocell®, Tencell®, Modal® and Rayon®). Synthetic polymer fibres are mainly petroleum-based. This group includes basic fibres such as acrylic (e.g., Dralon®), aramid (e.g., Kevlar®, Nomex®), polyamide (e.g., Nylon®), polyolefin and polyester (e.g., Trevira®, Diolen®), but also block-copolymers from polyurethane and polyethyleneglycole (e.g., Spandex®, Elastan®). There is a controlled fabrication process for the basic synthetic fibres, and the properties of the surfaces created should be carefully defined.

9.1.2 Structure and surface

The macroscopic surface structure of textile fabrics can vary, depending on the fabrication process and the end-use requirements. Fabrics are usually divided into the following types (Kastner, 2008):

- knitted
- woven
- non-woven
- microfibre.

Knitted fabrics are more elastic than woven or non-woven textiles. They are commonly used in clothing and textiles that require changes in shape. Knitted fabrics can stretch, depending on their material and knitting pattern, to up to 500% of their original size. This stretching behaviour must be taken into account when considering bonding possibilities for knitted textiles. Typical surfaces of knitted textiles are shown in Fig. 9.1a-c. It is clear that the structure and topography of the filaments and the resulting textiles differ between fabrics, according to the origin of the fibre (natural, synthetic). Good mechanical adhesion is possible with knitted fabrics because of their topography; however, the lightness of the textile can be disadvantageous, as viscous, low molecular adhesive systems can migrate more easily through the fabric.

A woven fabric is formed by weaving and has a more dense structure than a knitted fabric. Weaving is a textile craft in which two distinct sets of yarns, called the warp and the weft, are interlaced to form the textile. A woven fabric only stretches in the bias directions, unless the threads are elastic. Typical woven surfaces are shown in Figs 9.2 and 9.3. The plain

(a)

(b)

(c)

9.1 SEM pictures of knitted fabrics, comparable resolution (source: ifs, TU Braunschweig, Germany): (a) bleached and dyed cotton – visibly are loose fibres on the textile surface; (b) polytetrafluoroethylene (PTFE), EmmiTex® by Filtec, Düren (Germany) – visibly an even filament surface and a loose knitted structure; (c) polyamide, with parts of Spandex® – observable are contaminations between the filaments.

9.2 Laser scanning microscopy of polyester, plain weave (source: ifs, TU Braunschweig, Germany).

9.3 SEM picture of woven fabric, commercial cotton bed linen, satin weave (source: ifs, TU Braunschweig, Germany).

weave (Fig. 9.2) is the simplest of the three basic weaves. The satin weave (Fig. 9.3) is characterised by four or more weft yarns floating over a warp yarn, or vice versa, four warp yarns floating over a single weft yarn. Single floating fibres on the surface cause problems when the textile is bonded by adhesives, as the floating fibres are easy to remove, leading to a decrease in the strength of the final joint. Twill wovens have a pattern of diagonal parallel ribs. The weft thread is passed over one or more warp threads, and then under two or more warp threads and so on, with a 'step' or offset between rows to create the characteristic diagonal pattern.

Non-woven fabrics are defined as sheet or web structures bonded together by entangling fibres or filaments mechanically, thermally or chemically. They typically lack strength unless they are densified or reinforced by a backing. Non-wovens are flat, porous sheets, often containing a certain percentage of recycled fabrics (see Fig. 9.4) and oil-based materials. They provide specific functions such as absorbency, liquid repellence, stretch, softness, strength, filtering and sterility. Care must be taken to ensure that, when bonding non-wovens with adhesives, these functions are not destroyed by the joint.

A special topography is generated on the surface of a fabric when filaments with smaller diameters are used; microfibres are synthetic fibres such as polyesters, polyamides or a combination of these that measure less than one denier. The shape, size and combination of the polymers are selected for their specific characteristics like softness, durability, absorbency, breathability, filtering and water repellence. Microfibres are used to make woven, non-woven and knitted textiles.

9.1.3 Surface effects

In textile bonding, the topography of the fabric, along with the fibre properties and geometric factors such as yarn diameter, twist and fibre compactness, is

9.4 Commercial protective non-woven fabric with parts of recycling fibres (source: ifs, TU Braunschweig, Germany).

an important factor in creating good adhesion. However, the chemical surface behind the topography is not necessarily homogeneous; mostly processing chemicals and finishing products are used to modify or cover the original base material. Natural impurities are present on the surface of natural fibres, too, such as cotton wax or wool grease.

In order to efficiently manufacture yarns and fabrics of high quality, the fibres must be processed at high speed and with an efficient process control. Fibre lubricants are useful, when applied to the surface in order to avoid the build-up of undesirable friction or static electrification. Some commonly-used lubricants are aqueous formulations with lubricant, antistatic agents, emulsifier, water and biocide (Holme, 1999). Some fibres are washed in aqueous detergent solutions or solvents to remove applied fibre lubricants or natural contaminants. This procedure must be carried out uniformly; otherwise the surface will no longer be homogeneous, which can impair satisfactory adhesion.

To finish the fabric, different aqueous processing treatments can be carried out, such as bleaching, dyeing and finishing with optical bleach and/or softener. These modifications improve the visual and technical appearance of the textile, but also modify the electrochemical surface and affect the wetting, spreading and wicking of adhesives. The Washburn method is used for measuring the influence of such modification on a textile. This method is an established means of measuring the critical surface energy of porous substrates. Table 9.1 shows surface energies of cotton textiles with different kinds of finishing. The type of finishing selected obviously has a significant effect on the fibre surface (AiF, 2011). In washed cottons, the surface tension is low, but after treatment the surface tension increases dramatically. These results correspond to those discussed in Holme (1999: 458).

It is well known that low values of critical surface energy can lead to difficulties in wetting, spreading and wicking of adhesives, and that higher

Table 9.1 Surface energy of cotton textiles with different finishes (source: ifs, TU Braunschweig, Germany)

Substrate	Surface energy (total) (mN/m)	Dispersive component (mN/m)	Polar component (mN/m)
Cotton, washed	25.7	14.4	11.4
Cotton, washed, dyed	39.6	34.2	5.5
Cotton, bleached	41.8	39.3	2.4
Cotton, optical bleach	43.2	41.2	2.0
Cotton, bleached, silicone-finishing	28.6	23.2	5.5
Cotton bed linen, coloured print, finishing unknown	26.9	16.3	10.6

values are more suitable for effective adhesive bonding. A general overview of surface energy measurements is given by Mittal (1993). Additional physical pre-treatments of the textile surface such as corona discharge or atmospheric plasma (Ryu *et al.*, 1991), with or without additional process gas or even polymer deposition (Kan *et al.*, 2010), can modify the surface for – in the best case – improved wetting (Karahan and Ozdogan, 2008).

Additional functions of modern textiles, such as breathability and water resistance, have also been developed through surface modifications. These additional functions will become increasingly important for the future of textiles, particularly in areas such as sportswear. Coatings with polytetrafluoroethylene or hydrophobic treatments have a significant effect on the surface energies; consumer demand for functional textiles will influence the future of textiles and will lead to greater variation in the surfaces to be bonded. Overall, however, the most important aspects to consider in creating an effective adhesive joint are the selection of an adequate adhesive (base) in combination with the electrochemical nature of the finished textile surface and its topography.

9.1.4 Short survey of applications

Textile adhesives are extremely versatile and are used in a wide range of applications. One main field is apparel, especially lingerie, leisurewear and sports gear. Home textiles, upholstery and even blinds and awnings also use adhesive joining techniques. Applications in medical textiles have specific requirements, such as joint stability against hot washing or sterilisation. In many cases, the importance of adhesive bonds between textiles is such that life depends on them; this applies not only to paragliders, hot air balloons and climbing equipment, but also to adhesive bonds in airbags used in the automotive industry. Moreover, textile adhesives play an important role in the design of protective equipment used by the military, police and fire-fighters. Everyday products such as those used in personal hygiene are low-cost disposables, but consumers have specific requirements with respect to feel and function. Other products, such as those used in car or airplane interiors, should be long-lasting with outstanding performance and visual properties.

Many applications are in the field known as technical textiles. These include filters for cleaning air or fuels and oils, geotextiles for the protection of hillsides and cliffs, and for prevention against mudslides. Even in the construction industry, innovative methods of joining new textile skins with adhesives offer textile design effects for interior architecture and outdoor applications. More simply, they can also protect against dust and mechanical damage during construction.

9.2 Adhesives and adhesive applications

Adhesives in the textile industry can be classified on the basis of their chemical base polymer, use or physical/application form. The chemistry of the base polymers affects the performance of the joint while the physical form determines the processing characteristics (Fries, 1983; Holme, 1999; Habenicht, 2006).

9.2.1 Adhesives

A wide range of adhesive base polymers are in use in fabrics, such as polyurethanes, polyamides, polyester, polyolefins, ethylene-vinyl-acetate copolymers and acrylates. The physical form of these adhesives is variable.

Polyurethanes are available as waterborne or solvent-based thermoplastic (non-reactive) or crosslinkable types (with co-reactant), but also as 100% solid systems, as thermoplastic powders or in film form. Reactive types cure under atmospheric conditions with moisture from the atmosphere (reactive hot melts) or as two-component liquid systems or reactive film. They have a high adhesive strength on a wide range of (textile) surfaces. The advantages of polyurethane are good elongation and toughness of the joint, good water and solvent resistance (types with sterilisation properties have also been developed) and, in the case of reactive systems, sufficient heat resistance. One disadvantage is yellowing under UV conditions.

Polyamides are generally used as hot melt adhesives as granules or in film and powder form; post-curing types (with atmospheric moisture or via heat-cure with epoxide modified systems) are available. Polyamides were formerly restricted to fusible interlining applications (Fries, 1983), but nowadays the advantages of good heat and chemical (humidity, oil, fuel) resistance have led to their use in other applications, e.g., in the filter industry. Polyamides are often used in textiles where resistance against dry cleaning or even plasticisers is required.

Polyesters are used as hot melt systems (granules, film, powder) or solvent solutions. Chemically they are related to the fabric base polymers; polyester adhesives therefore show particularly good adhesion to polyester fabrics. They are used in interlining or applications where good heat resistance is required (such as in the filter industry).

Acrylates are available in solutions and as aqueous emulsions. Heat-curing systems or thermoplastic non-curing systems are common for both forms. As they offer good adhesion to many substrates, they are versatile and can be used in a range of applications. They are particularly in demand for the dry cleaning and washing resistance of the curing types.

Rubber-based, synthetic or natural, adhesives are available in aqueous dispersions or as hot melts, such as polyolefins and block-polymers (e.g. ethylene-vinyl-acetate). Some systems are available with post-curing, e.g. reactive polyolefin hot melts cure under atmospheric conditions with moisture. The flexibility of the resulting joint and the low costs are two of the main advantages of this type of adhesive. Polyolefin types are mainly mixtures of copolyolefins with maleic anhydride modified ethylene-vinyl-acetate, polyethylene or polypropylene as the adhesion promoter, but they can also be mixed with resins.

The performance of adhesive joints in a fabric depends on two main factors: the resulting adhesion between the textile substrates (based on topography of the textile and its surface energy and the surface wetting by the adhesive) and the cohesion of the adhesive itself. To ensure a good wetting the adhesive must be able to flow out onto the surface; a sufficiently low viscosity is therefore required, in combination with adequate pressure during the bonding process, taking into account the potential for undesirable wicking through the textile.

Good wetting is achieved by applying the adhesive dissolved or suspended in a vehicle such as a solvent or water. A disadvantage of this process is the need to then remove this vehicle, which incurs further costs, particularly for environmentally harmful organic solvents. Another possibility is to provide a lower molecular polymer with post-curing possibilities. The polymer itself has a liquid flow which disappears after the curing process. Thermoplastic systems obtain reduced viscosity by means of heat. It is possible to pre-apply the thermoplastic systems and reactivate the adhesive in a later bonding step. Inline fabrication, such as the lamination process, is also very common.

9.2.2 Viscously applied adhesive systems

Viscously applied adhesive systems comprise waterborne or solvent-based adhesives and 100% solid systems. Solid systems can be applied liquidly at room temperature, as mixed two-component adhesives or one-component hot melt systems using additional heat to reduce the viscosity of the polymer.

The kind of application varies widely: viscous adhesives can be sprayed, applied with different kinds of rollers, holohedral or open coating, or screen-printing systems. For waterborne or solvent-based systems a drying/evaporating step (inline/batch) at moderate temperatures is required after application of the adhesive; the other adhesives can move directly to lamination after the adhesive coating has been applied. Reactive adhesives need a self-contained oxygen- and humidity-free applicator system; otherwise, the reactions of the adhesive with the environment would lead to impurities.

Spray applications are used when a thin and even distribution of the adhesive is required even on three-dimensional surfaces like interior trim and on thermally fragile substrates. There are several different types of process and equipment that can be used in spray application of hot melts, including large and mini swirl, consistent or random pattern, meltblown, slot-like spray and elastic coating (see Fig 9.5).

Applying adhesives with engraved rollers (see Fig. 9.6) or using screen printing technology leads to defined adhesive structures on the textile surface. Rollers apply molten hot melt adhesives or adhesive dispersions onto textile substrates. This method of application is typically used for dispersions of pressure-sensitive formulations on textile structures, e.g. for interior use or for dispersions of hot melt powder formulations (see Section 9.2.4).

In the application of hot melts using rollers, granules or blocks of adhesive are melted in an extruder, applied with a doctor blade onto the engravings of an engraved roller through a slit die, and then applied to a substrate (Evonik, 2011). Even heat-sensitive textiles can be coated with dot or rhombic patterns using polyamide or polyester hot melt adhesives, which require high application temperatures. The extremely short contact time between the textile and the printing roll, which can reach temperatures of up

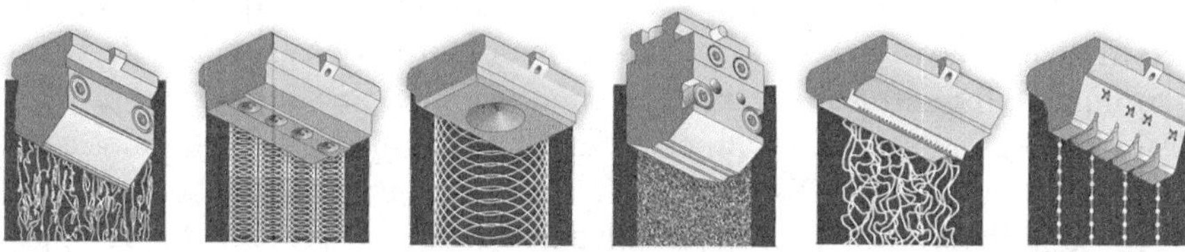

9.5 Different nozzles for spray applications (source: Nordson Deutschland GmbH, Germany).

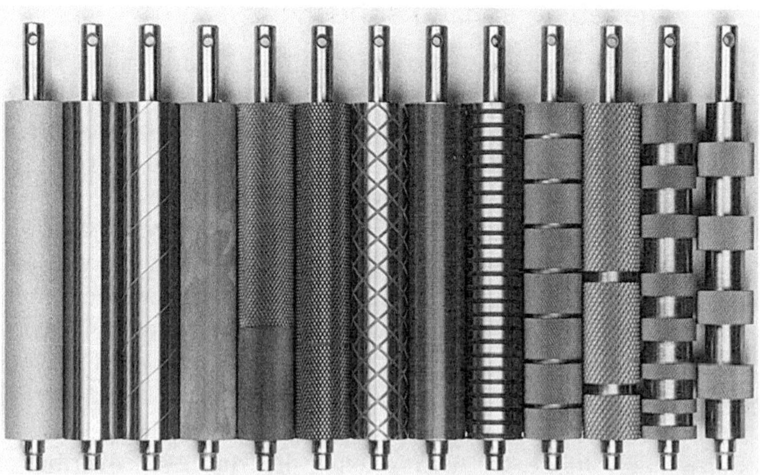

9.6 Engraved rollers – overview (source: Hardo Maschinenbau GmbH, Germany).

to 220°C, protects the substrate from thermal damage. Some standardised dot patterns have become accepted over time, varying according to fabric type and application. The dot-like or rhombic application of hot melt adhesives onto one of the two substrates is particularly economical, achieving high coefficients of adhesion and low coat weights. There is also no barrier layer to steam, which is a problem in the case of padding laminates, for example. Only a part of the intersections and junctions is coated, depending on the density and size of the adhesive dots. The coating consequently does not significantly affect the softness and elasticity of the fabric. For coatings and laminates in which the softness and elasticity of the composite are less important, any lattice and line pattern may be used.

The hot melt print process offers special advantages to the automotive industry in particular. Since this industry is attempting to manufacture a car that is almost completely recyclable, all components of the seats (covering fabric, hot melt adhesive, and non-woven fabric) are often required to be composed of the same chemical base. Since seat cover fabrics often consist of polyester, polyester non-woven fabrics are replacing the polyurethane foams previously used.

9.2.3 Adhesive films

Thermoplastic adhesive films (Collano, 2011) are manufactured by extrusion processes. The raw material is heated in an extruder and formed to a flat film using blow extrusion or cast extrusion on a chill roll. The result is an adhesive film with a thickness of 15–500 g/m^2, depending on the demands of the customer. Multilayer systems can be produced by co-(blow) extrusion of up to five different base polymers. In these systems, the properties of different polymers for coating and sealing can be combined; they are used mainly in textile sealing and are able to bond even incompatible substrates thanks to the systematic combination of various bonding agents. In addition, flat substrates such as textiles can also be coated and/or bonded directly on an extrusion installation.

The typical base polymers used for films are copolyolefins, copolyester, copolyamides and thermoplastic polyurethanes. Since the substrates used in the market are extremely varied and the bond requirements are very different (hard/soft; porous/plane; chemically stable/soluble), the full range of available raw materials is required to construct films that satisfy all possible requirements.

The bonding principle for films is quite simple. The thermoplastic adhesive film is laid between the two parts to be bonded, pressure is applied and the system is heated to 30°C above the melting point of the adhesive. The molten film wets the textile surface under low pressure and the final joint is formed as it cools. The speed of this process is determined by the time within which the heat penetrates into and out of the glue line. Very rapid heating is generated by high-frequency (for adhesives like polyamide, polyester and thermoplastic polyurethane) or ultrasonic techniques.

Reactive adhesive films for textile applications are known as 'latent-reactive' polyurethane adhesives (Büchner *et al.*, 2005). The films are rugged under standard conditions (10–40°C) for up to six months and can be laminated like any thermoplastic adhesive film. They are composed of polyurethane dispersions, mixed with a solid isocyanate such as aromatic toluylene-diisocyanate-dimer or aliphatic isophorone-diisocyanate-trimer. The dispersions are dried on a liner material at ambient temperatures. Crosslinking starts as a result of thermoactivation during the lamination process: the solid isocyanate dissolves at temperatures above 70°C in the molten polyurethane matrix and starts crosslinking reactions. The temperatures must be high enough to induce the solvation and diffusion process completely. After cooling, the crosslinking process proceeds in the same way as in two-component systems. Because of the thermal step between the melting of polyurethane soft segments at 45–50°C and the start of the crosslinking reaction above 70°C, the systems may be pre-applied. Reactive films are used when very high levels of performance are required in terms of heat and humidity stability.

Lamination processes for film bonding can be classified into two major groups: continuous and batch processes (Collano, 2011). In continuous processes the laminate is created by a transfer calender or flat bed laminator, or in flame-bonding machines. With a calender (Fig. 9.7), heat is applied onto the glue line by means of a heated roller. This method is ideal for thin, flexible substrates and achieves a production speed of 2–35 m/min. With flat bed laminators, thick and rigid substrates are heated and pressed between heating and cooling bands with a production speed of 2–25 m/min. The flat laminating installation is used to manufacture small scouring pads like sponges. An older technique that is still commonly used is flame bonding, in which foams (polyethylene or polyurethane) are bonded by means of scorching. Through the use of a thermoplastic adhesive film, this technique can be extended to substrates that cannot normally be activated by scorching.

Special bonding techniques that are used particularly in the textile industry are high-frequency, ultrasonic, hot-air and hot-wedge bonding (Collano, 2011). In principle, when welding, the substrates must be heated to, or above, the melting point, so that they can then fuse with one another. In many cases, this is not possible for a variety of reasons: the melting point of the substrate is too high; the softening range is too narrow (the polymer goes directly from the plastic state to the liquid state and loses its shape); or the technique used does not permit sufficient warming, as is the case when high frequency is employed with polyolefins. In all these cases, thermoplastic adhesive films can be used as welding aids. As a result, welding in combination with

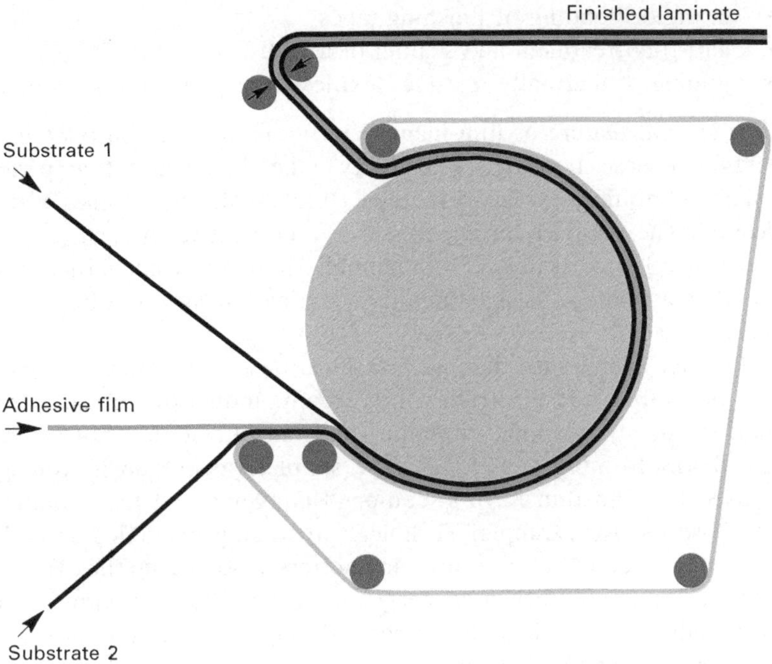

9.7 Calender: operating mode (source: ifs, TU Braunschweig, Germany/Collano Adhesives AG, Switzerland).

thermoplastic adhesive films becomes a process somewhere between welding and bonding. In most cases when using a thermoplastic adhesive film, the temperature necessary at the glue line is reduced.

High-frequency welding is used when polar adhesive molecules are placed in a high-frequency field due to their oscillation. In ultrasonic welding the substrates are made to vibrate by means of ultrasound. The resulting friction causes them to warm up. A certain stiffness of the substrate is a prerequisite, as otherwise sufficient stimulation to cause vibration of the textile by means of the ultrasonic probes (the 'loudspeakers') is impossible. Ultrasonic welding has been a standard process in joining techniques for 40 years. Here too, thermoplastic adhesive films have the primary task of reducing the welding/bonding temperature. Welding and bonding techniques are used in weather protection, work and sports clothes, protective helmets, car tarpaulins, sun blinds, all types of awnings, insulation and filter tubes, cable sheaths, dressing materials, foundation garments and diapers.

The following applications of thermoplastic adhesive films are particularly notable:

- fixing and welding/bonding of hems and overlapping seams
- sealing of hems and overlapping seams

- fixing or bonding of finishing tapes
- anti-fray treatment and sealing of textiles
- joining of thermally resistant textiles (e.g., glass fibres, carbon fibres).

A special feature of film manufacturing is the slit film (Collano, 2011). Slitting increases not only the flexibility of the film but also its air permeability, which is required when membrane ('breathable membranes') effects are desired. The regularly arranged slits can be opened by heat (see Fig. 9.8), whereby a lattice structure with naturally high air permeability is obtained. Another advantage is a bubble-free lamination and a softer hand of the resulting joint.

In batch techniques thermal presses are used. A variety of heat sources can be employed: electricity, oil, steam, induction, infrared radiation, high-frequency welding, vibration (friction, ultrasonic), hot air, hot gases and thermoforming techniques. The simplest technique is ironing, which is used to bond transfer-film-equipped emblems and logos onto sweaters and T-shirts, for example. Technical processing examples are side panels and other decorative parts for car interiors made of textiles (polyethylene terephthalate and other polyesters) and artificial leather with thermoplastic polyurethane, as well as headliners, sliding roofs, fitted carpets, airbags, various seat parts and shoes.

9.2.4 Adhesive powder

Many industrial applications require the use of hot melt adhesives in powder form. To obtain this powder, polymer granules are embrittled with liquid

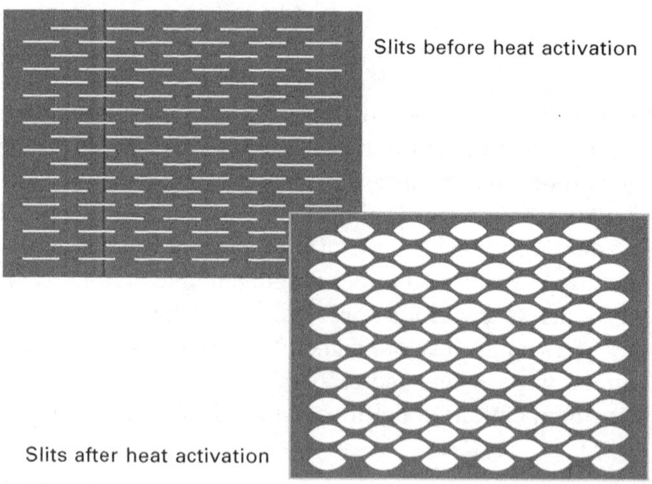

9.8 Slit films: operating mode (source: Collano Adhesives AG, Switzerland).

nitrogen. Common base polymers are polyamides containing laurolactam as an essential monomer. These hot melt adhesives have melting points that allow textiles to be fixed efficiently without causing damage; the resulting textiles are able to withstand washing and dry cleaning. As the adhesive is applied in powder form, air permeability of the joint is also achieved (Evonik, 2011).

Besides thermoplastics, reactive hot melt powders are also available. Epoxy-modified polyamide powders, such as those described in Kalbe *et al.* (2000) and latent reactive polyurethane powders, such as those described in Abend (1999), can be pre-applied at moderate temperatures, and can cure by heat at temperatures above 120°C via isocyanate crosslinking. A good resistance against solvents, e.g. steam and fuels, is one of the goals for heavy-duty applications such as industry filters.

Depending on the particle size, the applications of these powders can vary: particles up to 80 μm are used in light fabrics and light charmeuse with powder dot (fine mesh) or on non-wovens in paste dot applications; particle up to 120 μm in size are used in knitted and woven interlining as well as fabrics; and powders with particle size of between 300 and 500 μm are used in needle-punched non-wovens or for glass fabrics by means of powder scattering (Evonik, 2011).

The simplest and oldest industrial process for manufacturing fusible interlinings consists of scattering the thermoplastic adhesive powder onto the lining substrate (see Fig. 9.9). Today the scattering process is used mainly to manufacture shirt collar interlinings and industrial textiles. In the oldest design of industrial scattering machines, the powder falls from the supply hopper through a slit and onto an oblique vibrating conveyor, which uniformly distributes the powder to be scattered. Advanced units use a hopper mounted over a transfer roller. An oscillating brush brushes off the powder carried in the wells of the transfer roller. It is therefore important that the powders used do not have too high a flowability rate: this ensures that the powder will stick to the roller until it is brushed off. The powder then falls through one or two oscillating screens onto the web of lining material, which is then fed through a sintering tunnel with the least vibration possible. There it is sintered onto the surface by hot air or infrared radiation. The coating width production lines of scattering systems range from 0.3 to 5 m. If the machine parameters are optimally matched to the powder used, it is possible to attain coat weights of 10–200 g/m^2. This corresponds to production rates of 10–30 m/min. The scattering process is primarily used for industrial textiles that require large quantities to be applied, such as thermoformable carpeting for the automotive industry (Evonik, 2011).

The powder dot process is a gravure printing process (Fig. 9.10). The web of liner material is heated by winding it around a steel roller heated to 220°C. Together with the fabric, this roller is pressed against a hot printing roll,

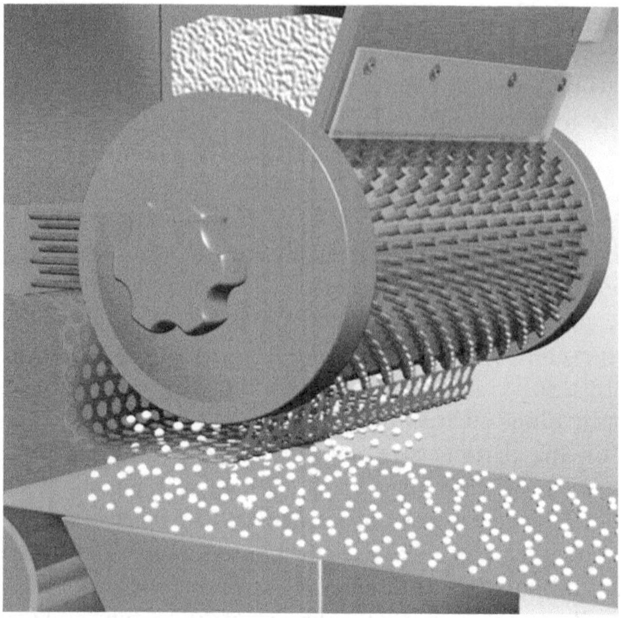

9.9 Scattering of adhesive powder operating mode (source: Evonik Degussa GmbH, Germany).

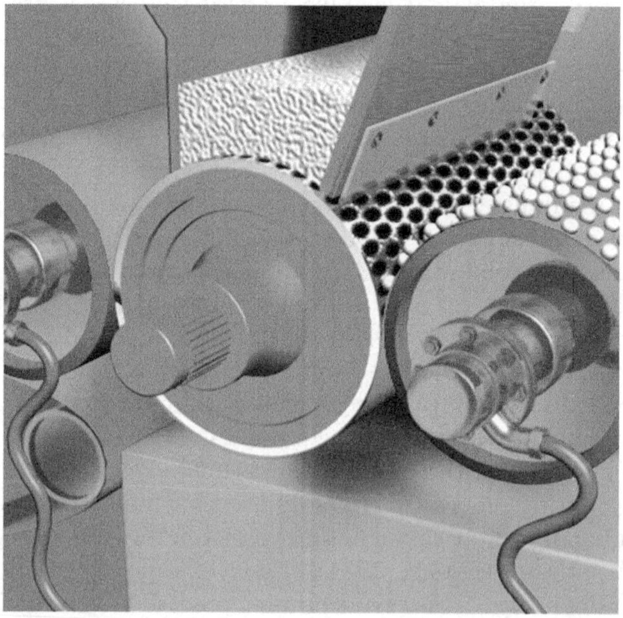

9.10 Powder dot process: operating mode (source: Evonik Degussa GmbH, Germany).

which contains the thermoplastic powder in its dot-like wells (dot row roller engraving). A doctor blade with a hopper applies the powder into the wells. The printing roll (engraved printing roller, dot roller, cup roller) immediately presses the powder in the wells onto the warm web of lining material, producing surface melting and agglomeration of the powder and emptying the cup-like wells of their powder. This agglomeration and/or subsequent irradiation with height-adjustable infrared radiators sinters the powder grains of each pile of powder, making it sticky and anchoring the powder grains firmly onto the web of lining. The infrared rays furthermore ensure that a smooth hemispheric surface will form on the dots of adhesive.

The powder dot process is used mainly to coat lining materials that are not sensitive to thermal stress, such as materials made of natural fibres. The adhesive powders must be carefully matched to the surface texture of the material, the dot image, and the final adhesive properties in order to obtain a good balance between them (Evonik, 2011).

The paste dot application (Fig. 9.11) runs as a rotary screen printing process, in which an aqueous suspension of finely divided thermoplastic powder adhesives and additives is pressed through the holes of a rotating, perforated cylinder (the screen stencil) onto a cold web of fabric. A typical paste contains the following ingredients:

- water as the matrix
- a thickener for controlling the paste viscosity

9.11 Paste dot process: operating mode (source: Evonik Degussa GmbH, Germany).

- a dispersing agent to ensure that the powder is uniformly distributed in the water
- a plasticiser to reduce the temperature of the joint and improve adhesion
- a wetting agent to improve adhesion on hydrophobic surfaces
- a slip additive to stabilise the aqueous phase paste; this keeps the water flow enhancer in the paste upon application and improves the dot geometry
- a hotmelt adhesive powder.

The paste dot application procedure is particularly gentle to the substrate, and the wide range of options for formulating the paste grants the user flexibility in the application procedure. Then the paste dots are dried, and circulating air or infrared radiators are used to sinter the textile web. The process is characterised by high and economical coating rates ranging from 30 to 60 m/min. In particularly favourable cases, rates of 90 m/min have been attained. This method of application has thus surpassed the dry processes and is now the most popular of the powder applications (about 60%) (Evonik, 2011).

A very special type of hot melt powder application is the double dot, shown in Fig. 9.12. A double dot consists of a highly viscous or crosslinked base dot and a low-viscous top dot. This allows light and open face materials to

9.12 Double dot process: operating mode (source: Evonik Degussa GmbH, Germany).

be bonded to equally open linings without strike-back, without the adhesive penetrating the fabric, and without the back of the material sticking to the machine. Other advantages include a low application weight, better adhesion to surfaces that are difficult to fuse, and a soft textile feel.

In recent years, new base materials such as polyester charmeuses (about 35 g/m^2) have come into use to close the gap between non-woven fabric and woven lining. Because these materials have open surfaces and are also very temperature-sensitive, paste dot or powder dot coatings are not suitable, as the coating dot is pressed too deeply into the lining, causing hardening. Powder dot coating has the additional disadvantage that the lining is subjected to too much thermal stress, impairing the product's elasticity.

The double dot process is ideal for coating and bonding these materials, because the double dot sinks only slightly into the lining compared to the other two processes, thanks to the high viscosity of the base dot, which is applied via rotary-screen printing. Then a powder adhesive is scattered onto the base dot; it adheres because the base dot is still wet. Any powder that does not stick is removed by aspiration in a later step. The water is removed via a dry tunnel and the base dot is sintered on the lining substrate; finally both dots are bonded together.

A double dot produced in this manner consists of approximately 3–5 g/m^2 base dot and 4–5 g/m^2 top-dot material.

Fashionable face fabrics, such as polyester and viscose georgette and microfibres, are provided with a special finish (siliconising or fluorocarbonising). Conventional systems fail to bring about adhesion in these materials because suitable low-viscous hot melt adhesives will seek the path of least resistance, causing strike back deep into the liner. In the double dot process, the base dot forms a highly viscous barrier layer that prevents strike back into the liner. The top dot's hot melt adhesive is forced to run in the direction of the face fabric; the resultant strong bonding to the base dot and excellent surface wetting lead to good adhesion. If all elements are correctly matched, running speeds of 50 m/min are possible (Evonik, 2011).

9.3 Properties achieved by adhesive joining

In textile fabrics, bonding applications usually involve disparate materials which cannot be joined by other techniques, or for which other techniques do not provide a satisfactory joint. The ongoing development of new, often functional materials and fabrics, together with the usage of textiles in new areas, poses particularly difficult challenges. The bonding process should be easy, quick and safe, and economical through all production steps. Filaments should not be damaged by sewing needles.

The adhesive and the resulting joint are subject to touch requirements, such as a flexible surface with a 'soft touch'. Adhesives and joints should either be water-resistant or should support water diffusion (according to the end-use requirements), breathable, elastic, wear-resistant, non-yellowing and non-brittle and, particularly for apparel, resistant to washing and dry cleaning.

The motivation for bonding textiles with adhesives is often the insufficient performance of a sewn seam: crimping, ladders and seam visibility, particularly in thin textiles such as those used in lingerie, are speeding up developments in the field. A sewn seam does not usually offer any functions besides 'joining'; as mentioned above, future joints will be required to fulfil multiple functions.

The requirements of the adhesive joint are as versatile as the testing methods are variable. In general, the cohesive resistance for textile laminates, determined according to DIN 53530 (DIN, 1981; ASTM, 1998), is a standard parameter. The absolute number can vary according to demand, depending, for example, on whether the adhesive bond is being used simply to fix two parts or to join textiles 'for a lifetime' in applications with high demands on their strength. Non-woven laminates, for instance, which are used as protective membranes in the building and construction industry, are bonded adhesively to polyolefin foils (as a protective foil against liquids) by hot melts; the cohesive resistance of such a laminate is about 0.5 N/mm with failure in the upper non-woven layers (see Fig. 9.13).

9.13 Fracture appearance of a non-woven foil laminate. Left: non-woven fabric; middle and right: foil with adhesive dots and loose fibres (source: ifs, TU Braunschweig, Germany).

In apparel uses, when changing from a 'stitched' seam to 'stitchless' seams (AiF, 2011), the cohesive resistance of the joint must be higher than 2 N/mm; this describes the strength 'men cannot tear apart by hand'. For example, Fig. 9.14 shows knitted cotton, laminated with a thermoplastic polyurethane and a peel strength of 3 N/mm. The failure in this case is caused by loose cotton fibres, pulled out of the textile under peel influence. In addition, a low bending stiffness is required in order to achieve a stitchless seam with a soft touch. Bending stiffness can be determined according to DIN 53121 (DIN, 2008; ASTM, 2002) or, more simply, using the old but straightforward cantilever procedure (Wegener and Sommer, 1960; Reumann, 2000).

Another important test, used for lingerie, pullovers and shirts, for example, is the simulation of the behaviour of the clothing. This is described by the elongation according to DIN 53835-T14 (DIN, 1992), 'knitted fabrics, single strain between two force limits'. Measurements simulate dressing, for instance, with 100% applied load over various cycles, and wearing, with 40% applied load over time. Typically, the elongation to break the adhesives used in textile joints is quite high; polyurethanes, reactive polyolefins or silicones have elongations to break of 500% up to 1,000%. However, the behaviour of the joint should be similar to the behaviour of the textile; otherwise, different types of resetting lead to insufficient visual performance of the joint, as shown in Fig. 9.15.

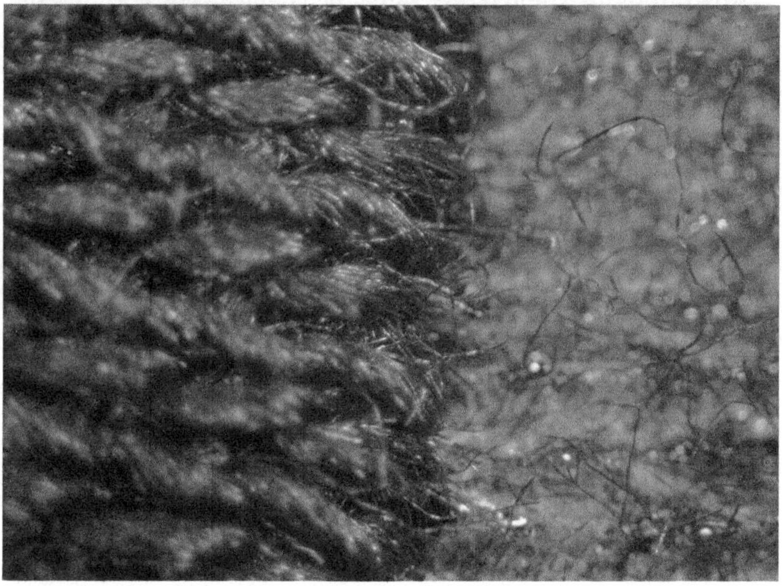

9.14 Fracture appearance of a knitted cotton fabric, bonded with polyurethane film Collano® 36.304. Left: knitted cotton; middle and right: adhesive foil with loose fibres (source: ifs, TU Braunschweig, Germany).

9.15 Insufficient optical performance because of different types of resetting, knitted cotton and hot melt adhesive (source: ifs, TU Braunschweig, Germany).

Apart from the strength, the aging behaviour is also of high importance. Apparel requires resistance to washing and dry cleaning; typically the fabrics of the test specimen are washed and sometimes also tumble-dried in various cycles at the required temperatures. Other fabrics such as technical textiles, particularly geotextiles or automotive interiors, require a completely different type of aging test; climate changing tests at environmental temperatures and humidity, sometimes with additional UV irradiation, are carried out; the parameters are specific to the application and can vary.

Fabrics with additional functions also require specific tests, according to the demands of the end-use (e.g. Reumann, 2000). In breathable membranes and in filters, air permeability should not be affected by the adhesive or by the bonding process. DIN EN ISO 9237 (DIN, 1995) determines the parameters for testing; the pressure difference in static tests for apparel is 100 Pa (1 mbar), for filter and technical fabrics 200 Pa (2 mbar). Airbag textiles need more realistic test parameters and equipment because of the intermittent load; in this case a developing curve of the air permeability as a function of pressure between 0 and 10^5 Pa (0 and 1 bar) is used to simulate the inflating of the airbag in the case of an accident.

A special quality of (apparel) textiles which can be transferred to the requirements for adhesives used in textiles is the Oeko-Tex® mark. Testing procedures, standardised by the International Association for Research and Testing in the Field of Textile Ecology, analyse the content of various parameters which could affect health and quality of life. Adhesives, used, for instance, in baby products (class 1; highest class in this specification), are tested by specific institutes (normally initiated by the adhesive supplier) in order to obtain the authorisation to use the Oeko-Tex® mark.

9.4 Examples of adhesive use

Table 9.2 shows a selection of adhesive base chemistries used in the textile industry, their performance and some typical applications. The chemical type of the base polymer employed in an adhesive formula generally determines its properties. The following examples show how the demands on adhesives and joints can vary.

9.4.1 High-tech, low-cost and mass production: personal hygiene products

There is a wide range of applications of non-woven textiles in the personal hygiene field: baby diapers, child training and swim pants, adult incontinence products and feminine hygiene pads and shields. Overall, these products combine many constructive components with excellent functional efficiency (Fornes, 2009; Gries and Klopp, 2006). Hot melts are usually used in these applications, not only for bonding the components but also for fit, breathability, retention, absorption and feel. In baby diapers (see Fig. 9.16), absorbency and fluid retention of the core and the acquisition layers can be affected by the application of the adhesive. Poor use of adhesives on elastic applications can translate into leakage and redness on the baby's legs.

Product function is one key requirement, but the manufacturing process must also be taken into account. Baby diapers are products which cost only a few cents to make; these low costs are only possible thanks to the use of thermoplastic hot melts with high efficiency and product performance. Production speed is the key to leveraging economies of scale. Typically, 800–1000 pieces are produced every minute. Each baby diaper contains a total of only 1 to 1.5 g adhesive, which means that any element in the production process that leads to waste or a faulty product will have a significant negative impact on the bottom line.

The choice of adhesive is determined by two main requirements: bond strength between the different layers, which is tested with peel strength tests, and effectiveness of the elastic bonding, measured by elastic creep tests. It is also important that the product's breathability (which allows air to pass through the product for comfort) and permeability (which allows fluid to flow away from the skin) of the product are not affected by the adhesives.

Synthetic rubber-based adhesives, such as styrene-butadiene-styrene or styrene-isoprene-styrene rubbers have traditionally been used for this type of

Table 9.2 Selection of adhesives for textile applications

Base polymer	Types available	Properties	Applications
Polyurethane	• Liquids (one or two component, waterborne, solvent-based) • Reactive hot melts • Films	• Good adhesion even to difficult substrates like PTFE • High strength • Good elongation • Toughness • Water and solvent resistance (washing and dry cleaning) • Some types with sterilisation properties • Heat resistance (reactive types) • Some types yellow on exposure to UV	• Polyether-polyurethane films for Spandex® and hydrophobic textiles • Reactive polyurethane hot melts for breathable membranes like Sympatex®, non-wovens and foils for apparel, home textiles and medical applications • Waterborne dispersions and hot melts for automotive interior • (Multilayer) films for textile seam sealing
Polyester	• Films • Solvent solutions	Good adhesion to polyester fabrics	• Interlining • Applications in filter industry • Automotive interior (recycling!)
Polyamide	• (Reactive) hot melts • Powder, granules • Films	• Good heat resistance • Good resistance against oil, fuel and humidity Resistance possible against: • dry cleaning • plasticisers	• Fusible interlining applications • Filter applications • Carpeting for automotive industry • Powders for outerwear
Polyolefin	• Films • (Reactive) hot melts	• Good adhesion even to non-treated polyolefins • Good resistance against humidity • Less adhesive amounts possible because of lower density	• Personal hygiene products • Textile/non-woven lamination
Silicone	(High temperature curing) one component systems in film and band form (extruded)	• Transparency • Good mechanical properties	• Bonding of silicone-containing surfaces • Seam sealing

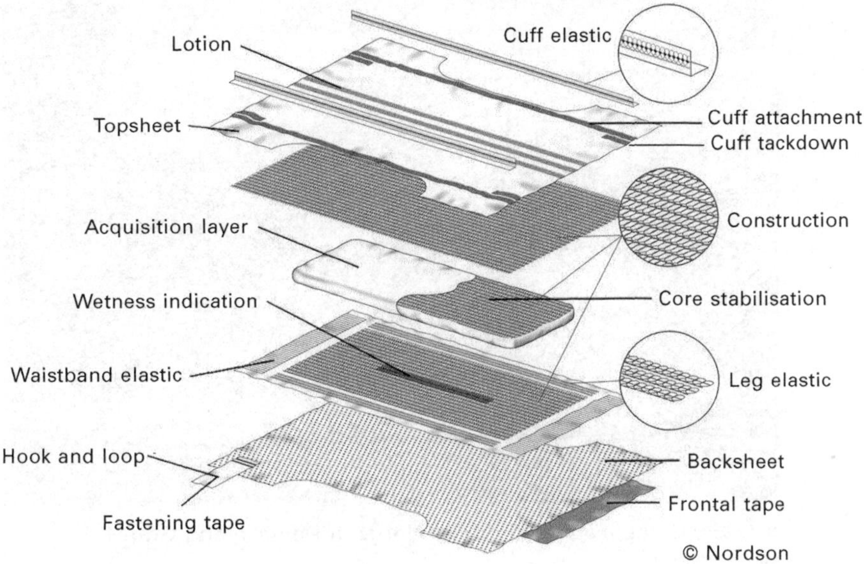

9.16 Assembly of a baby diaper (source: Nordson Deutschland GmbH, Germany).

application. These polymers need tackifiers like hydrogenated hydrocarbons or even white oil for improved tack. All polymers used are thoroughly cleaned as required by legislation and for consumer protection purposes. The adhesives have a good cohesive performance and flexibility and are easily applied in swirl processes with various nozzles (see Fig. 9.17 for an example of a mini-swirl application), especially for covering the elastic threads, which can range from 1 to 30 strands, depending on the product and feature type.

Nowadays, an economic alternative to the base polymers described above is atactic poly-alpha-olefins. Their sprayability is reduced, so they are commonly used as constructive adhesives in a diaper, applied with slots, as shown in Fig. 9.18. The constructive adhesive bonding covers about 70% of the whole adhesive requirement of a diaper; the lower costs of the adhesive could have a positive impact on the price, making the product more consumer-friendly, if the application speed of these alternative adhesives is able to compete with that of the traditionally used rubber systems.

9.4.2 Breathable membranes: (microfibre) textiles for sport and outdoor use

Textiles laminated with breathable membranes like Sympatex® or Gore-Tex® for sportswear, shoes and outdoor use are extremely hard-wearing, yet very

9.17 Swirl applications (source: Nordson Deutschland GmbH, Germany).

9.18 Slot applications (source: Nordson Deutschland GmbH, Germany).

comfortable and very functional with outstanding design possibilities. In Sympatex® fabrics, a 5–25 μm polyether-polyester copolymer membrane without pores is laminated onto a textile carrier by adhesive bonding. In contrast, Gore-Tex® laminates are made of thermo-mechanically expanded polytetrafluoroethylene and other fluoropolymer products, combined with inner and outer shell textiles. The adhesives used in these assemblies must therefore be able to bond even problematic surfaces without affecting their function; a solution for Gore-Tex® laminates was the development of the GORE-SEAM® tape-sealing technology.

Breathable membranes prevent water entering from external sources, but are permeable to water vapour diffusion from the inside (see Fig. 9.19). Every adhesive applied restricts the function of the membrane. Two factors are required to optimise the process: an open coating of the adhesive and improved breathability of the adhesive itself. Reactive polyurethane hot melt systems are commonly used for this purpose (Schindler *et al.*, 2009). Engraved rollers apply the adhesive in dots (Fig. 9.20) leading to a surface covering of 20–40%. One major factor affecting the covering rate is the viscosity: low-viscous systems spread more easily and lead to a higher covering rate of the membrane surface than specially formulated systems with optimised viscosity. Another goal is the formulation of adhesives which are breathable in themselves; new polyurethane systems have been developed that have a steam permeability up to 50 times higher than commonly used polyurethane adhesives.

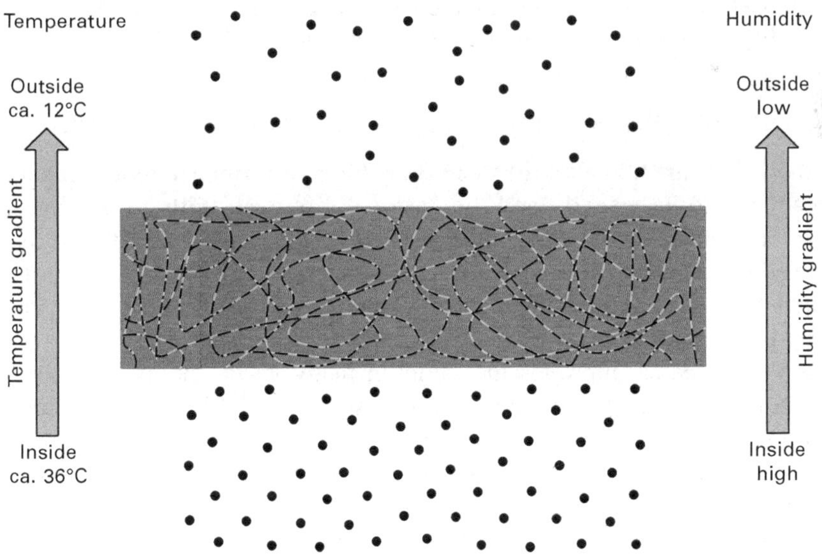

9.19 Breathable membranes: operating mode (source: ifs, TU Braunschweig, Germany/Jowat AG, Germany).

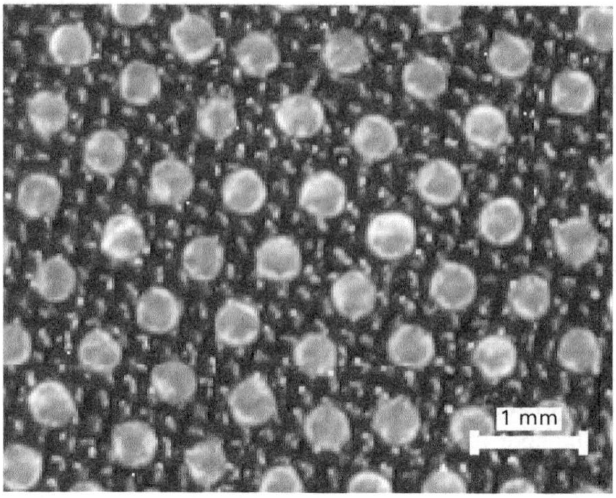

9.20 Dot application of polyurethane hot melt on breathable membranes (source: Jowat AG, Germany).

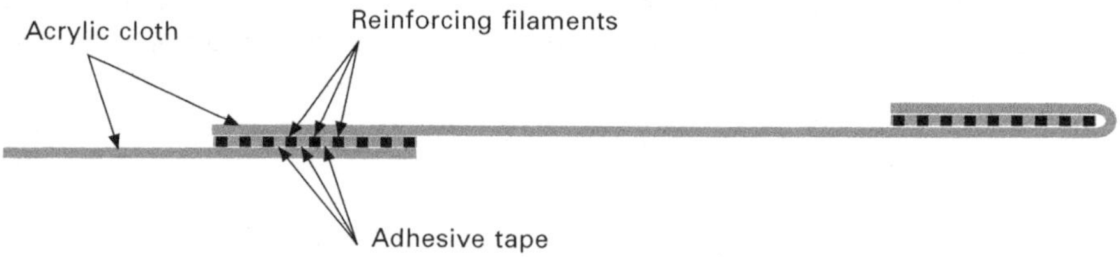

9.21 Glued seam with reinforced adhesive tape (source: ifs, TU Braunschweig, Germany/Jentschmann AG, Switzerland).

9.4.3 Stitchless seams, better performance: awnings

Sewed awnings are home textiles with very high performance requirements: the seam has to be visually and dimensionally perfect, with a good aging stability against environmental influences. However, it has been found that the sewed seams and side hems have a tendency to stretch, particularly after repeated rolling in and out cycles over some years of use; the result is a sagging of the side hems, which is unacceptable for customers.

Gluing the seams improves the joints in many ways; the awning cover has a more pleasant appearance with almost undetectable, thinner seams, which are water-tight with no dripping from needle holes (a requirement that is becoming particularly important as awnings are increasingly used as water-protective umbrellas for restaurants and cafes). However, gluing alone cannot provide significant seam stretching. This problem is instead solved by bonding with specially reinforced adhesive tapes (see Fig. 9.21) (Jentschmann, 2011; Gluetex, 2011).

Polyacrylic textiles are particularly susceptible to the sagging problems described above. The adhesive tapes used in this case are double-sided UV-stabilised hot melt tapes, reinforced with glass fibres or non-wovens. The adhesive is applied onto the textile and the seam is assembled using ultrasonic welding. The welding machine travels along the seam while the material is clamped on the machine table, as shown in Fig. 9.22. Guiding devices ensure that the seam or hem overlap is perfectly maintained, while the heat-activated adhesive tape is fed between the layers to be joined. In general, the tensile strength of an adhesive joint is quite high when reinforced polyester textiles are used, and is comparable to that of awning seams sewn with Tenara® (PTFE) yarn. Problems occur with dirt-reflecting surfaces of acrylic textiles; the heat-activated adhesive tapes do not adhere to these surfaces, necessitating a modification of the seam design, so that only uncoated sides are bonded. The result is that the hems are still glued in a conventional manner with the reinforced adhesive tape, but the seams are joined side-by-side with the adhesive tape and a textile tape with a coating underneath. Overall, the resulting reinforced seams are thinner than sewn seams because of the slight compression that occurs during the welding process, and have an optimal seam strength due to full surface bonding.

9.4.4 Developments in airbag applications

Driver and passenger airbags were first developed in 1951 (Linderer, 1953) with the aim of reducing injuries caused by a crash. These airbags had to

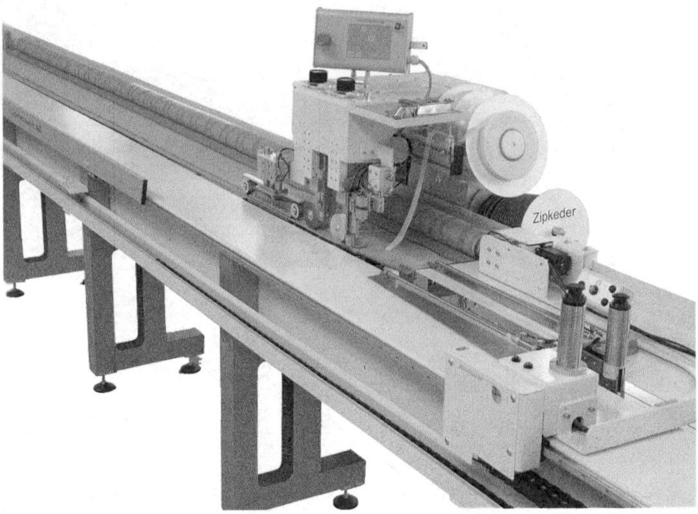

9.22 Ultrasonic welding machine for awning seams (source: Jentschmann AG, Switzerland).

inflate within 60 milliseconds and then collapse. One way of achieving this was to coat a sewn textile bag with a defined open area with silicone rubber. The silicone coating has a certain porosity which does not render it gas-tight, but allows the airbag textile to be impermeable enough to become inflated; the second aim was the reduction of the surface friction to achieve improved inflation. Airbags are usually made of multi-fibre filaments, based on polyamide 6.6 due to its high temperature stability. In 2003, 67,000 tons of polyamide yarns were produced for airbag textiles (Konrad, 2004) with an upward trend.

In the last few years, the movement towards rollover or side curtain airbags made a redesign of standard airbag systems necessary: a side curtain airbag has to stay inflated for a minimum of six seconds. This could only be achieved by optimising the textile coating and the sewing/bonding steps (Studt, 2006; Rychter, 2004). Sewing of airbag textiles always posed one significant problem: stitches lead to seam leakage, especially on coated fabrics, and damage to the textile fibres. One way of optimising the process, therefore, was to substitute (at least partially) the sewing step.

Studies (Rychter, 2004) have shown that the peel-strength required in the case of an accident, namely 8–10 N/mm, can be achieved through an optimised pre-treatment of the textile and, for instance, polyurethane hotmelts, as shown in Fig. 9.23. It is important to note that the adhesives used must be resistant to aging over their lifetime (10–15 years) under adverse conditions, such as environmental thermal influences in a close-flattened structure. Another

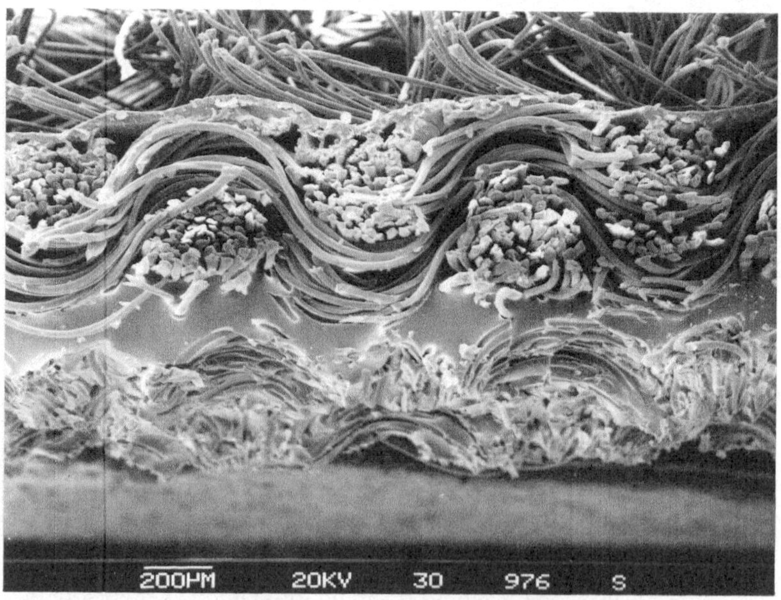

9.23 Glued airbag seam (source: ifs, TU Braunschweig, Germany).

method for solving the problems described above was the creation of a 'one piece woven', an airbag textile, woven by a special process that allows the production of bags with no seam (Studt, 2006). Coating textiles through lamination with a polymer film is a well-known procedure. Now, a perfect control of the gas permeability of the textile can be achieved by lamination with a two-layered film. The first layer, with a relatively low melting viscosity and orientated to the textile surface, is used as the 'adhesive'; it bonds the second layer, which is high-melting and gas-tight, onto the textile surface. Special thermoplastic polyurethane films were used as the adhesive layer, while a high melting copolyester with high elasticity was used as the top layer, so that surface friction and adhesion to the polyurethane layer were optimised. Airbags with this coating passed peel, surface friction, fire hazard and airbag shooting tests (Studt, 2006) and allow the design of a very compact, low weight package of side-curtains.

9.4.5 Textile reinforcement: concrete, composites and more

The reinforcement of bulk materials with textiles of inorganic or organic origins is a well-known procedure. A typical example is the application of textiles in rubbers such as natural rubbers or styrene-butadiene copolymers for tyres and heavy industry products (Wootton, 2001). However, the desire for light-weight and filigree substrates for construction applications has led to ongoing investigations into fibre- and textile-reinforced alternative materials in recent years. Fibre reinforced plastics have been in use for decades, and their properties have been studied in numerous publications. Thus, this chapter simply provides a few selected publications on specific topics.

The materials investigated vary widely, from reinforcement of concrete with, for example, glass fibres (Mund *et al.*, 2005) or polymers (Hartung *et al.*, 2010) on the one hand, to carbon-fibre reinforced plastics (Frauenhofer *et al.*, 2008). The aim, especially in the case of reinforced concrete, is to improve the tensile strength of the matrix material while maintaining good durability and ductile behaviour. Figure 9.24 shows a four-point bending test with a reinforced concrete specimen and the clear advantages in terms of 'elastic' behaviour.

Adhesives are mainly used as binders within textile structures to maintain the fibre alignment, to improve the load-bearing cross sections of the yarn and to strengthen the outer bond between textile and matrix, e.g. concrete, polymer or rubber. The textile reinforcement system can be used as a performance element; thus three-dimensional, light-weight structures become possible. Applications can vary with the diversity of matrices; reinforced concrete can substitute the state-of-the-art steel-reinforced applications, but is also available for thin-walled building applications or even in furniture. Reinforced

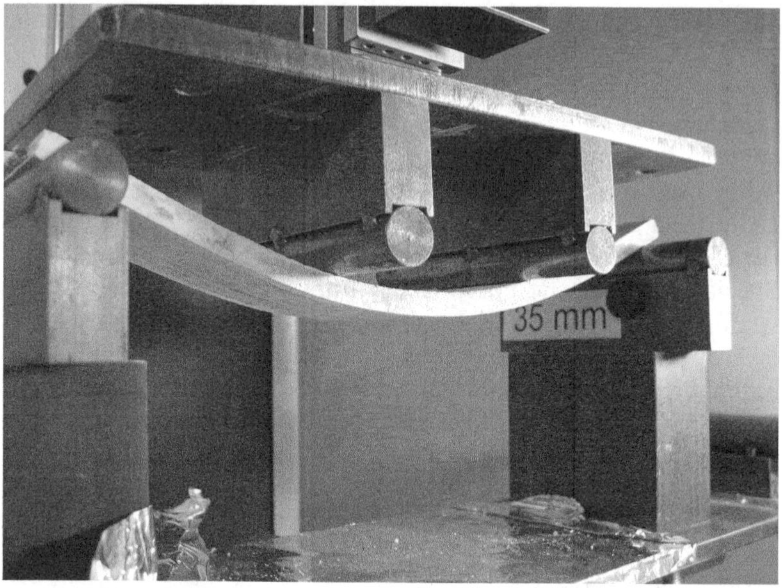

9.24 Four-point bending test of reinforced concrete (source: ifs, TU Braunschweig, Germany).

plastics are becoming increasingly important as low-weight elements for cars, especially those with electric drive, and in future developments in aircraft.

9.5 Future trends

The ongoing development of new materials with complex functionalities will in future place high demands on joining techniques; adhesive bonding has the capacity to solve some of the prospective problems. Disparate materials must be joined, without damaging the filaments by needle stitches, and insufficient performance such as crimping or joint visibility must be reduced. Consequently, the substitution of sewed seams by adhesively bonded seams is a major challenge for the future, especially for light-weight fabrics such as lingerie and functional membranes. The adhesive and the joint must fulfil touch requirements like soft touch and flexibility, usually through the involvement of continuative functions.

By functionalising the bond itself, smart textiles will become more common, especially in healthcare and medicine. Techniques will be improved to save cost, reduce waste and improve the resulting fabric; for example, direct coating with steam-permeable adhesives will lead to textiles that are no longer reduced in functionality by the bonding step. However, any future fruitful investigations will always be the result of effective collaborations between research institutes, adhesive suppliers, plant engineers and users. Interdisciplinary requirements will be an essential part of these studies.

9.6 Acknowledgements

The authors would like to thank the following institutions for their excellent co-operation and support (in alphabetical order):

- Collano Adhesives AG, Switzerland, for discussion, technical information and figures on the topic of adhesive films;
- Evonik Degussa GmbH, Germany, for technical information and figures on the topic of adhesive powder and hot melt application;
- Gluetex GmbH, Germany, and Jentschmann AG, Switzerland for information and figures on the topic of awnings;
- Hardo Maschinenbau GmbH, Germany, for Fig. 9.6;
- Jowat AG, Germany, for discussion, information and figures on the topic of hot melt adhesives, particularly joining breathable membranes;
- Nordson Deutschland GmbH, Germany, for technical information and figures on the topic of spray technology, particularly in the fabrication of diapers.

9.7 References

Abend T (1999), 'Verfahren zur Herstellung und Verwendung von lagerstabilen latentreaktiven Schichten oder Pulvern aus oberflächendesaktivierten festen Polyisocyanaten und Dispersionspolymeren mit funktionellen Gruppen', European Patent EP 0922720 A1, 16.6.1999.

AiF 15927 N (2011), Final report of the project 'Funktionalisierung von Nähten bei der Verarbeitung von Textilien durch klebtechnisches Fügen', 1.12.2008–30.11.2010, funded by the BMWi, Germany, March.

ASTM (1998) ASTM D413, 00-1998; Standard Test Method for Rubber Property – Adhesion to Flexible Substrate.

ASTM (2002) ASTM D828, 00-2002; Test Method for Tensile Properties of Paper and Paperboard Using constant rate of elongation apparatus.

Büchner J, Henning W, Stepanski H, Raffel B (2005), 'Lagerstabile latent-reaktive Klebfolien und ihre Einsatzchancen', *Adhäsion – Kleben und Dichten*, 7–8, 23–27.

Collano (2011), technical information, available at www.collano.com.

DIN (1981) DIN 53530, 02-1981; Testing of organic materials; Separation test on fabric plies bonded together.

DIN (1992) DIN 53835, 11-1992; Testing of textiles; tensile test for testing the elastic behaviour; knitted fabrics, single strain between two force limits.

DIN (1995) DIN EN ISO 9237, 12-1995; Textiles – Determination of permeability of fabrics to air.

DIN (2008) DIN 53121, 10-2008; Testing of paper and board – Determination of the bending stiffness by the beam method.

Evonik (2011), technical information, available at www.evonik.com.

Fornes M (2009), 'Holding it all together', Nonwovens Report, August.

Frauenhofer M, Ströhlein T, Fabig S, Boehm S, Herbeck L, Dilger K (2008), 'Induction technique in manufacturing preforms', *Mechanics of Composite Materials*, 44(5), 523–530.

Fries J A (1983), 'Adhesives for fabric lamination', *Journal of Coated Fabrics*, 12, 174–186.

Gluetex (2011), technical information, available at www.gluetex.de.

Gries T, Klopp K (eds) (2006), *Füge- und Oberflächentechnologien für Textilien*, Springer, Berlin, Heidelberg.

Habenicht G (2006), *Kleben – Grundlagen, Technologien, Anwendungen*, 6th edn, Springer, Berlin, Heidelberg.

Hartung I, Böhm S, Dilger K, Mott R, Brameshuber W (2010), 'Spacers for elements of textile reinforced concrete', *3rd fib International Congress Washington*, 29 May–2 June.

Holme I (1999), 'Adhesion to textile fibres and fabrics', *International Journal of Adhesives & Adhesion*, 19, 455–463.

Jentschmann (2011), technical information, available at www.jentschmann.ch.

Kalbe M, Wagener S, Grynaeus P (2000), 'Adhesive powder', WO 002000034406 A1, 15.6.2000.

Kan, C W, Yuen C W M, Tsoi W Y I, Tang T B (2010), 'Plasma pretreatment for polymer deposition – improving antifelting properties of wool', *IEEE Transactions on Plasma Science*, 38(6), 1505–1511.

Karahan H A, Ozdogan E (2008), 'Improvements of surface functionality of cotton fibers by atmospheric plasma treatment', *Fibers and Polymers*, 9(1), 21–26.

Kastner A (2008), *Faser- und Gewebekunde*, 15th edn, Verlag Handwerk und Technik, Hamburg.

Konrad B (2004), 'Bird's eye view of the global airbag market', *Bag&Belt 2004, 8. Internationales Symposium für Fahrzeuginsassen-Rückhaltesysteme*, 28–30 April, Cologne, Germany.

Linderer W (1953), 'Einrichtung zum Schutze von in Fahrzeugen befindlichen Personen gegen Verletzungen bei Zusammenstößen', German Patent No. 896312, Klasse 63e, Gruppe 70, 12.11.1953.

Mittal K L (ed.) (1993), *Contact Angle, Wettability and Adhesion, Festschrift in honor of Professor R J Good*, VSP, Utrecht, The Netherlands.

Mund F, Dilger K, Böhm S (2005), 'Improvement of durability and bond properties by use of adhesives for coating textiles in textile reinforced concrete elements', *Proceedings of the 28th Annual Meeting of the Adhesion Society*, Mobile, AL, The Adhesion Society, pp. 492–494.

Reumann R D (ed.) (2000), *Prüfverfahren in der Textil- und Bekleidungstechnik*, Springer, Berlin, Heidelberg.

Rychter J (2004), 'Kleben von technischen Textilien', *Bag&Belt 2004, 8. Internationales Symposium für Fahrzeuginsassen-Rückhaltesysteme*, 28–30 April, Cologne, Germany.

Ryu J, Wakida T, Takagishi T (1991), 'Effect of corona discharge on the surface of wool and its application for printing', *Textile Research Journal*, 61, 595–601.

Schindler M, Hellwig J, Klaes M (2009), 'Innovative Klebstoffe zur Membrankaschierung', *Technische Textilien*, 52(4), 195–197.

Studt H J (2006), 'Innovative use of thermoplastic film adhesives in automotive airbags', available at www.speautomotive.com/SPEA_CD/SPEA2006/pdf.

Wegener W, Sommer H (eds) (1960), *Die Prüfung der Textilien*, Springer, Berlin, Heidelberg.

Wootton D B (2001), *The Application of Textiles in Rubber*, Rapra Technology Ltd., Shrewsbury.

10
Bonding requirements in coating and laminating of textiles

E. SHIM, North Carolina State University, USA

DOI: 10.1533/9780857093967.2.309

Abstract: Coating and laminating are among the best known and oldest techniques to produce composite or hybrid textiles to achieve functionality and properties that cannot be achieved with single components. General characteristics of coated and laminated fabrics and several of their applications are also discussed.

Key words: coating, laminating, fabrics, adhesives.

10.1 Introduction

A coated/laminated fabric is one kind of composite or hybrid textile. Like other composite/hybrid textiles, it can provide functionality that cannot be achieved with a single component by combining multiple components. It is also capable of utilizing wide ranges of materials, processing techniques and creating numerous different structures targeted to deliver a specific performance. Because of their high functionality, versatility and ability to be customized, hybrid/composite textiles are better positioned to provide engineered solutions for diverse applications in traditional and non-traditional areas (Bitz, 2002). Coating and laminating are among the best known and oldest techniques to produce composite or hybrid textiles (American Chemcial Society, n.d.).

A coated fabric is composed of a fabric and any adherent material applied to one or both surface of the fabric (ASTM D123). Coating materials, which are mostly fluids or precursors of fluids (i.e., solid powders), are deposited on a substrate fabric, form a film, and then bond to the substrate surface. Both film formation of coating materials and bonding to the substrate are required to achieve quality coating.

A laminated fabric is defined as 'a layered fabric structure produced by joining several layers of *preformed* continuous sheet substrates in such a way that the identity of the continuous sheet material is retained' (ASTM D123). A laminating process uses fully formed sheet materials such as films, fabrics, and foams, while film formation is an essential part of a coating process. So, bonding of preformed material to a fabric substrate is the most critical aspect of a laminating process (Fung, 2005).

Despite their differences, both coating and laminating processes share many common principles. One of them is adhesion. Adhesion and interfacial strength directly affect performance and properties of coated/laminated fabrics, including tear strength, resistance to flex, and peel strength (Dartman & Shishoo, 1993). Adhesion is the attraction between two materials in contact and it can be described in terms of interfacial energy. The level of adhesion is determined by cohesion forces of bulk materials and the interactions between the layers, so the work of adhesion W_a, that is the energy required to separate two layers in contact, is defined as:

$$W_a = \gamma_1 + \gamma_2 - \gamma_{12} \qquad [10.1]$$

where γ_1 and γ_2 are the surface tension of component 1 and 2, respectively, and γ_{12} is the interfacial tension between components 1 and 2 (Dahlquist, 2001; Dartman & Shishoo, 1993; Wicks *et al.*, 1999; Yosomiya, 1990).

Equation [10.1] demonstrates that adhesion is the function of surface/interfacial energies of materials, which represent the nature of intermolecular interactions (Dartman & Shishoo, 1993). Generally speaking, the larger the work of adhesion, the better the adhesion is. The positive work of adhesion indicates the tendency of two surfaces to adhere, while the negative work of adhesion indicates a tendency towards delamination, as illustrated in Fig. 10.1 (Schonhorn, 1985). Low interfacial energy as the result of molecular affinity of two materials in contact leads to high work of adhesion. Chemical natures of molecules in coatings and substrates determine the work of adhesion, so their chemical affinities are most fundamental requirements to achieve good adhesion in coated and laminated fabrics. It is hard to achieve good adhesion between two layers lacking chemical affinity. In this case, interfacial adhesion can be improved by surface modification of one or both surfaces, such as grafting and plasma treatment (Paul, 2009). The addition of a third layer

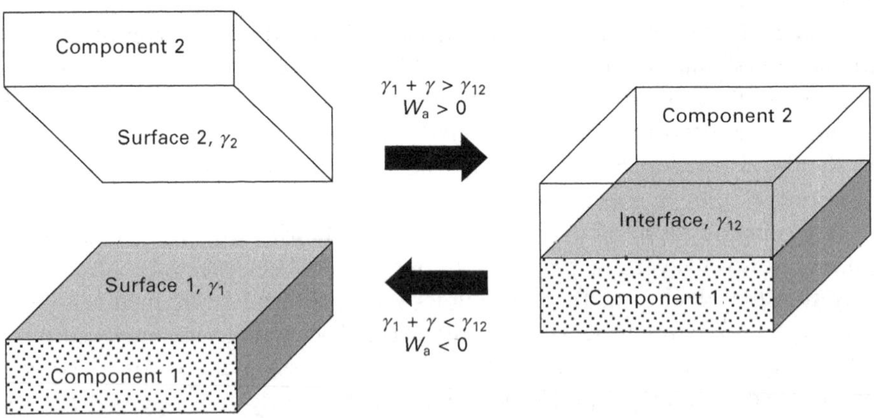

10.1 Work of adhesion.

having affinity to both layers, which is often called a tie-layer, can impart adhesiveness between materials otherwise chemically incompatible (Cole & Macosko, 2000).

When two materials have sufficient interactions and affinity, it is possible for polymer molecules to diffuse beyond the interface and intermingle with the other layer. This interfacial diffusion of molecules at the interface results in strong adhesion (Schonhorn, 1985). Having chemical affinity between two materials is one requirement to achieve interfacial diffusion, but interfacial diffusion is also governed by other factors, such as mobility of molecules, microstructure of materials, and contact condition – time, temperature, and pressure. When material is heated above its glass transition temperature, the polymer chains have mobility to penetrate deeper by diffusion, and adhesion strength increases if time permits (Wicks *et al.*, 1999).

Textile fabrics are highly porous so coating liquid or adhesive applied does not always stay on the surface but penetrates into the bulk of the fabric. Penetration increases the interfacial contact area of materials/fabric substrate and when penetrated liquid is solidified, mechanical locking is formed. Increasing the interfacial area and mechanical anchoring improve adhesion strength greatly, so adhesion is the function of coating material penetrations (Dartman & Shishoo, 1993; Wake, 1973; Yosomiya, 1990). However, penetration of coating into the structure of the fabric causes the internal part of fibers to lose their freedom of movement and, as a result, the fabric becomes rigid. Since the rigid structure cannot redistribute the stress applied, excessive penetration of coating in fabric deteriorates mechanical strength (Farboodmanesh *et al.*, 2005). Coating liquid penetration also causes reduction of porosity, so high penetration impacts permeability negatively. Therefore, the penetration depth of the coating liquid should be optimized to achieve good adhesion while minimizing impacts on fabric rigidity and permeability.

Coating liquid penetration into the fabric structure can be viewed as capillary intake of liquid. Then, it is governed by wettability of system, capillary and pore structure of fabric substrates, coating materials/adhesive rheology, and processing condition. Wettability is one of most basic requirements to achieve penetration. Wetting can be defined as the displacement of a fluid from a solid surface by a second immiscible fluid (Padday, 1978). In the coating process, wetting occurs when air in contact with fibers is replaced by a coating liquid. The most often used parameter describing wettability is a contact angle. A contact angle is defined as the angle between the liquid–air and solid–liquid interfaces when there is no relative movement of contact line (Fig. 10.2). The static contact angle is determined by the balance of interfacial tensions in the three-phase contact line as described in Young's equation (Young, 1805):

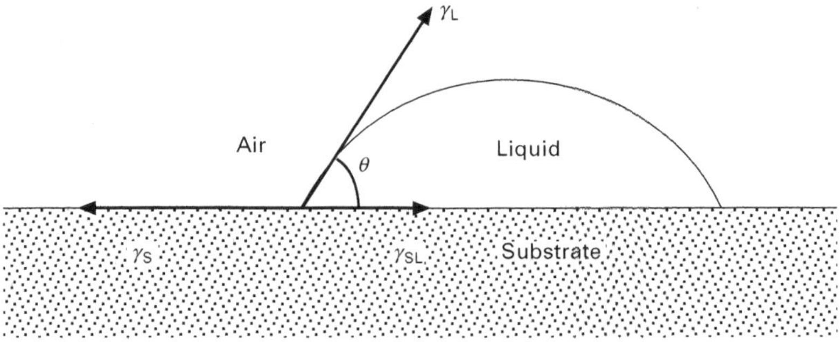

10.2 Interfacial tensions and the static contact angle.

$$\cos \quad = \frac{S - SL}{L} \qquad [10.2]$$

where θ is contact angle, γ_S is interfacial tension between solid and vapor, γ_{SL} is interfacial tension between solid and liquid, and γ_L is interfacial tension between liquid and vapor.

It relates the static contact angle and interfacial tensions of liquid–air, solid–air, and solid–liquid for conditions of thermodynamic equilibrium and a perfectly flat homogeneous solid surface. A low contact angle indicates good wettability of the system and when the contact angle between the coating liquid and the substrate is zero, the coating liquid spreads spontaneously without a limit (Berg, 1993; Dartman & Shishoo, 1993). In the case of a textiles substrate, the substrate surface is not flat but porous and the coating liquid penetrates by capillary wicking. Capillary pressure is determined by the Young–Laplace equation (Adamson & Gasy, 1997):

$$P = \frac{2 \quad \cos}{r} \qquad [10.3]$$

where ΔP is capillary pressure, γ is liquid surface tension, θ is contact angle, and r is the capillary radius.

The equation shows how contact angle governs liquid penetration. When $\theta <$ 90, ΔP becomes positive. Then the coating liquid spontaneously penetrates and fills the capillary. When $\theta > 90$, ΔP is negative, so the liquid cannot penetrate into the capillary of the substrate without external pressure. The capillary pressure is also governed by r, the capillary radius, which is determined by fabric structures and fiber size. So, fabric openness, construction type, and fiber/yarn size affect pore diameter and impact penetrations of liquid.

However, one should note that wetting and penetration during the coating process is a dynamic process, where the liquid/substrate contact line moves at high speed and both coating liquid and the substrate are under external forces. Therefore, the equilibrium contact angle that measures the tendency of spontaneous spreading and penetrating without the presence of external pressure has limitations (Dussan, 1979). The kinetics of wetting, governed both by liquid rheological properties and surface energy, influence interface formation and adhesion (Bluestein *et al.*, 1975; Mewes, 1989).

In addition to interfacial properties, interfacial energy, penetration and anchoring, molecular diffusion, and bulk properties also play a role in interlayer adhesion strength. It is hard for materials with poor mechanical strength to form strong interlayer adhesion even when mechanical anchoring is achieved by penetration. Differences in thermal shrinkage between the layers cause poor adhesion or delamination.

10.2 Materials and adhesives in coating and laminating

As pointed out previously, the advantages of coated and laminated fabrics are based on their ability to utilize many different materials in one structure where all components synergistically deliver unique properties and functionalities. There are numerous different possible combinations, but most common compositions are fabric/polymer coating, fabric/film laminates, fabric/foam laminates, fabric/fabric laminates and net/fabric laminates. Moreover, coating and laminating are very versatile processes and capable of incorporating non-traditional and novel materials into textile structures, such as inorganic particles, metal and other conductive layers, fiber bundles, electrospun nanofiber webs, and membranes. In this section, we will discuss the most basic and common components of coated and laminated fabrics.

10.2.1 Fabrics

A textile fabric is a carrier of other components in coated and laminated fabrics and provides mechanical strength and integrity to whole structures (Fung, 2002). Textile fabrics are a fibrous assembly where numerous fibers are held together and individual fibers possess some degree of mobility. Then, structures of fabrics are inherently porous and compliable (Hu, 2004). Depending on how fibers are aligned, structured and held together, fabric properties can be dramatically different even among fabrics composed of exactly the same fibers. Since fabric properties are governed by not only chemical composition of fibers but also fabric construction and structures, the selection of fabrics in coating and laminating has to be made based on comprehensive understandings of fabrics, including fiber chemical composition, fiber size and cross-sectional shape, fabric construction, and finish present on the fabrics.

Most common textile fibers are organic polymers including cotton, wool, rayon, polyester, polypropylene, nylon, and acrylics. For technical applications inorganic fibers including carbon, ceramic, and metal fibers can be used. The type of polymers and their chemical composition mostly determine affinity to other materials, i.e., adhesion and wettability. Cellulosic fibers such as cotton and viscous rayon are highly polar and hydrophilic, so aqueous media easily wet the surface and penetrate into fabrics. It has high affinity to polar substance, which will result in good adhesion with polar coating material/adhesives. In contrast, polypropylene and polyethylene are non-polar and oleophilic, so it is hard for the aqueous coating system to wet and penetrate fabrics made of these fibers without wetting agents. They also show little adhesion to polar adhesives and coating layers, which have better affinity to oleophilic materials. The chemical composition of fibers mostly determines their thermal properties and resistance to heat. Therefore, fiber chemical compositions need to be considered when setting the processing temperatures.

Fiber size and their cross-sectional shape also affect penetration of coating resin/adhesives and adhesion between layers. Smaller fibers create finer capillary networks and coating liquid tends to penetrate deeper and faster in materials with good wettability. In addition, smaller fibers have a larger surface area, which leads to a larger contact area between fibers to the coating layer even at the same penetration depth. This will enhance adhesion strength.

Fabrics consist of fibers, or yarns made of fibers. Fabric constructions are categorized into woven, knit, and nonwoven fabrics. Woven is the first fabric used in the production of coated and laminated fabrics and is still the most commonly used ('Cotton textiles for plastic laminates', 1944). Woven fabrics consist of warp and weft yarns interlaced at right angles (Joseph, 1981). Warp and weft density and intersection patterns determine density and porosity of fabric, which are key factors in laminated and coated fabrics. Woven construction creates a relatively rigid structure with little stretch and it can be used in the coating/laminating process to impart high stress and tension to fabrics (i.e., direct coating). The most frequently used woven construction in laminates and coatings is plain woven, but twill and basket constructions, and their modifications are also used (Fung, 2002).

A knitted fabric is produced by 'interloping one or more ends of yarn or comparable materials' (ASTM D123). It is a fast and low cost fabric production technique. Knitted fabrics have soft hand and good drapability, but tend to have high stretch and openness. Therefore, they are not desirable substrates for coating and laminating. Choices of coating and laminating methods for knitted fabrics are limited, and in most cases they cannot be directly coated (Fung, 2002).

Nonwovens, 'a textile structure produced by bonding or interlocking of fibers by mechanical, chemical, or thermal means' (ASTM D123), have become increasingly popular as substrates for coating and laminating with the advances in nonwoven technology. Nonwoven fabrics are often called engineered fabrics because of their versatility and unique ability to be customized to provide targeted properties. They have grown a broad array of multifunctional and high performance products (Pourdeyhimi, 2004). The growth of nonwovens is driven by high speed, low manufacturing cost as well as diverse processes capable of producing a wide range of structures and utilizing a variety of different raw materials. Nonwoven fabric technology continues to grow both in quantity and quality (Pourdeyhimi, 2004). The combination of nonwoven and coating and laminating processes provides even more possibility for creating highly functional products. Coating and laminating of nonwovens can be added as finishing step, simply to improve abrasion resistance or to create the unique aesthetics of artificial leather for highly functional selective barrier fabrics.

Regardless of type of fabric construction, their structural characteristics and properties impact the processability and performance of final composite fabrics. Fabric characteristics can limit selection of coating/laminating technique and affect optimum process conditions (temperature, speed, etc.). Openness, porosity, and pore structures of fabric substrates affect penetrations of resins and adhesives, coating film formation, and strength of adhesion between layers. When a fabric has an open structure, adhesive or coating resin penetrates more readily. High penetration can increase adhesion but excessive penetration may degrade mechanical properties and stiffen the structure by limiting fiber mobility inside. If fabric becomes too open, resins and adhesive cannot bridge the pores and fail to form a continuous film layer, which is critical in some applications, such as waterproof apparel (Fung, 2002). However, low resin penetration in tight structures may result in poor resin adhesion.

Fabric dimensional stability is another consideration, since fabric substrates need to withstand stress and heat imposed during the coating and laminating process. Fabrics that shrink and/or stretch excessively during the processes can be distorted and damaged. The capability of fabrics to withstand the amount of heat and stress without damage will limit the choice of process technique. For example, delicate fabrics are treated with techniques where process tension is minimized, such as transfer coating. In this aspect, fabric pre-treatments aimed at improving dimensional stability are desired. Heat setting at a temperature higher than the expected processing temperature is a common way to achieve better stability (Fung, 2005). Surface roughness and smoothness will be related to coating thickness uniformity. Certain coating methods, such as knife coating, produce a rough and irregular surface, frequently causing irregular coating thickness. Uniformity and defects of fabric are also important because the presence of small defects can be exaggerated during the process or can cause material failure during the process.

10.2.2 Film

A film is a thin continuous sheet and has no or limited porosity and significantly higher solidity than a fabric. As a result, a film generally has higher barrier properties than a fabric and can be completely impermeable to liquid and gases. However, its stiffness increases greatly as thickness increases, so it is hard to produce flexible materials with enough mechanical strength using film by itself.

The most frequently used film materials are organic polymers including PVC (polyvinyl chloride), polyurethane (PU), natural and synthetic rubbers. Specialty films, such as breathable barriers which permit moisture to transfer while acting as a barrier for liquids, and reflective films having a smooth surface that provides high reflectivity, are often laminated to fabrics to improve strength and stability. Film/fabric laminate products include sails, hot air balloons, packaging, medical gowns and drapes, and protective garments. Films can be a form of adhesive, too. Instead of applying liquid adhesive, a preformed adhesive film is inserted between the layers in the laminating process.

10.2.3 Foams

Plastic foams, or cellular plastics, have a unique structure comprising a solid polymer matrix and voids containing air phase (Okoroafor & Frisch, 1995). According to the connectivity of voids, foam structures can be open or closed. Open cell foams are composed of predominantly interconnected cellular voids, so air can pass freely between the cells, while closed cell foams consists of mostly individual non-interconnecting cells where air is entrapped (ASTM C168; Mills, 2007).

Due to presence of air voids, foams in general have good insulation properties and energy absorption properties. Foams can be produced with a great variety in density, flexibility, and materials and exhibit a wide range of properties. Cell type – open or closed – as well as cell geometry – size and shape – greatly affect foam properties. For example, closed cell foams are most suitable for thermal insulation and open cell foams are better for acoustic insulation (Okoroafor & Frisch, 1995). Polymer foams laminated to fabrics are used in seating, car interior, protective packaging, and safety gear (Mills, 2007).

Typical polymer foam materials are PU, and other thermoplastics such as polyethylene (PE). PU foams are most common in fabric laminates. PU foam can be made with different density and porosity and both in open cell and closed cell structure. There are different types of PU chemistry present. Polyester-based PU can be easily laminated by the flame lamination technique but has limited hydrolysis resistance. Polyethylene-based PU foam provides better hydrolysis resistance but flame lamination cannot be employed. It is also possible to impart physical chemical functionality such as flame retardancy in PU chemistry.

Polyolefin foams, such as polypropylene (PP) and polyethylene (PE) foam, possess excellent microbial, oil, solvent, and chemical resistance. These can be used in car components such as headliner, dashboard, and door casing. They have shock and energy absorbing properties and provide heat, noise, and vibration insulation. Therefore they are also used in body protection clothing, luggage, and underfloor insulation (Mills, 2007; Okoroafor & Frisch, 1995).

10.2.4 Coating materials and adhesives

In a coated fabric, coating layers deliver added functionality which fabric itself cannot easily provide. Examples are barrier properties, impermeability, flame retardancy, insulation properties, aesthetics, hand, and printability. The most desirable properties coating materials deliver are governed mostly by the specific application. Adhesives are applied between substrate layers to be laminated and provide adhesion. So, the main requirement of adhesives is providing good adhesion to substrates. In addition to imparting desired properties to the product, coating/adhesive materials should be compatible with the coating process and have good adhesion properties on the substrate fabric (Shim, 2010). To satisfy these requirements, most coating materials and adhesives are complex mixtures and highly formulated with multiple components. The typical formulations include polymeric resin, medium, and additives (Wicks *et al.*, 1999).

Polymeric resin

The main component of coatings is polymer or polymer precursor to be polymerized during the coating/laminating process. A continuous film of polymers is formed and bonded to the substrate when the medium evaporates or melt solidifies. It provides bonding among other components and the substrate, so determines the adhesion properties and governs the mechanical properties of the coating layer. Types of polymers used in the coating process are poly (vinyl acetate) (PVA) and its copolymer, acrylic, polyvinylchloride (PVC), polyurethanes (PU) and natural and synthetic rubbers. Other special formulations such as fluoropolymer, silicone, and PVDC (polyvinylidene chloride) are also used. In many formulations, crosslinking between polymer chains can be formed to provide durability and abrasion resistance. Crosslinking reactions mostly occur after coating materials are deposited on the substrate, when a reaction condition is reached by elevated temperatures, mixing of two components, or UV irradiation. The degree of crosslinking affects many coating properties, such as tackiness, wearing properties, aesthetics of handle, drape, and softness.

PU polymers are condensation reaction products of polyols and diisocyanates as illustrated in Fig. 10.3(a). They consist of soft segments of long polyol chains and hard segments from diisocyanate. PU chemistry can be easily tailored to provide a wide range of properties. Lengths of hard and soft segments affect adhesive properties, modulus, tensile strength, hardness, tear strength, elongation, and glass transition temperatures. Hard segments form a pseudo crosslink network. So the long hard segment leads to high glass transient temperatures, modulus, hardness, tear strength, and resistant to compression and extension (Lay & Cranley, 2003).

Crosslinking can be easily introduced both in soft and hard segments and it results in high modules, hardness and few elastomeric properties. Polyether-based PU made from polyether polyol is most widely used but polyester-based PU is also used. PU has excellent flexibility, good elongation and abrasion properties, and many grades are available in solvents, dispersion, and hot melt

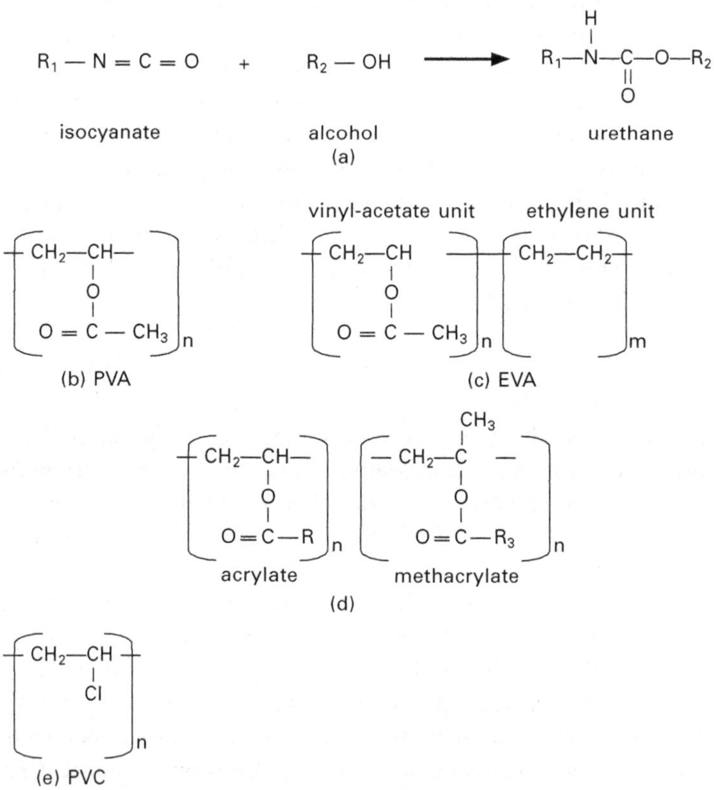

10.3 Chemical structures of coating resins: (a) basic urethane chemistry, (b) PVA, (c) EVA, (d) acrylate and methacryalte, and (e) PVC.

adhesives (Fung 2002). They are used for barrier coating, artificial leather, and adhesives for laminates (Lay & Cranley, 2003). Polyurethane reactive adhesives usually consist of oligomers which are converted to polymer later (Stepanski & Nebe, 1997).

Polyvinyl acetate (PVA) is prepared by emulsion polymerization through a free radical polymerization mechanism (Fig. 10.3(b)). EVA (ethylene-vinyl acetate) (Fig. 10.3(c)) is a copolymer consisting of polar, amorphous PVA block and partially crystalline polyethylene block. EVA properties are governed by vinyl acetate content and melt flow index (Martinez-Garcia *et al.*, 2004). They are mostly used as the form of water suspension (Geddes, 2003). They can be easily molded by heat, are relatively inexpensive, and can provide good adhesion properties and strength. They are sensitive to water, so exhibit poor washing resistance.

Acrylic-based latex (Fig. 10.3(d)) was introduced in the 1950s. Its basic formulation includes acrylic monomers and initiators, which undergo radical polymerization during curing. Selecting proper monomers among numerous available monomers and initiators allows tailoring properties. The formulation also contains polymers and organic/inorganic modifiers (Damico, 2003). One of the most interesting developments is energy curable acrylic formulations (Fung, 2002). Most other coating formulations are cured by heat, and it frequently requires drying prior to curing, which consumes energy and occupies space. Energy curable formulations contain photo-initiators, and radical polymerization is initiated when they are exposed to UV or E-beam energy. This is a fast process and requires less energy.

Unsaturated polyester-based chemistry is mostly used for adhesives. It is the product of the reaction between dicarboxylic acids or anhydrides and diols. The curing temperature can be controlled by introducing a catalyst into formulations. Other parameters such as molecular weight, viscosity, type, and concentration of monomers affect coating properties. Increasing molecular weight improves hardness, tensile and flexural strength and heat distortion temperature (HDT). Density of crosslinking is governed by monomer types, and high crosslinking density increases hardness and heat resistance but reduces flexural strength (Pizzi, 2003a).

Natural rubber was the first material used to coat fabrics (American Chemcial Society, n.d.). Natural and synthetic rubbers are elastomers that can be stretched many times their original size without failure and almost instantly recover from deformation when stress is removed. Natural rubber is obtained from the rubber plant and mostly composed of polyisopropen. There are many different types of synthetic rubbers and they include styrene-butadiene rubber (SBR), isoprene-isobutylene rubber (IIR, or butyl rubber), ethylene-propylyene polymers (EPM), polychlorpren rubber (CR), nitril-butadiene rubber (NBR) and silicone rubber. Rubber has a low glass transition temperature, so retains a mostly rubbery status at room temperature.

Vulcanization forms crosslinks to impart elastomeric properties (Sen, 2008).

Polyvinyl chloride (PVC) (Fig. 10.3(e)) is a commonly used coating material because of its low cost, excellent physical properties, and easy processability. It is produced by free radical polymerization of vinyl chloride. It has a very low crystal content and is considered an amorphous polymer. Most PVC is plasticized by addition of plasticizer to improve processability and to impart softness, flexibility, and extensibility (Sen, 2008).

Coating medium

Most coating and laminating systems require adequate flow properties of coating materials and these are provided by the medium – the carrier of other coating components. They ensure coating materials are transferred onto the substrate fabric and distributed uniformly. The medium also controls the coating material's rheology and wettability, which impact film formation and penetration during the coating process. They are eventually evaporated or removed from the product at the end of the process. Depending on the medium used, coating materials are classified as solvent-borne, waterborne, and solid. Solvent-borne coating materials were the most popular until 1945, as it is easy to evaporate them after application. However, their use has diminished because of environmental concerns and government regulations regarding high VOC (volatile organic compound) emission and toxicity of solvents. Water-based coatings do not produce harmful gas but it is not easy to remove water. Energy intensive drying units occupying large floor space are required in waterborne coating systems. Waterborne coating formulations – solutions, emulsions, and dispersions – are prepared through mixing of components, de-aeration, and filtering into one-part or two-part systems. One-part coating materials are supplied in ready to apply form by coating formulation suppliers having all the necessary components, but two-part coating systems need to be mixed just before application, as mixing the components initiates reactions such as polymerization and crosslinking (Licari, 2003). In most two-part systems, mixing and dispensing must be carried out immediately before the application process (Licari, 2003). Regardless of whether a one-part or two-part system is used, improper preparation and storage of coating formulations may cause changes in liquid rheology, adhesion properties, and reduction in process reproducibility, resulting in degradation of coating quality and uniformity.

Solid coating materials, such as powder and hot melt, use no medium but materials melt at processing temperature to provide flow properties. Materials for hot melt coating are supplied in the form of pellets or powder. Melting and filtering units are connected directly to the applicator unit that delivers hot melt at the right temperature and rheology. Powder coating consists of binders and additives without a liquid medium, so does not require de-aeration or filtration (Fung, 2002; Waelde, 2001).

Additives

Additives and pigment components are also incorporated into basic coating formulations of binders and media. They can impart color, opacity, antistatic properties, and processability. Catalysts, flow modifiers, emulsifiers, and wetting agents are common additives used in coating formulations (Wicks *et al.*, 1999). Some additives are used to improve functionality, including UV stabilizers, FR chemicals, heat stabilizers, antioxidant, fillers for mechanical property improvement, filler for economy, pigments, and coupling or bonding agent to promote adhesion.

10.3 Coating process

In the textile coating process, one or more layers of coating materials are deposited on the surface of fabric substrates to produce a uniform and stable coating layer (Shim, 2010). This is achieved by a three-step process: metering, transferring, and fixing of the coating materials to the substrate (Greer & Stelling, 1994; Scott, 1995). Control of these processes, in addition to material characteristics, determines the amount of coating materials applied to the substrate, uniformity of the coating layer, adhesion between the coating layer and the substrate, penetration of the coating material into the substrate, and eventually performance and quality of coated fabrics (Scott, 1995).

Metering controls the amount of coating material to be applied to the fabric substrate. Coating processes can be categorized by whether and when the metering occurs in free meniscus coating, pre-metering coating, or post-metering coating. In free meniscus coating, coating thickness is determined solely by the rheological properties of coating fluid and substrate movement without metering. It is the simplest coating application method, but its production rate and coating thickness is too dependent on material characteristics. Therefore, most commercial coating techniques employ some method of metering. Pre-metering coating techniques use a metering device, e.g. a displacement pump and a metering roll to control the amount of coating materials applied on the substrates. In post-metering coating, an excess amount of coating material is applied to the substrate, followed by removal of the excess material. Examples of post-metering coatings are knife coating and air knife coating. In some coating techniques, both pre-metering and post-metering are used.

Transferring is the process where the coating material is deposited on the substrate and forms a coating layer. Achieving a uniform coating layer transfer on to the substrate with controlled penetration is desirable and is affected by wettability and rheology of the coating liquid and coating applicator geometry. Series of rollers or calendars, spray, direct extrusion to the substrate, or immersion of the substrate in a coating bath are ways to transfer coating materials to the substrate. The final step of the coating process is fixing, where transferred and metered coating layers are fixed to the substrate through drying, curing or solidifying. In this step, liquid film on the substrate becomes a solid coating layer and evaporation of media or cooling down of melt occurs (Geddes, 2003). Polymerization reaction of polymer precursors and crosslinking reactions also occur.

Many different types of coating techniques have been developed to adapt characteristics of a wide range of coating materials. Even though the majority of coating materials are liquid, solid or gas has also been used and each requires different processing methods to form layers on the substrate. The characteristics of the substrates also have an impact on coating processes. A highly delicate stretchable fabric requires completely different coating techniques and conditions from the substrate with dense structure and high strength. In addition to material properties, cost, production rates, reproducibility, environmental impacts, versatility and fast adjustment are also important criteria for selection of coating technique (Shim, 2010).

10.3.1 Immersion/saturation coating

Immersion or saturation coating is one of the simplest coating techniques. The fabric substrate passes through a bath of coating liquid and picks up coating materials on it. The amount of coating liquid applied to the substrate and their penetration are the result of processing dynamics, fabric structure, coating liquid rheology, interfacial wetting tendency of the coating fluid and fabric surfaces (Kistler & Schweizer, 1997). Excess coating liquid can be removed by passing through a squeeze roll or nip rolls (Fig. 10.4) (Prud'homme &

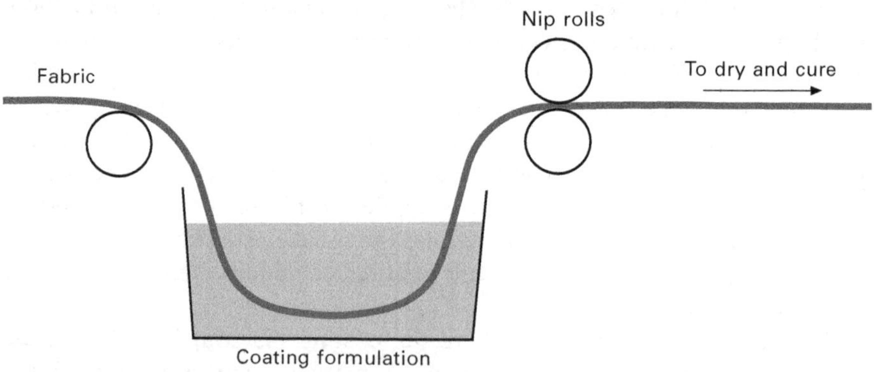

10.4 Saturation coating.

Khan, 1996; Whiteman, 1993). Coating materials tend to penetrate deep and generally require high drying requirement.

10.3.2 Knife or blade coating

Knife coating is a process where an excess of coating material is applied to the substrate and removed by a metering blade to achieve the desired coating thickness (Fig. 10.5). There are different knife configurations, which are floating knife, knife-over-roll, knife-over-fixed-table and knife-over-conveyor (Herrera, 1991). In a floating knife coating or a blade over air, the knife actually touches the surface of the fabric, which is only supported by tension between two rolls. The doctor blade pressures the upper surface of the fabric substrate and forces the coating formulations to penetrate into the substrate. Fabric tension and the depression of the knife mainly control the amount of coating applied and the depth of penetration, but sharpness of the blade and knife angle alignment also affect them (Woodruff, 1992). This method is usually used for applying relatively small amounts of coating material to the fabric (Fung & Hardcastle, 2000; Scott, 1995).

A more common technique is knife-over-roll coating where the knife blade is suspended above a roller and does not touch the substrate directly. In this configuration, the coating layer thickness is controlled by a gap between the substrate and the knife by setting the gap between the roller and the blade. This enables precise thickness control, but variations in substrate thickness

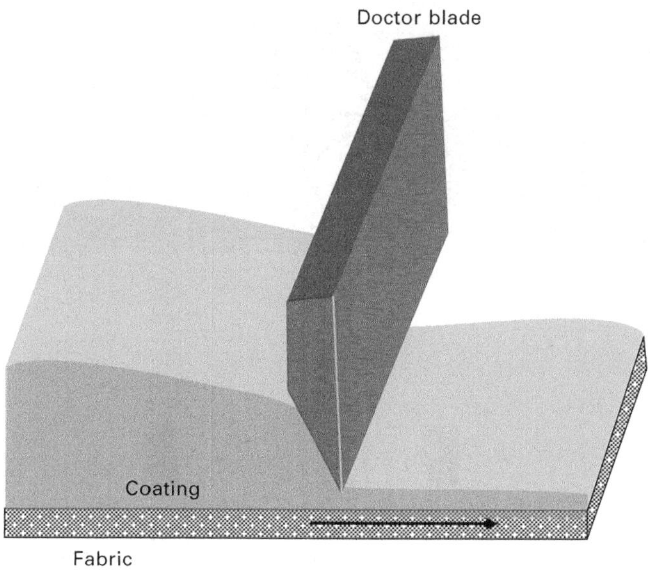

10.5 Knife coating.

may result in non-uniform coating thickness (Fung & Hardcastle, 2000; Herrera, 1991; Scott, 1995; Woodruff, 1992). Blade geometry (the shape and angle of the blade), blade flexibility, and the rheology of the coating fluid also affect the amount of coating applied and the extent of penetration into the substrate (Schindler & Hauser, 2004). Air knife coating uses an air jet to blow off excess coating materials and to control coating thickness instead of using a knife blade.

10.3.3 Metering rod coating

Metering rod coating uses a wired wound metering rod to control the amount of coating materials applied to the substrates. A metering rod is constructed with a stainless steel core tightly wound with polished stainless steel wires and is also called an applicator rod, a Mayer bar, an equalizer bar, a coating rod, or a doctor rod. As illustrated in Fig. 10.6, it allows coating material to pass through only between the wires, so the amount of coating applied to the fabric substrate is governed by wire construction. The cross-sectional area of the grooves between the wire coils of the rod directly determines the amount of coating materials that pass through it, and thus wet coating thickness. Using metering rods wound with large diameter wires produces a thick coating, while metering rods with fine wire results in a thin coating

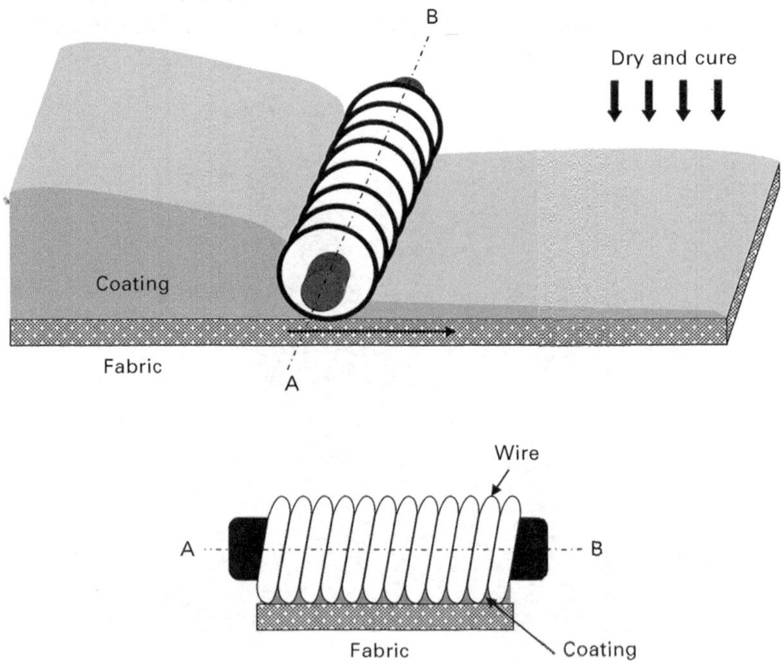

10.6 Metering wire coating.

layer (MacLeod, 2001). Coating thickness is easily altered by replacing the metering bar, which is quick and easy. Metering rod coating provides precise coating thickness control at a relatively low cost. In order to precisly control coating thickness, the metering rod must maintain intimate contact with the substrate fabric, but not to cause substrate deformation.

One limitation of the metering rod coating is striations on the coating layer. A wire wound bar leaves striations on the coating liquid surface immediately after metering, and these will eventually be leveled and smoothened by liquid surface tension. However, this is not an immediate process, and if the coating material surface cannot be leveled before drying, a smooth coating surface cannot be achieved. High viscosity liquid takes a long time to level, which limits coating speed. Coating with a metering rod works best with low viscosity liquid, since it flows easily between the wire windings and levels quickly.

10.3.4 Roll coating

Roll coating is a pre-metered coating and uses a series of rollers to meter and apply coating liquid on a substrate. A metered film of coating liquid is first formed on the roller surface before it is applied to the substrate, so the amount of coating material delivered to the substrate is nearly independent of the fabric properties and structures. Mostly determined by the rheology of the fluid and the relative speed of two rotation surfaces, precise control is possible (Alonso, 2003; Greer, 1995; Abbott *et al.*, 1972). Varying substrate thickness does not result in uneven coating thickness (Scott, 1995; Glawe *et al.*, 2003; 'Laminating and coating: flexible future', 2002).

The simplest roll coating set-up uses a single rotating roller. The bottom half of the roller is immersed in a coating liquid bath and the upper part of the roller is in contact with the fabric substrate (Fig. 10.7). As it rotates, the coating liquid forms a film on the roller surface and part of the liquid film is transferred from the roller surface to a fabric substrate. The amount of coating on the substrate is governed by hydrodynamics. The rotation speed of the roller, the substrate speed, and rheological properties of coating fluid (surface tension, viscosity, and density) are factors determining coating thickness (Wright, 1981). In this set-up, one roller is a metering device as well as an application device. More precise control is achieved by addition of more rollers.

Three-roll coating uses a metering roller, an applicator roller and a back-up roller. Common three-roll configurations are nip feed coating and L-head coating. In three-roll nip feed coating (Fig. 10.8(a)), the nip formed by a metering roller and an applicator roller is flooded with coating liquid and functions as a reservoir (Greer, 1995; Grant, 1978). The applicator roller picks up the coating liquid from the nip and the amount of coating liquid

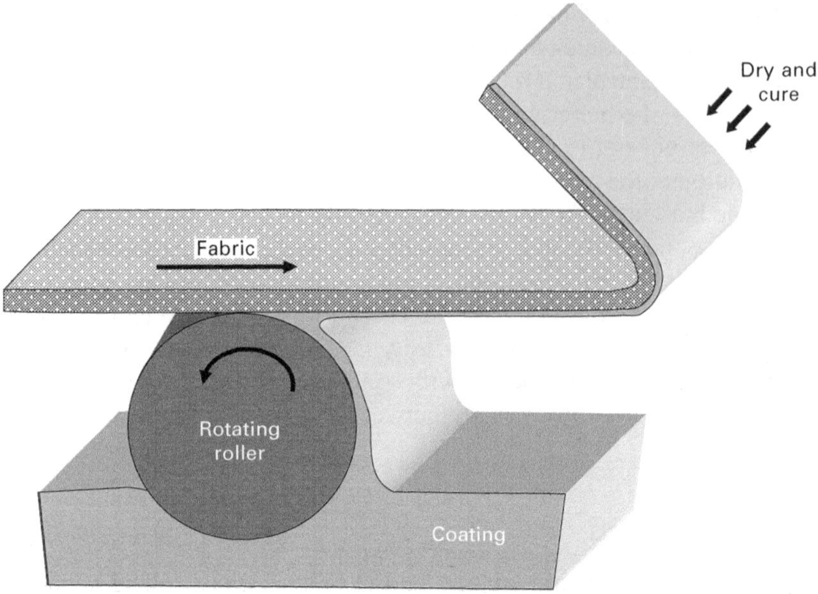

10.7 Roll coating (one roll reverse applications).

delivered to the fabric substrate is metered by the metering roller rotating in a reverse direction to the applicator roller. After coating liquid transfers from the metering roller to the applicator roller, any coating liquid remaining on the metering roller surface is cleaned by a doctor blade, otherwise it will result in coating defects, such as streaks or film roughness. Film formed on the applicator roller is deposited onto the substrate surface supported by a back-up roller. This configuration needs only a minimal amount of coating fluid, but a drawback is possible coating fluid leakage. It can be problematic when coating liquid has low viscosity. In three-roll pan feed or L-head coating (Fig. 10.8(b)), a liquid film is formed on the applicator roller rotating through the coating liquid, metered by a metering roller, and deposited on the substrate fabric on a back-up roller. To increase coating speed, one more roller – a pick-up roller running at a reduced speed – can be added, and this system is called a four-roll pan-fed coating system (Greer, 1995; Grant, 1978, 1981).

Configurations shown in Fig. 10.8 are called reverse metering, since applicator and metering rollers rotate in opposite directions. When they rotate in the same direction, it is called forward metering (Hannachi & Mitsoulis, 1990). Reverse metering produces a smoother film with better stability, while forward roll metering is prone to generate unstable, non-uniform films. Therefore reverse roll coating is more commonly used (Greer, 1995; Grant, 1978). Roller coating can use water-based solutions, solvent-based coating materials as well as hot melts (Grant, 1981). In hot melt roller coating, solid

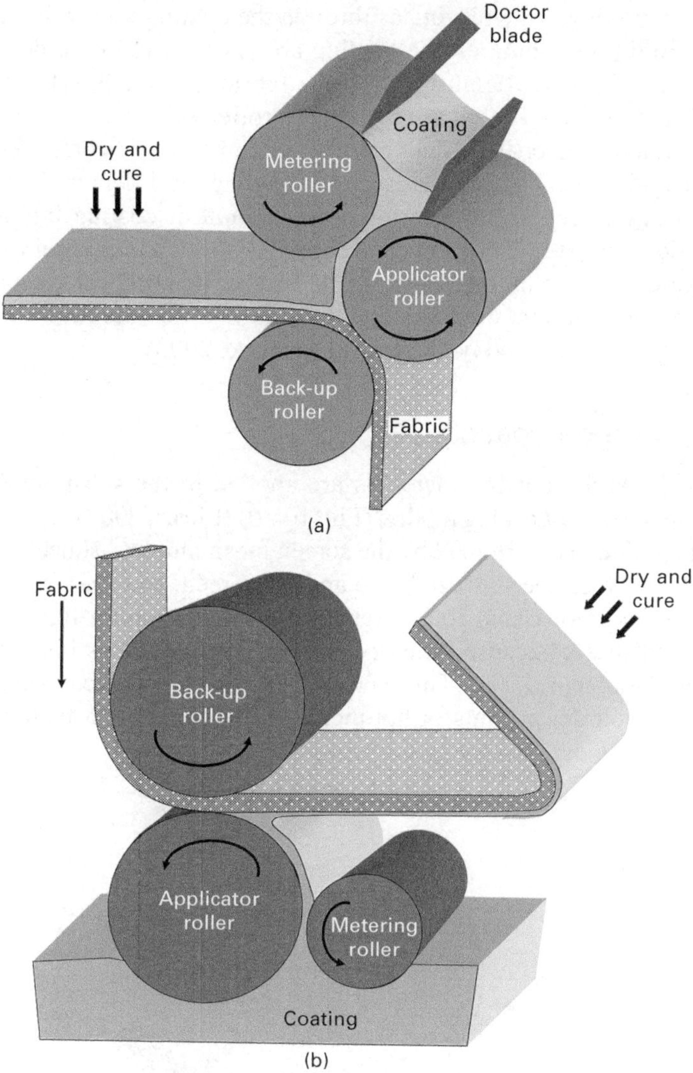

10.8 Three-roll coatings: (a) nip feed coating and (b) L-head coating.

pellet is melted between the heated melt rollers, forming a melt film and deposited on a substrate. The substrate fabric is usually preheated before hot melt is applied (Zickler, 1978).

10.3.5 Gravure or engraved roll coating

Instead of a smooth roll surface, the coating roll surface can have an engraved pattern to carry the coating liquid. In engraved roll coating, coating liquid fills the engravings when it rotates through the coating reservoir and excess coating liquid remaining on the landing area is removed by a doctor blade before the roll presses against a substrate fabric (Fig. 10.9) (Hewson *et al.*, 2006; Grant, 1981). The engraving on the roller acts as a metering device and the amount of coating delivered to the substrate is controlled mainly by the engravings. The land areas, cell opening, cell depths, cell volume, cell angles, and cell space determine the amount of coating liquid applied to a fabric substrate (Robinson & Marrick, 2006). Coating liquid viscosity, application pressure on the substrate, and substrate structure may influence actual delivery amount ('Laminating and coating: flexible future', 2002). It can also create patterned coating (Stukenbrock, 2003).

10.3.6 Screen coating

In screen coating, coating materials are applied to the substrate through a mesh screen by a squeezing motion (Fig. 10.10) (Licari, 2003). The amount of coating applied is determined by the screen mesh number, squeeze pressure, the angle between the squeeze blade and the screen, and the viscosity of the coating liquid (Goossens, 2001). Screen coating is ideal for lightweight and delicate substrate, because little or no friction or tension is imposed on the substrate (Goossens, 2001). This method is also easily adapted to use hot melt and in hot melt dot printing or hot melt printing where pre-melted hot melt

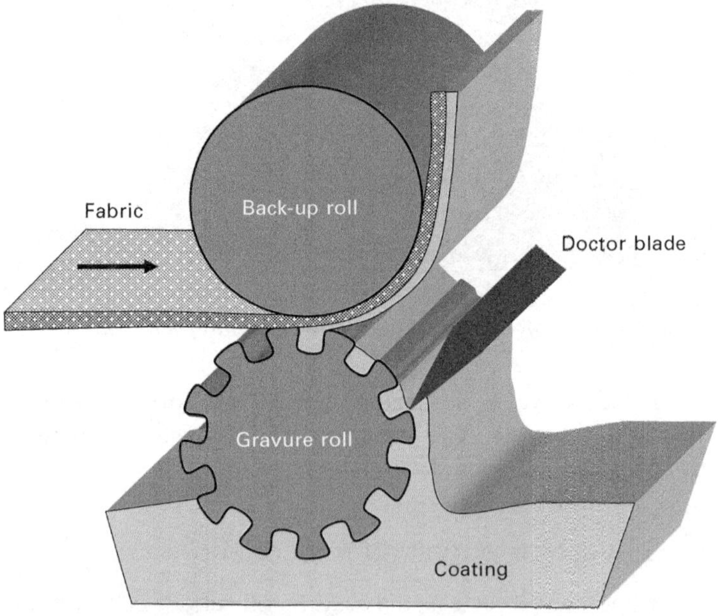

10.9 Gravure roll coating.

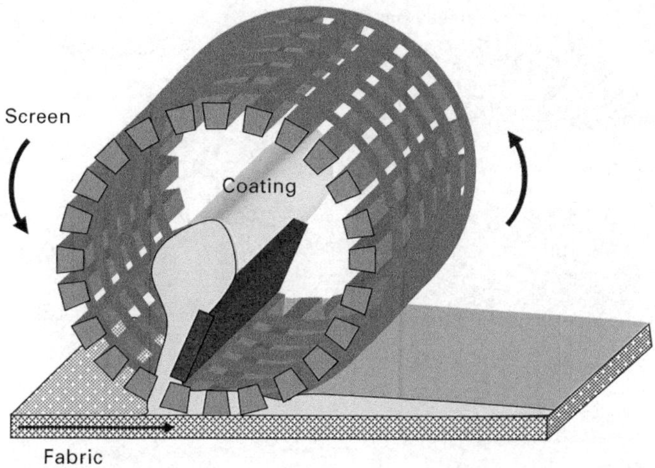

10.10 Rotary screen coating.

resin is deposited on the substrate through the screen. Advantages of using screen printing over engraved roll coating are relatively low cost of screen, shorter exchange times, lower friction (Goossens, 2001; 'Laminating and coating: flexible future', 2002) as well as good pattern definition. Since it is capable of providing a good pattern definition and precise coating thickness control, screen coating is also used for electrical applications, such as circuit board printing (Licari, 2003; Matthews, 1993).

10.3.7 Transfer coating

Direct application of coating material on a delicate substrate is challenging since tension required in the process often deforms or damages fabrics. It is also hard to prevent excessive penetration of coating liquid into a fabric substrate with very open structure. One solution to this problem is a sequential coating process where the coating material is first applied to a release paper and later transferred to the fabric substrate (Fig. 10.11). This process is called transfer coating (Scott, 1995; Fung, 2002).

A desired amount of coating material is first applied to a silicone release paper through one or other direct coating method and then dried. Then, this coated silicone release paper and the fabric substrate pass through the laminating rollers under heat and pressure, so coating layers backed by a release paper forms a bond with the substrate. After peeling off the silicone release paper, coating layers remain on the surface of the substrate fabric (Scott, 1995; Fung, 2002). It is more expensive than direct coating methods, but it can produce a flexible coated fabric because the process imposes little or no tension and penetration of coating formulations is low (Fung, 2002; Keeley, 1991).

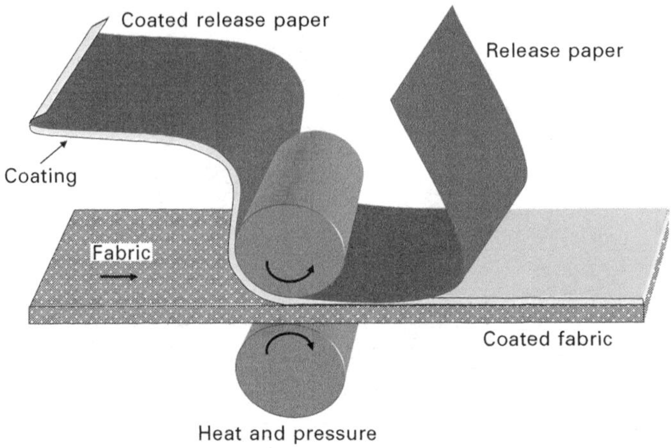

10.11 Transfer coating.

10.3.8 Spray coating

A spray coating is where a coating material is sprayed directly onto the substrate surface. Spray generation techniques are pressed air vaporization, airless pressure spray, hot flame spray, hot vapor impelled spray, electrostatic spray, and dry powder resin spray (Licari, 2003; Waelde, 2001), but the most common method is pressed air vaporization, where air and coating materials are pressed out through a nozzle (Glawe *et al.*, 2003). This is another method to cause minimal tension on the substrate, but it is difficult and expensive to achieve a uniform coating on a wide width of textile substrates (Woodruff, 1997).

10.3.9 Slot die and extrusion coating

Slot die coating is the coating process where a plasticized coating compound or a coating liquid is pressed through a die and transferred directly onto the surface of the fabric substrate (Whiteman, 1993; Zickler, 1978). The traditional slot die configuration is 'a closed system' where the die lip is in contact with a substrate backed by the roller, and coating materials are not exposed to air during the process (Glawe *et al.*, 2003). In this configuration, the substrate is often under high shear pressure, so it must be able to withstand it. Some modern slot die systems can have a gap between a die lip and the substrate fabric to impart less tension ('Laminating and coating: flexible future', 2002). The die configuration, the lip shape and lip opening, and die-to-roller gap, flow rate and flow distribution influence coating quality. A balance between die pressure and substrate tension should be achieved for good quality coating.

Extrusion coating mostly refers to a coating method that extrudes a polymer onto a substrate and it mostly has a gap between a die tip and a substrate where extruded polymers are self-supporting. The configuration of common extrusion coating is shown in Fig. 10.12 (Harold *et al.*, 2005). Molten polymers are extruded by an extruder pump through a die and deposited on the substrate – the substrate may be preheated – and the substrate and deposited coating layer passed through pressure nip rollers to improve adhesion. When an additional substrate is inserted through the nip rollers with the coated substrate, a laminated fabric can be produced.

In slot die and extrusion coating, coating weight is principally controlled by pump delivery rate and the substrate line speed (Glawe *et al.*, 2003; Lippert, 2001; 'Laminating and coating: flexible future', 2002). One major application is hot melt coating since it can handle a high viscosity of hot melt with relatively high line speed ('Lamination and coating; nonwovens', 2002). Parameters that affect adhesion between melt and the substrate are the compatibility of the coating materials and substrates, together with other processing factors, such as extrusion melt temperature, air gap, film thickness, coating nip pressure, substrate construction, resin melt index, coating line speed, and coating roller temperatures (Mamish, 1990). Frequent maintenance of the die is required and it makes extrusion coating an expensive option.

10.3.10 Scatter coating

Scattering coating is the process where coating powders are spread evenly on the substrate surface with a rotating scatter roller. The dosing roller, its rotation speed, and the substrate line speed control depositon of the coating power on the substrate fabric (Jarrell, 1992; Glawe *et al.*, 2003). Pre-prepared coating

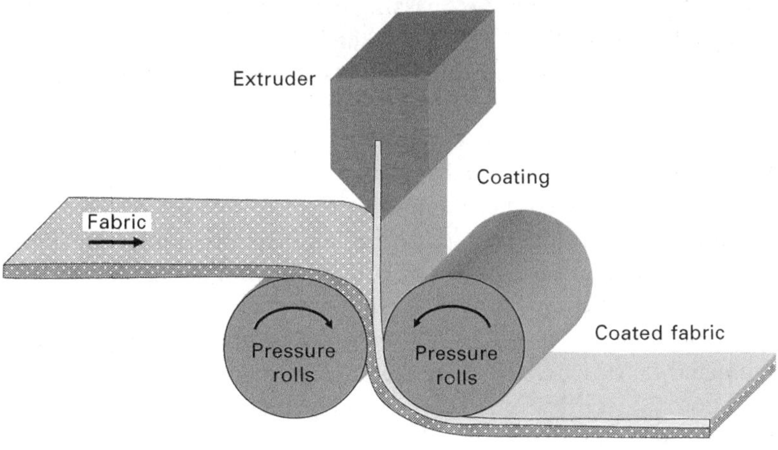

10.12 Extrusion coating.

powder in solid form is deposited directly onto the substrate without using a liquid medium. Utilizing 100% solid without medium reduces environmental impact. No solvent use results in little VOC release and no water use renders a drying process unnecessary. It generates few waste materials or hazardous chemicals (Waelde, 2001). Beyond its environmental attractiveness, scatter coating also has unique characteristics because the particles have a high surface area. It is able to create rich coatings with minimal penetration.

10.4 Laminating process

Laminating is the process that combines multiple substrates together and produces the stable multilayer structures by adhesives, heat, pressure, and mechanical bonding (Nair & Pandian, 2006; Scott, 1995). Good adhesion between layers is one of the essential requirements of the laminating process. In adhesive-based lamination, a substrate is coated with adhesive by the coating process followed by bonding with the other substrate (Schindler & Hauser, 2004). In this case, the adhesive coating is critical to the process of laminating (Swedberg, 1998). There are several different criteria to categorize the laminating process; substrate type combinations, the number of layers, and the method used. Here, laminating processes are classified by bonding methods and adhesive type into flame lamination, wet adhesive lamination, hot melt lamination, and dry heat lamination.

10.4.1 Flame lamination

Flame lamination is a simple and easy laminating process with high production rate and low cost. A preformed thin foam sheet is passed over an open flame to create a thin layer of molten polymer on the foam surface and subsequently bonded to fabric substrate by passing through a nip of calenders (Fig. 10.13) (Mansfield, 2003a; Scott, 1995). Molten layers of foam surface act as adhesive and form stable bonds to the fabric as it cools. Polyurethane foam is the most frequently used material in flame lamination.

Flame intensity should be optimally adjusted to ensure sufficient melting without burning or loss of excess foam layer thickness (Gillessen, 2000) and it is affected by gas type, flame height and spread. Constant line speed is also required to produce good quality laminate pressure (Mansfield, 2003a). Triple-ply lamination also can be produced using a dual burner or two-pass lamination (Gillessen, 2000). As indicated above, this is an inexpensive laminating process and does not require drying or a curing oven. But flame laminated fabrics are stiff since the process produces a solid bond layer with little porosity (Gillessen, 2000). Another main drawback is environmental unfriendliness, as it generates harmful emissions during the process.

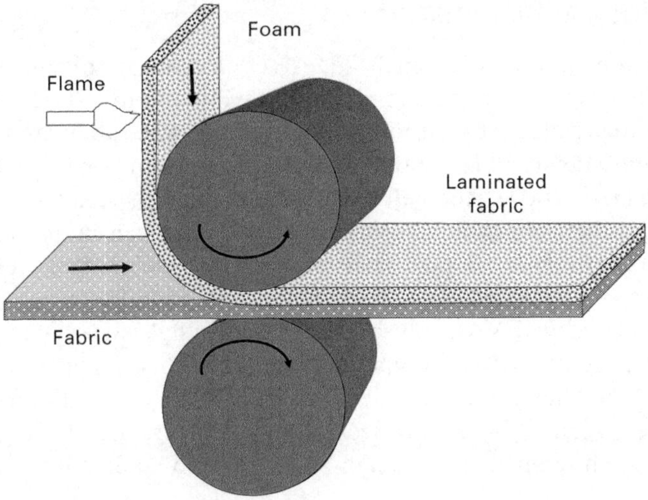

10.13 Flame laminating.

10.4.2 Wet adhesive lamination

Wet adhesive lamination is the process which uses liquid adhesives to bond layers together (Walker, 1994). Adhesives are first applied to one or more substrate surface by a coating technique, such as roll coating, spray coating, knife coating, etc. Then, the adhesive coated substrate is bonded with the other substrate under pressure, and dried and/or cured in the oven (Grant, 1981; Swedberg, 1998; Stukenbrock, 2003).

Properties of laminated fabrics are highly dependent on the adhesives used. Adhesive chemistry and rheology (viscosity and other flow characteristics) affect adhesion strength, chemical resistance, set-time required during the process, durability, resistance to heat, and penetrations (Crabtree, 1999). Penetration depth of adhesive must be controlled to achieve good adhesion while minimizing loss of structural flexibility. Some penetration is necessary to achieve good adhesion, but deep penetration of adhesive may generate stiff structures and cause loss of drapability, hand, tear strengths, and breathability (Crabtree, 1999; Stukenbrock, 2003).

A solvent-based adhesive dries fast, but it generates high VOC emissions and hazardous waste. Due to the environmental concerns and government regulations, conventional solvent adhesives are used only when it is coupled with expensive incineration or recovery equipment (Gillessen, 2000; Halbmaier, 1992). Water-based adhesive does not use toxic solvents, but requires a drying unit that consumes a lot of energy and occupies a large floor space (Gillessen, 2000).

10.4.3 Hot melt lamination

Hot melt lamination uses hot melt adhesives instead of solvent or water-based adhesives ('Laminating and coating: flexible future', 2002). They are 100% solid materials that melt to the liquid phase when they are heated to the temperature ranges of 80–200°C and solidify to form bonding when they cool (Glawe *et al.*, 2003). Hot melt adhesives are mostly specially formulated polymers and their compounds (Nussli, 2001). They can be thermoplastic or reactive (Glawe *et al.*, 2003). Thermoplastic hot melt softens at elevated temperatures, is deposited on the substrate, and solidifies when it cools. Ethylene vinyl acetate (EVA), polyamide (PA), polyethylene, polyvinylchloride (PVC), and polyester-based compounds are basic thermoplastic hot melts (Crabtree, 1999; Glawe *et al.*, 2003; Pizzi, 2003b). Since it is thermoplastic, it will soften again if it is exposed to high temperature. It is generally sensitive to high temperatures, steam and water, so application ranges are limited.

Reactive hot melt adhesives are also deposited onto the substrate in molten state. In addition to solidifying when it cools, it also undergoes reactions such as crosslinking. Crosslinking renders materials to thermoset, so after the crosslinking reaction has finished, heat does not soften the adhesives. It includes moisture crosslinking polyurethanes. They are highly durable, with good boiling and climate resistance, but they are rather expensive (Mansfield, 2003a; Woodruff, 2002b).

Hot melt adhesive is normally melted in an off-line melting unit and delivered to a hot melt coating unit (Gillessen, 2000). Many conventional coating processes, including roll coating, gravure roll coating, extrusion coating, screen coating, and knife-over-roll coating, are used to apply hot melt adhesive onto the substrate. However, coating units for hot melt adhesives must be able to handle a high viscosity of melt as well as the high temperature, required of this type of application. The substrate coated with hot melt adhesives bonded with another substrate and hot melt solidifies by cooling and/or crosslinking reaction to create a laminated fabric.

Selection of hot melt adhesives is based on both processing and performance. In addition to good adhesion properties, appropriate melt viscosity, melting temperature, and curing time are required to provide good peel, shear, and tensile strength. Depending on applications, breathability, lightness, porosity, and washing durability are also considered (Mansfield, 2003a; Nussli, 2001; 'Laminating and coating: flexible future', 2002).

One advantage of hot melt adhesives over water- and solvent-based system is their environmentally friendly operating condition – free of solvent and water use (Glawe *et al.*, 2003). VOC emission is minimal and they do not require a drying step ('Laminating and coating: flexible future', 2002) (Glawe *et al.*, 2003). Bonds are formed almost instantly without drying, so operating speeds are not limited by the rate of drying, and high productivity can be achieved (Halbmaier, 1992). In addition, the shelf life of hot melt adhesive is longer than that of wet adhesives. Laminating can be achieved at low coating level, so it can produce soft handle and flexibility ('Laminating and coating: flexible future', 2002). Compared to dry heat lamination, the substrate is not exposed directly to high temperature, so it can be used in the lamination of heat-sensitive substrates (Woodruff, 2002a). However, a change of adhesives type requires cleaning of pre-melt and application system, and they are relatively expensive (Glawe *et al.*, 2003).

10.4.4 Dry heat lamination

Dry heat lamination is the process where adhesive is deposited onto the substrate as 100% solid without medium or not being melted and later activated by heat and pressure to provide adhesion between the layers it bonds. Like hot melt adhesives, high solid content without harmful emissions and no need for drying are advantages of this process from an environmental point of view (Gillessen, 2000). But energy requirements for melting of dry adhesive are relatively high and the substrate is exposed to high temperature, so it is limited to substrates stable at process temperature (Gillessen, 2000).

Dry adhesives include polyester, polyamide, EVA, polyethylene and thermoplastic PU, and are produced in the form of powder, webs, or films (Gillessen, 2000; 'Lamination and coating', 2002). Powder adhesives are cryogenically ground to a fine texture and applied to a substrate by scatter coating or powder dot coating. When melted, they form a discontinuous bond between two layers, imparting bonding while providing high softness, drape, and permeability. Even though they do not produce any harmful emission, the fine particles may generate airborne dust (Crabtree, 1999; Gillessen, 2000).

Another type of dry adhesive is adhesive rolls either in the form of a nonwoven web or a film. It can be inserted between or over the substrate to be laminated (Jarrell, 1992). It is an ideal method to laminate substrates with open structures where powder adhesives tend to fill inside the void instead of staying on the surface (Crabtree, 1999; Gillessen, 2000). Lower add-on can also be achieved. However, they are more expensive than powder adhesive and changing application width and add-on requires use of different adhesives rolls (Gillessen, 2000).

Flat bed dry heat lamination is one method to achieve dry heat lamination. Dry adhesives are sandwiched between two substrates and transported through the heating tunnel by an upper and lower conveyer under pressure, where the adhesive melts and forms the bonding when it cools (Fig. 10.14) (Field, 2001; Smith, 2001). It is a high volume production technology in a continuous operation and this process also enables dry adhesive to be used at

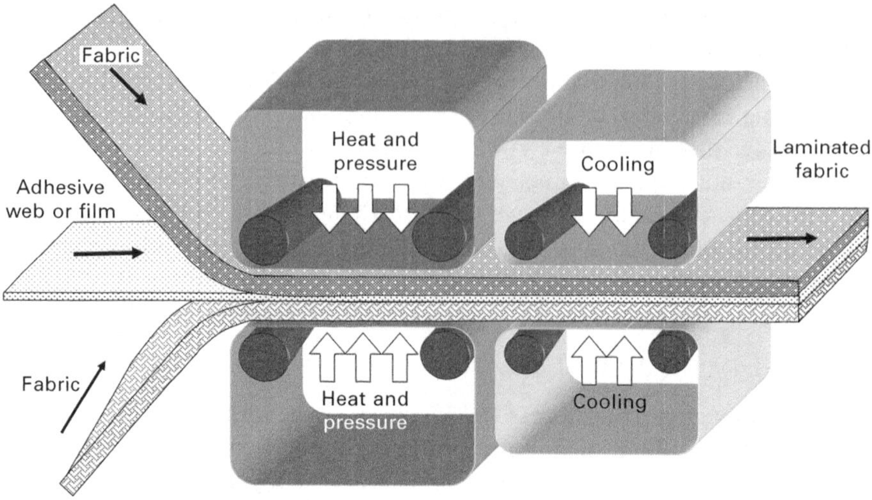

10.14 Flat bed laminating.

relatively low temperature, so reducing problems of substrate heat exposure (Field, 2000).

10.5 Properties and applications of coated and laminated fabrics

The reason for creating coated and laminated fabrics is to produce fabrics with characteristics and functionality that cannot be achieved by the fabric itself. In coated and laminated fabrics, individual components synergistically contribute properties and functionality to the whole system. The ultimate properties of coated and laminated fabrics are determined by structures created by many different components – how each layer is aligned and bonded, the order and ratio of layers, and adhesion between them (Fung, 2005). Even among laminates consisting of the same fabrics and coating materials, the performance of the final coated/laminated fabric can be drastically different depending on how the laminate structures are constructed and what process is used to produce them. As a result, coated and laminated fabric delivers properties beyond what traditional textiles can achieve and it is easier to customize based on final product requirements. Therefore, coated and laminated fabrics are used in a wide range of applications from traditional apparel to industrial technical textiles. These include diapers, incontinence pads, feminine hygiene, filter media, geotextiles, protective clothing, footwear, furniture upholstery, automotive interiors, membranes, tents, sheet roofing, bandages, surgical dressings and gowns, composites for automotives and aerospace, sails, many industrial specialty fabrics, packaging, and inflatables (Aberly,

2005; Hopkins, 1995; Mansfield, 2003a, 2003b). General characteristics of coated and laminated fabrics and a few examples of coated/laminated products are discussed further in this section.

10.5.1 General characteristics of coated and laminated fabrics

The properties and performance of coated/laminated fabrics are the result of a combination of effects of multiple components, but these are not the simple sum of the properties of the individual components (Farboodmanesh *et al.*, 2005). Rather, individual components synergistically contribute to the final properties, which are highly dependent not only on the properties of each component, but also on the structures created during the coating/laminating process (Fung, 2002, 2005). Figure 10.15 schematically illustrates factors affecting the final properties of coated fabrics through multiple processing steps. The properties of laminated fabrics are also determined in similar, complex ways. Several components including fabrics, adhesives, films, and foams are processed through multiple steps to produce final products, and the properties of each component and process parameters collectively determine the structures of the coated/laminated fabrics and their final properties.

Fabric components mostly impart integrity to the system. Fabrics consist of fibers and yarns and their chemical composition controls affinity between coating materials and fabrics. In addition to their fineness and morphology, fiber chemical properties are key factors determining adhesion. Fabric strength, governed by fabric structures and fiber/yarn strength will dominate their strength-bearing capabilities, and fabric openness will directly influence their ability to function as a carrier for adhesives and coating materials and affect penetration depths.

Another major component of coated and laminated fabrics is polymeric materials used as coating layers or adhesives. They are mostly designed to provide application-specific functions such as barrier properties, aesthetics, bonding between layers and flame resistance. But one cannot ignore the influence of their properties on basic properties such as durability, strength, and hand. Their rheological properties govern penetration behaviors into fabric substrates and their hardness, extensibility and strength also impact the strength and hand of the whole system.

In addition to the properties of individual components, the structures of coated and laminated fabrics formed through coating and laminating influence the final product properties and functionality greatly: thickness of coating and adhesive layers, their depth of penetration to the fabric, uniformity and distribution of these materials, order of layers, adhesion between them, etc. One of the most influential structural characteristics is the distribution and penetration depth of coating materials (Abbott *et al.*, 1972). Generally

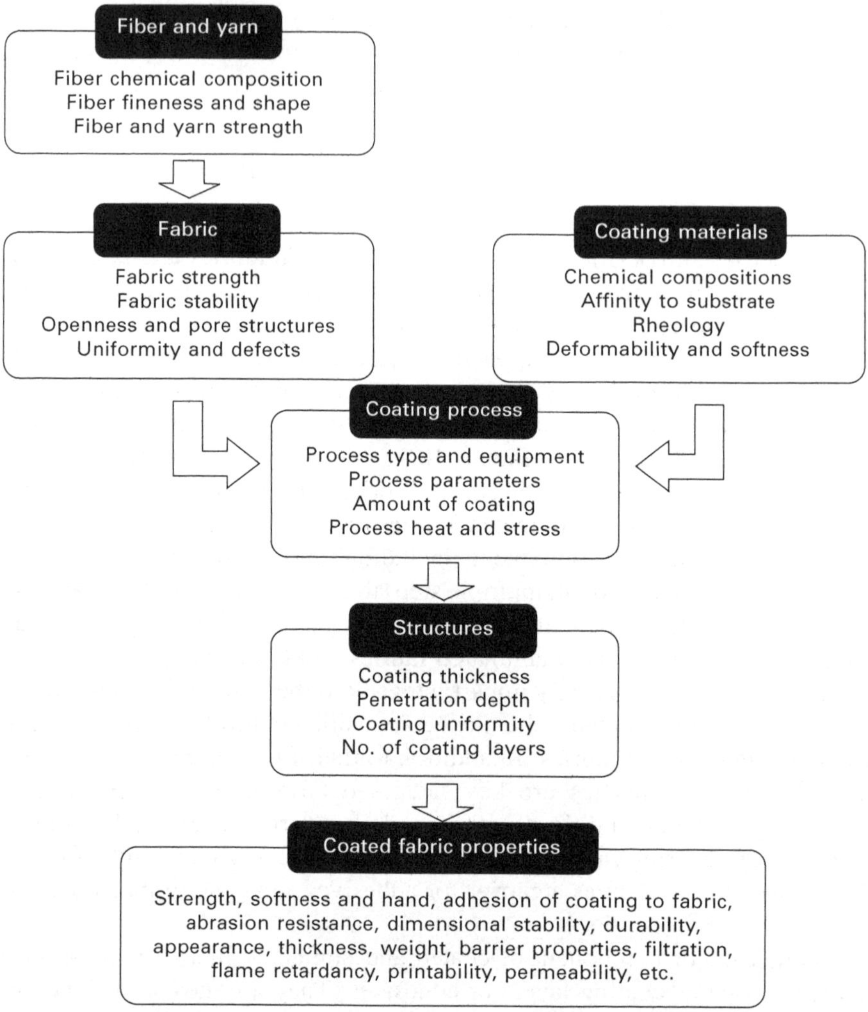

10.15 Factors affecting properties of coated fabrics.

speaking, there is an optimum depth of penetration where enough adhesion is provided between layers without causing excess loss of yarn and fiber mobility. Too much penetration of coating/adhesive materials results in the loss of yarn mobility, which makes fabrics stiff and deteriorates fabric strength. Soft handle can be achieved by minimum penetration, but it also causes poor adhesion. These structural components are governed by processing method and parameters. For example, a transfer coating method tends to create a thin and flexible coating, while an immersion coating produces a high penetration of coating polymers.

The desirable properties of coated laminated fabrics are highly dependent on the application areas where they are used. However, some basic properties are generally required regardless of application areas. These include strengths, hand, appearance, abrasion resistance, delamination resistance and dimensional stability (ASTM D3135, ASTM D2724, ASTM D4851). Resistance to delamination is one of the basic strength and durability requirements in coated/laminated fabrics. Adhesion is often recognized as the single most important factor determining strength, delamination resistance, abrasion resistance and coated/laminated fabric appearance. It is governed both by materials parameters such as chemical compatibility between layers, surface roughness, strength and hardness of components, and by structural characteristics such as penetration depth.

Fabric strength, such as resistance to tensile failure, tear propagation and burst strength are important in most applications and in particular technical applications, such as shelters, and architectural applications (Sen, 2008). Fabric components often dominate the strength of coated/laminated fabrics, but properties of the coating layer, distribution of coating materials, and adhesion between base materials are also important. Coating material can reinforce the base substrate and increase tensile and tear strength by penetrating between fibers, bunching them together, and making them bear higher applied force than individual fibers can do. But when it exceeds a certain critical level of penetration, it hinders movement of yarns and fibers and makes structures rigid, which does not allow for redistribution of stress concentrations. It results in losses of tensile and tear strength (Farboodmanesh *et al.*, 2005). Reduction of yarn mobility in knit and woven structure also hinders bundling of yarn, so applied force breaks individual yarns instead of moving and bundling them (Fung, 2002). Properties of coating materials also affect redistribution of applied forces and coated fabric strengths. When coating materials are deformable, higher tensile and tear strength are possible since deformability of coating materials allows applied stress to be redistributed (Sen, 2008).

Another consequence of limited freedom of fiber movements by coating penetration is reduction in hand and drapability of coated and laminated fabrics. During coating/laminating, coating materials or adhesives fill the space between fibers and yarns and constrain their movement (Y. Chen *et al.*, 1995). This results in significant changes in their bending, shear and tensile deformation behavior which are highly related to fabric hand (Weedall *et al.*, 1995; S. Chen *et al.*, 2007; Y. Chen *et al.*, 1995). Penetration of coating material increases fabric rigidity and creates stiff hand. Generally, the higher the penetration depth, the greater the flexural rigidity of coated fabrics (Farboodmanesh *et al.*, 2005). Again, flexibility of coated materials plays a great role in achieving better and softer hand. Another key factor is effectiveness of coating materials and adhesive to deliver functionality. Material which can deliver better functionality and adhesion at low thickness and penetration level can provide better hand without compromising functionality requirements.

10.5.2 Medical applications

Medical disposable fabrics ranging from medical drapes to dental bibs are mostly laminated, multilayer fabrics (Aberly, 2005). The use of laminating/coating is essential in medical applications. One of the most fundamental requirements of these medical fabrics is infection control. In this aspect, medical barrier fabric should provide protection by stopping transmission of blood or any bodily fluid and microbial organism.

Even though protection is most important in the medical barrier fabric, comfort should not be ignored. For example, a surgical gown should not interfere with surgical procedures by restricting or causing discomfort to surgeons. Therefore, medical barrier fabrics must be designed to meet the delicate balance of comfort and protection. Breathable barrier fabrics that allow dissipating moisture vapor from wearers while blocking water or other liquid penetration can achieve two of the most sought-after attributes in protective garments: comfort and protection. Typically, breathable barrier fabrics have three-layer laminated structures – breathable barrier film sandwiched between two fabrics. Nonwoven fabrics are frequently used as the fabric of choice to insure sterility and achieve disposability. Other properties required are light weight, soft hand, flexibility, tear and tensile strength, abrasion resistance, dimensional stability, resistance to delaminating, general durability to flexing/aging, and easy care. Trilaminate is a suitable construction to achieve all these performance requirements (Bitz, 2002). An outer layer resists abrasion and puncture failure, and a middle layer composed of breathable barrier film prevents fluid penetrations while permitting water vapor dissipation. An inner layer is designed to use next to the skin and have soft feel.

A middle layer is where the breathable barrier technology is used. There are three different approaches to produce breathable barrier fabrics: water-repellent treated densely woven fabric, microporous film, and monolithic hydrophilic film. Microporous breathable barrier film has micropores of controlled size and uses a size exclusion mechanism to block any relatively large water droplets while vapors pass through pore structures (Fig. 10.16(a)). The monolithic barrier film does not have any pore but is hydrophilic. Water vapor permeates through the hydrophilic groups in the polymer structure – a process called 'active diffusion' (Fig. 10.16(b)) (Paul, 2009).

10.5.3 Apparel

The first use of coated fabric in apparel dated back to ancient times. For example, Mayan Indians applied gum from rubber trees on clothing and

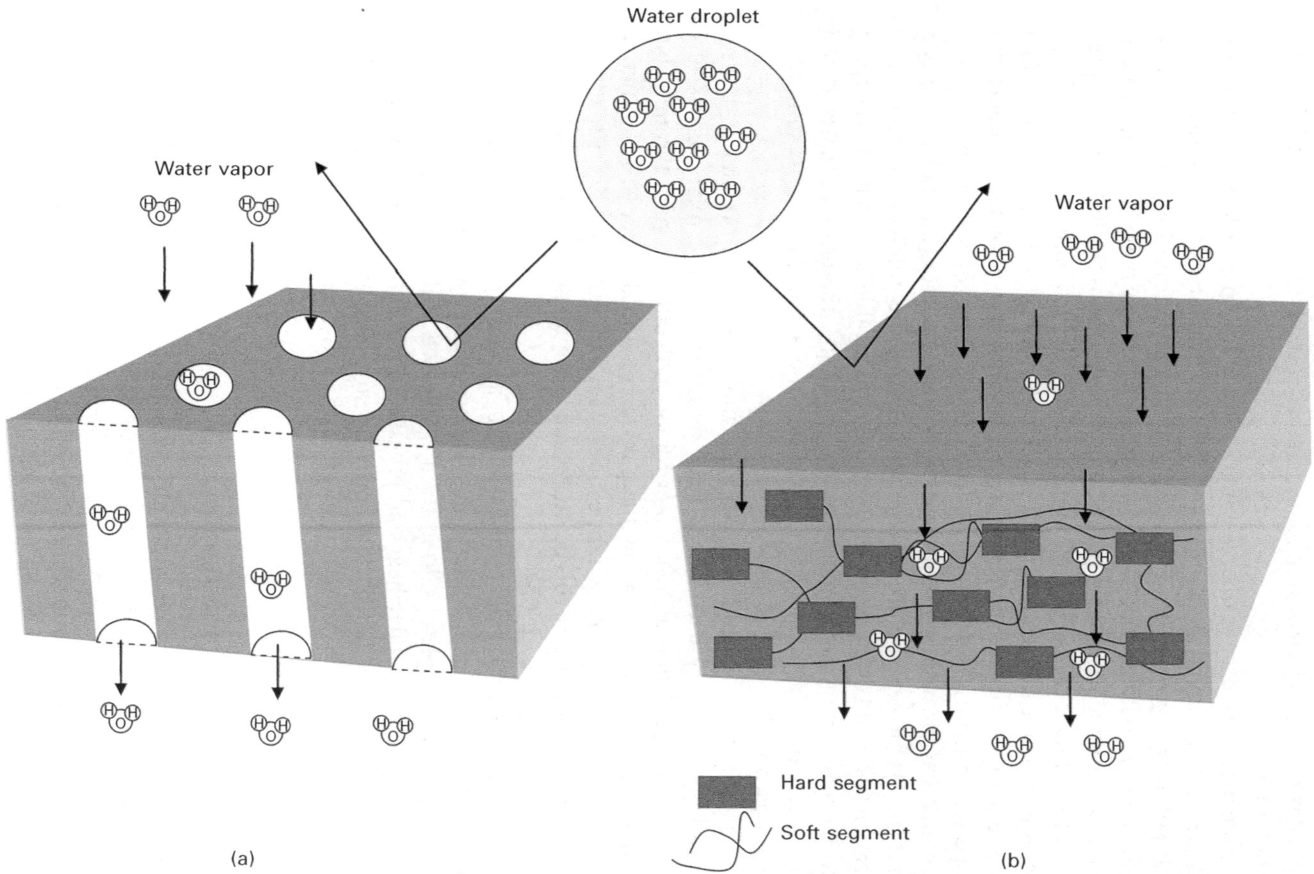

10.16 Mechanism of breathable barrier film: (a) microporous (pore structure is simplified) and (b) monolithic, hydrophilic film.

footwear to make them waterproof (Sen, 2008). The modern coating industry also started with a rubber-coated waterproof garment when Charles Macintosh applied rubber on a fabric to produce a raincoat in 1823 (American Chemcial Society, n.d.).

Protection against weather – rain, snow, wind, and cold temperature – can be best achieved by multilayer fabrics. An outer layer fabric coated with polymeric coating provides waterproofness and an inner layer can be designed to maximize insulation properties. In the early stage of waterproof fabric development, water repellency is achieved mostly by blocking pores of textiles material by use of compact polymeric coating. PU, PVC, or neoprene coated fabrics are used. PU has advantages of light weight and excellent low temperature flexibility, while PVC has poor low temperature properties and low solvent reissuance. These impermeable coating based garments can provide good protection against weather, but fail to provide comfort. Achieving comfort requires keeping the micro-climate formed between body and cloth layer constant and comfortable, which means keeping it to a narrow range of temperature and humidity. In addition, these should not restrict body movement, so this should be achieved with as minimum weight as possible (Fung, 2005).

Because of these needs for comfort and protection, breathable barrier technology is widely used. Gore-Tex® fabric consists of microporous film made from PTFE that allows water vapor molecules to pass through, but the pores are too small to allow the passage of liquid water. It is laminated between the outer garment fabric and a second scrim fabric. Breathable barrier technology has evolved to be more sophisticated, as described in Section 10.5.1 (Fung, 2005).

Another coated fabric application in apparel is fusible interlinings (Goossens, 1994). Fusible interlinings are produced by coating with adhesive and shipped in completely dried form. Adhesive will melt and form bonds with another fabric under heat and pressure during the garment production process to reinforce the fabric to provide stiffness. Powder coating, either scatter coating or powder dot coating, is the most used process to produce fusible interlinings.

10.5.4 Synthetic leather

Leather has unique properties both aesthetically and functionally – hand, drape, and surface texture as well as good moisture absorption and high air and moisture vapor permeability. It has been recognized as a most desirable material for footwear, upholstery, and other apparel accessories. As demand for leather grows, the limited availability of raw hide and skins has driven the development of synthetic leather materials (Lunden & Schmel, 1984; Sen, 2008).

Synthetic or artificial leather is divided into two groups: compact coated fabrics and Poromerics. Compact coated fabrics are produced by transfer coating and consist of base fabric, foam coat and top coat. It gives leather-like handle and drapes through a cellular foam layer created mostly by a blowing agent. The top coat imparts water resistance and leather-like texture by embossed release paper. PVC and PU are the most used coating materials, but PVC has low vapor permeability and low temperature flexibility, and tends to crack quickly due to the migration of plasticizers. PU-based fabric has better properties. They are widely used for upholstery, soft luggage, handbags, shoes, and seat covers (Morley, 1984; Sen, 2008).

Another class of more advanced synthetic leather is called 'Poromerics'. Poromerics has not only leather-like aesthetics (hand and feel) but also leather-like properties: good air and water permeability. Excellent permeability is the result of a microporous structure following a coagulation process (Durst, 1984; Merkle & Tackenberg, 1984). The typical process is that the base fabric is impregnated in PU solution, dissolved in solvent and PU precipitate, then successively dipped in water, forming porous structures. Some synthetic leather uses microfiber nonwovens created from a sea-island configuration to resemble the feel and properties of genuine leather. Microfiber nonwoven is produced by spinning of bicomponent fibers in a sea-island configuration, removing the sea components resulting in individual island-form microfibers (Fig. 10.17). It is an excellent material for footwear where high permeability is an essential requirement (Sen, 2008).

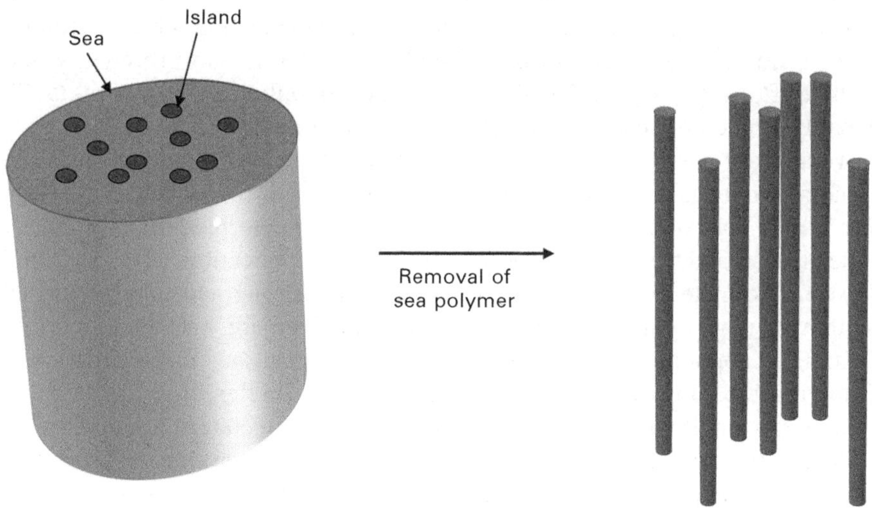

10.17 Island in the sea microfiber formation.

10.5.5 Automotive applications

Automotive seat covers and other interior fabrics are mostly laminated fabrics. Car seat covers should have durability, high tear and tensile strength, high resistance to abrasion, UV and mildew growth. In addition to these functional requirements, car seat covers must have an appealing and decorative appearance, and soft handle, so it is hard to deliver all the desired properties with a single-layer textile fabric (Sen, 2008; Woodruff, 1993).

Most automotive seat covers are trilaminates consisting of a face fabric, foam layer and scrim backing. A face fabric provides an attractive appearance that lasts throughout the car's lifetime. Therefore, lightfastness, soil resistance and abrasion resistance are required (Fung & Hardcastle, 2000). The foam layer, mostly PU with thickness ranging from 1 to 10 mm, is laminated between the face fabric and the scrim backing and it provides a soft, cushioning feel (Sen, 2008). A scrim backing fabric controls stretch properties and dimensional stability. An extra coating may be applied to the trilaminate to improve abrasion resistance, flame retardancy, and barrier properties (Fung & Hardcastle, 2000).

Door casing and headliner fabrics are also based on laminated fabrics to provide a soft touch and appealing look. Acoustic and thermal insulation of car headliner is also important, so multiple layers consisting of PU foam, fiber glass and other components, and a face fabric are laminated together (Sen, 2008).

Air bag fabrics are another important automotive application. As concerns over safety increases, demands on air bags are rising. During the last few decades, air bags have evolved from one type – the driver side air bag – to diverse forms and shapes including side impact, knee bolster, side curtain air bag, inflatable restraints, etc., where requirements are subtly different. Driver side air bags need to be inflated within 60 ms, absorb energy and deflate after 120 ms (Sen, 2008) and withstand an inflation pressure of 35–70 Kpa. It should not permeate any chemicals and hot gas. Side curtain air bags have appropriate air hold capacity to remain inflated for about 3–12 sec. In general, air bags are required to have high strength to withstand inflation force and controlled air permeability, from complete blocking for driver side air bag, and some release properties in other applications. The preferred base fabric material is nylon 66 because it is a strong and tough material with high tenacity and breaking strain. For a driver side air bag, a polymer coating is applied to prevent hot gas penetration – it has to be completely impermeable. Good adhesion, maintaining long-term flexibility while being stored in a small compartment and experiencing temperature fluctuations of cold and hot conditions for about 10 years or so is also required. Elastomeric materials – neoprene and silicone-based material – are mostly used for coating. Silicone-based coating has the advantages of creating a smoother surface and reducing secondary injury such as abrasion, scrapes, and bruises by friction reduction (Crouch, 1994).

10.5.6 Architectural applications

Coated materials are used in constructing shelters. High strength and environmental resistance – resistance against wind, rain, snow, sunlight, and biological degradation – are desirable. One of these applications is fabrics for tents. Historically, tent fabrics are cotton canvas coated with wax to impart water repellency. Current tents mostly use high strength, hydrophobic synthetic fiber-based fabric coated with polymeric coating. Tent fabric must have high strength, creep resistance, resistance to temperature changes, durability and water-repellency, impermeability to air, light transmission and flame-retardancy. PVC, PTEF or silicone, PU, and acrylic materials are used as coating material. In general, coating is preferred over laminating since coating has less delamination problems, but fluoropolymer film lamination is sometimes used (Sen, 2008).

Wall covering is another coated/laminated fabric in architectural applications. In this application, aesthetic aspects are most important. Fabrics laminated with a vinyl decorated surface, fabric laminated to paper backing, or coated nonwoven fabrics are all used as wall covering materials. They are mostly used to improve and maintain aesthetic aspects – touch and appearance – but fabric-based wall coverings can provide the added benefits of improved acoustic and thermal insulation (Reagan & Villasi, 1982).

10.6 Conclusion

Coating and laminating processes have unique abilities to use and combine multiple materials, even with drastically different properties and structures. They have routinely provided solutions to many challenges in textiles applications and will continue to serve as tools for innovation in the textile industry.

Innovations are driven by increasing demands for high functionality. Textile fabrics have become indispensible tools for protection from micro-organism, fire and heat, while the desire for comfort is mounting. This demand can only be achieved by hybrid and composite fabrics and coated/laminated fabrics are becoming more and more sophisticated to satisfy these demands. The growth of technical and industrial textiles, such as geotextiles, filtration, automotive interior, and hygiene also require better performance at lower cost (Scott, 1995). Coating and laminating techniques are well positioned to respond to this request since they can construct structures by synergistically utilizing many different components.

Another interesting development is increasing interest in the use of textile structure as a platform technology. Fibrous materials with their high surface area, sophisticated pore structure and high flexibility have unique advantages over other structures such as film or plastics when one incorporates functionality – conductivity, shape memory, drug delivery, for example – to develop integrated system responses to the environmental stimuli and to fulfill the engineered function (Yonenaga, 2001). Transformations from commodity textiles to highly engineered products with tailored properties are on the way, and coating and lamination is a key technology in this transfer.

Environmental concerns pose both challenges and opportunities for the coating and laminating industry. Many traditional methods, such as flame lamination, face environmental challenges, while coating and laminating processes with less energy use, low pollution and hazardous waste generation have been developed (Woodruff, 2002b). This drives increasing use of high solid content coating – powder coating and hot melt coating where there are low VOC emissions and no drying requirements. The rise of energy curable coating formulations is partly driven by this demand because UV curable formation does not require solvent or a drying process and uses less energy and occupies less floor space. Coating materials from renewable resources or high recyclability have been active research areas.

In summary, coating and laminating developments are driven by demand for high and multi-functionality, growth of technical textiles, and development of environmentally friendly processes. Capability of multiple coating processes, the quick response to demand, and process flexibility become more and more important. Precise coating thickness control becomes more critical in some technical textile applications such as membrane formation. The ability to process a wide range of materials is also needed as innovations in materials science, nano-technology, and biotechnology bring novel materials to the coating and laminating process.

10.7 Further reading

Brockmann W (2009), Adhesive Bonding: Materials, Applications and Technology, Weinheim: Wiley-VCH.

Fernando R H & Sung L (2009), Nanotechnology Applications in Coatings, ACS symposium series, Washington, DC: American Chemical Society.

Fourche G (1995), 'An overview of the basic aspects of polymer adhesion. Part I: Fundamentals', *Polymer Engineering and Science*, 35(12), 957–967.

Glawe A, Reuscher R, Koppe R & Kolbusch T (2003), 'Hot-melt application for functional compounds on technical textiles', *Journal of Industrial Textiles*, 33(2), 85–92.

Lee S & Obendorf S K (2005), 'Barrier effectiveness and thermal comfort of protective clothing materials', *Journal of the Textile Institute*, 98(2), 87–98.

Mukhopadhyay A & Midha V K (2008), 'A review on designing the waterproof breathable fabrics. Part II: Construction and suitability of breathable fabrics for different uses', *Journal of Industrial Textiles*, 38(1), 17–41.

Packham D E (2005), *Handbook of Adhesion*, 2nd edn, Chichester: John Wiley & Sons.

Schwartz M M (2006), *New Materials, Processes, and Methods Technology*, Boca Raton, FL: Taylor & Francis/CRC Press.

Wicks Z W (2007), *Organic Coatings: Science and Technology*, 3rd edn, Hoboken, NJ: Wiley-Interscience,

Weinstein S J & Ruschak K J (2004), 'Coating flows', *Annual Review of Fluid Mechanics*, 36(1), 29–53.

Zarras P (2007), *New Developments in Coatings Technology*, Oxford: Oxford University Press.

10.8 References

Abbott N J, Lannefeld T E, Barish L & Brysson R J (1972), 'A study of tearing coated cotton fabrics. Part III: The influence of coating application techniques', *Journal of Coated Fibrous Materials*, 1, 130–149.

Aberly D (2005), 'Converting nonwovens for improved product features', *Nonwovens Industry*, November.

Adamson A W & Gasy A P (1997), *Physical Chemistry of Surfaces,* New York: John Wiley & Sons.

Alonso S B (2003), 'A torque-based analysis of the reverse roll coating process', *Chemical Engineering Science*, 58, 1831–1837.

American Chemcial Society (n.d.), *United States Synthetic Rubber Program, 1939–1945*. National Historic Chemical Landmarks: http://acswebcontent.acs.org/landmarks/landmarks/rbb/rbb_natural.html (accessed August 24, 2012).

ASTM D123 Standard terminology related to textiles, 2009.

ASTM C168 Standard terminology relating to thermal insulation, 2010.

ASTM D2724 Standard Test Methods for bonded, fused, and laminated apparel fabrics, 2007.

ASTM D3135 Standard specification for performance of bonded, fused, and laminated apparel fabrics, 1987 (R2004).

ASTM D4851 Standard Test Methods for coated and laminated fabrics for architectural use, 2007.

Berg J C (1993), *Wettability*, New York: Marcel Dekker.

Bitz K (2002), 'The best of both worlds', *Nonwovens Industry*, August.

Bluestein C, Loewrigkeit P, McGimpsey T T & Van Dyk K A (1975), 'Factors influencing latex tie-coat adhesion in urethane skin coat to fabric bonding', *Journal of Coated Fabrics*, 4, 158–165.

Chen S, Ding X & Yi H (2007), 'On the anisotropic tensile behaviors of flexible polyvinyl chloride-coated fabrics', *Textile Research Journal*, 77(6), 369–374.

Chen Y, Lloyd D W & Harlock S C (1995), 'Mechanical characteristics of coated fabrics', *Journal of the Textile Institute*, 86(4), 690–700.

Cole P J & Macosko C W (2000), 'Polymer–polymer adhesion in melt-processed layered structures', *Journal of Plastic Film and Sheeting*, 16, 213–222.

'Cotton textiles for plastic laminates', (1944), *Textile Research Journal*, 14, 11–14.
Crabtree A (1999), 'Hot-melt adhesives for textile laminates', *Technical Textiles International*, 8(2), 11.
Crouch E T (1994), 'Evoluiton of coated fabric for automotive air bags', *Journal of Coated Fabrics*, 23, 202.
Dahlquist C A (2001), 'The theory of adhesion', in Satas D & Tracton A A, *Coatings Technology Handbook*, New York: Marcel Dekker, 47–56.
Damico D J (2003), 'Reactive acrylic adhesives', in Pizzi A & Mittal K L, *Handbook of Adhesive Technology*, New York: Marcel Dekker, 747.
Dartman T & Shishoo R (1993), 'Studies of adhesion mechanisms between PVC coating and different textile substrates', *Journal of Coated Fabrics*, 22, 317–335.
Durst P (1984), 'The PU coagulation process and its sucess in PUCFs', *Journal of Industrial Textiles*, 13, 175.
Dussan E B (1979), 'Spreading of liquids on solid surfaces: static and dynamic contact lines', *Annual Review of Fluid Mechanics*, 11, 371–400.
Farboodmanesh S, Chen J, Mean J L, White K D, Yesilalan H E, Laoulache R & Warner S B (2005), 'Effect of coating thickness and penetration on shear behavior of coating fabrics', *Journal of Elastomers and Plastics*, 37, 197–227.
Field I (2000), 'Why is the industry turning to dry heat lamination technology?', *Technical Textiles International*, 9(7), 21.
Field I (2001), 'Dry heat lamination technology', *International Dyer*, 186(9), 39–43.
Fung W (2002), *Coated and Laminated Textiles*, Cambridge: Woodhead Publishing.
Fung W (2005), 'Coated and laminated textiles in sportswear', in Shishoo R, *Textiles in Sport*, Cambridge: Woodhead Publishing.
Fung W & Hardcastle M (2000), *Textiles in Automotive Engineering*, Cambridge: Woodhead Publishing.
Geddes K (2003), 'Polyvinyl and ethylene-vinyl acetates', in Mittal, K L, *Handbook of Adhesive Technology,* New York: Marcel Dekker, 719.
Gillessen G (2000), 'Flame, dry or hot-melt', *International Dyer*, 185(11), 34.
Glawe A, Reuscher R, Koppe R & Kolbusch T (2003), 'Hot melt application for functional compounds on technical textiles', *Journal of Industrial Textiles*, 33, 85–92.
Goossens F (1994), 'Quality improvement in the paste dot coating of fusible interlinings', *International Textile Bulletin: Nonwoven Industrial Textiles*, 40(1), 6.
Goossens F A (2001), 'Rotary screen coating', in Satas D & Tracton A A, *Coatings Technology Handbook,* New York: Marcel Dekker, 179–183.
Grant R R (1978), 'Coating techniques for the manufacture of continuous web products', *Journal of Industrial Textiles*, 8, 171–182.
Grant R R (1981), 'Coating and laminating applied to new product development', *Journal of Industrial Textiles*, 10, 232.
Greer R (1995), 'The reverse roll coater', *Journal of Coated Fabrics*, 24, 287.
Greer R & Stelling P (1994), 'Application equipment for radiation curable coatings', *Journal of Coated Fabrics*, 24, 40.
Halbmaier J (1992), 'Overview of hot melt adhesives application equipment for coating and laminating full-width fabrics', *Journal of Coated Fabrics*, 21, 301–310.
Hannachi A & Mitsoulis E (1990), 'Numerical analysis of multilayer forward roll coating', *Journal of Plastic Film and Sheeting*, 6, 170–190.
Harold G, Wagner J R & Eldrige M (2005), *Extrusion – The Definitive Processing Guide and Handbook*, Burlington, MA: William Andrew Publishing.
Herrera A (1991), 'Versatility in coating operations – knife coatings', *Journal of Coated Fabrics*, 20, 289.

Hewson R W, Kapur N & Gaskell P H (2006), 'A theoretical and experimental investigation of tri-heilcal gravure roll coating', *Chemical Engineering Science*, 61, 5487–5499.

Hopkins J (1995), 'A comprarative analysis of laminating automotive textiles to foam', *Journal of Coated Fabrics*, 24, 250.

Hu J-L (2004), *Structure and Mechanics of Woven Fabrics*, Cambridge: Woodhead Publishing.

Jarrell C (1992), 'A new process for coating and laminating face-finished fabrics', *Journal of Industrial Textiles*, 21, 212.

Joseph M L (1981), *Introductory Textile Science*, New York: CBS College Publishing.

Keeley V E (1991), 'Transfer coating with polyurethanes', *Journal of Coated Fabrics*, 20, 176.

Kistler S F & Schweizer P M (1997), 'Coating science and technology: an overview', in Kistler S F & Schweizer P M, *Liquid Film Coating: Scientific Principles and their Technological Implications*, London: Chapman & Hall, 3–16.

'Laminating and coating: flexible future' (2002), *Textile Month*, March, 24.

'Lamination and coating' (2002), *Textile Month*, March, 22.

'Lamination and coating; nonwovens' (2002), *Textile Month*, March, 22.

Lay D G & Cranley P (2003), 'Polyurethane adehsives', in Pizzi A & Mittal K L, *Handbook of Adhesive Technology*, New York: Marcel Dekker, 695.

Licari J J (2003), *Coating Materials for Electronic Applications – Polymers, Processes, Reliability, Testing*. Norwich' NY: William Andrew Publishing.

Lippert H G (2001), 'Slot die coating for low viscosity fluids', in Satas D & Tracton A A, *Coatings Technology Handbook,* New York: Marcel Dekker, 139–155.

Lunden B O & Schmel F (1984), 'Soft leather substitute materials and their impact on the international leather and leather product trade', *Journal of Industrial Textiles*, 14, 9.

MacLeod D M (2001), 'Wire-wound rod coating', in Satas D & Tracton A A, *Coatings Technology Handbook*, New York: Marcel Dekker, 129–137.

Mamish A L (1990), 'Co-extrusion coating of woven fabrics', *Journal of Coated Fabrics*, 20, 108.

Mansfield R G (2003a), 'Combining nonwovens by lamination and other methods', *Textile World*, 153(1), 22–25.

Mansfield R G (2003b), 'Expanding industrial fabric uses'. *Textile World*, September, 93.

Martinez-Garcia A, Reche A S & Martin-Martinez J M (2004), 'Improved adhesion of EVAs with different vinyl acetate contents treated with sulphuric acid', *Journal of Adhesion Science and Technology*, 18(9), 967–982.

Matthews T (1993), 'Polymer coatings for textiles', *Technical Textiles International*, June, 14–20.

Merkle R & Tackenberg W (1984), 'Alternatives to leather: Laif Poromerics A new family of polyurethance coated fabrics', *Journal of Industrial Textiles*, 13, 228.

Mewes H (1989), 'Adhesion and tear resistance of coated fabrics from polyester and nylon', *Journal of Coated Fabrics*, 19, 112.

Mills N J (2007), *Polymer Foams Handbook: Engineering and Biomechanics Applications and Design Guide*, Oxford: Elsevier.

Morley D (1984), 'Coated fabrics for upholstery', *Journal of Coated Fabrics*, 14, 46.

Nair G P & Pandian S P (2006), 'Spotlight on laminating machines', *Colourage*, 53(10), 96–112.

Nussli R (2001), 'The hot-melt coating and laminating of industrial textiles', *International Textile Bulletin: Nonwovens – Industrial Textiles*, 47(4), 14.

Okoroafor M O & Frisch K C (1995), 'Introduciton to foams and foam formation', in Landrock A H, *Handbook of Plastic Foams*, Norwich, NY: William Andrew Publishing, 1–10.

Padday J F (1978), *Wetting, Spreading and Adhesion*, London: Academic Press.

Paul S (2009), 'Surface modificaitons of polypropylene nonwovens to improve adhesion to elastomers', PhD thesis, Raleigh, NC: North Carolina State University.

Pizzi A (2003a), 'Unsaturated polyester adhesives', in Pizzi A & Mittal K L, *Handbook of Adhesive Technology*, New York: Marcel Dekker, 731.

Pizzi A (2003b), 'Hot-Melt adhesives', in Pizzi A & Mittal K L, *Handbook of Adhesive Technology*, New York: Marcel Dekker, 739.

Pourdeyhimi B (2004), 'Nonwovens and traditional textile know-how', *Textile World*, August, 42–45.

Prud'homme K & Khan S A (1996), *Foams: Theory, Measurement and Applications*, New York: Marcel Dekker.

Reagan B M & Villasi L (1982), 'Thermal properties of wall covering materials', *Textile Reserach Journal*, 52, 703.

Robinson J & Marrick K (2006), 'Printing techniques for enhancing nonwovens', *Nonwovens World*, 15(2), 36–44.

Schindler W D & Hauser P J (2004), *Chemical Finishing of Textiles*, Cambridge: Woodhead Publishing.

Schonhorn H (1985), 'Adhesion and adhesive: interactions at interface', *Journal of Plastic Film and Sheeting*, 1, 163–179.

Scott R A (1995), 'Coated and laminated fabrics', in Carr C M, *Chemistry of the Textile Industry*, New York: Springer, 210–248.

Sen A K (2008), Coated Textiles: Principles and Applications, Boca Raton, FL: CRC Press.

Shim E (2010), 'Coating and laminating processes and techniques for textiles', in Smith W C, *Smart Textile Coating and Laminates*, Cambridge: Woodhead Publishing.

Smith W C (2001), 'Laminate faster, more competitive', *Textile World*, November, p. 55.

Stepanski H & Nebe V (1997), 'Polyurethane adhesive for textiles', *Journal of Inudstrial Textiles*, 27, 27.

Stukenbrock K H (2003), 'Chemical finishing', in Albrecht W, Fuchs H, Kittelmann W & Lünenschloss J, *Nonwoven Fabrics*, Weinheim: Wiley-VCH, 435–454.

Swedberg J (1998), 'Sticking to it: learning about laminates', *Industrial Fabric Products Review*, 75(8), 30.

Waelde L R (2001), 'Thermoset powder coatings', in Satas D & Tracton A A, *Coatings Technology Handbook*, New York: Marcel Dekker, 691–705.

Wake W C (1973), 'The adhesion of polymers to fibrous masses', *Journal of Coated Fabrics*, 3, 84.

Walker R (1994), 'Thermoplastic powder adhesive for lamination', *Technical Textiles International*, 3(9), 28.

Weedall P J, Harwood R J & Shaw N (1995), 'An assessment of the Kawabata transformation equations for primary-hand values', *Journal of the Textile Institute*, 86(3), 470.

Whiteman R (1993), 'On-line gauging, control and benefits for the coating/laminating process', *Journal of Coated Fabrics*, 23, 87–104.

Wicks Z W, Jones F N & Pappas S P (1999), *Organic Coatings: Science and Technology*, New York: John Wiley & Sons.

Woodruff F A (1992), 'Environmentally friendly coating and laminating: new processes and techniques', *Journal of Coated Fabrics*, 21, 24.

Woodruff F A (1993), 'Coated fabrics for air bags', *Journal of Coated Fabrics*, 23, 14.

Woodruff F A (1997), 'Advances in powder and other thermoplastic coating and laminating systems', *Journal of Coated Fabrics*, 26, 308.

Woodruff F A (2002a), 'Laminating and coating: advanced composites', *Textile Month*, March, 23.

Woodruff F A (2002b), 'Exciting prospect for PUR compounds', *International Dyer*, 187(11), 34.

Wright E L (1981), 'Low wet pick-up finishing: a review of commercial experience', *Textile Research Journal*, 51(4), 251.

Yonenaga A (2001), 'Solvent-free industrial textile coatings', *International Textile Bulletin: Nonwovens – Industrial Textiles*, 47(4), 17–18.

Yosomiya R (1990), *Adhesion and Bonding in Composites*, New York: Marcel Dekker.

Young T (1805), 'An essay on the cohesion of fluids', *Philosophical Transactions of the Royal Society of London*, 95, 65–87.

Zickler D (1978), 'Coating industrial fabrics by means of the melt roll process', *Journal of Coated Fabrics*, 8, 121.

Part III

Welding technologies

11
The use of heat sealing, hot air and hot wedge to join textile materials

I. JONES, TWI Ltd, UK

DOI: 10.1533/9780857093967.3.355

Abstract: This chapter reviews heat sealing and hot air/wedge welding techniques that are used to join textile materials. In each case, the chapter discusses key principles, advantages and disadvantages of the technique, the equipment used, factors affecting quality, and the applications of each technique.

Key words: textiles, fabrics, heat sealing, hot press, hot bar, impulse, heated belt, rollers, lamination, fusing, bonding, hot air, hot wedge.

11.1 Heat sealing of textiles: introduction

This technique uses an external heat source, either a heated bar, roller or plate. As the hot tool is pressed against the fabric layers, they are softened and a joint can be made. The heating is by conduction through one of the fabric layers. The process is rapid, taking a few seconds when a bar is used or at a rate of a few metres per minute when using rollers. The outer surface of the fabric is melted when this process is used for welding two similar layers together. However, it is more commonly used to fuse a hot melt adhesive film between two layers of fabric. It can also be used to melt a film to a fabric. The film layers can be melted before the fabric melts either due to use of polymer film with a lower melting point than the fabric or because of the insulating effect of the fabric layer. The heat sealing technique is used in laminating (see Chapter 10), seam waterproofing, garment making and for medical dressings amongst other areas.

The main advantages of using heat sealing over other joining processes are:

- short weld times (typically a few seconds) or a few metres per minute
- can be used for discreet parts or areas, or as a continuous process in lamination
- simple technique; can be manually manipulated (heated tool or fabrics), and straightforward to put into automated production
- flexible process applicable to films, fabrics and combinations of material and may be used for small or large parts

- low cost equipment and processing (though adhesive film consumable may be required)
- low energy consumption, especially when impulse heating is used
- an impermeable seam can be achieved
- a fusible film can be melted at the joint interface, avoiding use of stitching and providing a soft, stretchy, flexible joint; it is also lower weight and less bulky than a stitched seam, and can seal the edges of the fabric even when a simple overlap seam design is used.

The main disadvantages of heat sealing are:

- it relies on conduction to heat the materials, therefore is not effective for thick or low thermal conductivity materials
- there can be heat marking of the outer surfaces
- there can be inconsistent heating or pressure at different parts of the roll, bar or plate
- there is potential to contaminate the outer surfaces of the materials if melt products stick to the heated tool surface.

11.2 Equipment for heat sealing

Heat sealing equipment comes in many forms. However, they all have the following basic components:

- a tool containing a heating element; this may be in the form of a narrow bar, a wide rectangular platen or shaped iron, or a roller
- a cool (typically) base plate for the heated tool to work against; this may be actively cooled to improve consistency
- pressure application method
- process control; a timer for the heating and cooling cycle or thermostatic controls for repeated or continuous procedures.

11.2.1 Hot bar and impulse sealing

The hot bar process (Fig. 11.1) is based on the principle that if thermoplastic materials are pressed against a heated metal bar, they will soften and a joint can be made. Since the technique relies on conduction of heat through one of the fabrics, this limits the thickness that can be joined to less than 0.5 mm.

The equipment comprises one or two metal bars, which are generally electrically heated. One of the bars is hinged to allow the placement and removal of the fabrics. The weld pressure is applied manually or via pneumatic cylinders. A coating of PTFE is often applied to the bars to prevent the softened or molten material from adhering to them. The process is controlled

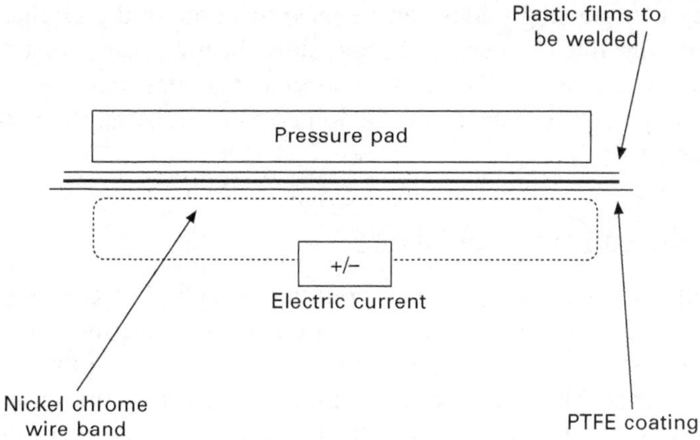

11.1 Main features of impulse sealing.

by the bar temperature, pressure, and heating and cooling times. A weld in thin materials can be completed in 1–3 seconds.

Impulse welding is a variation on hot bar welding where the heating is carried out using a resistive nichrome wire, giving a rapid heating and cooling cycle. In this way the fabrics are given a well-controlled welding cycle whilst still being held under pressure.

One or both of the upper and lower bars may be heated. In the case of impulse welding equipment, the bars may also be force cooled with a liquid flow after the weld time has been completed. Complete cooling prior to release of the fabrics helps to eliminate wrinkles.

11.2.2 Ironing

This is a manual process, which relies on the operator to control pressure and time. The temperature is variable. With polyamide resins, bond formation is assisted by providing steam. The result is better if a damp cloth is used, rather than a steam iron, because the steam is more uniformly distributed.

11.2.3 Hot press

A hot press consists of two metal plates, one or both of which are heated, which press and heat the fabrics to be joined. The bottom platen is sometimes covered with a sponge pad to aid even application of the pressure. With this type of equipment, pressure, temperature and time are accurately controlled. A relatively high pressure is required, which is applied either pneumatically or hydraulically. It is typically used where two fabrics or a hem are sealed using a hot melt adhesive film between the two layers. The process can be applied to uncoated, coated or laminated fabrics to provide a sealed seam.

Double-sided heating plates can be used to improve the production rate or when joining thick materials. Single-sided heating can be used for thin material, or where one of the parts to be joined is heat sensitive. There are examples in which thin fabrics can be joined to foam by applying heat only on the fabric side.

11.2.4 Heated belts or bands

Heated belts can be used to provide a continuous sealing process in some thin material joining applications. The belts are made of a thermally conducting material supported by driven rollers at either end. The belts move over fixed heating elements. They may be used single sided with fabric moving with the belt, or as two moving belts with the fabric between them. They can form a stand-alone machine or part of seaming machine with other forms of heating. In the latter case, the heated belt can provide part of the controlled cooling cycle, enabling consolidation and completion of the sealing whilst the parts are pressed together for a more extended period than can be achieved with a roller alone.

11.2.5 Heated rollers/laminator

Heated rollers may form part of a laminating machine typically with a smooth surface, a calendering machine where the surfaces are patterned to put a textured finish on a film or fabric, or as part of a product manufacturing machine where the rollers may be smooth or embossed to give a pattern to the joints in the product. The latter type of equipment has been proposed for making nappies from non-woven fabrics. The fabrics proposed contain a proportion of low melting point fibres which act as a fusing mechanism and can be melted without requiring the whole of the fabric to be melted. These are layered and positioned as required in the product and sealed in a pattern defined by the embossing on the rollers in a continuous process (Mukai *et al.*, 2009).

11.2.6 Continuous fusing press

The fabrics (potentially adhesive coated) and fusing interlayer films are laid up as a multiple layer system. They enter a heating oven followed by a press to consolidate the seals and provide control over the cooling cycle. This type of equipment is typically used in clothing manufacturing, where the fusing press can be used to bond an interlining to an outer textile fabric by means of an adhesive, often previously applied to the interlining and which melts under certain conditions of temperature, time and pressure. They are also used for applying foam layers and for leather backing.

Fusible interlinings are textiles coated with an adhesive. The textile structure can be woven, knitted or non-woven. The resin is usually a polymeric material, which melts and flows within a defined temperature range. The adhesive can be applied to the textile in a variety of ways, e.g., scatter coating, dot printing, paste printing, laminating, melt coating, etc.

Temperature, pressure and time are infinitely variable within the working ranges. Short pressing times give a relatively gentle process. Fusing can be achieved with a significantly lower temperature of the top fabric, which helps to avoid heat stress. They are capable of fusing multiple layers together. There are different types of fusing machines on the market. Figure 11.2 shows one of them. Using the appropriate pressure and keeping the temperature constant are two critical points of the fusing process. The touch screen control allows the adjustment and control of all required parameters. The transport system employs a continuous belt conveyor that feeds the fabrics into the fusing machine.

The quality of the bond, in terms of resistance to washing, ironing and dry-cleaning, depends on the interlining, the adhesive, the outer layer fabric and the fusing conditions. These have to be matched to the specific requirements of a garment. The bond should be formed without modifying dramatically the appearance, the structure, the comfort and the property of the garment. Different materials react differently to heat, pressure and moisture. They may either shrink, become shiny, melt or burn. The bond made between the outer fabric and the interlining must not be weakened by maintenance operations.

11.3 Factors affecting the quality of heat sealing

11.3.1 Sealing procedures

The basic parameters to be controlled in heat sealing are:

11.2 Continuous fusing press (Veit).

- temperature
- pressure
- dwell time
- cooling time.

These are discussed below.

Temperature

The temperature should be set to melt material at the joint line. This may require melting through the base fabric or of an adhesive film at the joint. The temperature also controls the viscosity of the melt materials. Higher temperatures lead to greater flow of the melt into the surrounding fabric. Overheating causes excess melt flow and ultimately degradation of the plastics or scorching of natural fibres. The optimal place to measure the temperature is at the joint line. This can be achieved using thermocouples or flat film thermometers, but it is generally more convenient to monitor the temperature in the heated plate or tool. The temperature set here needs to be significantly higher (e.g. 20°C) than the required temperature at the joint line, which in turn is higher than the melting point of the film or fabric to be melted. It will generally be necessary to carry out tests at a range of temperatures until the required joint performance is achieved.

Pressure

Typically the pressure required at the joint is 0.1–1.0 N/mm^2. Thin, tightly structured fabrics are processed at higher pressure than more open, thicker materials. Very low pressures may be used for foams.

Dwell time

This is generally in the range of a few seconds (2–15 sec). Enough time must be allowed for heat to conduct through to the joint line. In theory, shorter times can be used if the tool temperature is increased; however, this can risk overheating the outer surfaces of the fabrics, leading to scorching, shrinkage or glazing. For continuous processing, the dwell time is controlled by the speed of processing and the time in contact with the heated tool.

Cooling time

Sufficient cooling time (typically 2–10 sec) is required before any parting stresses are applied to the joint to allow the melted materials to solidify. Preferably the joint will be held under pressure for this period.

Impulse heat sealing performance in plastic films was examined by Raymond *et al.* (2005). For a constant film thickness combination, the impulse seal is characterised by an increase in strength up to an optimal duration/temperature setting; beyond this point, breaking load remains constant or deteriorates. It was also found to be possible to define duration/temperature settings to achieve less strong joints that could be opened easily by hand if required by the application.

The heat sealing procedures can be monitored and controlled to provide greater consistency and reproducibility of processing (Trillick, 2005). The temperature of the heating element must be constantly monitored and controlled such that each and every heating cycle is the same. To ensure that the joint reaches a high and consistent strength, the cooling cycle is also important. Preferably the jaws of the sealing equipment are opened at a preset cooling temperature after the melted materials have solidified and reached sufficient strength to support the joint in further processing.

The bar or platen design can also affect the joint performance. The heat-seal band is designed with tapered edges that reduce thinning and tearing along the edge of the seal. In addition, care must be taken to ensure the bars and platens are straight and flat to assure uniform contact and application pressure across the face of the seal. Aluminium jaw bars and platens are preferred as the higher thermal conductivity aids in the cooling process. The pressure of the bar can be monitored and controlled using pressure sensors.

11.3.2 Adhesive film

Where the sealing process incorporates an adhesive film or adhesive coating, the adhesive selection is important in defining the quality, performance, feel, stretch and longevity of the joint made. The selection may be based on the polymer or natural fibre type(s), the fabric type and tightness of the weave or knit, the weight of the fabrics, the mechanical property requirements of the joint (strength and elongation to failure), and the environmental conditions of use. The adhesives can also be supplied in a variety of forms including film, web, tape, multilayer, fabric backed and release film backed. They may also be pre-applied to the fabric in the form of dots or non-woven layer, or built into the structure of the fabric as a low melting point fibre amongst higher melting point material.

Processing conditions may also be important in selecting an adhesive, including the melting point, which can be selected based on the ease of processing compromised with the upper service temperature of the product. As a general guide, the following adhesives have been proposed for the given fabric type, though as stated above, there are many considerations leading to the final selection (Bemis, 2005; Xiro, 2000):

- polyester fabrics – polyurethane, polyamide, polyolefin and polyester-based adhesive films
- polyamide (nylon) fabrics – some polyurethane films, polyamide and polyester films
- acrylic fabrics – polyamide films
- acetate fabrics – some polyamide, polyester and polyolefin films
- aramid fabrics – polyurethane films
- cotton fabrics – some polyurethane, polyamide, polyester, polyolefin and vinyl films
- wool fabrics – some polyurethane, copolyester and polyamide films
- elastic fabrics (containing lycra) – polyurethane and polyamide films have been designed that retain their elastic properties after the bonding process
- polyurethane foams – polyurethane and copolyester films
- PVC foams – polyurethane, polyester and vinyl films
- polyethylene foams – polyolefin films.

Typically the film thickness used is in the range 25–125 microns. For tightly woven materials and lightweight fabrics, a thin film is used (<75 microns), and for heavier fabrics or open structured fabrics, a thicker film is used (up to 200 microns).

11.4 Applications of heat sealing in textiles

- There are a range of applications of heat sealing in textiles:
- Apparel – stitch-free seams in intimate garments, swimwear and outdoor clothing use heat sealing and adhesive films. Nappies can be made using embossed heated rollers and fabrics containing low melting point fibres. Embroidered badges are applied using heat sealing.
- Automotive – soft tops and some seating fabrics use these techniques.
- Architectural – awnings, inflatable structures and roofing materials are joined with sealed seams using heat sealing.
- Medical – gowns, dressings.

11.5 Hot air wedge and hot wedge welding of textiles: introduction

Wedge welding is a continuous process in which a thermostatically controlled heated wedge or a perforated wedge or nozzle delivering a heated air stream is positioned between the fabrics to be joined. The inner surfaces of the fabrics are heated as they move over the heated surfaces or air stream and they are drawn through and compressed by rollers running at a controlled speed to consolidate the weld. In the case of using a hot air stream, again thermostatically controlled, the fabrics are heated without direct contact being made with the hot surfaces. The process will operate at a few m/min and can produce strong joints that do not melt the outer surfaces of the fabrics. The hot air wedge is also commonly used to apply a sealing tape to the surface of a previously stitched seam. The equipment is used in a wide range of applications from elastic fabrics to tarpaulins and geomembranes.

The main advantages of using hot air/wedge welding over other joining processes are:

- applicable to thin and thick materials (including foams)
- heat is applied directly at the joint line, enabling high speed, energy efficient joints to be made without affecting the outer surfaces of the fabrics
- sealed joints can be tested non-destructively when a dual-track machine is used
- manual or automated machines available
- consistent performance once process is stabilised
- relatively low cost (consumables such as needles, solvents, adhesives are not required)
- an impermeable seam can be achieved
- a fusible film can be melted at the joint over a previously stitched seam to provide sealing
- edges can be sealed and there are no stitch holes
- the air jet blows contaminants away from the joint
- the simple overlap seam provides an alternative to stitching that is lower weight and less bulky. With suitable film adhesive it can be a soft, stretchy, flexible joint.

The main disadvantages of hot air/wedge welding are:

- the wedge or air temperature cannot be altered quickly, so care is needed when changing process speed to go round more complex parts of a seam to avoid melting through the fabrics
- pre-placement of the fabrics with the joint in contact before welding is not possible
- the seam must be open at the end so that the wedge can be removed from between the fabrics; a heated press is sometimes used to close seams
- seam designs are restricted to overlap, peel and taped.

11.6 Equipment for hot air/wedge welding

Hot air/wedge welding equipment has the following basic components:

- A heated wedge or air knife delivering heated air, which is mounted on a bar to deliver heat between the fabrics being joined.

- A drive mechanism either to pull the fabrics through the machine or to drive the machine along the fabrics (in large-scale applications). The drive wheels also have the function of pressing the materials together after heating and consolidating the joint as it cools.
- Process control. Thermostatic controls on the air heater or the wedge, combined with speed control.

11.6.1 Hot air wedge welding

This equipment uses a shaped or perforated nozzle to deliver hot air to heat one or both of the fabric surfaces before they are pressed between driven rollers (Fig. 11.3) to apply pressure and complete the weld between the two surfaces. The hot air method of welding makes use of a device consisting of a resistance heater, a blower and a temperature controller to blow hot air between the fabrics. The hot air wedges and nozzles are typically 10–40 mm wide.

The same equipment can often be used for applying sealing tape to a stitched seam. A tape feed is added to deliver tape into the rollers over the previously stitched seam. The hot air stream is directed to heat the lower surface of the tape before it enters the rollers (Fig. 11.4). Tapes in the range 10–26 mm width can typically be processed. The air nozzle and rollers are specified to be at least 3 mm wider than the tape to ensure effective sealing. The lower roller surface may be grooved for seam taping to allow the stitching thickness to be accommodated and present the flattest possible surface to the tape.

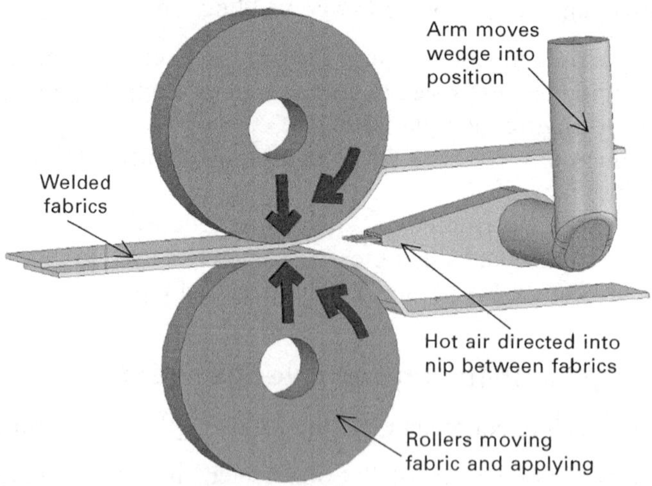

11.3 Hot air wedge welding equipment.

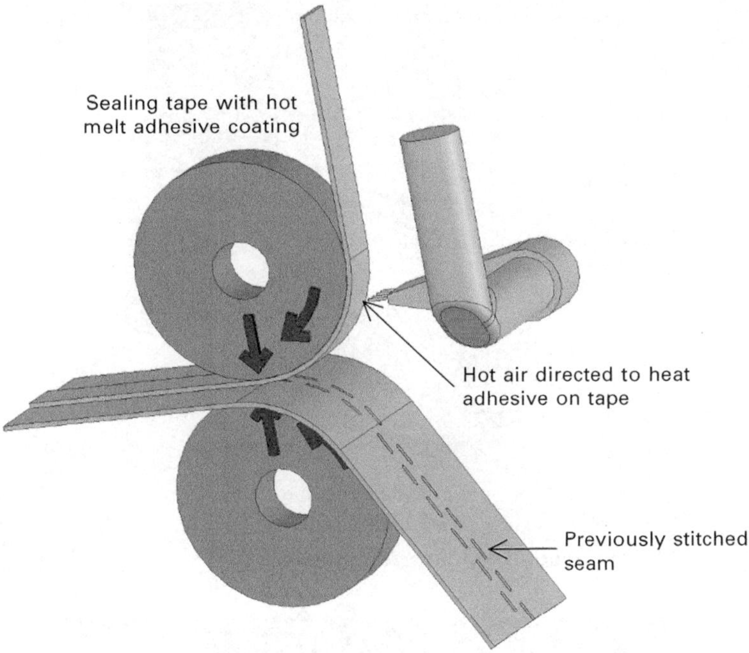

11.4 Hot air seam taping and sealing equipment.

11.6.2 Hot wedge welding

Hot wedge welding uses an electrically heated wedge, which contains one or more cartridge heaters. The temperature is controlled by a closed loop feedback using a thermocouple positioned in the wedge. This maintains the wedge temperature typically to within ±10°C of the set temperature during welding. The principle of electrically heated wedge welding is that both fabric surfaces to be welded pass over, and are forced into intimate contact with, the hot surface of the heated wedge. In some systems, for heavy duty applications, there are additional wedge rollers to hold the fabrics against the heated wedge. The drive rollers or consolidating rollers pull the fabric through the machine and press the heated surfaces together (Fig. 11.5). The hot wedge is typically 7–50 mm wide for bench top machines and 25–75 mm wide for floor or ground-based machines.

Seams can be made as a single or double track. Double-track welding requires a heated wedge with a cartridge heater at each edge and a gap in the centre. This leaves an unwelded channel that can be pressurised with air after the completion of the weld in order to non-destructively test the integrity of the joint (BGA, 2006).

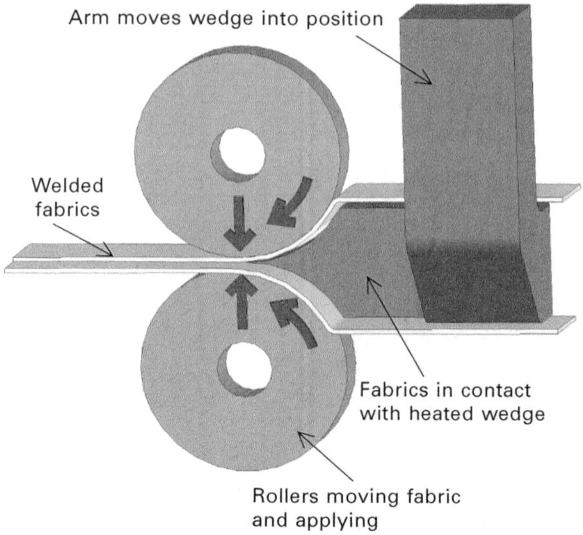

11.5 Hot wedge welding equipment.

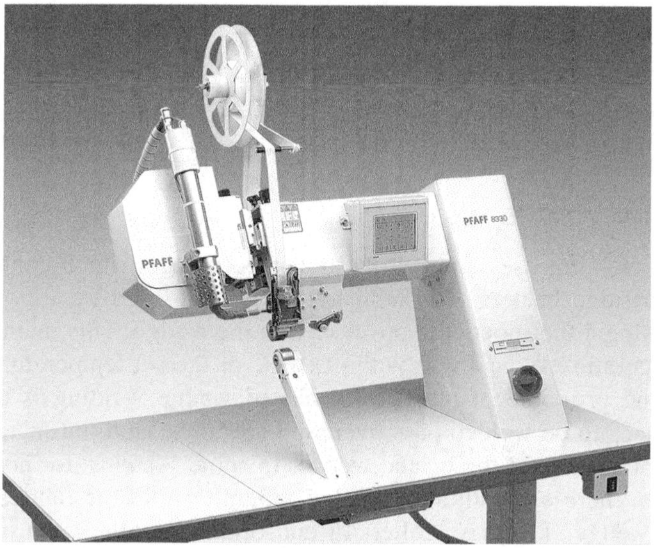

11.6 Bench-top sewing machine type equipment for hot air/wedge welding and seam taping (Pfaff).

11.6.3 Bench-top sewing machine type equipment

These machines are bench mounted such that the user manually manipulates the fabrics through the machines assisted by the driven consolidation rollers (Fig. 11.6). They may use a hot air wedge or a hot wedge as the heating mechanism. In some cases this may be combined with a heated belt to more effectively control the heating/cooling cycle for the process.

For seam taping or for the use of tape between the fabrics, a tape feed mechanism is attached, generally to hot air wedge machines. Alternatively, the fabric may be pre-coated with adhesive in the region to be joined. This assists in the procedure for hemming, binding and other more complex seams. It is also possible to use a range of fabric folding tools and guides to shape and support these more complex seams.

Machines may include edge trimming, tape cutting or an automated reverse motion at the end of a seam to ensure that the seams are heated and consolidated at the seam end. When adhesive tape feed is used, it may be pre-heated before it is fed into the joint, with edge trimming of the adhesive tape used to ensure there is no under- or overrun at the bonded edge.

For heavier or larger applications this equipment is designed with a long upper and lower arm (in place of a flat table base), typically up to a metre in length. This allows fabric to be gathered on either side of the welding tool.

11.6.4 Floor or ground-based equipment

For large-scale applications, indoors and outdoors, the wedge welding equipment is mounted in equipment that drives itself along the ground, over the sheets being welded, with the heating tool sliding along between the edges of the sheets (Fig. 11.7). There is a variation on this type of equipment that injects an extruded bead of plastic at the joint line in addition to the

11.7 Floor-based hot wedge welding equipment (Leister).

hot air wedge or hot wedge, which is present as a pre-heat for the surface. The extruded bead is generated in the welding equipment by delivering plastic pellets through an extruder barrel. This is described as extrusion flatwelding or extru-wedge flatwelding and is typically used for geotextiles (Koerner, 1990).

11.6.5 Gantry-mounted automated equipment

Hot wedge and hot air wedge equipment is mounted on single axis and potentially two-axis overhead gantry systems, which moves the welding head over a flat worktable of various sizes. This enables processing of large products and heavy fabrics of many metres in length and width (Fig. 11.8). Such equipment is used for seaming and hemming of banners and tarpaulins. Like the bench-top equipment, the welding head has a consolidating roller which follows the hot wedge, and tape can also be dispensed for over taping, taped butt welds and lap seals with cover strips. The fabrics are pre-placed and fixed into position using bar clamps or vacuum clamping (for impervious materials). Equipment can also be adapted with a cutting wheel to allow cutting before or after welding.

11.7 Factors affecting the quality of hot air/wedge welding

The basic parameters that need to be controlled in wedge welding are:

- seam design and preparation
- the wedge or air temperature
- consolidating rotor pressure
- process speed.

Seam design/preparation

The seam designs used in wedge welding are restricted due to the need to insert the wedge between the fabric layers during welding. Most commonly

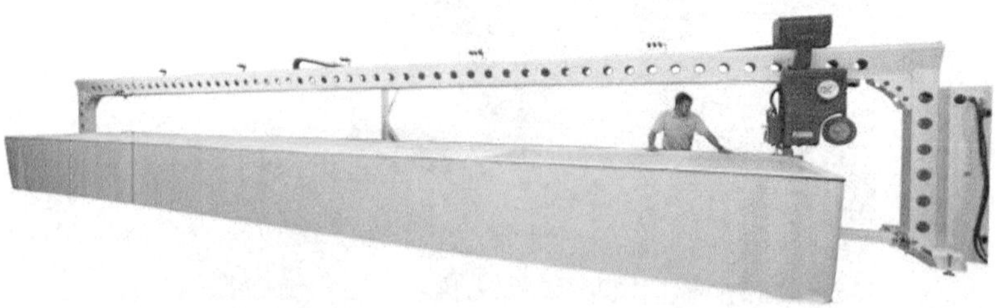

11.8 Gantry-mounted automated equipment (Weldmaster).

either an overlap seam or a taped seam is used, but it is also possible to use a peel seam or double tape seam for extra strength (Fig. 11.9). The equipment may be fitted with guides to control the size of the overlap at the joint or the position of fabrics during taped seaming. Guides may also be fitted to fold the fabrics as they enter the welding zone, allowing hemming and other more complex features to be carried out (hem with rope, hem with pocket, welds with channels, etc.). Accessories for tube welding for sleeves or other applications are available. Taped seams can also be made without previously stitching the two pieces together, with precise position of the two fabric pieces and the overlying tape.

Particularly when welding coated fabrics and plastic and when using adhesive films or tapes, it is important that the area of the seam to be welded is clean and free from dirt and dust particles. This includes the parts of the welding machine that come into contact with the fabric or the adhesive. Sometimes it may be necessary to use a water-based detergent solution to clean the materials. When using the hot air wedge, the air stream is effective in removing loose dust from the joint region.

The gap at the rollers and between the wedge and the rollers is set to correctly accommodate the thickness of fabric being welded. If only one side of the joint is to be heated, such as in bonding with a pre-coated adhesive, the hot air stream is positioned accordingly (Fig. 11.4). Prior to welding it is important to check that the heat source is centred in line with the rollers. The position of the hot air stream in relation to the fabric and the tape for seam taping is important. It should be positioned to heat the tape preferentially to allow the adhesive to be melted and then pressed into the fabric surface structure and not to damage the sensitive backings of multilayer waterproof fabrics.

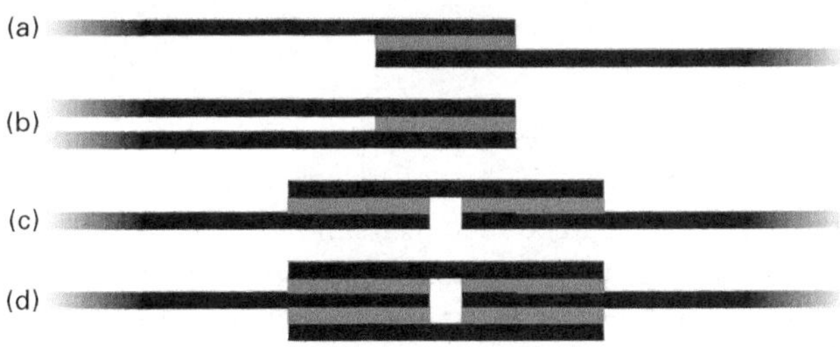

11.9 Some seam designs suitable for wedge welding. Various hemming types are also possible such as hems with a channel and hems with rope as well as plain hems: (a) ASTM D6193 designation LSa – Simple lap seam; (b) ASTM D6193 designation SSa – Simple peel seam; (c) ASTM D6193 designation LSp – Taped seam. This may also be over a previously stitched seam; (d) No ASTM D6193 designation – Double taped seam carried out in two stages.

Wedge or air temperature

The temperature of the air stream or solid wedge should be set to melt the material at the joint line. This is either the base fabric, plastic coating or an adhesive film at the joint. The temperature also controls the viscosity of the melt materials. Higher temperatures lead to greater flow of the melt into the surrounding fabric. Overheating causes excess melt flow and leads to holes melted through the fabric layers, or degradation or carburising of the materials, severely reducing the joint strength. The temperature is measured using a sensor in the wedge or in the air stream. The temperature set here needs to be significantly higher (100–250°C) than the material melting point. Hot air systems typically allow operation at up to 600–750°C, hot wedge at up to 500°C. It will generally be necessary to carry out tests at a range of temperatures until the required joint performance is achieved. For hot air processing the air flow rate and/or pressure must also be controlled to ensure suitable transfer of heat. It should also be noted that the ambient temperature of the working zone (such as working outdoors), or effects of the sun will affect the optimal welding parameters. If the ambient conditions are not controlled, then it is advisable to carry out test welds and assess the weld strength in advance of production.

Clamping pressure

The pressure at the rollers is controlled by a pneumatic actuator that can typically be controlled by setting the air pressure in the range 0–7 bar. The pressure and gap setting are varied depending on the thickness and resilience of the materials to be welded. Thin, tightly structured fabrics are processed with higher pressure than more open thicker materials. Very low pressures are used for foams.

Roller gap

It is usually possible to set the roller gap independently of the roller pressure, otherwise called the depth stop. It should be set to less than the combined thickness of the materials and can be set to zero allowing the weld consolidation to be fully controlled by the pressure setting. The gap is opened to work on foam backed material such as neoprene to prevent stretching.

Process speed

This is set at the drive rollers, generally in the range 0–20 m/min. Typically 0.7–4.0 m/min is used for thick materials and faster for thinner fabrics. If the speed is too low, or reduces as the operator is negotiating a corner in the product, there is a danger of melting through the fabrics or overheating the materials. Some equipment has automated adjustment of the heating power to compensate for changes in the processing speed. If the speed is too high, the fabric or adhesive materials will be insufficiently softened resulting in a low strength joint. Some welding equipment allows differential adjustment of the welding speed between the upper and lower rollers. This helps to control puckering at the seam.

After welding is completed, in the case of coated fabrics or plastic sheet, it is common to see a small melt bead at the weld edge. This is a reliable indication that proper seaming procedures have been used. Excessive melt beads or wrinkling of the sheets suggests that excessive heating or pressure may have been applied.

When double-track welding is used, in geomembrane welding for example, the unwelded centre channel can be non-destructively tested to ensure that the welds are sealed. It is also possible to cut mechanical test specimens from the surplus at the ends of welds.

11.8 Applications of hot air/wedge welding in textiles

The fabrics that may be joined using wedge welding are shown in Table 11.1:

- Hot air wedge welding is used in the healthcare and personal protective product sectors for making seams in non-woven and coated non-woven fabrics, also in welding of neoprene wet or dry suits
- Hot wedge welding is used in joining of heavy fabrics and films in outdoor applications, such as lining of swimming pools, reservoirs and landfill sites.

The bench-top hot air and wedge welding equipment is used for sealing coated fabrics in a wide range of applications including boat covers, soft tops, inflatable boats and toys and sealed bags. There are also applications in light fabrics for lingerie hems and swimwear. These can use tape fed into the joint or fabric pre-coated with hot melt adhesive at the joint region.

Hot air taping is used to seal the stitch holes in previously stitched seams in applications requiring waterproofing. Applications include outerwear, NBC suits, other protective workwear, tents, waders, footwear, sportswear, rucksacks and sealed bags. It is also used in producing sealed seams for seating for industrial and open-top vehicles, hospital furniture and motorcycle seats.

The gantry and floor-mounted equipment is used in a wide range of industrial applications using coated and heavy fabrics such as tarpaulins,

Table 11.1 Fabrics that may be joined using wedge welding

Fabric group	Sub-group	Suitability of wedge welding			Comments
		Yes	Some cases	No	
Natural					
Cotton			x		Using adhesive film
Cellulose			x		Using adhesive film
Wool			x		Using adhesive film
Silk			x		Using adhesive film
Glass			x		Using adhesive film
Carbon fibre			x		Using adhesive film
Synthetic					
Acrylic	Woven, knitted		x		Using adhesive film
Cellulose	Acetate		x		Using adhesive film
	Viscose		x		Using adhesive film
Nylon	Woven, knitted, fleece, non-woven	x			With or without adhesive film
Polyester	Woven, knitted, fleece, non-woven	x			With or without adhesive film
Polypropylene	Woven, knitted, fleece, non-woven	x			With or without adhesive film
Polyethylene	Woven, knitted, fleece, non-woven	x			With or without adhesive film
Coated	PU, PC	x			
	PVC	x			
	PTFE			x	
Laminated		x			Preferably with a film interlayer
Elastic fabrics		x			With an elastic film interlayer
Natural to synthetic		x			Synthetic melts into natural fibres or use film layer

Note: Foam materials can be joined to fabrics using the hot air wedge method.

inflatable boats and other inflatables, airships and aerostats, flexible roofing and tents, signs, banners and awnings.

11.9 References and further reading

Adhesive Films Inc. (2002), *Instruction Manual and Guide for Effective Waterproofing of Sewn Seams.*

Bemis (2005), *Bemis Thermoplastic Adhesive Selector Guide*, January.

BGA (British Geomembrane Association) (2006), *Geomembrane Welding*.
Koerner R M (1990), *The Seaming of Geosynthetics*, London: Elsevier.
Mukai H *et al.* (2009), 'Stretchable non-woven fabric, absorbent article and method of producing absorbent article', European Patent EP2 090 684 A1, August.
Pfaff (2011), *Pfaff Machine Guide*, November.
Raymond M, Leczynski A, Iovanna J, Tayebi A (2005), 'Effect of impulse heat sealing process parameters on bond strength of low density polyethylene films', *ANTEC 2005*, Boston, MA: Society of Plastics Engineers, pp. 2925–2929.
Trillick C (2005), 'Process control improves heat seal quality', *Pharmaceutical and Medical Packaging News*, February.
Wise R, *Thermal Welding of Polymers*, Cambridge: Woodhead Publishing.
Xiro (2000), *Xiro Adhesive Selector Guide*, November.

12
Ultrasonic and dielectric welding of textiles

I. JONES, TWI Ltd, UK

DOI: 10.1533/9780857093967.3.374

Abstract: This chapter reviews two major welding techniques for textiles: ultrasonic welding and dielectric welding (also known as high frequency (HF) or radio frequency (RF) welding). For each technique, the chapter discusses the key principles, their advantages and disadvantages, the equipment used, factors affecting quality and applications.

Key words: textile, fabrics, welding, bonding, ultrasonic plunge, wheel, roller, continuous, dielectric, RF, HF.

12.1 Ultrasonic welding: an introduction

Ultrasonic welding is a process that uses mechanical vibrations to soften or melt a thermoplastic material at the joint line. The fabrics to be joined are held together under pressure and subjected to ultrasonic vibrations, usually at a frequency of 20–40 kHz. The mechanical energy is converted to thermal energy due to intermolecular and surface friction. The mechanisms responsible for generating heat at the joint line are not fully understood. However, Benatar and Gutowski (1989) have summarised the key steps as:

- the mechanics and vibration of the parts
- the viscoelastic heating of the thermoplastics
- heat transfer
- flow and wetting
- intermolecular diffusion.

Studies suggest that the ultrasonic bonding mechanism is a physical process and not a chemical process, i.e. when the process is carried out within the normal operating range, no new chemical groups or bonds are formed. Other studies of the mechanisms of ultrasonic welding include Tolunay (1982), Shi and Little (2000) and Mao and Goswami (2001). The ultrasonic process can be used for seaming, cutting, slitting, trimming, tacking, embossing, quilting, or simultaneous cutting and sealing. The process is fast (200 m/min has been demonstrated) and energy efficient. The equipment may be bench or table mounted, or integrated into an automated line.

The main advantages of using ultrasonic welding over other welding processes are:

- short weld times (typically less than one second) – suits high volume production of discreet parts
- high continuous welding speeds (>30 m/min possible)
- simple technique – straightforward to put into production
- consistent performance (once weld parameters are optimised) – a reliable process; precise automated assembly is possible using CAM technology
- clean process – can be used in a clean-room environment
- minimal thermal damage to fibres: heat energy is generated within the fibres using ultrasonic energy at the joint site; this heat energy is not conducted through the fibres as is the case in thermal bonding
- relatively low cost (consumables such as needles, solvents, adhesives, etc., are not required)
- low energy consumption
- an impermeable seam can be achieved (essential for certain technical textile applications such as clothing for contaminated environments)
- recycling is relatively easy since other yarns have not been used in seam formation
- a fusible film can be melted at the joint interface providing a seal without affecting the fabric layers
- edges are sealed and there are no stitch holes.

The main disadvantages of ultrasonic welding are:

- all aspects of the process must be considered carefully together to ensure successful welding
- particles of flash are sometimes generated
- vibrations may affect sensitive components
- when plunge welding is used (i.e. not continuous), the maximum size of component is limited to 250 mm; this size is reduced when using materials that require high energy for welding, for example, semi-crystalline materials.

12.2 Equipment for ultrasonic welding

A typical ultrasonic welding machine for plunge welding (Fig. 12.1) consists of five main components (Froment, 1999):

- ultrasonic generator
- microprocessor control system, and user interface
- machine stand, consisting of the base-plate, pneumatic system and welding press
- welding stack, which incorporates the ultrasonic transducer, booster and welding horn
- component holding fixture.

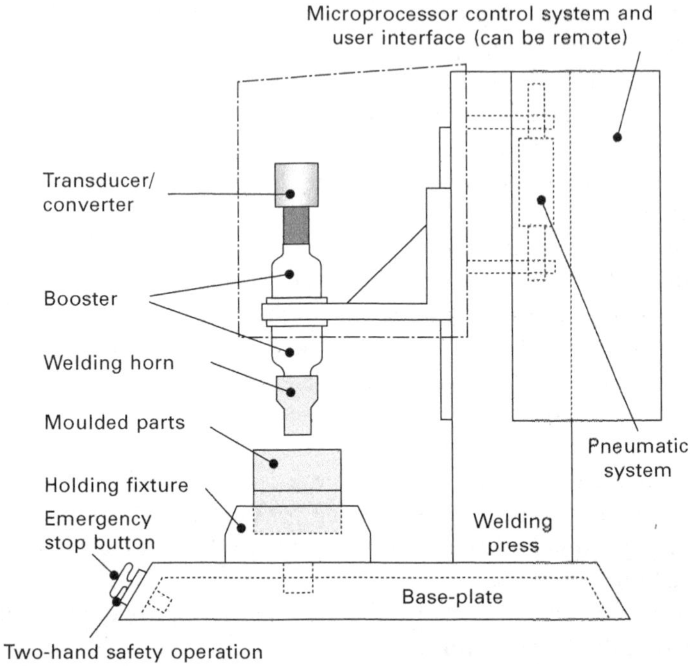

12.1 Schematic of a typical ultrasonic plunge welding machine.

Figure 12.1 shows a breakdown of the individual parts of an ultrasonic welding machine. A generator converts electrical power from the single-phase mains to the correct frequency and voltage for the transducer to convert into mechanical vibrations. The generator thus provides the ultrasonic energy to the welding stack. The welding stack is a tuned resonator, rather like a musical instrument tuning fork. In order to function, the resonant frequency of the tuned welding stack must closely match the frequency of the electrical signal from the generator (within 30 Hz). The microprocessor unit controls the welding cycle and feeds back key welding information to the user, via the user interface. The user interface also allows the operator to enter the required welding parameters.

The machine stand is designed to hold the welding stack and apply the force necessary for welding. It consists of a base-plate, to hold the tooling jig, and a pneumatic cylinder to apply the force. The machine will have a pressure gauge and regulator for adjustment of the welding force. It should be noted that a particular gauge pressure set on one piece of ultrasonic welding equipment will not necessarily provide the same welding force on another. Welding force should be calibrated using a load cell so that direct comparison of welding forces can be made from equipment to equipment. There is also a flow control valve to allow adjustment of the speed at which the welding head approaches the component being welded. Some equipment manufacturers have introduced an electromagnetic force application system in place of the traditional pneumatic cylinder. This gives better control of the approach rate and can be beneficial when welding small or delicate components.

Finally, the welding stack is the part of the machine that provides the ultrasonic mechanical vibrations to effect the welding operation. This is generally a three-part unit consisting of:

- transducer
- booster
- welding horn.

These are mounted on the welding press at the centre-point of the booster section.

The transducer, also known as the converter, is the key component of the ultrasonic welding system. The transducer converts the electrical energy from the generator to the mechanical vibrations used for the welding process. The booster section of the welding stack serves two purposes. Its primary purpose is to amplify the mechanical vibrations produced at the tip of the transducer. Its secondary purpose is to provide a place where it can be mounted on the welding press and a point where the load can be transferred from the press to the components being welded. The booster expands and contracts as the transducer applies the ultrasonic energy.

The welding horn (or 'sonotrode' as it is also known) is the element of the welding stack which supplies energy to the component being welded. The horn is designed to resonate at the frequency of the ultrasonic system. When the horn vibrates, it stretches and shrinks in length by a small amount; this motion is referred to as the amplitude of the horn (Devine 1998). Amplitude is typically between 30 and 120 μm. Design of the welding horn is critical to successful welding. The welding horn, like the booster element, is a tuned device which, in the majority of applications, also provides mechanical gain. The length of the welding horn must be a minimum of one half of a wavelength of ultrasound in the material from which it is manufactured, but can also be multiples of one half wavelength. This ensures that there will be sufficient amplitude at the end of the welding horn to effect welding. The horn must be made from a material with good acoustic and mechanical properties. These properties are usually found in low-density materials such as aluminium or titanium. If wear is likely to be a problem, aluminium can be chrome plated and titanium may be carbide coated.

There are three basic designs of welding horn: exponential, step and catenoidal. The catenoidal welding horn is used in the majority of applications. This design, which combines a gentle curve with a modest step at the node, gives more gain than the exponential design but without the stress concentration associated with the stepped horn. The tip of the welding horn delivers the ultrasonic energy to the component being welded. In the majority of applications, the tip of the horn is profiled to provide a pattern on the fabric and to focus the energy delivered, although some applications use a simple flat-ended welding horn. The pattern ensures that maximum energy transfer between the horn and the component is achieved. A variation on the process uses a two-part horn with a gap between the tips of 0.15–0.55 mm, allowing independent vibration of the two (Tsujino *et al.*, 1986). The result is a weld of a similar size to the gap between the horn parts, which is effective for micro-welding of thin materials over a limited area of a plastic part.

In the case of textiles, there are two main welding techniques:

- plunge mode welding
- rotary/continuous welding using a wheel or a wide drum and static horn.

The general principles of the continuous welding equipment are very similar to the plunge welding equipment. In continuous welding the ultrasonic vibrating horn can be wheel shaped, and rotates during processing. The anvil is a rotating wheel or drum, rather than being static as in plunge welding.

In plunge mode welding, the fabrics are held under pressure onto an anvil by a vibrating horn, the process typically taking around one second to complete. Either the anvil or the horn may be patterned.

In rotary/continuous mode, the fabric is fed between two wheels, one of which is ultrasonically vibrated while the other acts as a rolling anvil (Figs 12.2 and 12.3), or between a fixed vibrating horn and a rotating drum anvil (Fig. 12.4) to provide a continuous welding process. The wheels are normally patterned to provide local intensification of the ultrasonic energy and to provide a patterned seam. They may also incorporate a cutting edge to enable simultaneous 'cut and seal' processing. Seaming rates of at least 30 m/min are possible. The drum type machines can be extended to a full fabric width (up to 3 m) to provide continuous quilting (Fig. 12.5) at rates up to 10 times faster than a stitch quilting machine and offering a different variety of quilting patterns defined by the patterns on the drum. These machines can have tens of horns, positioned to provide a pattern over the full width of the fabric.

A horn with a sharp edge may be used for ultrasonic slitting. This is a generally a continuous process with either the horn or the fabric moving. Ultrasonic slitting produces an edge that is smooth, durable and clean, with no discoloration of the fabric. Ultrasonic edge sealing prevents unravelling of knitted or woven materials. The sealed edge is tapered without a bead that would add unnecessary bulk.

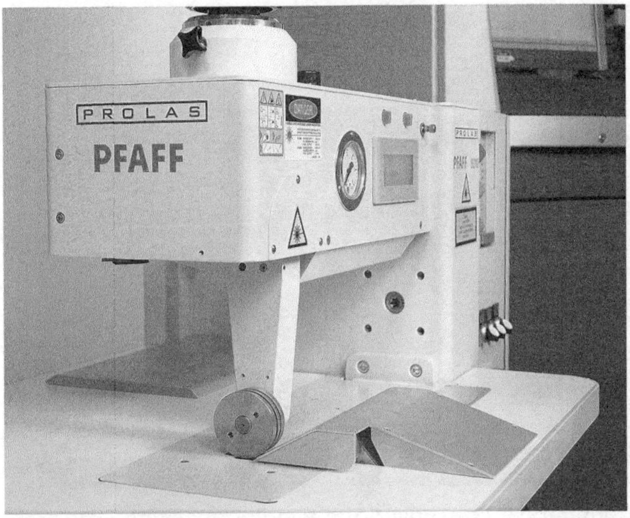

12.2 Continuous ultrasonic wheel welding and cutting machine (note this machine also includes laser welding).

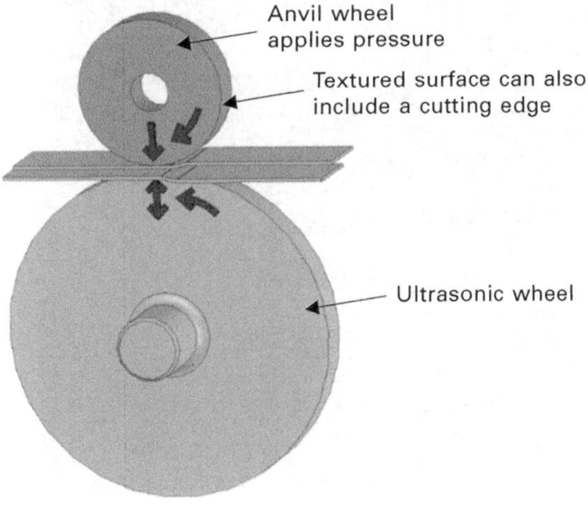

12.3 Continuous ultrasonic wheel welding.

12.3 Factors affecting the quality of ultrasonic welding

To achieve a good ultrasonically welded joint, designers are advised to consider the whole process carefully. The following sequence serves as a guide to successful ultrasonic welding:

- Select the right material or combination of materials and interlayer films.

12.4 Continuous ultrasonic drum welding machine.

12.5 Wide format (up to 3 m) ultrasonic drum welding machine used for quilting.

- Select the relevant seam design for the material being welded.
- Select the most suitable type of equipment, based on the size, shape and contents of the component being welded. This includes the correct horn design, booster type and support fixture.
- Evaluate designs with prototype components. Selection of weld parameters should take place at this stage. Welding parameters should be set in relation to the desired performance of the product, for example, to

maximise failure strength. Use of designed experiments to set the welding parameters is recommended.

- Make any necessary changes to the design and then put the components into production.

Faults occurring in production due to poorly designed components can prove costly and time-consuming to rectify. It is wise to be aware of typical welding faults before designing a product. This helps designers to avoid faults.

12.3.1 Materials

Both modes of welding (plunge and continuous) can be used to join two virgin fabrics without coatings or interlayers. In this case the fabrics will be melted through their full thickness. Alternatively, the ultrasonic energy can be used to melt either a coating or a hot melt adhesive film interlayer. The film can be pre-placed or delivered between the fabrics from a roll during the welding process. Only the film or the coating is melted and infiltrates the fibres in the fabric, creating a strong joint that remains flexible and often retaining elasticity if the film is of a suitable type.

The ability to ultrasonically weld textiles depends on their thermoplastic content. As a minimum, the material must have a thermoplastic content of 65%. The weldability of the main fabric groups is listed in Table 12.1. As a general rule, natural fabrics do not melt and require a hot melt adhesive film to be placed between them before welding. This will melt into the natural fibres creating a strong joint. Synthetic thermoplastic materials can be welded without an additional film interlayer, although as with most other welding methods, a low melting point film can be used that will melt preferentially to the fabrics and provide a joint that leaves the original fabrics unaffected by the process. This is especially useful for elastic fabrics and laminated fabrics, where a seal can be achieved without damaging the barrier layer in the laminate.

The fabric structure, thickness and base material affect the performance of the ultrasonic welding process. Factors affecting weldability include the following (Branson 2011):

- Wovens: Yarn density, thermoplastic content, tightness of weave and uniformity of material thickness. Weld strength may vary according to the orientation of yarns or filaments.
- Non-wovens: Uniformity of material thickness and thermoplastic content. Random orientation of fibres gives non-wovens excellent strength.
- Knits: Style of knit, thermoplastic content and elasticity of construction.
- Coated materials: Coating material, thickness and substrate characteristics.

Table 12.1 Fabrics that may be joined using ultrasonic welding

Fabric group	Sub-group	Suitability of ultrasonic welding			Comments
		Yes	Some cases	No	
Natural					
Cotton			x		Requires adhesive film
Cellulose			x		Requires adhesive film
Wool			x		Requires adhesive film
Silk			x		Requires adhesive film
Glass			x		Requires adhesive film
Carbon fibre			x		Requires adhesive film
Synthetic					
Acrylic	Woven, knitted		x		Can be cut, but bonding requires adhesive
Cellulose	Acetate		x		Requires adhesive film
	Viscose		x		Requires adhesive film
Nylon	Woven, knitted, fleece, non-woven	x			
Polyester	Woven, knitted, fleece, non-woven	x			
Polypropylene	Woven, knitted, fleece, non-woven	x			
Polyethylene	Woven, knitted, fleece, non-woven	x			
Coated	PU, PC	x			
	PVC		x		Welding inhibited by plasticisers
	PTFE		x		Spray coats can be welded, thick coats need a film interlayer
Laminated			x		Preferably with a film interlayer
Elastic fabrics			x		With an elastic film interlayer
Natural to synthetic		x			Synthetic melts into natural fibres or use film layer

Notes: For best results the mixed natural/synthetic fabric to be welded should contain 50%, and preferably over 65%, thermoplastic synthetic fibres. Foam materials weld better in plunge mode than continuous, so that they can be held under pressure during cooling.

- Laminates: Thermoplastic type at weld interface and laminate structure configuration.

The type of thermoplastic is also important in defining welding performance. Amorphous and semi-crystalline thermoplastic materials will behave differently when ultrasonically welded as bulk plastic material. However, the characteristics of fabrics usually ensure that all the processing is near-field, i.e. the ultrasonic horn is less than 6 mm from the joint interface. The differences between amorphous and semi-crystalline polymer processing therefore become less distinct. Often, it is found that soft, elastomeric materials require more energy for ultrasonic welding than more rigid materials because the ultrasonic vibrations are damped by the soft material.

Semi-crystalline materials have a well-defined melting point. High heat energy is required to break down the crystalline structure before melting can occur. The molecular chains are then mobile and inter-diffusion can begin to occur and form the weld. Generally, semi-crystalline plastics require more energy to weld ultrasonically than amorphous plastics. Solidification of the material also occurs more rapidly compared with amorphous plastics. Higher horn tip amplitudes are required to ultrasonically weld semi-crystalline materials. This ensures that the required high heat energy is achieved. In contrast, amorphous materials do not have a defined melting point. When the materials are heated, they gradually soften as they pass through the glass transition temperature of the material, and move into a rubbery state. These materials require less energy for welding than the semi-crystalline materials and hence lower horn tip amplitudes can be used.

There are other properties of thermoplastic materials which govern their ability to be ultrasonically welded successfully, including the presence of additives such as flame retardants. Typical flame retardant additives such as boron, antimony, halogens, nitrogen and phosphorous reduce the strength of the welded component and often require higher ultrasonic amplitude to make the weld. Some polymer materials, for example nylon, have a tendency to absorb moisture from the atmosphere. In the case of nylons, moisture absorption can be as high as 4% of the material weight when fully saturated and this has a detrimental effect on welding. Moisture tends to boil off during the rapid ultrasonic heating process, leaving voids at the weld interface. This is a particular problem when the joint is required to produce a hermetic seal. If materials have been stored in a humid atmosphere, they should be dried thoroughly. Components which are likely to experience problems with moisture may need to be stored in polyethylene bags, along with a suitable moisture absorption medium such as silica gel. Yarn density, tightness of weave, elasticity of material and style of knit are other factors which can influence textile weldability.

An important issue when considering weldability is the compatibility of different materials. The general rule is that very few materials are considered compatible for welding. A further practical consideration is the thickness of material through which the ultrasonic energy needs to travel between the welding horn and the joint. As a minimum, the material must have uniform thickness.

12.3.2 Welding parameters

Once the welding parameters have been optimised, ultrasonic welding is a consistent and reliable process. Key welding parameters include:

- amplitude
- power
- welding mode (including hold time in plunge welding)
- pressure and wheel or anvil gap (continuous welding).

These must all be correctly set to achieve the required joint quality.

Amplitude

Amplitude selection depends on the type of material being welded, amorphous or semi-crystalline. In general, semi-crystalline materials require more energy and, therefore, more horn tip amplitude than amorphous materials. The gain in the horn and in the booster are altered to set the amplitude. Depending on the material being welded, the amplitude may be set at 20–100 μm.

Power

Typically, a product produced in an easy-to-weld material will require a machine with a maximum power output of 800 W, though some machines can deliver over 2 kW per horn, and may have multiple horns. During the welding cycle, the machine may never draw the maximum power available. Most ultrasonic welding equipment has the ability to limit the amount of power consumed during the welding cycle.

Welding mode

Time, energy or distance are the three key variables in the welding mode for plunge welding. The ultrasonic welding cycle consists of two phases: the 'ultrasonics on' period and the hold time. Throughout the welding cycle, the welding horn remains in contact with the components being welded and an axial force is applied. The period when the ultrasonic energy is applied can be controlled by a time setting (typically 0.1–0.3 s for small parts), energy applied or the movement (displacement) measured at the joint. During the hold time, the parts are supported under pressure for typically 0.3–0.5 s to allow the joint to cool.

Pressure

The force applied to the joint must be enough to hold the parts together but not overdampen the vibrations. For parts with less than 20 mm weld length, typically forces less than 200 N are required.

Wheel or anvil gap

For continuous welding, the relation between the fabric thickness and the gap at the wheel or anvil is important in controlling the weld properties. Typically a gap of 0.1 mm may be set for thin materials, but this may be increased for thicker materials to limit the thinning at the weld zone.

Shi and Little (2000) have explored the importance of these parameters in welding seams to incorporate optical fibres into garments. Their research suggested an increase of weld strength with an increase in pressure and time until a point at which strength peaked and then began to drop due to polymer orientation. Weld strength also increased when amplitude was increased from 40 to 60 μm.

12.4 Applications of ultrasonic welding in textiles

The ultrasonic welding process is typically used where short weld times and a hermetic seal are required. The process can be easily automated. It is therefore well placed to replace the two-stage stitch and sealing process with a cheaper one-stage process for sealed seams. The potential for automation is particularly important for garment making where there is considerable pressure to speed up production whilst still allowing the flexibility to provide for design and sizing variations. Specific applications include hook and loop straps (Branson, 2011) and belt loops. The ability of ultrasonic welding to provide a hermetic seal also means it is particularly suitable for the manufacture of medical textiles. Ultrasonic welding is used, for example, to cut and weld non-woven materials for face masks, filters and clean room capes. In the plunge mode, the ultrasonic horn operates perpendicularly to the material, and fuses the layers together in the pattern of the stationary anvil. This technique can be used to simultaneously cut and seal the edges of material. Typical applications using this mode include bra straps and buckles, buttonholes, collar stays, darts, strapping, embossing, grommets, belt loops, filter media (woven and non-woven), zipper stops, and vertical blinds.

In the continuous mode, typical applications include protective garments, disposable hospital gowns, shoe covers, face masks, infants' nursery garments, nappies, filters, bags, curtains, sails, and web splicing. Sealed edges and seams with no stitch holes prevent penetration of chemicals, liquids, blood-borne pathogens, or particulates. The rotating drum operation is used for quilting, medical disposables and some bag materials.

Ghosh and Reddy (2009) studied the effect of ultrasonic welding on polyethylene terephthalate (PET) and Spectra fabrics. In particular, they investigated the effect of weld pressure and weld time on the resultant seam strength. Results showed that these parameters need to be carefully controlled in order to maximise seam strength; the optimum value of these parameters also varied for the different fabrics. If weld times were too short, the thermoplastic fibres did not soften sufficiently and inadequate bonding between the two layers of fabric occurred. However, if weld pressures were too high, excessive melting of the fibres occurred and the bonds formed between fabric layers were weaker. Table 12.2 shows the seam efficiencies of ultrasonic bonded PET and Spectra fabrics; efficiencies of traditionally sewn seams are included for comparison.

Hayes and McLoughlin (2007) compared ultrasonically welded, sewn and sewn/tape-sealed seams used for waterproof polyester fabric. The sewn/tape-sealed seam was found to withstand a higher load than the welded seam, probably because of the double mechanism of the sewn seam and tape in maintaining seam integrity. The sewn and tape-sealed seam had a longer bending length than the welded seam, making it less flexible. Welded seams were found to have a greater load carrying capacity than sewn seams. There was no significant difference in bending length between welded and sewn seams. Table 12.3 summarises the relative performance of the three seam types.

Volkov (2002) described the use of ultrasonic welding to produce filters suitable for the filtration of sterile solutions. The filters investigated were formed from four layers of Lavsan knitted fabric using an ultrasound profile press welding method. Volkov concluded that this method could be used successfully to produce components for filtration. However, the strength of the welded joint and the strength of the individual layers in the fabric varied

Table 12.2 Comparison of the strength of ultrasonic bonded and sewn seams (from Ghosh and Reddy, 2009)

Fabric breaking load (lbf)	Ultrasonic seam breaking load (lbf)	Traditional seam breaking load (lbf)	Seam efficiency (%)	
			Ultrasonic	Traditional
PET fabrics				
112.6	26.1	31.5	23.2	28.0
Spectra fabrics				
450.0	22.1	22.1	4.09	4.09

Table 12.3 Relative performance of ultrasonically welded, sewn/tape-sealed and sewn seams (from Hayes and McLoughlin, 2007)

	Welded	Sewn and tape-sealed	Sewn
Load	Next best	Best	Worst
Extension	Worst	Best	Next best
Bending length	Best	Worst	Best

due to scattering and attenuation of the ultrasonic energy and the generation of differing amounts of heat between the layers. Optimisation of welding time, welding pressure and the amplitude of oscillations was again found to be essential.

There have been a number of studies of ultrasonic welding of nonwovens. Ozaltun *et al.* (2006) reported on the ultrasonic plunge welding of polypropylene blown microfibres with spunbonded cover web made from polypropylene and polyester fibres. Three different weld profiles were used: a rectangular solid profile; a diamond-shaped, repeated pattern; and a star-shaped repeated pattern. The rectangular weld profile produced the best bonding results, with the diamond-shaped pattern outperforming the star-shaped pattern. This was attributed to the higher weld area of the rectangular weld pattern. Ozaltun *et al.* (2006) also investigated the effects of three variables – weld time, vibration amplitude and weld pressure – on the resultant bond quality. The authors concluded that the optimum settings were 27.31 μm, 75 ms and 206.84 kPa for amplitude, weld time and weld pressure respectively; with vibration amplitude and weld time both having a significant effect on weld strength. Further increases in amplitude and weld pressure could produce a stronger seam but also resulted in apparel damage and marks on the fabric. For the lowest vibration amplitude investigated (19.09 μm), the weld strength was low because insufficient melting and fusion occurred.

Mao and Goswami (2001) described the ultrasonic bonding of bicomponent PET fibre batts and a spunbonded PET fabric, and the development of a model to predict the temperature change in webs during the bonding process. The authors used various combinations of weld pressure (20, 25 and 30 psi) and speed (0.048, 0.070 and 0.083 m/s). SEM analysis of the bonds showed that the degree of bonding was correlated with the temperature changes and the processing conditions. The slower speed resulted in fabrics that were stiffer, stronger and had a higher degree of bonding; as weld speed increased, the degree of bonding was reduced.

Medical applications using ultrasonic welding have also been explored. Bonds formed by ultrasonic welding have the advantage that they have no stitch holes and therefore form an effective barrier to chemicals, bloodborne pathogens and particulates. As a result, they are employed for the joining of textiles in many medical applications such as in medical gowns, caps, face masks, hygiene products, surgical drapes, etc. Patil *et al.* (2010) reported on trials to develop an alternative joining process for endovascular graft-stent devices. These devices consist of a polyester textile graft material and a supporting wire stent. The complex design of these stents means that current manufacturing processes involve hand sewing with resultant high cost, long production times and quality control issues.

In earlier research, Patil *et al.* (2008) investigated two types of ultrasonic welding using a plunge-type welding machine and a wheel-type machine. In both machines the two fabrics to be joined were placed in a peel joint configuration (one on top of the other). In an alternative arrangement, and with the aim of increasing weld strength, a copolymer adhesive film was placed between the two fabrics. For the plunge-type process, the frictional heat generated between the two fabrics was not high enough to melt and fuse them together due to the fabrics being tightly woven, very smooth and only 0.1 mm thick. When the adhesive film (with a relatively lower melting point) was used, however, good welds were created for both the plunge-type and wheel-type machines. Although the welds initially appeared satisfactory, SEM images showed the fabrics had melted through the entire thickness which resulted in limited weld strength. The authors concluded that the complex design of the stent-graft systems and the thin fabrics involved mean that it is not yet feasible to use ultrasonic welding in this application.

Shi and Little (2000) investigated the use of ultrasonic bonding to create smart seams by incorporating optic fibres in military uniforms. Experiments were carried out to determine whether the technical performance of the optic fibres was affected adversely by the ultrasonic welding process. Owing to the potentially fast material processing speeds that can be obtained, ultrasonic bonding has applications in the automation of certain manufacturing processes. Longerich *et al.* (2003) reported on efforts to use CAD technology to combine ultrasonic cutting and welding processes on a single workstation to improve manufacturing efficiencies. The materials to be joined were foils and PU-coated textile films. The authors developed a manufacturing method whereby the pattern pieces to be processed were created in the CAD system and an optimal cutting plan created. The physical cutting and welding processes were then carried out using ultrasonic techniques. The cutting and welding parameters (such as weld energy, weld speed, pressure on the material) could be adjusted for each operation. The period between creating the blueprint based on CAD and actual fabrication was less than 30 minutes. The authors report a reduction in manufacturing time of about 80%.

12.5 Dielectric welding: an introduction

Dielectric welding is also known as high frequency (HF) welding and radio frequency (RF) heating or welding. Dielectric welding generates heat in a polymer as a result of the application of a high frequency alternating electric field (around 27 MHz). The oscillating field generates motion in the molecules of certain polymer types which is converted to heat. This process is often described as bulk heating in contrast to heat sealing which is purely conductive heating.

Dielectric welding is only applicable to materials with:

- polarisable or polarised macromolecules with comparatively short chains
- good mobility of the molecule segments
- occurrence of inducible dipoles.

If these properties are present, the molecules start to oscillate as soon as the alternating field is applied, secondary valency forces weaken and molecules can interchange sites. Typical materials with these properties include poly(vinyl chloride) (PVC or vinyl), poly(ethylene-co-vinyl acetate) (EVA) and thermoplastic polyurethane (TPU). Polyamide (nylon) can be welded in a dielectric field if a pre-heat to approximately 50°C is applied. This is normally achieved by heating the electrode used in the process. Polyolefin fibres can be used either if a copolymer of ethylene with a polar comonomer is used or, alternatively, a multi-layer laminate film is employed using polyolefin and a polar polymer (Anon., 1995). The blending of linear low density polyethylene (LLDPE) to produce a material suitable for dielectric welding has been described, for example, by Yoon *et al.* (2008). Hollande *et al.* (1998) have reviewed dielectric welding of polyurethane elastomer-coated fabric. Table 12.4 shows how suited a range of materials are to dielectric welding.

The advantages of dielectric welding include the following:

- It is a clean process.
- Heating is localised, minimising fabric distortion.
- Material is 'bulk' heated rather than conduction heated, which reduces cycle times and allows the electrodes to be run cool, improving consistency compared to heat sealing.
- It is possible to heat contact surfaces to provide a strong weld without damaging the outer surfaces of the materials being joined.
- It produces a hermetic seal.
- It is easily controlled and automated.
- It is a relatively quick process (a few seconds to complete a weld) which needs little warm-up time to reach consistent processing.

Disadvantages include the following:

- It is restricted to a limited range of materials such as polyurethane and PVC fibres or coatings.
- The process is not continuous but limited to platens or electrode bars

Table 12.4 Suitability of textile materials for dielectric welding

Material	Response				
	Excellent	Good	Fair	Poor	None
ABS polymers		×			
Acrylics			×		
Cellulose acetate (clear)		×			
Cellulose acetate (colour)		×			
Cellulose acetate butyrate		×			
Cellulose nitrate			×		
Cellulose triacetate			×		
CPET					×
Ethyl cellulose					×
EVA (Ethyl vinyl acetate)		×			
EVOH (Ethyl vinyl alcohol)			×		
Nylon (Polyamide)			×		
PET (Polyethylene terephthalate)		×			
PETG (Polyethylene terephthalate glycol)	×				
Pliofilm (Rubber hydrochloride)	×				
Polychlorotrifluoroethylene				×	
Polyester				×	
Polyethylene (All)					×
Polymethyl (Methacrylate)			×		
Polypropylene					×
Polystyrene					×
Polyurethane			×		
Polyurethane foam				×	
Polyurethane-vinyl film		×			
Polyvinyl acetate		×			
Polyvinyl chloride (PVC) (flexible, clear)	×				
Polyvinyl chloride (PVC) (colour)	×				
Polyvinyl chloride (PVC) (opaque)		×			
Polyvinyl chloride (PVC) (semi rigid)		×			
Polyvinyl chloride (PVC) (rigid)			×		
Polyvinyl chloride (PVC) (flexible, glass-bonded)	×				
Polvvinvl chloride (PVC) (coated cloth & paper)	×				
Polyvinyl chloride (PVC) (adhesive emulsions)	×				
Saran (Polyvinylidene chloride)	×				

typically less than 2m in dimension. Longer seams can be made using repeated indexing of the material under a bar.

- PVC emits volatile organic compounds (VOCs) when heated.
- There is a potential risk to operators or surrounding equipment from high-frequency radiation. Dielectric welding equipment is usually shielded.

12.6 Equipment for dielectric welding

A typical dielectric welding machine is shown in Fig. 12.6. It uses two electrodes (also known as platens):

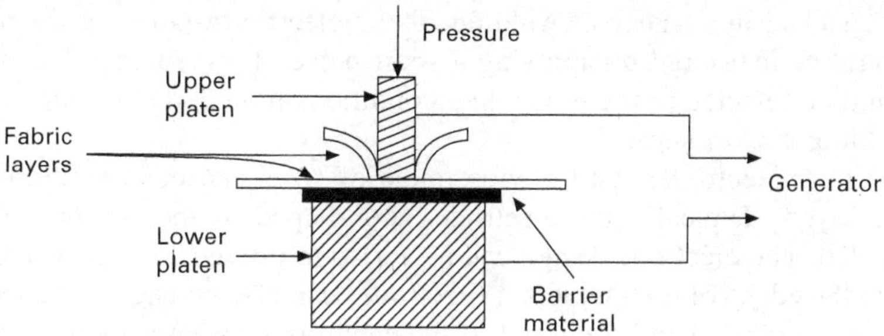

12.6 A typical dielectric welding machine.

- upper platen
- lower platen.

The platens are connected to the dielectric supply. The upper platen is movable to allow the electrode to be brought into contact with the fabric. The weld is created by the combination of the effect of the alternating electric field generated by the electrodes and the downward pressure of the upper platen. The lower platen is usually fixed in position and earthed. The lower platen usually serves as the work surface over which the fabric is spread. The temperature of both platens is often controlled to counteract uncontrolled heating of the platens caused by the dielectric supply. A heated platen reduces heat loss from the fabric, allowing a larger area to be welded for a given dielectric power rating and a more consistent weld. Heated platens are also used to allow welding of materials which do not heat from cold using a dielectric field.

When high frequency power is passed between the electrodes, an alternating electric field is generated through the materials. As has been noted, the switching electric field agitates the molecules in the material, causing it to heat up. The process generates heating throughout the thickness of the materials. However, the heating effect is at a maximum at a point half way between the electrodes at the junction of the two materials to be joined.

When the electrodes sink into the fabric pieces to be joined, they require more power from the generator. When the fabric reaches melting point, an electrical change occurs which causes power to reduce. This reduction is often used to mark the completion of a weld. Once the materials have fused, the dielectric power is switched off to prevent overheating of the fabric pieces and the joined fabric is allowed to cool. Typically cooling time is 20% the length of the weld time. It is evident that it is possible to monitor the weld by measuring the changes in the dielectric properties of the material being welded. This is an output that can be sensed directly from the process electrodes, and can be used to control the power applied and heating cycle of the weld. A further extension of this concept would allow the materials to be checked in advance of welding. The dielectric response of the materials would be measured by applying a short pulse of RF energy. The materials could be rejected or the processing conditions modified to ensure consistent welding performance.

Upper electrodes with a wide range of sizes and configurations can be employed. Typically the electrodes are shaped to the components to be welded. The electrode design can be varied to produce different seams or to trim the edges of textile pieces. Electrodes can also be engraved to emboss a pattern on the seam if required. If the electrodes are long enough, dielectric welding can produce whole seams in one continuous process. With well-designed electrodes, the temperature of the electrodes themselves seldom needs to exceed 40°C.

Where either very thin layers or layers of unequal thickness are to be welded, it is essential to include a dielectric barrier material of up to 0.5 mm thickness between the electrodes and the fabric to ensure that welding is uniformly applied. Barriers are also essential in tear-seal welding (see below) to prevent electrical contact between the cutting edge of the upper platen and the lower platen, which would cause an arc, which is evident as burnt patches in the welded seam, possibly puncturing the materials. Typical barrier materials are composites including:

- glass-fibre fabric coated with polytetraflourethylene
- natural and silicon rubbers (with fabric backing)
- poly(ethylene terephthalate) film (with paper backing)
- varnished cotton fabric.

The generator produces the high frequency electricity required for welding. The maximum output power level required varies, depending on such factors as the type and thickness of the weld materials and the area of the required weld. The power rating is one of the most important parts of a machine's specification, varying from a few hundred watts to tens of kilowatts. A single valve or electronic oscillator is typically employed working at a pre-set output frequency. The oscillator is supplied with a high-voltage direct current from a transformer-rectifier power pack unit. The tuning of the electrical circuit is usually achieved with a variable capacitor.

There are two main types of dielectric welding:

- plain welding
- tear-seal welding.

These are illustrated in Fig. 12.7. Tear-seal welding involves the simultaneous welding and cutting of a material. This is achieved by placing a cutting edge adjacent to the welding edge. The cutting edge compresses the material so that it can be separated into two. The ratio of the compressed thickness of the welded area to the combined thickness of the original two layers is known

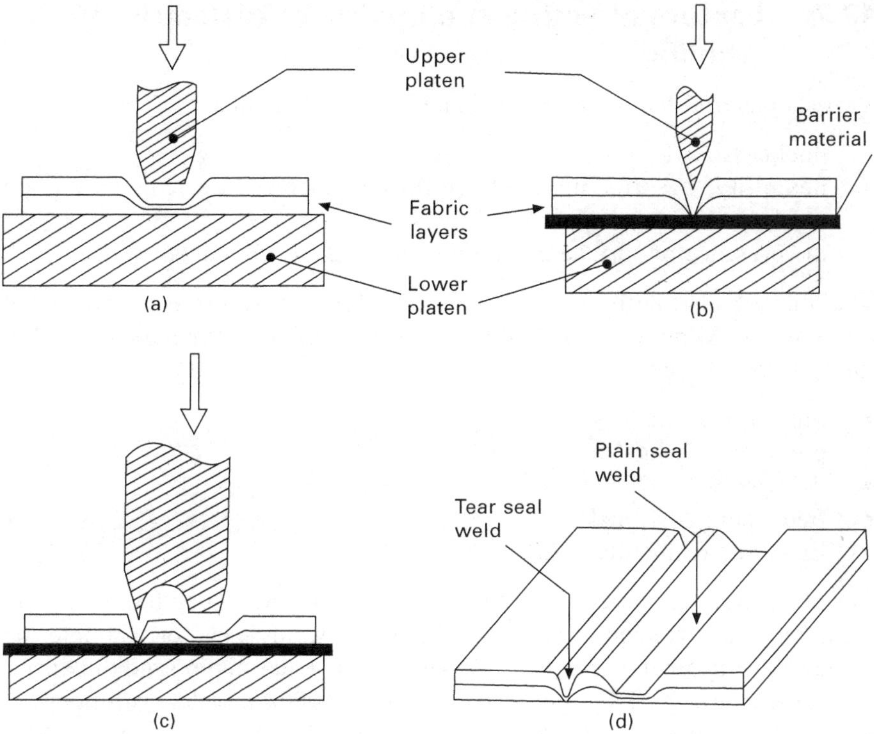

12.7 Types of dielectric welding: (a) plain-seal welding; (b) tear-seal welding; (c) combined plain and tear-seal welding; (d) combined weld from (c).

as the welding differential. Plain and tear-seal welds can be combined to provide a particularly strong bond together with a sharp edge to the fabric. There are also more specialised dielectric welding techniques. In appliqué dielectric welding, a flexible vinyl of a different colour is superimposed and welded over a substrate. The surplus material is removed to leave a decorative surface with contrasting colours and surface variation.

There are various ways of handling materials for welding:

- manual: fabrics are fed, positioned and removed by hand
- linear: a conveyor belt feeds fabrics to the machine, positions them (a process known as 'indexing') and removes them after welding
- rotary: a circular table rotates to position fabrics under the welding unit; this configuration allows operators to assemble more complex workpieces before they are welded and then carry out further operations (e.g., stripping off surplus material from tear-seal welding) as required
- rail: large components (e.g. tarpaulins) may need to be held in a stationary position whilst the welding unit is moved along a rail to make repeated welds.

12.7 Factors affecting the quality of dielectric welding

Various material factors affect weldability. These include:

- thickness
- flexibility: a softer material requires less power to achieve a good weld
- the presence of surface contaminants, films or lubricants.

The impact of coatings on dielectric welding has, for example, been discussed by Skripnik *et al.* (1984). Typical welding parameters are within the following ranges:

- time: 1–6 seconds
- pressure: 1–3 kgf/cm^2
- voltage: 3–15 kV
- frequency: 27.12 MHz
- Electrode temperature: 20–40°C.

The key process parameters are time, pressure (including depth of sink) and high-frequency voltage (Anon., 1995). Faster welds can, for example, be achieved using greater pressures. Depending on material type and thickness, pressures as low as 1 kgf/cm^2 can be used to obtain a weld. An important aspect of pressure is controlling depth of sink. The sink control limits how far the electrode will sink into the fabric. Adjusting the depth of sink can help in achieving maximum weld strength. Excessive depth of sink will lead to a thinner and weaker weld. The depth of sink can also help to control the melt profile at the edges of the weld, which is important in defining the tensile and fatigue performance of the joints made. In terms of power applied, a rough rule of thumb is 1 kW of RF power per 30 cm^2 of weld area, depending on the fabric.

12.8 Applications of dielectric welding

Dielectric welding is used for cutting and joining PVC films or sheets. It is suitable for both woven and non-woven fabrics. Dielectric welding is used in packaging in such areas as sachets and blister packs. Because of the water and airtight quality of the seal, dielectric welding is used to join PVC and polyurethane-coated fabrics in such applications as inflatable life jackets, life rafts and other buoyancy equipment, as well as air beds and inflatable toys. It is also used to join materials for inflatable temporary structures, e.g. for sporting and leisure events, tents, marquees and tarpaulins.

Dielectric welding is also well suited to the manufacture of medical products, e.g. containers such as blood or colostomy bags. The use of the technique for implanted medical devices is discussed by Yussuf *et al.* (2007). Specific applications in medical textiles include:

- blood bags, blood separation systems and blood pressure cuffs
- urinary drainage, colostomy and enema bags
- *in vitro* feeding bags
- wound evacuation pumps and bags
- pulsating air and water mattresses (e.g., for pressure relief in preventing sores)
- stretcher and bed covers and pads (including vacuum mattresses for immobilising patients during transport)
- oxygen tents and mobile clean rooms.

Dielectric is used in protective clothing, e.g. rainwear and waterproof one- or two-piece polyurethane clothing for motorcyclists, high-visibility clothing for the emergency services, chemical or biological protective clothing. Specific non-medical applications include:

- HAZMAT and radiation suits
- survival suits (e.g., for naval personnel)
- industrial gloves
- rainwear
- life jackets/vests and buoyancy aids
- inflatable boats
- tents and tarpaulins
- conveyor belts
- gel and foam-filled inserts for shoes
- slippers.

12.9 References and further reading

Anon. (1995), *High Frequency Welding Handbook*, Federation of High Frequency Welders.

Benatar A and Gutowski T G (1989), Ultrasonic welding of PEEK graphite APC-2 composites. *Polymer Engineering and Science*, 29, 1705–1721.

Branson Ultrasonic Corporation (2011), Technical Information Sheet PW-7.

Chen J-W and Zykbo J M (2005), Diode laser bonding of planar microfluidic devices, MOEMS, bioMEMS, diagnostic chips and microarrays. *Proc SPIE*, 5718, 92–98.

Curtis *et al.* (2004), The uses and advantages of laser sealing in high density assay plates. *Society of Biomolecular Sciences 10th Conf.*, 11–15 September.

Devine J (1998), Ultrasonic bonding of plastics and textiles for medical and other devices. In: *Joining Applications in Electronics and Medical Devices. Proceedings, ICAWT '98 International Conference on Advances in Welding Technology*, 30 September–2 October, Columbus, OH: Edison Welding Institute.

Froment I D (1999), *Ultrasonic Welding of Thermoplastics*, Cambridge: TWI.

Ghosh S and Reddy R (2009), Ultrasonic sealing of polyester and Spectra fabrics using

thermoplastic properties. *Journal of Applied Polymer Science*, 113, 1082–1089.
Grewell D and Benatar A (2003), Experiments in micro-welding of polycarbonate with laser diodes. *Proc. ANTEC conference of the Society of Plastics Engineers*, May, 1039–1044.
Haire K R and Windle A H (2001), Monte Carlo simulation of polymer welding. *Computational and Theoretical Polymer Science*, 11(3), 227–240.
Hayes S G and McLoughlin J (2007), Welded and sewn seams: a comparative analysis of their mechanical behaviour. In: Ariadurai S A & Wimalaweera W A (eds) *The Textile Institute 85th World Conference*, 1–3 May 2007, Colombo, Sri Lanka, pp. 131–142.
Hollande S, Laurent J-L and Lebey T (1998), High-frequency welding of an industrial thermoplastic polyurethane elastomer-coated fabric. *Polymer*, 39(22), 5343–5349.
Hustedt M *et al.* (2006), Laser micro welding of polymeric components for dental prostheses. *Proc. SPIE*, 6107.
Jones I A (2006), Precision laser welding of plastic components. TWI members report.
Jones I A and Olden E (2000), 'A thermal model for transmission laser welding of thermoplastic polymers', TWI Members report 708/2000, July.
Jones IA and Taylor NS (1994), High speed welding of plastics using lasers. *ANTEC '94 conference proceedings*, 1–5 May, San Francisco, CA.
Klotzbuecher T *et al.* (2004), Microclear – a novel method for diode laser welding of transparent microstructured polymer chips. *Proceedings of the 23rd International Congress on Applications of Lasers and Electro-Optics*.
Longerich B, Clausnitzer B and CompMess GmbH & Co (2003), XY plot editing table with ultrasonic cutting and ultrasonic welding system for foil and coated textile film. *Technical Textiles*, 46.
Mao Z and Goswami B C (2001), Studies on the process of ultrasonic bonding of nonwovens: Part 2: Experiments and results. *International Nonwovens Journal*, Fall, 17–28.
McCrum N G, Buckley C P and Buckley C B (1988), *Principles of Polymer Engineering*, Oxford: Oxford University Press.
Ozaltun H, Benatar A and Nayar S K (2006), Ultrasonic plunge welding of polypropylene nonwovens, ANTEC 2006, *Annual Technical Conference of the Society for Plastics Engineers*, pp. 2184–2188.
Patil A, Jones I, Ashton T and Tavakoli M (2008), 'Investigation into novel joining techniques for rapid and automated manufacturing of a new generation of endovascular grafts', *Proceedings of Health Technologies KTN/University of Brighton Annual Conference*, 3 April.
Patil A, Jones I and Ashton T (2010), 'New manufacturing methods for vascular grafts using laser welding and laser cutting techniques', *Joining Plastics 2010* – 2nd DVS/WJS Conference, Dusseldorf, Germany, November.
Pfleging W *et al.* (2005), Laser assisted welding of transparent polymers for micro-chemical engineering and life science. *Proc SPIE*, 5713 479–488.
Shi W and Little T (2000), Mechanisms of ultrasonic joining of textile materials. *International Journal of Clothing Science and Technology,* 12(5), 331–350.
Simpson W (1995), Sealing and welding of thermoplastic materials. In *Plastics: Surface and Finish*, Cambridge: Royal Society of Chemistry.
Skripnik V, Radzieski V and Butov V (1984), Materials with a fibrous structure welded in a high-frequency current field. *Automatic Welding*, 2, 54–56.
Tolunay M N (1982), An experimental study of the bonding mechanism in the ultrasonic welding of thermoplastics. PhD thesis, Cornell University.

Truckenmuller *et al.* (2006), Micro ultrasonic welding: joining of chemically inert microparts for single microfluidic components and systems. *Microsyst Technol*, 12, 1027–1029.

Tsujino J, Hongoh M, Tanaka R, Onoguchi R and Ueoka T (1986), Ultrasonic plastic welding with a welding tip pair. *Japanese Journal of Applied Physics*, 25, Supplement, 168–170.

Ussing T *et al.* (2005), Micro laser welding of polymer microstructures using low power laser diodes. *Proc Int Conf on multi-material micro manufacture*, Karlsruhe, 291–293.

Volkov S S (2002), Ultrasonic welding of knitted Lavsan fabric filters. *Welding International*, 16, 664–665.

Warwick C M and Gordon M (2006), Application studies using through-transmission laser welding of polymers. *Proc. Joining Plastics 2006*, 25–26 April, NPL, London.

Wise R J (1999), *Thermal Welding of Polymers*, Cambridge: Abington Publishing.

Wool R P (1995), *Polymer Interfaces: Structure and Strength*, Carl Hanser Verlag.

Yoon C, Lee E, Yang M, Song C, Yang C, Kim K and Kim B (2008), High frequency welding of thermoplastic LLDPE/PA 6/PE-*g*-MAH ternary blends. *Journal of Applied Polymer Science*, 109, 3355–3360.

Yussuf A, Sbarski I, Solomon M, Tran N and Hayes J (2007), Sealing of polymeric-microfluidic devices by using a high frequency electromagnetic field and screen printing technique. *Journal of Materials Processing Technology*, 189(1–3), 401–408.

13
Laser seaming of fabrics

I. JONES, TWI Ltd, UK and A. PATIL, Vascutek Ltd, UK

DOI: 10.1533/9780857093967.3.398

Abstract: Laser welding provides a fast, easily automated technique for joining textiles. It can be used with or without added polymer film at the joint, and allows accurate control over the position of welding and the amount of heat applied. Heating also starts from the inside of the joint, so in most cases it is possible to make a weld without affecting the outward appearance, and retaining the flexibility more than other welding processes. The process can be used to join fabrics to rigid plastics. Rapid developments have been seen since 2000 and early applications are in place, but as yet the process is not widely used.

Key words: textiles, fabrics, welding, bonding, seaming, laser, automation, manual, robotic, joint design, monitoring, quality control, performance, sealing.

13.1 Introduction

The use of lasers in materials processing is growing rapidly. Speed, accuracy, flexibility and the potential for automation are just a few benefits of this advanced manufacturing technique. The acceptance of laser welding of plastics by various industry sectors has recently increased due to the development of reliable, efficient and cost effective laser systems. Many new materials processing developments are now being pursued using various types of lasers. There are many applications that have already moved into industry and there are many more that are in the early stages of evaluation. Of these applications, a large number can be classified as plastics joining applications (which include more recent developments in textiles joining using laser systems). This chapter explores the process of joining fabrics using the laser welding technique. First, the process is described briefly and the applicability of the process to textile fabrics is discussed. Then, critical process parameters are presented along with the materials that can be welded and possible joint designs. Finally, a discussion on potential applications and future trends are presented.

13.2 The laser welding process

Welding is a process in which materials (including fabrics) of the same type or class are joined together through the formation of primary (and occasionally, secondary) bonds under the combined action of heat and pressure. These bonds are the natural consequence of bringing two similar materials close enough together to allow intermolecular diffusion.

Techniques currently used for welding fabrics can be divided into three groups: those in which heat is generated by mechanical movement; those in which heat is conducted to the joint from an external heat source; and those using electromagnetic radiation directly. Laser welding falls into the third group.

The methods for laser welding of fabrics fall into two distinct groups defined by the type and, more specifically, the wavelength of the laser being used for the process:

- Direct welding using CO_2 lasers (10,600 nm wavelength). Welding occurs when the laser beam is absorbed directly at the interface by the top surface of the fabric. The laser is often directed at the nip between two rollers so that heating and clamping can be achieved almost simultaneously.
- Transmission laser welding using Nd: YAG, diode or fibre lasers that, unlike CO_2 lasers have wavelengths that transmit through unpigmented thermoplastics (400–1500 nm). To use these lasers successfully for joining fabrics, the laser energy must be properly harnessed to create the heating and intermolecular diffusion critical for weld formation. The location of the weld is controlled by placing an absorber of the laser radiation at the joint interface. The absorber heats to generate melting locally.

13.2.1 Direct laser welding

In direct laser welding (Fig. 13.1) the materials are heated from the outer surface, and generally an additional absorber applied to the fabrics is not required. A laser source is chosen that has a wavelength that is absorbed by the materials. Laser sources from 0.8–10.6 μm wavelength have been used. At 10.6 μm (CO_2 laser), radiation is strongly absorbed by plastic and fabric surfaces independently of any colouration or additive, allowing high speed joints to be made. If diode or fibre lasers in the range 0.8–1.1 μm wavelength are used, then the fabrics must either have a dark carbon-based pigmentation or an absorptive coating is applied as used in transmission laser welding. A diagram of direct laser welding is shown in Fig. 13.1.

13.2.2 Through-transmission laser welding (TTLW)

Through-transmission laser welding of fabrics involves the use of laser energy to produce the heat necessary to generate a weld. Traditionally this was accomplished by using an additive in one of the substrates to absorb the laser

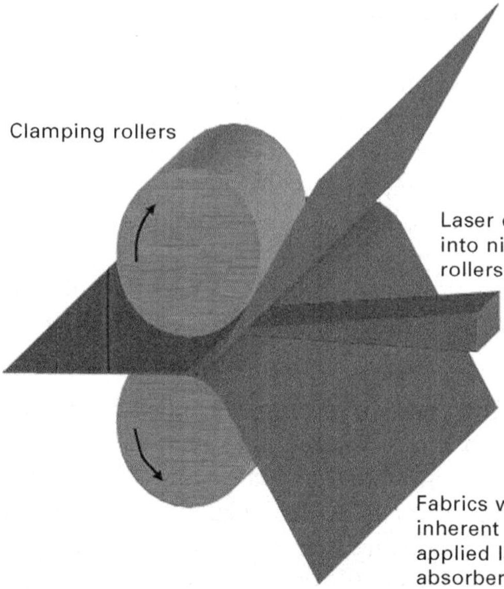

13.1 Direct laser welding.

energy. Most thermoplastic fabrics and some natural fabrics transmit near-infrared laser energy. Therefore, one substrate must remain in its natural state, while the second substrate is altered, typically by adding carbon black, in order to make it absorbent to the laser energy. The laser beam transmits through the top surface until it reaches the bottom substrate where it is absorbed. Under favourable welding parameters, the heating conducts into the top substrate allowing the two materials to melt and flow together forming a weld. The main advantages of the laser welding process are that it does not involve vibration, is non-contact, does not impart mechanical damage and is a clean process. The main disadvantage is the use of conventional laser absorbing materials such as carbon black, as it virtually rules out using this process for any applications in which aesthetics and lack of colour are important. A diagram of transmission laser welding is shown in in Fig. 13.2.

13.2.3 Colour independence

A significant process advancement to TTLW has been the development of an innovative process called the Clearweld which offers all the advantages of conventional laser welding without the use of opaque materials or addition of unwanted colour (Jones *et al.*, 1999; Hilton *et al.*, 2002). The process uses near-infrared (NIR) absorbing material systems to convert laser energy into heat. A thin layer of these materials applied at the interface of two pieces of fabrics to be joined absorbs light, acting as a focal point for the laser.

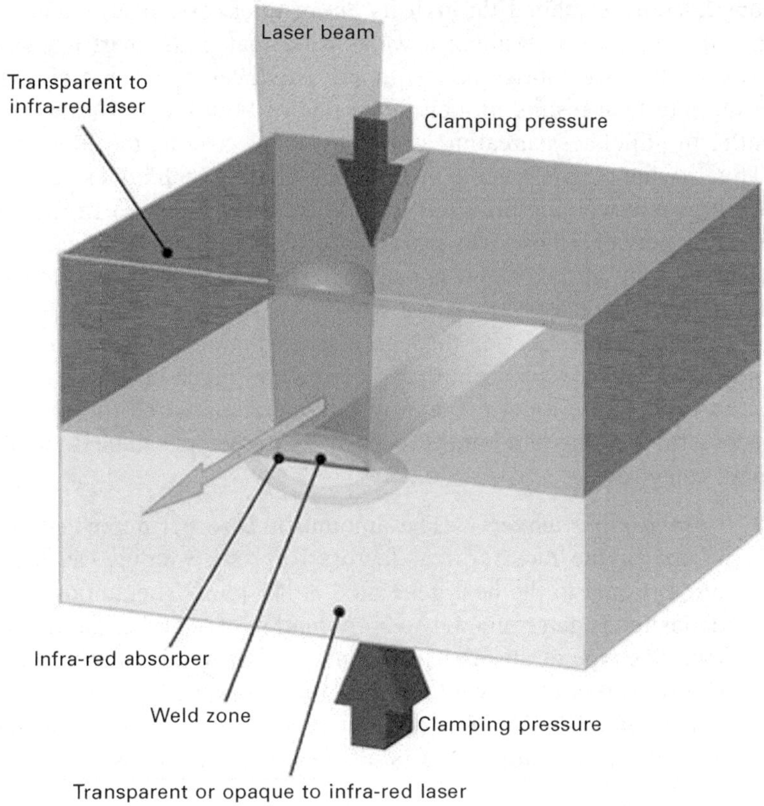

13.2 Transmission laser welding.

Localised heating of the substrates occurs at the joint interface, producing clean, optically clear joints with no particulates and visible colour. The Clearweld process was invented by TWI Ltd; a UK-based industrial research and development organisation, and has been developed for commercial use by Gentex Corporation (Jones and Wise, 1998). Clearweld consists of a series of materials capable of powerful absorption in the near infrared spectrum, while remaining virtually colourless.

13.3 The main processing parameters in laser welding and their effects

The key to producing a successful laser welded seam is to use the correct process parameters for the fabrics being joined (Jones and Olden, 2000; Rostami and Jones, 2003). The heat put into the seam must be sufficient to melt enough of the material to accommodate any surface roughness and to form a weld over the whole area, but not so much that one or both of the substrates have melted through its entire thickness. Insufficient heat input into the seam will result in a weak weld that pulls apart at the interface between the two fabrics being joined. However, too much heat generated at the interface results in melting of one or both the substrates through its entire full thickness, creating a line of weakness along the edge of the weld. (This is what occurs when conventional welding techniques such as hot bar or ultrasonic welding are used to produce fabric seams.) In the processing window between these two extremes, it is possible to produce seams with high strength and excellent appearance. Figure 13.3 shows a cross section through a satisfactory seam in a woven nylon fabric. For comparison, Fig. 13.4 shows a cross section through a seam that did not receive enough heat and Fig. 13.5 shows a seam that received too much heat.

A number of critical processing parameters must be followed to ensure successful welding of a joint. Processing parameters can be divided into two main categories:

1. *Absorber parameters* – The amount of absorber deposited at the joint is one of the most critical factors for laser welding, as it is directly proportional to the heat generated at the joint. The amount of absorber deposited is determined by a combination of absorber parameters which includes type of absorber, concentration of the dye in the solvent, flow rate, absorber spray width and application speed.
2. *Laser parameters* – The amount of heat generated at the interface is an equally critical process parameter which depends on the amount of

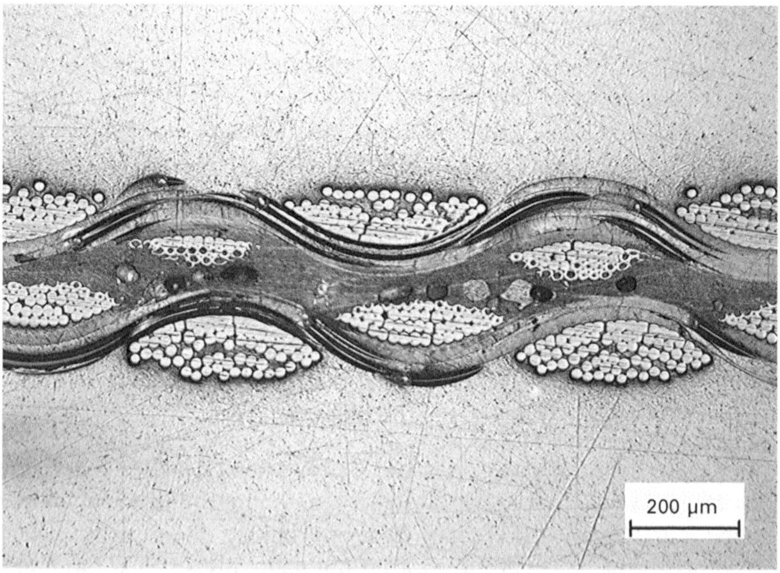

13.3 Cross section through satisfactory seam.

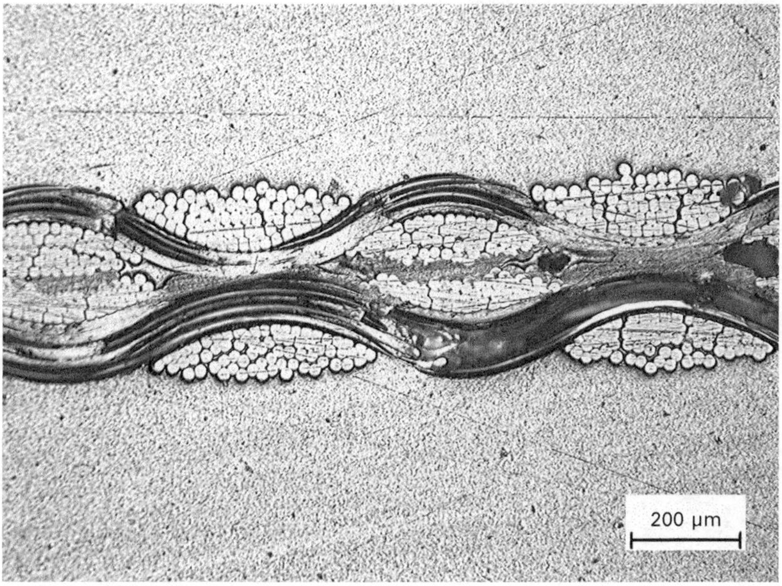

13.4 Cross section through seam with insufficient heat.

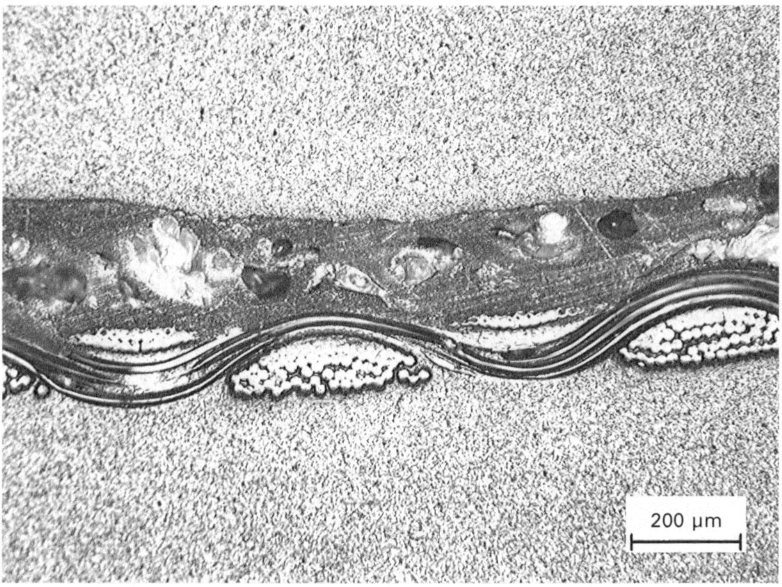

13.5 Cross section through seam with too much heat.

absorber present in combination with the laser processing parameters such as laser power, beam size, weld speed and the amount of clamping pressure applied during or just after the weld.

The amount of heat put into the seam per unit area (and hence the resulting weld performance) is determined by four critical parameters:

- the transmission properties of the upper fabric
- the power density of the laser
- the welding speed
- the quantity of laser-absorbing material system at the seam.

Transmission properties

The fraction of incident NIR radiation by the upper fabric must be at least 10% in order for the through-transmission laser welding process to be used successfully. This should not vary significantly from region to region in the textile as this will alter the amount of heat generated at the seam and therefore affect the quality and performance of the seam.

Power density

The power density of the laser is determined by the total power of the beam and by the beam spot size used for welding. As the total power increases, the power density increases in direct proportion, but as the spot size increases, the power density falls in proportion to the square of the spot size. The spot size will usually be determined by the width of the seam required, so it is usual to adjust the power density simply by altering the total power.

Welding speed

Use of low welding speed will result in more heat generated at the interface. It might be expected that with all the other parameters kept the same, doubling the laser power and doubling the welding speed would result in a weld of the same quality. However, in reality, when the time dependent conduction of heat away from the seam is considered, theoretical models predict that doubling the power should allow the speed to be increased by a factor of four.

Absorber coating material system density

To maximise efficiency and to minimise any residual colour in the seam, the minimum amount of absorber material system that will give the required weld should be used.

In practice, for textiles welding, required seam width and minimum required welding speed will often be known, so experimentation will only be required to find the correct absorber material system density and laser power to give a satisfactory weld.

13.4 Equipment in laser welding

13.4.1 Laser types

The following laser types may be used for textiles welding:

- CO_2 – 10,600 nm
- Fibre – 1,070–2,000 nm
- Nd:YAG – 1,064 nm
- Diode – 635–670, 808 nm, 940 nm, 980 nm.

The beams are either pulsed or continuous wave.

For transmission laser welding a laser with output in the region of 1 micron is selected so that the beam is transmitted through to the absorber coating at the joint. An absorber coating is selected to absorb the laser. Carbon-based absorbers will absorb over a wide range of visible and IR wavelengths. This is why they appear black. The absorber used in Clearweld material systems has been designed for optimum absorption in the wavelength range 940–1,064 nm. Continuous wave lasers available in this range include Nd:YAG, fibre and diode lasers (see Fig. 13.6).

Direct laser welding is carried out using a laser that will absorb in the fabrics directly. So a CO_2 laser can be selected which has an output wavelength

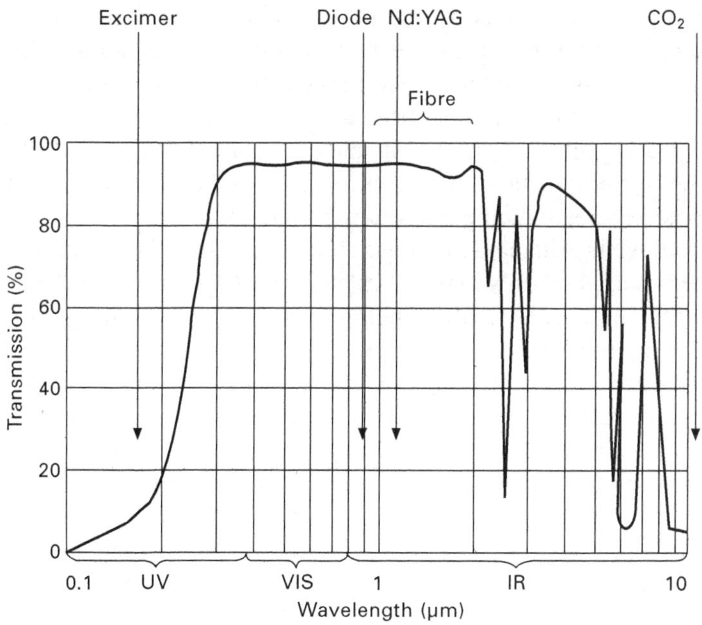

13.6 Transmission properties for a typical unfilled polymer at UV, VIS and IR wavelengths, showing also the lasers and their output wavelengths.

of 10.6 microns where there is strong absorption in polymers in general. The lasers with output at around 1 micron wavelength can only be used for direct laser welding if they contain an additive that is inherently absorbing at this wavelength. This is sometimes present as a carbon black pigment or can be deliberately added as a low colour IR absorber.

Textile welding has been carried out using 940 nm direct diode lasers with power ranging from 25 to 1000 W, and at speeds between 0.25 and 30.0 m/min, demonstrating the huge flexibility of this process. A high power laser may be needed to achieve a high welding speed, particularly because of the relatively low transmission properties of many textiles. However, it should be noted that the purchase and running costs of a laser increase as its maximum power increases. For a cost-effective textile joining process it is therefore important to understand the application requirements and to select the appropriate equipment.

13.4.2 Fabric positioning and clamping

The laser welding process, like other welding techniques, requires the two textiles to be joined under pressure. Clamping pressure is required for several reasons:

- To provide intimate contact between the two surfaces to be welded. To obtain intimate contact of the substrates at the weld interface, a certain amount of clamping pressure is required for most joints (the notable exception being an interference fit, as clamping may not be necessary). The amount of pressure required depends on the materials being joined, the specific joint design chosen, and the quality of the surface conditions at the weld interface. Surfaces of fabrics have bumps, valleys and inconsistencies which prevent close contact.
- To achieve heating and melting in both substrates. The absorber material system is applied only to one textile surface (in most cases). The absorber generates heat as a result of being exposed to laser energy. In order to obtain a strong weld, both textiles must melt. Intimate contact between the surfaces allows this heating and melting to occur.
- To allow diffusion of polymer chains between surfaces. A weld is created in plastics by intermingling of the polymer chains from both substrates. When the plastic melts, the polymer chains are able to move. Clamping pressure brings the surfaces in contact to enable diffusion. In addition, the pressure provides the force to push the chains together.
- To prevent separation of the fabrics during the cooling phase. All materials expand upon heating and contract upon cooling to some degree. Once the laser energy is removed the material begins to cool and hence contracts. If clamping pressure is not applied during cooling, the fabric layers may

separate. When joining most materials using the laser welding process, only a small localised region of substrate is melted. The melted plastic cools rapidly, therefore shorter hold times are required.

As with other welding methods, it is important to have a clean surface free of dirt and other contamination for assuring consistent weld performance. The presence of dirt could result in high laser absorption leading to burn marks.

13.4.3 Laser or fabric manipulation

Three generic machine concepts have been identified: moving-bed, flying optics and sewing machine arrangement. These concepts are discussed in more detail in the following section.

Moving-bed scheme

This arrangement, shown in Figs 13.7 and 13.8, is possibly the simplest and most easily implemented form of laser-based machine and, as such, is one that is often employed for work in the laboratory environment. It consists of a structure mounted over CNC operated tables. This arrangement consists of fixing the laser head or beam delivery optics (depending on the type of laser and beam delivery method) above movable tables. The structure used to mount the laser can be constructed from simple off-the-shelf framing components, with reasonable stiffness. A bolted construction may be adequate. For a modest sized work envelope, standard x-y tables with a controller would have sufficient accuracy and speed capability for the absorber application. By virtue of its simplicity, the fixed-optics–moving-bed concept will probably be the lowest cost option of those presented here.

A possible drawback of this arrangement is the physical size of the tables if a large working envelope is required. The presence of the table

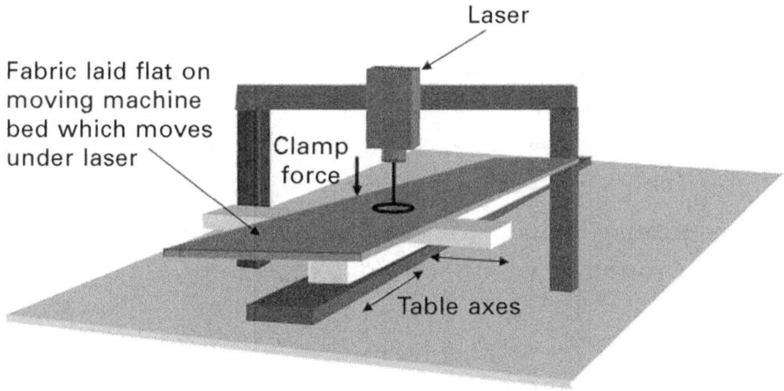

13.7 Moving-bed equipment scheme.

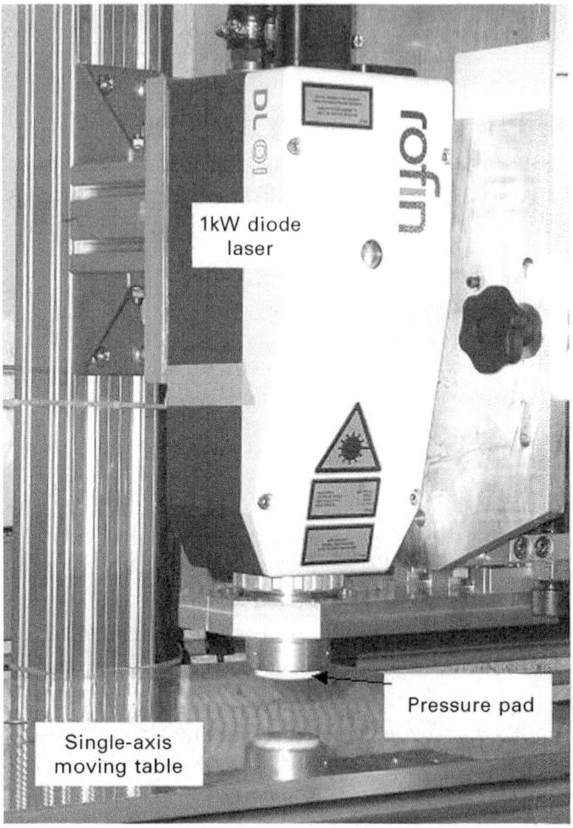

13.8 Fixed laser and ring shaped pressure pad slider working over a moving flat-bed table.

mechanism (motor and lead screw) beneath the tables may restrict access or increase the inertia of the system which could decrease the acceleration capability of the tables. This may not be important during the welding cycle, which requires constant speed, but may be an important consideration in a production environment where minimising indexing time between welds may be an important consideration.

Flying optics or robotic scheme

This approach is similar to the moving-bed scheme but with the laser and optics mounted on moving axes above a fixed table. This type of system is often referred to as a 'flying optics' configuration and is shown in Figs 13.9–13.12. A flying optics arrangement can be achieved in a number of ways; mounting the laser source adjacent to the machine and using mirrors and lenses to deliver the beam to the work area, the use of fibre optic beam

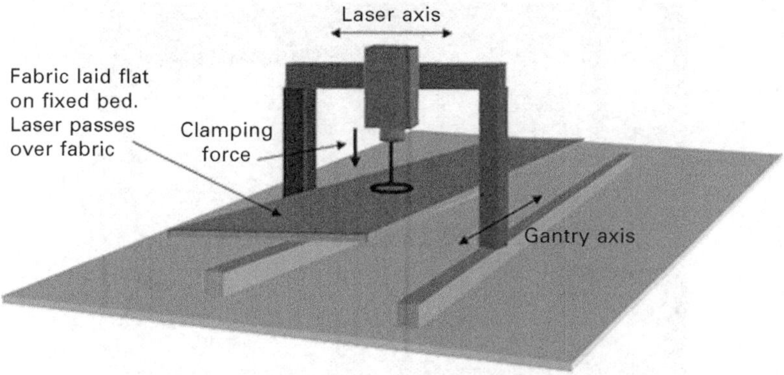

13.9 Fixed bed moving laser equipment scheme.

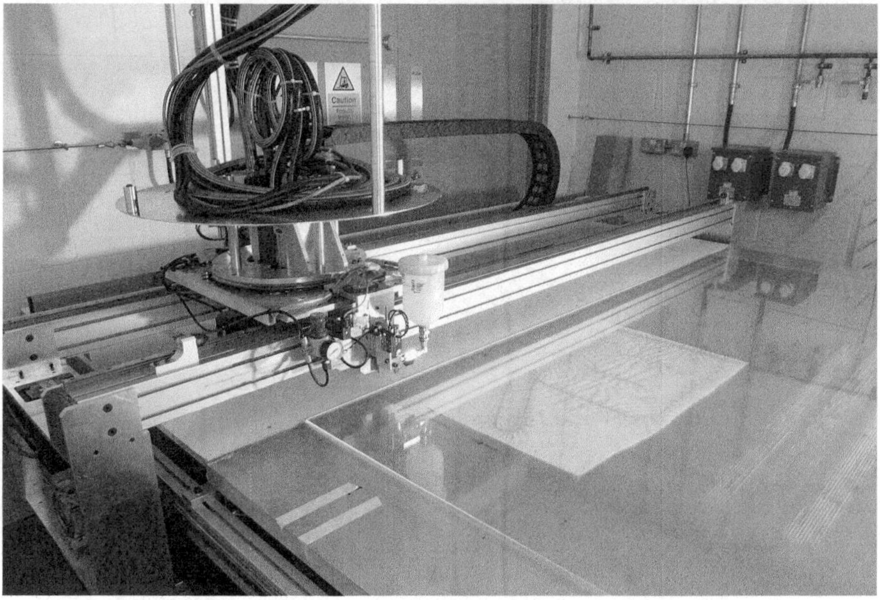

13.10 Flat bed moving laser demonstration equipment.

delivery or by mounting the laser head directly on to the moving axes. The relatively light weight (~10–15 kg) of the direct diode laser makes it a practical proposition to mount the laser head and optics on moving axes to access the area to be welded. One of the main advantages of a flying optics system is the potential for large component sizes to be addressed. The disadvantage is the requirement for greater complexity in the design of overhead gantry to include the moving axes and accommodate the reaction forces necessary for clamping.

The flying optics concept can be extended to three dimensions by attaching

13.11 Spray unit and reservoir.

13.12 Clamping unit and slider.

the laser or a fibre optic delivery system directly to a multi-axis robotic arm as shown in Figs 13.13 and 13.14. The textile handling and support become more complex in this case. A shaped support would be required, which could include vacuum clamping and may also incorporate methods of adjustment for different sized or shaped products. If vacuum is not used or is augmented with other clamping, then the reaction forces need to be accommodated by both the robot and the support structure.

Sewing machine scheme

This approach is similar to the principle of sewing machine operation and is shown in Figs 13.15 and 13.16. With this arrangement, the laser head is fixed and the fabric is fed into the machine under a processing head using a feed mechanism. Alternatively for very large constructions the processing head could be moved along the seam line, possibly with the machine mounted on wheels or rails. The processing head is likely to be more complex than in the other systems described because of the need to incorporate both a clamping and feed mechanisms. The sewing machine approach offers the potential for tremendous flexibility to address a variety of applications. However,

13.13 Six-axis robotic system for laser welding of fabrics.

13.14 Six-axis robotic system for laser welding of fabrics.

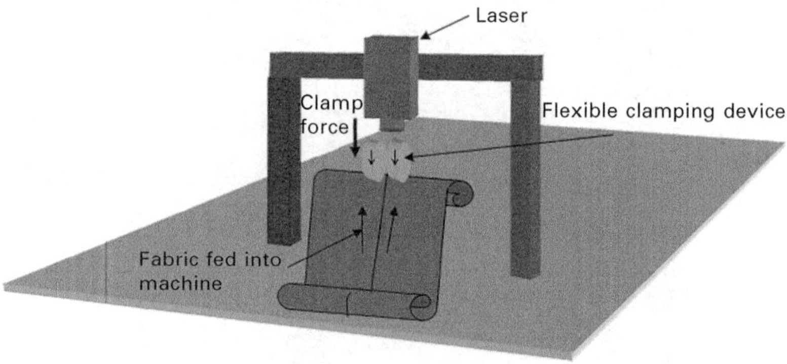

13.15 Sewing machine equipment scheme.

this flexibility does rely on the presence of a skilled operator controlling the delivery of the fabric into the machine. This must raise concerns about operator safety, particularly in respect to exposure to laser radiation and entrapment within moving parts.

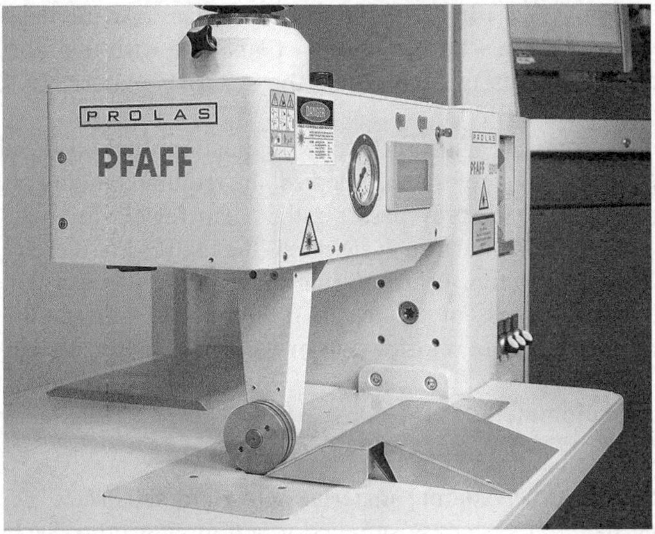

13.16 Multifunctional sewing machine type system including transmission laser welding, direct laser welding and ultrasonic welding in the same machine.

Table 13.1 Summary of the main advantages and disadvantages of the machine schemes discussed

Features/ type	Easy material handling?	Small fabric items?	Large fabric items?	Hollow sections?	Small footprint?	Ease of safety set up?	Compatibility with existing practice?
'Flying optics'	✓	✓	✓	✓	✓	✓	✓
Moving bed	✓	✓	✓	×	×	✓	×
'Sewing machine'	×	✓	×*	✓	✓	×	✓

Note: *long arm machines can be used to work with larger items.

Table 13.1 presents a summary of the main advantages and disadvantages of the machine schemes discussed.

13.4.4 Absorber deposition methods (for TTLW)

The transmission laser welding process depends on accurate and repeatable deposition of the NIR absorbing layer at the localised joint interface or in the lower piece of material. The absorbing material system may be applied in the form of a low-viscosity liquid, which dries rapidly to leave a deposit of the absorber on the surface of the textile. It may also be deposited on the surface of an interleaving film that is compatible with the fabrics being joined. A third option is to incorporate the absorber material system in a compatible interleaving film.

Spray deposition is the most effective application method for absorbent surfaces. In general, it is preferable to deposit the required amount of absorber using several high speed spray passes rather than fewer slow passes. This limits penetration of the material system into the textile and thereby helps to restrict melting only to the surface during the welding process. This avoids melting the full thickness of the textile, retaining the surface appearance and avoiding the introduction of a line of weakness along the edge of the weld.

Other application methods include:

- Needle dispensing – This technique has been extensively used for depositing absorber coating material onto rigid substrates. A constant flow rate material system is supplied to a hollow needle dispenser that is traversed over the substrate by an automated robot. This produces a highly reproducible deposit of the material system.
- Brushing – Brushing of the absorber onto textiles has also been used to create seams. Liquid dispensing equipment can be adapted to use a brush tip rather than needles. Brushing can apply wider lines than normally deposited by liquid dispensing techniques.
- Felt tip applicators – Felt tip applicators have also been successfully used to deposit absorber onto fabrics. Since the flow through a felt tip would be less than from a needle or a brush, the deposits would be less dense than normal liquid dispensing methods. However, for some applications, the method might provide a more uniform deposit.
- Films – Incorporation of the absorbing material into film has also been validated. The film then serves as a carrier for the absorbing material in the welding operation. This method of application would have to be tailored to each end use.

13.5 Textile materials that can be laser welded

Laser welding, like the other welding techniques, is mainly applicable to the thermoplastic synthetic materials (polyester, nylon, polypropylene and polyethylene). Joints can be made without the need for adhesive film between the layers of fabric, and generally leave the outer surfaces of the fabrics unaffected because all the heating is concentrated at a thin coating of absorber on the surface of one or both fabrics. The laser process can also be used to melt a film adhesive at the joint between a wider range of fabrics, including natural fabrics. The laser energy will pass through most fabrics to heat the film layer preferentially. The film is treated or coated with laser absorber to allow this. Table 13.2 indicates the types of fabrics that can be joined using the laser welding process.

Fabrics with thermoplastic coatings (e.g. PC, PU or PVC) can be very effectively joined and provide joints with performance limited only by the adhesion performance between the coating and the base fabric. Laminated

Table 13.2 Fabrics that may be joined using laser welding

Fabric group	Sub-group	Suitability of laser welding			Comments
		Yes	Some cases	No	
Natural					
Cotton			×		Requires adhesive film
Cellulose			×		Requires adhesive film
Wool			×		Requires adhesive film
Silk			×		Requires adhesive film
Glass			×		Requires adhesive film
Carbon fibre			×		Requires adhesive film and only by direct process
Synthetic					
Cellulose	Acetate		×		Requires adhesive film
	Viscose		×		Requires adhesive film
Nylon	Woven, knitted, fleece, non-woven	×			
Polyester	Woven, knitted, fleece, non-woven	×			
Polypropylene	Woven, knitted, fleece, non-woven	×			
Polyethylene	Woven, knitted, fleece, non-woven	×			
Coated	PU, PC, PVC	×			
	PTFE		×		Spray coats can be welded, thick coats need a film interlayer
Laminated		×			
Elastic fabrics			×		Special adaptations needed
Natural to synthetic		×			Synthetic melts into natural fibres

multi-layer fabrics can be welded such that only the sides of the fabric in contact are welded together without adhesive film, providing a seal without affecting the external appearance of the fabric. An external sealing tape is also not required.

Natural fabrics can be joined to synthetic fabrics by melting the surface of the synthetic material such that it flows into and interlocks with the natural fibres, again without using adhesive. Natural fabric to natural fabric and thermoset synthetic fabrics such as aramids (e.g. Kevlar) and cellulosics cannot be welded because they do not melt. Adhesive methods must be applied in these cases.

Stretchy materials do not generally respond well to welding. Nylon fabrics in particular become brittle when melted, and if the weld is stretched along its length, it will crack. However, in some cases, a fabric with two components, e.g. elastic fibres within a matrix of other fibres, can be welded and made to retain elastic properties. There are also alternative techniques using an array of dots, or a curved or zig-zag form to the weld which also retain elasticity. A laser source and scanning mirror system can easily generate any design along the seam line.

It is important to note that fabrics coloured with carbon-based pigments cannot be welded using the transmission laser process, but can be welded using the direct process where the laser beam does not have to pass through the fabric.

13.6 Joint designs in laser welding

The industrial standard, ASTM D6193-97, 'Standard practice for stitches and seams' covers the requirements and characteristics of stitches and seams in the fabrication of sewn items. It lists 55 types of seam in which the fabrics are superimposed and 101 types of seam in which the fabrics are lapped. These categories are also known as peel and lap seams, respectively. The standard also contains 18 types of bound seam, six types of flat seam, eight types of ornamental seam and 32 types of edge finishing seam. Many of these seams are complex and not well suited for through-transmission laser welding processes, for example because they require transmission of the laser through a large number of layers of textile. Some of the simpler types of seam, which are suitable for through-transmission laser welding are illustrated in Fig. 13.17.

The most common type of seam is LSa, the simple lap seam. This is readily manufactured using the laser welding process. The SSa simple peel seam is also readily manufactured using laser welding, but places higher demands on the weld in service. For a given fabric, the strongest SSa seam will be substantially weaker than the strongest LSa seam. Therefore, if strength is a primary consideration, the LSa seam is preferred.

(a)

(b)

(c)

(d)

(e)

(f)

(g)

(h)

(i)

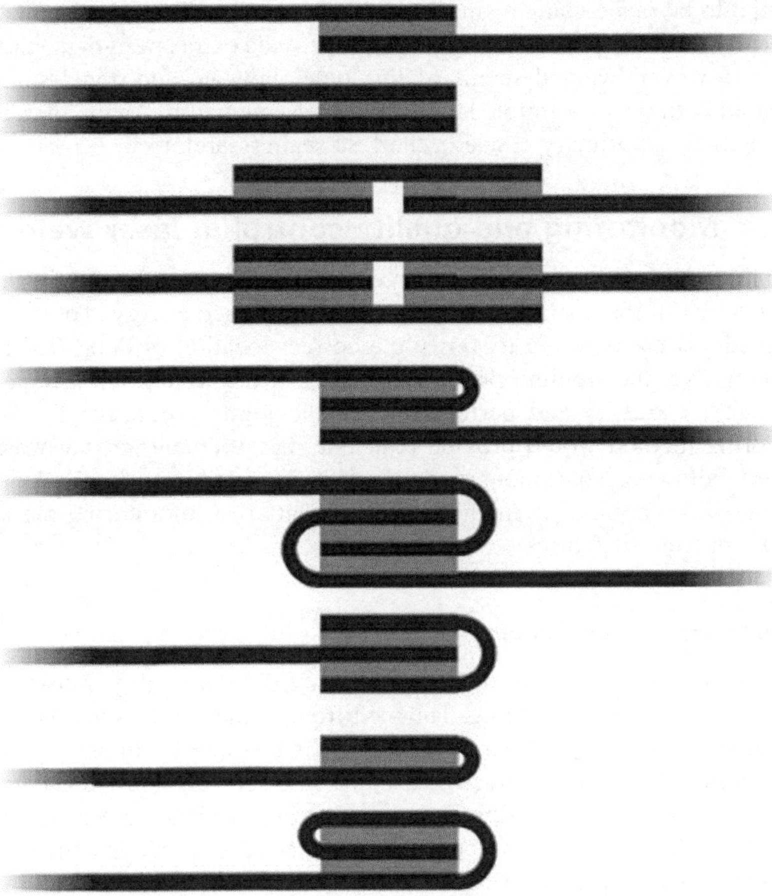

13.17 Some seam designs suitable for welding processes: (a) ASTM D6193 designation LSa – Simple lap seam; (b) ASTM D6193 designation SSa – Simple peel seam; (c) ASTM D6193 designation LSp – Taped seam; (d) No ASTM D6193 designation – Double taped seam; (e) ASTM D6193 designation LSb or LSq – Mock run & fell; (f) ASTM D6193 designation LSc – Run & fell; (g) ASTM D6193 designation BSa – Simple bound seam; (h) ASTM D6193 designation EFa – Simple edge finishing seam; (i) ASTM D6193 designation EFb – Overlocked edge finishing seam.

The LSp taped seam and the double-taped seam offer the benefit that they can be produced by applying the absorber to the tape only, therefore avoiding additional handling of the material on either side of the seam. If the laminate is constructed to include an absorber at the joint layer, it is possible to avoid a further consumable. In certain cases, if the textiles are sufficiently transmissive, then it may be feasible to fabricate a double-taped seam with a single welding operation.

The LSc (run & fell) and LSb (mock run & fell) seams also have the potential to be made using a single welding operation. These types of seam are widely used for aesthetic reasons and to provide extra strength in stitched seams. However, welded seams of this type show no additional strength compared with the LSa simple lap seam of the same width. Again, therefore, if strength is the primary concern, the LSa seam is preferred.

13.7 Monitoring and quality control in laser welding

Using laser radiation as the energy source enables precisely controlled welding both in terms of the position and the amount of applied energy. To optimise this, methods are required to verify the correct location of deposited laser absorbing dye, the welding process and to view the weld after completion to ensure the quality and performance of the joints produced. The ideal monitoring method would provide real-time measurements of the weld as it forms, allowing continuous control of the process in a feedback loop. The following monitoring methods offer potential for monitoring the laser welding process of fabrics.

Real-time visible light imaging

This is a suitable method for monitoring the welding of visibly transmissive materials at the time of welding. The melt front of the welds is visible in the video images and the pictures are relatively free from interference from the laser beam. This method can also be used when carbon black filled material is used as the lower fabric. Visible light imaging uses a video camera sensitive to visible light. It should be noted that some solid state CCD cameras are sensitive in the infrared as well as visible spectrum. If this is the case, then a filter can be used to block any radiation from the laser that would interfere with the images. In addition, an image analyser can be used to provide automated feedback of the items being viewed. Visible light imaging has been described when welding clear to black parts (Hierl and Hofmann, 2003), where variations in intensity of the dark mark at the joint indicate changes in the quality of the joint. In theory this can be extended to welds between two clear materials if the changes in reflection from the surfaces to be joined can be identified. Visible light imaging is appropriate when transparent and translucent materials are used as the top layer. It is a low cost option that uses off-the-shelf equipment. However, this method would be unsuitable when visibly opaque materials are used as the top layer.

Infrared thermography and temperature measurement

Thermographic techniques remotely measure the radiation emitted from a hot surface by virtue of its temperature. The temperature of the surface can be calculated from the spectrum and intensity of the radiation emitted. The strength of the signal available is the product of the spectral radiance and the material's emissivity at a given wavelength. Emissivity values range from 1.0 for lampblack down to 0.02 for polished silver. Thermography is used effectively for monitoring laser welding where carbon black is used as the absorber. Use of the method presents difficulties if the emissivity of the heated region is low.

The wavelengths at which infrared thermometers are sensitive define the range of temperatures that can be measured. To make measurements of thermal radiance in the melting range of plastics (150–400°C), it is necessary to monitor at wavelengths longer than approximately 1,800 nm where the radiance is at measurable levels (>0.1 lumens). If the materials being measured have low emissivity, then a longer measurement wavelength would be preferred to make use of the more intense region of the spectrum.

For transmission laser welding, Table 13.3 indicates that a short wavelength thermographic system is required to measure the weld interface temperature. It is preferable to monitor at infrared wavelengths shorter than 2,200 nm, where the PMMA transmission is greatest, and at wavelengths longer than 1,800 nm to read temperatures as low as 150°C.

Near infrared (NIR) imaging

Infrared imaging is carried out in the same way as visible light imaging, with cameras sensitive to the near infrared range of wavelengths. In principle, if wavelengths of the thermal emission spectrum from the weld are used, then a thermal picture of the weld can be seen. However, this procedure is subject to the same limitations as described above for infrared thermography. A method of using an IR camera to image a complete weld line immediately after the welding process has finished has been described when welding clear to black polycarbonate parts (Russek, 2003). Regions of differing temperature could be identified and could indicate variations in weld quality along the joint.

Alternatively, a camera can be selected that is sensitive to the absorption wavelength of the infrared absorber used in the welding process. The camera

Table 13.3 Infrared wavelength and detectable temperature range for commercially available thermographic equipment

Infrared wavelength (nm)	Temperature range (°C)	Applicability to textiles
8,000–14,000	–50 to +500	Low transmission; can only read temperature at top surface
1,000–2,000	+150 to +3,000	High transmission; can read temperature within component

would then be used to monitor the position of the coating before, during and after welding. The infrared absorption of the coating gradually decreases during welding, so potentially this could offer a good monitoring method. The difficulty in applying this method at the time of welding is that it is difficult to filter out interference from the laser wavelength, as this is selected to match the absorber. Because of this, the technique has only been applied to monitor the weld zone before and after welding. Image analysis may also be used in conjunction with infrared images taken.

Spectrophotometric measurement of pre- and post-weld samples

In this method, a spectrophotometer is used to measure the absorption properties of the IR absorber and substrate materials, over a range of wavelengths. This procedure may be used before and after the coating deposition to check the position and amount of absorber present. As a result of the laser heating, some or all of the dye decomposes during the welding process into products that have a different absorption signature. Measurements made using a spectrophotometer before and after welding can therefore reveal information about the heat generated during welding.

This method is currently limited to use before and after welding for similar reasons to infrared imaging. To use this method at the time of welding, in addition to filters for the laser wavelength, processing of the recorded spectrum would have to be very rapid.

13.8 Comparison between stitched and laser welded seams

A study was carried out by TWI to compare the local deformation (strain) under stress and the ultimate tensile stress of welded and stitched seams. The assessment was carried out using tensile testing equipment (BS 13935-1) and electronic speckle pattern interferometry (ESPI) to identify and map local strain in the region of the joints.

ESPI generates an area map of the local strains generated in a surface. This was carried out on selected welded and stitched seams to show the zones of high and low strain (and therefore stress) generated during small tensile load change. The ESPI equipment and test arrangement are shown in Fig. 13.18. Areas of high strain are expected to indicate the ultimate location of failure. The peak local strains for different samples are shown in Table 13.4, and the strain maps for a stitched seam, and welded seams in peel and lap configuration are shown in Figs 13.19–13.21.

The measurements indicate that peak local strains or displacements are greatest in welded seams in peel loading, and lowest and delocalised in welded seams under lap/shear loading. Stitched seams both in lap and peel

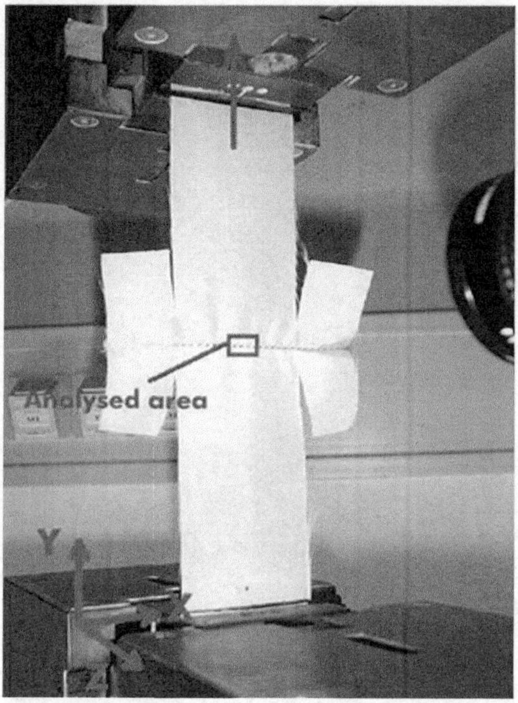

13.18 Analysed area and sensor coordinates system related to the specimen.

Table 13.4 Peak local strain in the y-axis measured using ESPI at different pre-loads for an incremental cross-head movement of 0.1 mm

Specimen code	Stitched or welded	Type of seam	Absorber	Preload (N)	Peak Strain (μm/mm)	Location
ST2-V	Stitched	Peel	None	7.8	3.59	Around stitches
				71.0	0.86	Between stitches
13-I	Welded	Peel	Spray	10.0	12.70	At weld opening
				31.7	19.90	At weld opening
19-I	Welded	Lap	Tape	30.0	1.19	Beyond edge of weld
				77.0	1.57	Beyond edge of weld
ST1-V	Stitched	Lap	None	25.0	1.97	Around one side of stitches
				137.0	5.42	Around one side of stitches
3-I	Welded	Lap	Spray	30.0	2.47	Beyond edge of weld

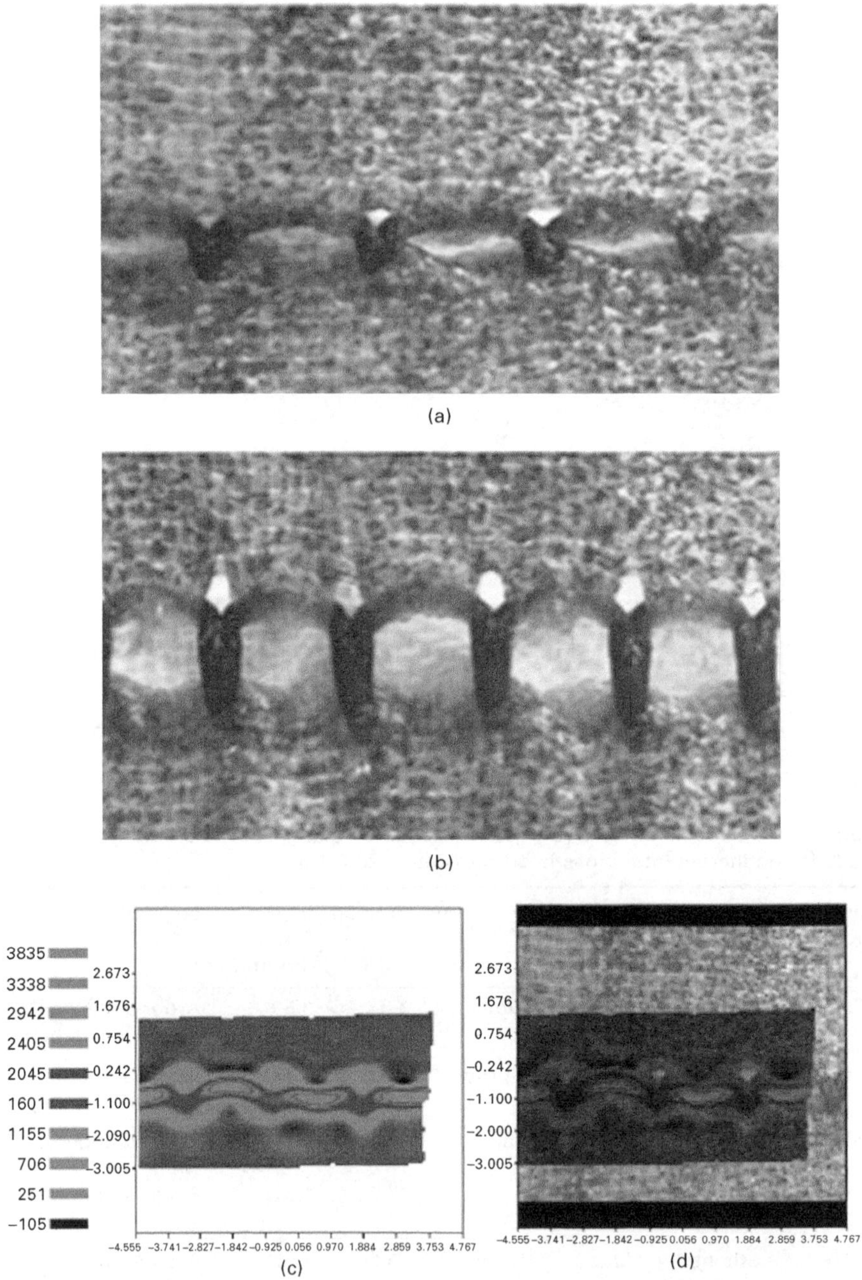

13.19 Sample ST2-V, showing strain by mapping between low and high loads, gaping at the stitches is visible: (a) live image at reference load of 66 N; (b) live image at load of 455 N; (c) strain in y-direction between 66 N and 455 N; (d) strain overlaid on live image.

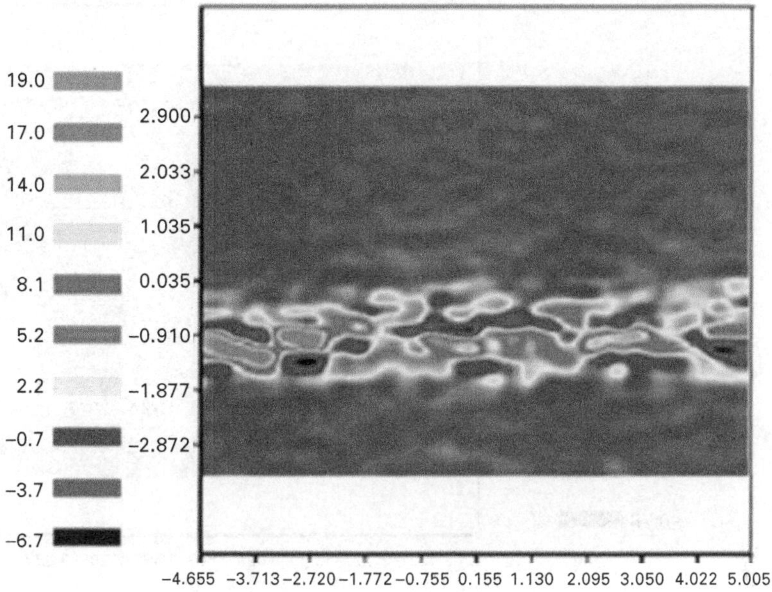

(a)

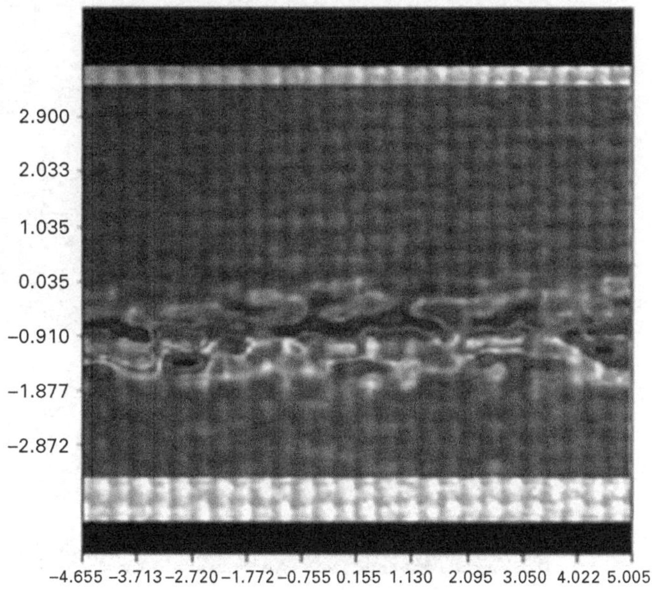

(b)

13.20 Welded sample in peel (13-I). Weld opening in peel near the centre of the picture. Preload 31.7 N: (a) strain (μm/mm) in y-direction; (b) overlaid on live image.

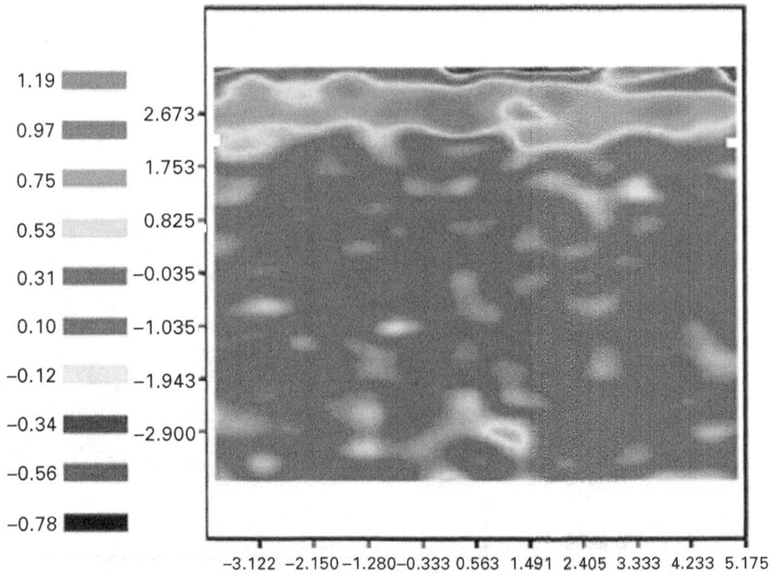

(a)

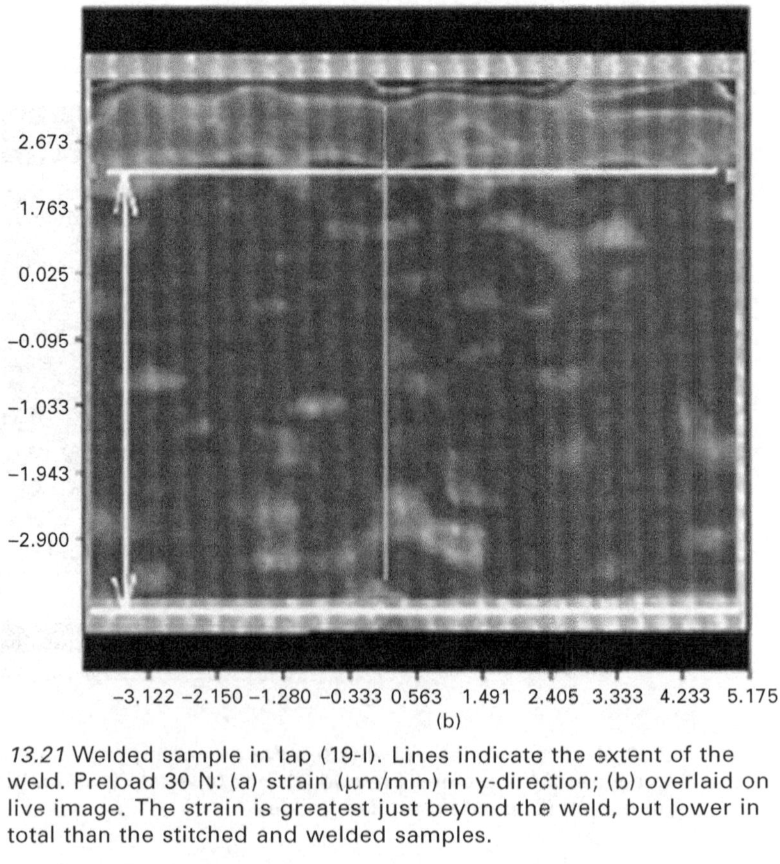

(b)

13.21 Welded sample in lap (19-I). Lines indicate the extent of the weld. Preload 30 N: (a) strain (μm/mm) in y-direction; (b) overlaid on live image. The strain is greatest just beyond the weld, but lower in total than the stitched and welded samples.

show highly localised strains between these two, with peel seams indicating strains that are higher than lap seams.

The peak local strain measurements were not necessarily an indicator of the ultimate tensile load of a seam. In particular in the welded lap seam, failure occurred at lower loads than in stitched seams even though the local strains are low. Table 13.5 summarises the results from this analysis.

13.9 Applications of laser welding

Some examples of applications tested are given in Table 13.6. The application to waterproof garments and a medical device are discussed in detail below.

13.9.1 Waterproof garments

Laser welding allows sealed seams to be made for applications such as outdoor clothing, with the potential for a reduced number of process steps and improved reliability, consistency and security compared to alternative techniques. The process is also suited to automation and would potentially fit into a fully automated production line for garment or other textile product manufacture (Carosio *et al.*, 2004).

The fabrics tested included three-layer laminated materials with polyester face, nylon back and a breathable membrane, polyurethane coated materials and bi-adhesive laminated materials. The results have shown that high integrity seams can be made which exceed the current industry requirements for resistance to water penetration (at least 200 cm water head required to

Table 13.5 Samples assessed in tension to failure. The peak load and the failure mode are also noted

Stitched or welded	Type of seam	Absorber	Peak load (N)	Peak strain, (μm/mm)	Failure mode
Stitched	Peel	None	1,050	3,835 at 455 N	Fabric gapes at stitches, pulls through and ultimately fabric yarns break
Welded	Peel	Spray	240	4,137 at 92 N	Peel seam opens in steps as melt detaches from weave
Welded	Lap	Tape	480	171 at 478 N	Fabric tears at edge of weld
Stitched	Lap	None	approx. 1,100 N	1,221 at approx. 300 N	Fabric gapes at stitches, pulls through and ultimately fabric yarns break

Table 13.6 Examples of tested applications of laser welding

Industry	Product	Fabric	Seam type	Individual seam length (m)	Product seam length (m)	Speed (target) (m/min)
Automotive	Air bags	Woven nylon	Overlap	Up to 3	Up to 100	>1
Garments	General apparel	Polyester, polypropylene, nylon	Peel and multiple-overlap	Up to 2	Up to 24	>2
	Protective clothing and waterproofs	PU-coated nylon Laminated 2/3 layer with microporous membrane	Peel and multiple-overlap	Up to 2	Up to 24	>2
Aerospace	Helium balloons	PU-or PC-coated nylon or polyester	Overlap or Butt with weld tape	Up to 30	Over 1,000	1–4
Civil engineering	Inflatable structures	Coated nylon	Overlap or Butt with weld tape	Up to 30	Over 1,000	1–4
	Tentage	PU-coated woven nylon	Peel or overlap	Up to 5	Up to 100	>3
	Flexible roofing	Polymer films	Overlap and multiple-overlap	Up to 10	Over 1,000	3
Boats	Sails	Woven nylon Spectra (PE) Polymer films	Overlap and multiple-overlap	Up to 4	Up to 1,000	>1
Medical	Vascular grafts	Knitted polyester	Overlap	Up to 0.1	Up to 3	1

ISO 811), and survive a 50-cycle wash test without detriment (to ISO 105-C01/89). The seam strengths are also very good with many existing fabrics, easily meeting the 120 N industry requirement (to ASTM D1683-04).

A summary of results of tests on seam strength and resistance to water penetration is shown in Table 13.7. The results for a single-layer fabric, similar to that used in the bi-adhesive material, are shown for comparison. This fabric shows higher strength than the others (it was not designed to be waterproof), and serves to demonstrate that the seam strengths in the laminated materials were limited by the lamination strength in the parent fabrics. This result was further reinforced by examination of the failure mechanisms for the tensile tests.

Equipment development

Equipment has been developed to allow construction of complex textile products in a wide range of sizes and designs. A reconfigurable table to support the fabric parts in the required location and shape has been made. This has been integrated with a laser beam delivery unit to provide controlled laser heating and pressure application along the textile seams. The diode laser is delivered by a fibre optic and the head is manipulated with six-axis robot to provide a functioning laser welding workstation.

The final welding head incorporates sensors to monitor the weld temperature and provide continuous control of the laser power allowing very high weld consistency at all times. This work has already resulted in a patent application. Two support systems have been developed to deal with either 2D or 3D seaming procedures. The reconfigurable table assembly for 2D procedures

Table 13.7 Laser welded seam performance for different fabric groups

Material	Seam strength Lap seam ASTM D1683-04			Resistance to water penetration ISO 811
Industry requirement	120 N			200 cm
	FF	FB	BB	
Nylon with polyurethane transfer coating, 222 g/m^2	274 N	278 N	289 N	>200 cm
Three-layer laminate, polyester/PTFE/nylon, 190 g/m^2	447 N	247 N	243 N	>200 cm
Two-layer polyester laminated with 14% PU biadhesive, 240 g/m^2	256 N	245 N	100 N	>200 cm
Single-layer polyester/lycra, Teflon coated, 250 g/m^2	547 N			33 cm

Notes: FF = face to face lap joint, FB = face to back, BB = back to back.

uses driven rollers that can operate jointly with the air suction system to ensure no wrinkle or crease formation on fabric. For 3D seams, the curved shape of the seam may be defined by a shaped substrate. The feasibility of welding a curved 3D seam has been demonstrated on a rigid mould, and a small-scale prototype equipment has been made with a rubber membrane and actuated pin elements able to precisely reproduce the 3D shape required. This indicates a route forward to a fully flexible system.

The robotic laser welding equipment has been constructed and integrated and a simulation model for the integration of the welding unit into a fully automated textile product factory has been developed. The equipment is shown in Fig. 13.22.

Waterproof jacket and 3D hood

An existing jacket design was modified to have seam designs and forms more suited to a laser welding fabrication (Fig. 13.23). Additionally, the total number of seams was reduced. A series of fabrication steps was developed starting from a pattern for the trunk followed by attachment of cuffs, pockets, collar and lapels. The welds were completed at a speed of 3 m/min using

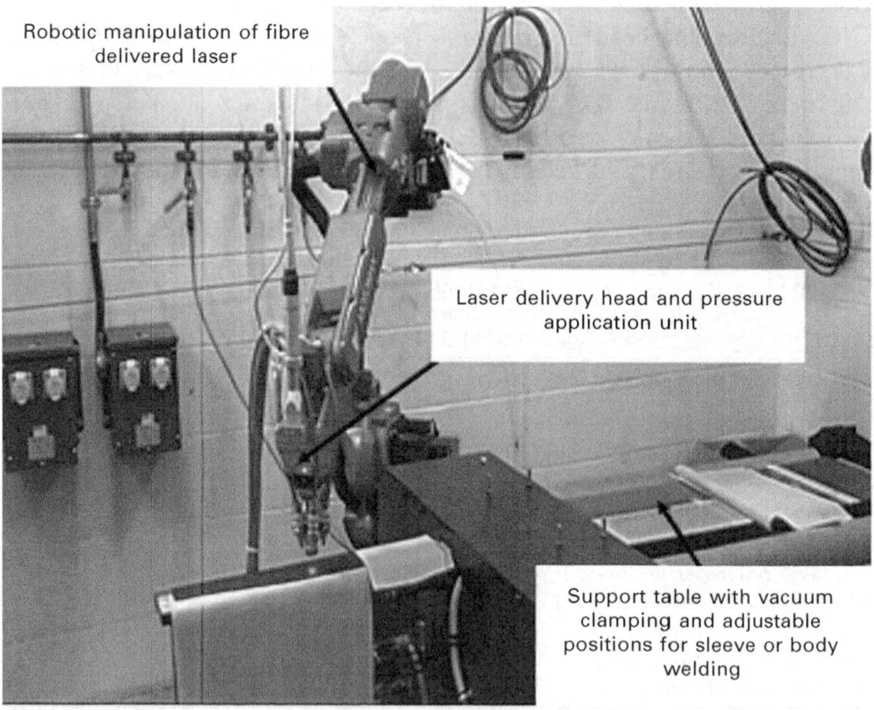

13.22 Automated laser welding equipment.

13.23 Laser welded waterproof jacket.

13.24 3D processing for a hood seam.

a laser power of around 75 W. The equipment has been evaluated for the manufacturing of 3D seams. The hood for the waterproof jacket was chosen to demonstrate 3D processing as shown in Fig. 13.24.

Laser welding of textile can be used for the manufacturing of simple or complex shapes, if the appropriate clamping and support systems are used.

Automated welding has been demonstrated in specific application areas. However, to take full advantage of the automated procedures to provide time and labour savings, the equipment needs to be integrated with an automated production line that would include preparation and handling of the fabrics. Preparation of the fabrics such as cutting, spraying of absorber and positioning could be done automatically using a combination of flat-bed tables and robots. The laser could also be coupled to several fibres so that welding of several jackets, for example, could be made simultaneously. A simulated automated manufacturing environment is shown in Fig. 13.25.

It is estimated that the laser welding operation of the jacket would take around 10 minutes compared to 45 minutes for the manual stitching and taping process. Therefore, an automated procedure would provide a higher production rate per manufacturing unit, whilst also providing a higher quality product (Jones, 2008).

13.9.2 Vascular graft stents

Abdominal aortic aneurysms (AAA) are a type of cardiovascular disease, a life-threatening condition that occurs when a section of the abdominal aorta, the body's main circulatory vessel, weakens and bulges outwards, as shown in Fig. 13.26, to form a fragile, balloon-like swelling called an aneurysm which is prone to rupture.

13.25 Simulated example of automation using laser welding of textiles.

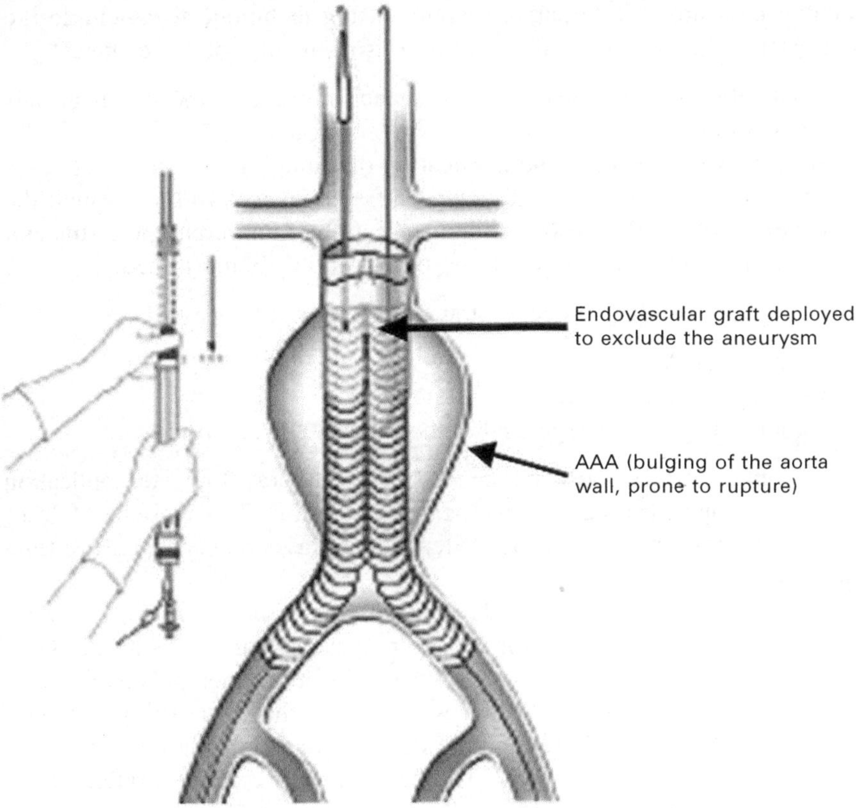

13.26 Schematic illustration of the AAA along with the Anaconda endovascular device manufactured by Vascutek Ltd (courtesy of Vascutek Ltd).

An endovascular stent-graft system can be used to treat AAA and is made from polyester fabric (used as the graft material) and supporting Nitinol wires (as individual ring-stents). Currently, weaving is used to produce the textile graft material, and hand-sewing is used to add the Nitinol ring-stents to the graft, which is very complex and highly labour intensive with quality dependent on the operator.

Some of the joining techniques that have been considered in place of manual stitching are as follows:

- automated sewing machines
- adhesive bonding
- laser welding
- ultrasonic welding
- hot-air welding.

After a series of preliminary feasibility trials, laser welding showed tremendous potential as a novel, alternative manufacturing technique to assemble the stent-grafts. The laser welding process consists mainly of three stages:

1. Application of laser absorber dye at the interface of the two materials being welded.
2. Assembly of the seam and application of clamping pressure.
3. Irradiation of the seam with a near infrared laser (940 nm) to melt the material where the absorber has been applied and create a permanent weld at the interface between the two materials being joined.

The main laser welding process parameters are as follows:

- laser power (W)
- welding speed (m/min)
- amount of absorber deposited at the joint interface (nl/mm^2).

The amount of absorber present is dependent on the spray flow rate, application speed, and concentration of absorber dye solution. The amount of laser energy input into the weld location depends on laser power, exposure time and laser beam size.

The main requirement is to hold the nitinol ring-stents securely to the graft fabric, which is currently performed by polyester sutures. Using the laser welding technique, this is achieved by welding the inner polyester fabric and an outer polyester film layer on either side of the Nitinol rings, thereby sandwiching the rings in between the two layers and hence holding the rings securely to the inner graft fabric. The heat affected zone generated during the laser welding process is localised and restricted to the interface between the graft layers, and the Nitinol ring-stents are not affected by the heat. A thin layer of polymer is melted in each graft layer and the application of clamping pressure brings the melted fibres in contact. The clamping pressure is maintained as the melted fibres cool and solidify to produce a permanent weld.

An ultrasonic spray system was identified as the favoured method for absorber application, as it produced an atomised spray of minute particles and tended to restrict the absorber to the surface of the outer graft layer (without wicking through its full thickness). The amount of absorber deposited can be precisely controlled by adjusting the spray system and XY CNC table timings.

Laser welding using a polyester fabric and polymer film configuration resulted in an ideal-case scenario where the absorber was confined to the interface causing some fibres to melt on either side. The main advantage of using a film as the other graft layer is its ability to prevent the absorber wicking through its thickness. During tensile tests, most of the samples failed in the parent material rather than at weld location or along the edge of the weld, indicating a very strong weld. Furthermore, this material combination resulted in minimal bulk (thickness) at the joint location and hence was suitable for use in the final product configuration.

The current hand-sewing process is compared to the laser welding process against the user requirement specifications and the results are summarised in Table 13.8. We can only expect the trend towards less-invasive techniques to continue, and perhaps most importantly automated manufacturing of medical devices will see significant growth in the future. The complex design of these stent-grafts limits the use of automated sewing machines, adhesive bonding, hot-air welding and hot-wedge welding to attach the Nitinol support wires to the graft. As a result of extensive feasibility studies, it is clear as that laser welding could potentially replace hand-sewing as it offers many functional advantages over conventional joining techniques, thereby reducing production times, costs and improving the quality and durability of the products (Patil *et al.*, 2010).

13.10 Future trends

Laser welding of fabrics can lead to greater automation, increased productivity and improved quality, offering manufacturers a competitive advantage and reducing the incentive to relocate production to regions with low labour costs. Producing finished goods close to where they are sold also reduces shipping costs. Additionally, the process reduces noise levels and injuries in the workplace. Further developments in material handling, clamping and fabric selection promise even greater benefits in terms of process speed, automation and quality improvements. Laser welding is already being used successfully in some simple applications, and it is expected that ever

Table 13.8 Comparison of current hand-sewing process with laser welding technique against user requirements

Product requirements	Current sewing process	Laser welding process
Joint strength (N/mm)	5.3–7.8	5.0–8.7
Joint appearance	OK, but chance of holes in fabric	Good – weld is at interface only
Bulk of the device	Minimal	Minimal
Flexibility of the device	High	High
Abrasion resistance	Average but acceptable	Good
Durability	Good	To be confirmed
Effect of ethylene oxide sterilisation	No effect of sterilisation	No effect of sterilisation
Time taken for 1 module	8 hours (only sewing process)	1 hour (estimation including setup)
Potential cost savings	Not applicable	£350,000 per year

more complicated articles will be manufactured using this technique. The widespread introduction of laser welding in textile product production has been limited to date, probably by the high cost of implementation. This is likely to change in the future as the cost of laser systems continues to fall and their electrical efficiency (now exceeding 50% in some diode lasers) continues to increase.

Future developments are likely to include studies on making stretchy welded seams. This has been difficult using other welding processes (though easy with stitching). The potential for laser welding just to heat and melt selected parts or certain fibres in a fabric means that elastic fibres or an elastic layer can be left untouched or selectively heated. This means that a seam can be made retaining the flexibility and stretchiness. Early studies have demonstrated this in certain materials, but more work is required to define the scope of application.

13.11 References

Carosio S, Molfino R M, Monero A, Pagliai F, Terentjev E M and Walter L: 'Reengineering the clothing manufacturing system through the cooperation of advanced robotics and multifunctional materials'. IMS International Forum 2004, Cernobbio, Italy, 17–19 May.

Hierl S and Hofmann A: 'Innovative solutions for laser plastics welding', *Proc 2nd Int Conf on Lasers in Manufacturing*, Munich, June 2003, pp 385–389.

Hilton P A, Jones I A and Kennish Y: 'Transmission laser welding of plastics', *International Congress on Laser Advanced Materials Processing (LAMP 2002)*, May 2002.

Jones I A; 'ALTEX final public report', March 2008. Available at: cordis.europa.eu/

Jones I A and Olden E: 'A thermal model for transmission laser welding of thermoplastic polymers'. TWI members report 708/2000, July 2000.

Jones I A and Wise R J: 'Welding method'. Patent WO 00/20157, 1 October 1998.

Jones I A, Hilton P A, Sallavanti R and Griffiths J: 'Use of infrared dyes for transmission laser welding of plastics', *ICALEO 1999*, Vol 87, Section B 1999, pp 71–79.

Patil A, Jones I and Ashton T: 'New manufacturing methods for vascular grafts using laser welding and laser cutting techniques', *Joining Plastics 2010 – 2nd DVS/WJS Conference*, Dusseldorf, Germany, November 2010.

Rostami S and Jones I: 'Process guidance and software for Clearweld®'. TWI Report 772/2003, August 2003.

Russek U A: 'Innovative trends in laser beam welding of thermoplastics', *Proc 2nd Int Conf on Lasers in Manufacturing*, Munich, June 2003, pp 105–111.

14 Properties and performance of welded or bonded seams

E. VUJASINOVIC and D. ROGALE,
University of Zagreb, Croatia

DOI: 10.1533/9780857093967.3.435

Abstract: The twenty-first century has witnessed an intensive application of new techniques in fabric joining. Conventional techniques of joining fabric pieces by sewing are faced with numerous innovative competitors who are bringing considerable changes to the area of garment manufacture, and textiles in general. It may be possible to implement contemporary joining techniques with the new and demanding high performance materials, a radical change in seam visual identity and achieving selected properties of end-use products, such as breathability, waterproofing and drag reduction. A comprehensive knowledge, primarily in the fields of materials science and machine construction, is however, necessary to implement these new joining techniques successfully, together with the notion of the importance of introducing the systems of objective measurement and evaluating the quality of new welds. Objective evaluation includes performance, conditions and properties. Although every seam analysis begins with checking its basic properties, different seams have different requirements that should be met, relating to their application and function. This chapter deals with contemporary methods of joining textiles, and is aimed at defining the differences between welded joints, and systematizing the knowledge within this field. The chapter aims to define key factors within the necessary and possible future research trends in the field.

Key words: welded seam property, seam performance, seam appearance, seam permeability, future trends in welding.

14.1 Introduction

In the past ten years there has been intensive implementation of new techniques in fabric welding. The conventional technique of sewing is now faced with numerous competitors, which has brought significant changes to the area of ready-made garments, and textiles in general. Modern techniques such as welding textiles using lasers, heat (hot air flow and heat sealing bar), ultrasound and high-frequency electromagnetic field, have introduced a number of new properties that cannot be achieved using conventional techniques. These new techniques are particularly efficient in joining new and demanding contemporary materials. In order to implement these techniques, it is necessary to have a specialized knowledge in the field of physics, machine construction and machine workings, as well as being familiar with the technological properties of these machines. These preconditions will determine the level of performance of the resultant seams, such as the appearance, porosity, mechanical properties, etc.

Contemporary welding techniques significantly broaden the range of seam properties that can be achieved, both aesthetically and with regards to technical quality. For this reason, it is necessary to develop adequate quality evaluation methods, which should be more comprehensive, more diverse and more demanding than the ones conventionally used in this area. However, it should be noted that conventional methods for evaluation are still widely used, with adequate modifications introduced in order to assess new techniques. New methods of evaluation are constantly being developed for particular welds with specific purposes, together with the measuring methods aimed at determining technical and technological seam properties. There is a growing need to modify machines and equipment, as well as the processes performed by them, whilst, at the same time, there is a pressure to introduce new control mechanisms for monitoring and regulating energy effects of new energy sources (laser beam energy, thermal energy, ultrasound energy, high-frequency energy, etc.) on the seam produced and its properties. This chapter deals with the contemporary methods of joining textiles, with the aim of differentiating the properties of bonded and welded pieces. The final aim is to determine key ideas and future directions for scientific research in the field.

14.2 Performance properties of seams

An old definition says that a seam is a line where two or more textile parts are joined into one by sewing. However, this definition is no longer valid. Some of the reasons for this are as follows (ASTM, 2000; Babic *et al.*, 1995; McQuaid, 2005; O'Mahony & Braddock, 2002; Saville, 2000; Seymour, 2008; Tao, 2001):

- modern technologies such as laser, ultrasonic and high frequency welding
- new materials, such as high performance, shape memory, smart or intelligent textiles that are widely present in our environment (architectural structures, geotextiles, agrotextiles, moby-textiles, sports textiles, medical textiles, etc.)
- socio-cultural and design engineering concepts (protective or reinforcing seams, decorative, conductive seam, etc.).

With this in mind, a seam can be defined as the area where two or more panels are permanently joined together by sewing, gluing, fusing, or using another mechanical manner of joining, with the aim of fulfilling engineering design requirements in the manufacture and end-use property requirements. Scientific research into the area of bonded or welded seam quality is limited. There are some EU associated projects, such as ALTEX, CLET and POLYBRIGHT. The majority of researchers agree that the method by which a seam is formed affects the performance of the seam assembly, and consequently the overall structural performance of the product (Frank, 1987; Hustedt *et al.*, 2008; Jones, 2005; Luders, 2000; Vujasinovic *et al.*, 2007b). This section will try to present the appearance and properties of welded and/or bonded seams, relevant for product end-use and its quality.

14.2.1 Seam appearance

Macroscopic seam appearance is one of the aesthetic quality parameters used, particularly with regard to garments (Fan & Liu, 2000; Kang, 2005; Zavec Pavlinic *et al.*, 2006). However, the requirement for visually perfect seams is often also applied to various types of technical textiles as well, e.g. building structures and sports textiles. Figure 14.1 shows the appearance of some welded seams, as well as the change in seam appearance caused by varying processing parameters and/or fabric type.

Laser welding can be used to produce welds in either lap or peel configurations, depending on the design and required appearance. The procedure mainly used in textiles is based on the Clearweld® method of transmission laser welding, which employs special materials to absorb infrared radiation. Traditional broad-spectrum absorbers, such as carbon black, add colour to the weld, which severely restricts the use of transmission laser welding in the applications where appearance matters. Clearweld is used to weld clear, coloured and opaque thermoplastics, where welds need to be made with no unwanted colour addition (Jones, 2005).

Figure 14.1 shows that PA 6.6 laser welded (LW) without a polymeric interlayer material may result in both textile layers becoming completely molten and re-solidified in the welding region, so that the fibre structure is completely destroyed. Whilst the seam region is more or less vitreous and brittle, with the advance application of laser welding parameters, better seam appearance can be achieved. The macroscopic appearance of the LW PA 6.6 textile layers with an absorbing polymeric interlayer looks much better than when no absorbing interlayer is used, and the fabric structure is completely retained at the outer sides of the samples.

According to Hustedt *et al.* (2008), this result cannot be achieved by ultrasonic welding (US), where at least the upper fabric layer is always molten across the whole thickness, similar to laser welding without an interlayer. The research performed at the University of Zagreb, Faculty of Textile Technology, indicates that this need not always be true. It is possible to preserve the original appearance of the fabric using ultrasound welding (Fig.

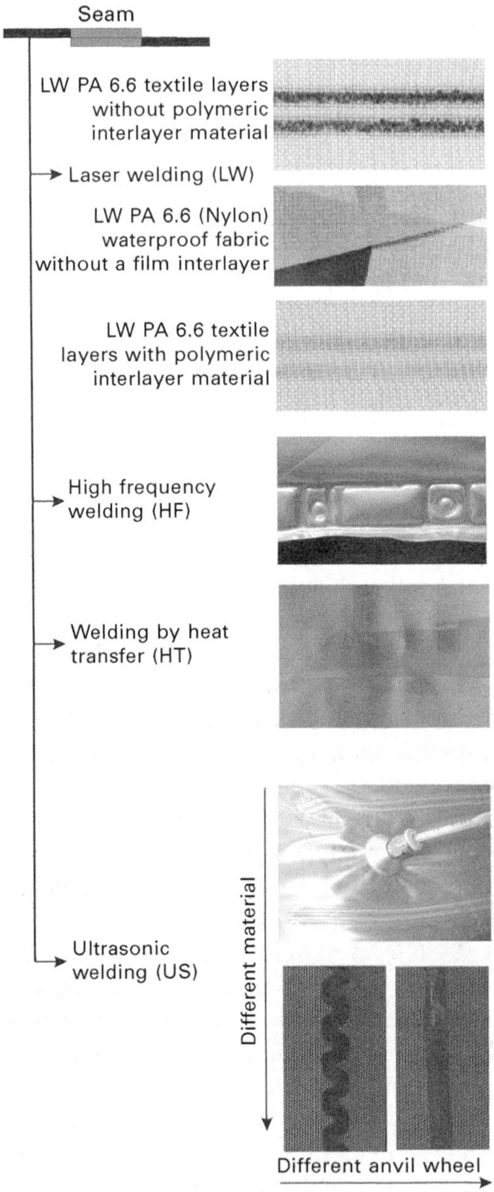

14.1 Macroscopic appearances of welded seams.

14.2), provided the high strength of the weld is not of utmost concern (e.g. in some medical textiles or geo- and agrotextiles). Similar conclusions were found in the paper by Shi and Little (2000), which used SEM to investigate US-welded seams incorporating optic fibres. The seams incorporating optic fibres, made at weld times of 0.2 seconds, weld pressure 0.2 MPa

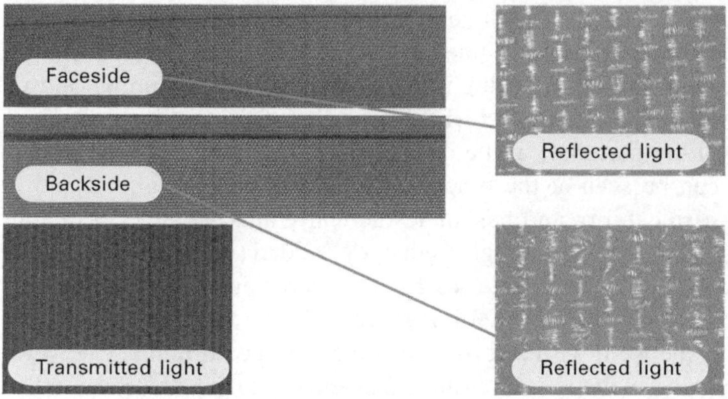

14.2 The appearance and microscopic image of an ultrasound weld (anvil wheel with point engraving, welding speed of 30 $dmmin^{-1}$ and amplitude of 200 W or 50%).

and amplitude of vibration 42 μm, showed that the two fabric layers were separated from each other with no obvious change in the optic fibre, while the interface of the two fabric layers began to fuse with an increase of the weld time. At a weld time of 1.5 seconds the optic fibre was damaged and a crack could be seen along the interface of the core and cladding, although a good weld resulted.

Welding textiles using ultrasound is also characterized by a pressing mark that corresponds to the embossing on the embossed counter roll. This type of welding offers proper joining of straight and gently curved contours; however, similar problems occur when joining more heavily curved contours as in thermal welding. Problems can be caused in this type of weld when a high welding energy is used, causing the fabric to melt at the beginning and end of the weld, as well as at spots of breaking in the weld.

High-temperature welding (HT) uses a convection method (heating the material by a hot-air stream between the fabrics) or a conduction method (using a contact heat transfer from a 'hot wedge' between the fabrics or a 'hot bar' over the fabrics), which has a key impact on the manner and rate of the fabric heating up. However, it does not determine the appearance of the weld. Seam appearance is mostly defined by pressure rollers used on the material after heating. These rollers are generally made of stainless steel or silicone rubber. When steel rollers are used, the quality of the pressing mark and its visibility depend on the force used on the rollers and the fabric. If silicone rollers are used, the pressing mark they make is hardly visible. When lower forces are used it is almost invisible (Rogale & Dragcevic, 2002). Still, with hot bar welding, in particular, the outer surface of the fabric is melted in this process (Jones & Wise, 2005). The machines for HT welding are also used to join straight and gently curved contours to those with a high curvature radius. The contours with smaller curvature radii are more difficult to join and aesthetically usually uneven and unacceptable. The second problem is associated with using a high energy flow at the beginning and the end of the seam and at the points where breaking of the weld is necessary in order to level the material, in the case of manual control. Molten material of the weld can be seen at these spots, which can lead to vertical and horizontal weld permeability and has an aesthetically unacceptable weld appearance.

The appearance of a high-frequency welded (HF) seam is characterized by a pressed and quite visible seam. This characteristic is due to the nature and functioning of the HF welding process. In the beginning, the HF electrode presses the weld with a particular force after positioning. The welding starts by bringing in the HF electromagnetic energy to the electrode, with the fabric positioned between the electrode and the grounded counter-plate, which causes the heat to develop within the material to be joined. The process generates melting throughout the thickness of the materials. As the material becomes softer with the rising temperature, the electrode increasingly presses the weld and penetrates the material. The result is a characteristic pressing mark of the bottom part of the electrode on the joined fabric pieces.

Mistakes are also possible in HF welding, which can be seen on the welded part of the seam. A number of HF machines are designed so as to cause the welding electrode to be lowered with the aid of pneumatic cylinders. If the mechanism for lowering the electrode is not properly adjusted, the electrode is not lowered parallel with the grounded counter-plate, which results in a deeper pressing mark at one part and somewhat shallower at the other. Apart from the uneven pressing mark, the shallower area bears the risk of electric breakthrough, since high-frequency tension is generated at the electrode (most often in the upper region of the short-wave radiofrequency spectrum, around 27 MHz), the amplitude of which, depending upon the power, often reaches as much as a few kilovolts of tension. Frequencies and tensions of this magnitude can lead to high dielectric stresses in polymer materials, which can result in material structure decomposition and an electric arc which additionally burns the material at the spot on the electric arc breakthrough. Permanent damage can occur at the breakthrough point, resulting in a lower weld strength, with vertical and horizontal porosity, as well as visual defects. HF welding is also limited to polyurethane and PVC-coated materials unless the electrodes are preheated.

All of this testifies to the fact that the appearance of welded or bonded seams depends primarily on the type of welding (Fig. 14.1) and processing parameters of the method used (Fig. 14.3). Seam geometry, its thickness and roughness, is determined primarily by the type of fabric to be welded, i.e. its construction characteristics, but also by the processing parameters of the welding method used. Thus, in US-welded seams of a PES sailcloth (extremely tight and compact fabric), for example, seam thickness and roughness are the

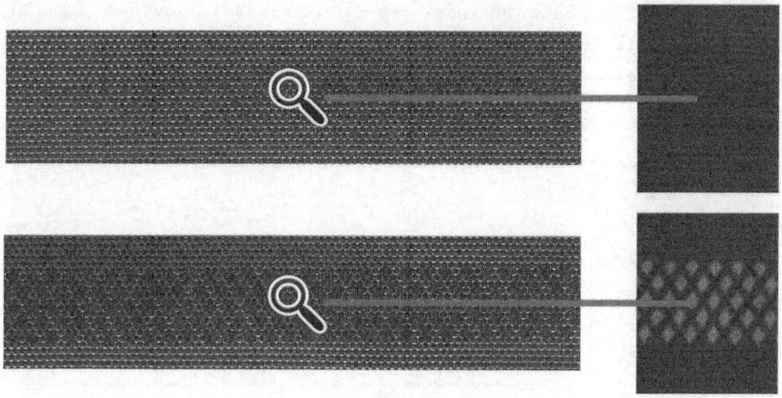

14.3 The appearance and microscopic image of an ultrasound weld; top – welding speed 30 dmmin^{-1} and amplitude of 200 W (50%); bottom – welding speed of 6 dmmin^{-1} and amplitude of 400 W (100%).

function of anvil wheel geometry and processing parameters of US welding, such as speed and/or amplitude, as well as sonotrode pressure during the welding process (Fig. 14.4; Vujasinovic *et al.*, 2006a).

Figure 14.4 shows that a higher amplitude results in a thicker weld for each of the alternating welding speeds, whilst constant amplitude with a higher welding speed reduces seam thickness. Accordingly, the thinnest weld is created at the highest speed and the lowest amplitude, and the thickest weld is created at the minimum speed and the highest amplitude. It can be seen that at an amplitude of 200 W (50%) seam thickness grows linearly with the reduction of welding speed, while at an amplitude of 400 W (100%) this correlation exhibits a completely non-linear character.

As opposed to seam thickness, the roughness of the seam is a property that can be hard to objectively measure and evaluate. Recent investigations (Vujasinovic *et al.*, 2007a) have shown that seam roughness can be qualitatively characterized by measuring the intensity of the transmitted light through the seam (Fig. 14.5). The graph of the intensity of the transmitted light passing through (I) the sample, shows the variability of the intensity on the measuring length, which is the result of:

- Sample roughness due to fabric construction (plain weave), where light intensity varies between 59 and 79, at the distance from 0 to 3 mm, and from 7.5 to 10.5 mm,
- Sample roughness due to weld geometry, exhibited as higher intensity of the transmitted light (up to 102).

Three such extremes can be seen at the sample length of 3 to 7.5 mm, thanks to the anvil wheel with point engraving used. Between the extremes, the

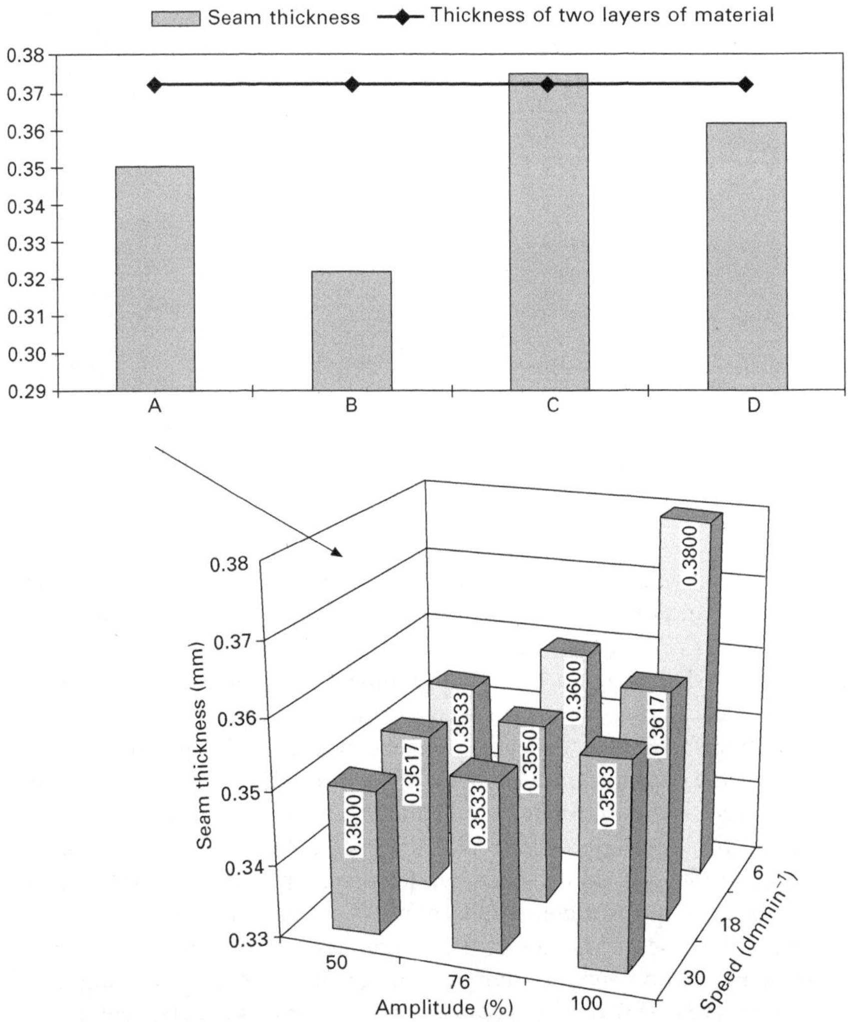

14.4 Comparative presentation of classical seam thickness and ultrasound weld thickness for anvil wheel of various gravures (A – point engraving; B – zigzag; C – one line; D – three lines) as well as variations of seam thicknesses for welding by anvil wheel with point engraving (A) and with weld alternating speed and amplitude.

intensity of the light transmitted through assumes the value of the intensity of the light passing through the fabric itself. The intensity of the transmitted light at the point of welding is higher (~29%), which means thickness is lower than the double thickness of the material, which is normally not the case. It can be assumed that the fabric is partially damaged (molten), which is augmented by the action of sonotode pressure and anvil wheel pressure exerted on the fabric.

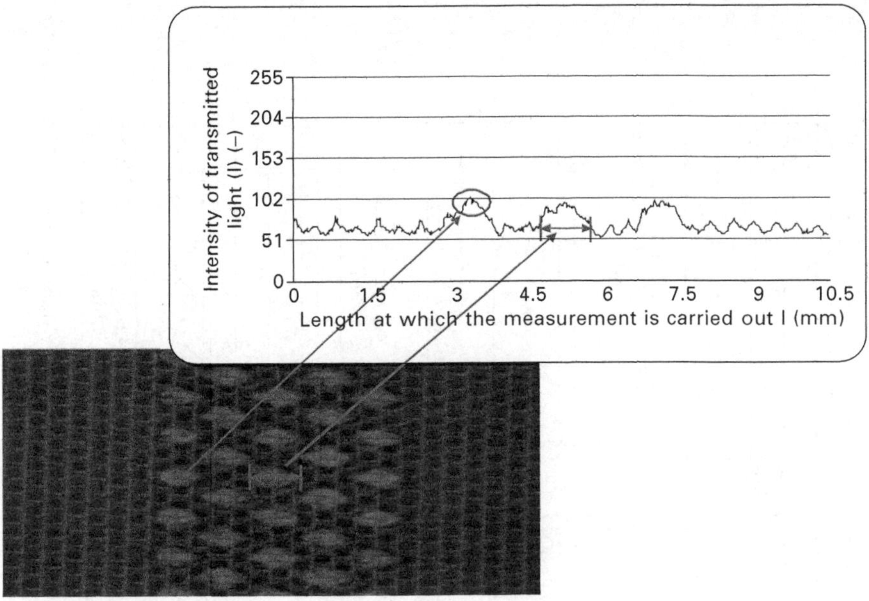

14.5 Microscopic image of the weld (obtained by anvil wheel with point engraving, welding speed of 6 dmmin^{-1} and amplitude 200 W (50%) and the related intensity of the light passing through the sample.

Figure 14.6 shows all the graphs recording the intensity of the transmitted light through welds, obtained at various speeds and amplitudes of US welding. It is evident that a higher amplitude results in a rougher weld for each of the welding speeds used, while constant amplitude results in reduced roughness under the conditions of increased welding speed. The intensity of the transmitted light passed through the sample at the lowest amplitude (200 W; 50%) and highest speed (30 dmmin^{-1}) reaches values from 54 to 78, which is the value determined by fabric construction. Obviously, such a weld will not increase product roughness, but it is questionable whether a strong enough bond is established between the two material layers. The weld created with the highest amplitude and lowest speed results in a high amount of roughness, and the intensity of the transmitted light through the weld is higher by around 70% than that passed through the fabric itself. The intensity of the transmitted light through the other tested welds varies between these two extremes, thus the selection of the working speed and the amplitude of the ultrasound welding machine should be optimized, in order to achieve minimum roughness and adequate strength.

Seam smoothness is often an imperative for numerous textile products, which can be illustrated by sports sails, where smooth seams ensure adequate laminar air-flow around the sails, which increases buoyancy and therefore

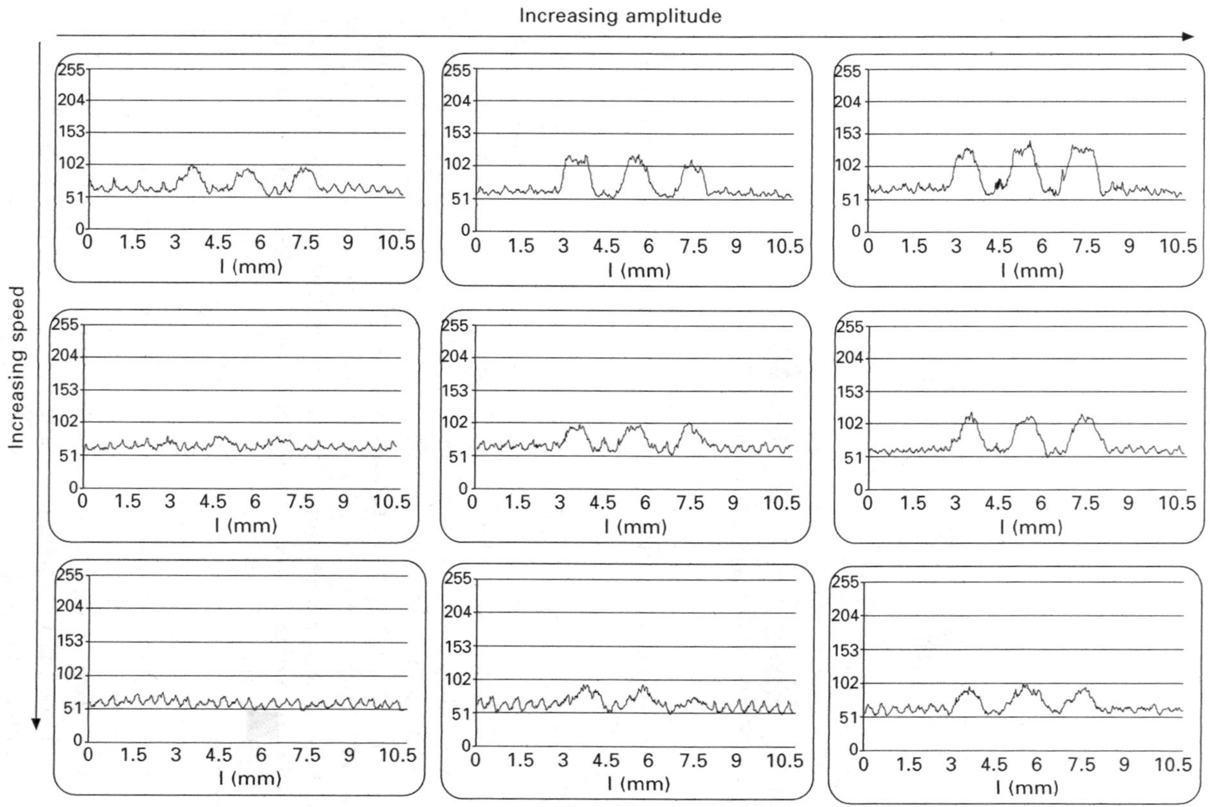

14.6 Changes of the intensity of transmitted light passed through the ultrasonic weld (the y axis represents the intensity of the transmitted light, and the x axis is measurement length).

the speed of sailing. Another example is swimming suits for top swimmers, where high performance fabrics and smooth ultrasonically welded seams (i.e., Speedo swimming suits, the LZR Racer) reduce the drag on body movement through water by 5%, which is certainly a significant contribution to the speed of swimming (NASA, 2009; Tucker & Dugas, 2008).

14.2.2 Seam permeability

There are two types of seam permeability/porosity in conventional sewing techniques. Vertical porosity is perpendicular to the surface of the fabrics joined and is caused by the needle penetrating the fabric, while horizontal porosity is parallel to the fabric surface and is caused by inadequate pressure of the thread in machine sewing stitch onto fabric layers. It is usually taken for granted that new joining techniques (welding, bonding, etc.) ensure seam impermeability, but this is not always so. Generally, there is no horizontal or vertical seam leakage in both types of fabric welding, provided impermeable fabrics are used in the process. The same is true for HF and laser welding. The only instance where horizontal permeability could occur is in ultrasound welding, if during the welding process an engraved roller is used with a discontinuous pressing surface (dotted weld, weld with a series of discontinuous lines). Horizontal leakage is avoided if engraved anvil wheels with a single or multiple continuous lines are used, or an anvil wheel with an engraved sinusoid or zigzag pattern. However, to make a weld vertically impermeable it is necessary to optimize processing parameters (Fig. 14.7) which, if inappropriate, could cause micro-porosity and make the product inadequate for some of the intended end-uses (protective clothing, tents, reservoir, etc.). Seam leakage is determined quickly and efficiently using the apparatuses (Fig. 14.8) for testing weld permeability with the aid of compressed air or water under a predetermined pressure.

As opposed to air- and water-permeability, seam resistance to the penetration of various chemicals is defined primarily by the type of fabric used, its raw material composition and associate chemical inertness, and only then by seam geometry (Racheel, 1995). The integrity of seams as well as closures or other clothing material interfaces is easily evaluated using penetration resistance testing (ASTM, 2010; Berardinelli & Cottingham, 1986; Racheel, 1995). Penetration is defined as physical transport of a chemical from one side of the material to the other, such as through imperfections, holes, tears, etc. A pass/fail test determines if a material is porous to a potentially hazardous liquid at a given pressure. Berardinelli and Cottingham (1986) demonstrated the utility of this test on a number of materials and seam samples. A part of it for Tyvek® (a registered trademark of DuPont) is shown in Table 14.1.

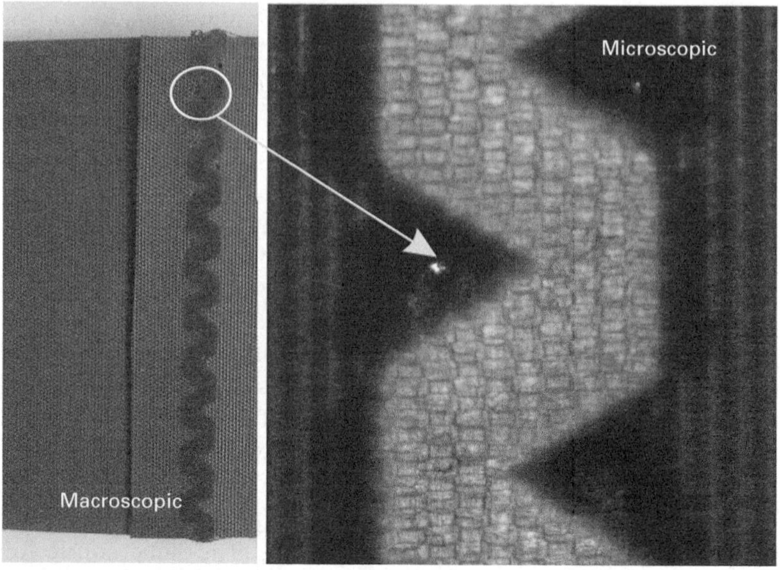

14.7 Macro- and micro-impermeability of ultrasonic weld.

14.8 Apparatus for determining seam permeability.

14.2.3 Seam mechanical properties

When it comes to making sure the final garments or textile products are fit for the end-use, seam mechanical properties, like strength, resistance to shear, elasticity etc., may become key performance factors. One of the major reasons for the consumer being dissatisfied with a garment, after poor colourfastness and dimensional stability, is when the seams in a garment or product fail due to seam breakdown. A seam should offer at least the same performance as the material used, otherwise the performance classification should be limited by the weakest element of the garment/textile product design. Having this in mind, as well as the fact that welded and bonded

Table 14.1 Penetration resistance data for Tyvek®

Material	Component	Chemical	Result 5 min*	Result 10 min**
Tyvek®	Fabric	Water	Pass	Pass
		Isooctane	Pass	Pass
		MEK	Pass	Pass
		TCE	Pass	Pass
	Stitched seam	Water	Fail	Fail
	Bonded seam	Water	Pass	Pass
		Isooctane	Pass	Pass
		MEK	Fail	Fail

Note: Based on overall 15-min exposure with *first 5 min at ambient pressure and **second 10 min at 13.8 kPa (ASTM, 2010). If liquid droplets are seen on the outside of the fabric, the material has failed the test.
MEK - methyl ethyl ketone; TCE - trichloroethylene.
Adapted from Berardinelli and Cottingham (1986).

seams have been used for years, mostly in high performance or extreme clothing and textile products (like spacesuits, protective clothing, medical textiles, architectural structures, sports clothing, etc.), it is surprising to note that the investigations of mechanical properties of such seams still present a challenge.

Most of the previous work was focused on ultrasonic welding of bulky polymers and thermal bonding of non-wovens (Shi & Little, 2000). The results of those studies can be summarized as follows.

- The highest weld strength of lap welds between polymers is attained at an intermediate value of weld force (Matsyuk & Bogdashevskii, 1960).
- Ultrasonic bonding strength of polyethylene increases within a range of welding time and welding force at first, and then decreases beyond that range (Mordvinteseva & Druzhinin, 1964).
- Ultrasonically welded polyethylene indicates that no new chemical group or bond is formed in the area of the seam (Mozgovoi *et al.*, 1968). The weld strength depends on the degree of fusion of the material at the interface for ultrasonically welded acrylonitrile-butadiene-styrene copolymer.
- Frankel and Wang (1980) studied the acrylonitrile-butadiene-styrene copolymer and recorded the highest weld strength in the case of longest weld time and lowest weld force, among nine different welding conditions, involving three levels of force and time.
- Ultrasonic welding of textile materials also depends on their thermoplastic content, fabricating type (woven or knit) and desired end results (Branson Ultrasonic Corp., 2009).
- In the work of Simmons *et al.* (1999), the effect of welding parameters

on seam strength of thermoplastic polyolefin roofing membranes was investigated, and the influence of surface texture of the membrane exhibited the highest impact among the optimal parameters.

- The main objective of the Shi and Little paper (2000) was to correlate the resulting bond strength with welding conditions and examine the nature of heat evolution in fabrics during ultrasonic welding. One of their conclusions was that increasing the weld pressure (up to 0.2 MPa) increased the weld strength. The tested specimens failed in three modes: fabric damage, bond damage and a combination of fabric and bond damage. The second mode was the most desirable because the seam was strong and flexible, and the yield point was observed in this mode.
- In the work of Vujasinovic *et al.* (2006b), the key result was that ultrasonic sailcloth bonding could successfully replace conventional sewing if an anvil wheel with appropriate engraving and optimal welding parameters (speed and amplitude) was selected. It not only increased bond strength in comparison to a conventional seam, but also provided sail air impermeability, being one of the basic aerodynamic requirements for sail making.
- Recent investigations by Cubric *et al.* (2010), performed using the Kawabata Evaluation System (KES FB system), indicate that the seam formed by ultrasound welding considerably impacts the mechanical properties of the welded polyurethane films. Significant changes have been established in deformation work and elongation as tensile loads for samples of ultrasound-welded films, while the changes of bending stiffness and shear rigidity are lower, compared to polyurethane film foil samples consisting of two layers before ultrasound welding. However, the investigations of shear resistance indicate that the weld created by ultrasound is rather stiff and prevents shear between the layers (Fig. 14.9).
- The work of Jakubčionienė and Masteikaitė (2010) analyses the strength of the textile hot bar bonded seams with respect to bonding temperature, pressure and duration. Optimal parameters of film transfer are established as 160°C and 10 seconds, while optimal parameters of layer bonding are 180°C and 30 seconds. In the case of film transfer temperature which is too low, the film soaks insufficiently into the lower layer of the specimen, whereas in the case where the temperature is too high it soaks excessively.

The end of the last century witnessed intensified research aimed at determining the impact of welding processing parameters on seam strength and behaviour under the conditions of mechanical stresses. This was partly due to the rapid development of new materials, development of machines, new applications for fabrics, but also to designing systems of objective measuring and

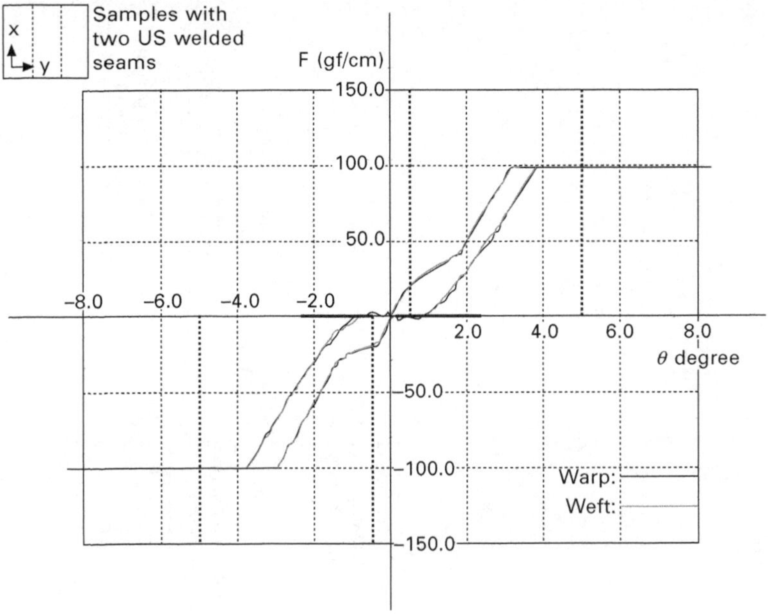

14.9 Shear deformation hysteresis for the tested sample, at the shear angle θ between ± 0 and 8°.

evaluation of textiles. Generally, every product has a characteristic number of elements which, taken together, describe and define its applicability. The properties describing these elements can de defined subjectively, but their numerical values can also be determined through objective measurements (Sommerville, 1998). The data obtained by these measurements can be used to predict product behaviour in end-use, to calculate the optimal behaviour, to compare the product with similar alternative products and to maintain and monitor quality. All of this can help to improve productivity, whilst at the same time reducing manufacturing costs. Apart from this, objective measurements offer important production data and can be used as guidelines for proper manufacturing, product and/or process design, with the aim of optimizing quality and ensuring a constant quality level of particular products of interest for the company (Humphries & Jackson, 1996).

To objectively evaluate any particular textile product manufactured from multiple parts, joined by a seam, it is necessary to include seam control in the process of measuring and evaluation (regardless of the technique of its making – conventional sewn seam, welded or bonded). A good illustration of this is the example of making sails. The attempt to establish whether an ultrasound-welded seam on a sail could replace a conventional sewn one resulted in a series of measurements involving ultrasound seam strength (Fig. 14.10; Vujasinovic *et al.*, 2006b).

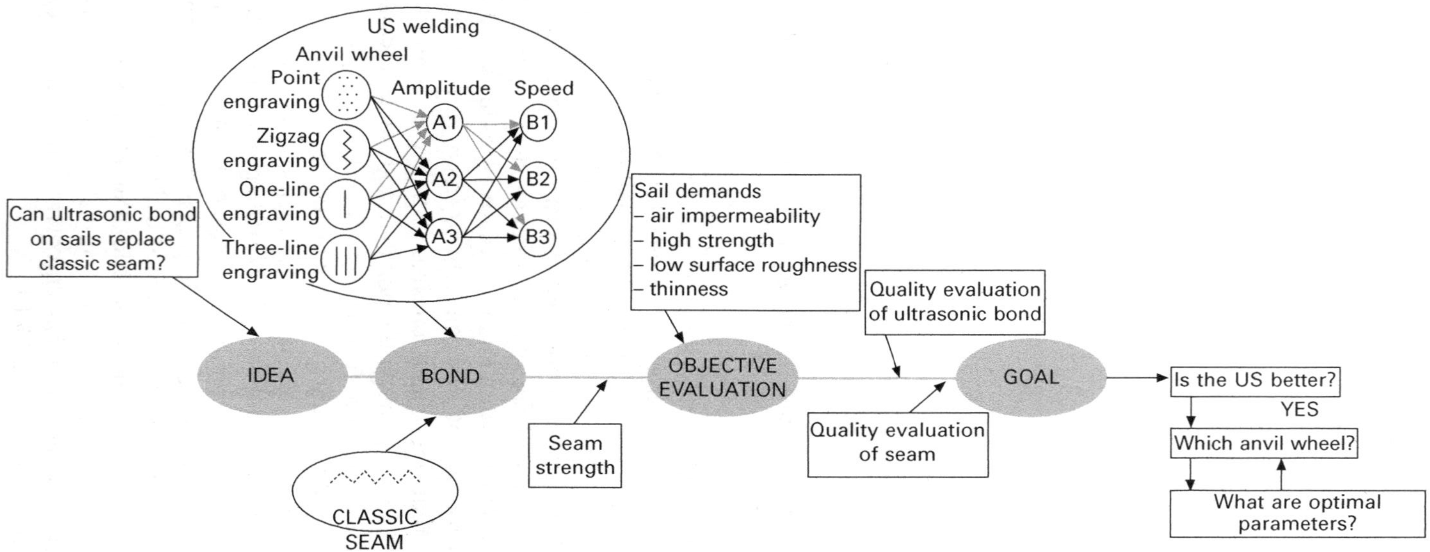

14.10 Objective quality evaluation of the sailcloth seam – research plan.

For the purposes of this work, PES Dacron® sailcloth samples were joined on the Pfaff 8310 Seamsonic ultrasonic welding machine, with a 400 W ultrasonic generator, frequency of 30 kHz, titanium sonotrodes 104 mm in diameter and a maximum weld width of 10 mm, welding speed ranging from 6 to 136 dmmin^{-1}, and welding pressure ranging from 0 to 800 N (5 bar), with the possibility of the amplitude and the gap between the sonotrode and anvil wheel to be regulated. The appearance of the ultrasonically welded seam, i.e. the impression of an engraved pattern, was determined by the shape of the anvil wheels (four of them, Fig. 14.11). Welding parameters were varied for each anvil wheel so that samples were joined at different welding speeds of 6, 18 and 30 dmmin^{-1}, and different amplitudes of 200, 304 and 400 W or 50, 76 and 100%. The gap between the sonotrode and the anvil wheel, as well as the pressure, were constant, amounting to 0.18 mm and 384 N (2.4 bar).

Tables 14.2 and 14.3 and Figs 14.12 to 14.15 show the results of determining sail seam resistance to tensile force applied. Figure 14.12 displays that the tensile strength of the ultrasonic bonds obtained by the use of the anvil wheel with point engraving (A) ranges from 0.44 to 31.77 Nmm^{-2}, whereby the tensile strength of the bond obtained at the medium amplitude (304 W; 76%) and welding speed (18 dmmin^{-1}) is the highest, whilst the lowest is the one of the bond obtained at the lowest amplitude (200 W; 50%) and the highest speed (30 dmmin^{-1}). In the sample bonded using the lowest welding speed (6 dmmin^{-1}), the tensile strength decreases with the rising amplitude, whereas in the samples bonded by using other speeds (18 and 30 dmmin^{-1}) the tensile strength of the bonds using medium amplitude (304 W; 76%) is the highest, with the 200 W (50%) amplitude the lowest. The speed affects bond strength most noticeably when bonding at the lowest amplitude, where the difference between the highest and the lowest value of tensile strength obtained using the highest and lowest welding speeds amounts to approximately 98.5%. With other applied amplitudes, the impact of the speed on bond strength is not so pronounced.

Figure 14.13 shows that tensile strength of the ultrasonic bonds obtained using the anvil wheel with a zigzag engraving (B) ranges from 2.31 to 13.48 Nmm^{-2}, whereby tensile strength of the bond obtained at the medium amplitude (304 W; 76%) and the highest welding speed (30 dmmin^{-1}) is the highest, the lowest being of the bond obtained at the highest amplitude (400 W; 100%) and the lowest speed (6 dmmin^{-1}). At the speeds of 6 and 18 dmmin^{-1} the inversely proportional dependence of the tensile strength of the amplitude is clearly seen; however, this is not the case at a speed of 30 dmmin^{-1}. In the samples bonded at the 304 W (76%) amplitude, a proportional dependence of tensile strength on the welding speed is pronounced, which is not the case for the lowest and highest amplitudes applied (200 and 400 W; 50 and 100%). It can be concluded that there is no general regularity,

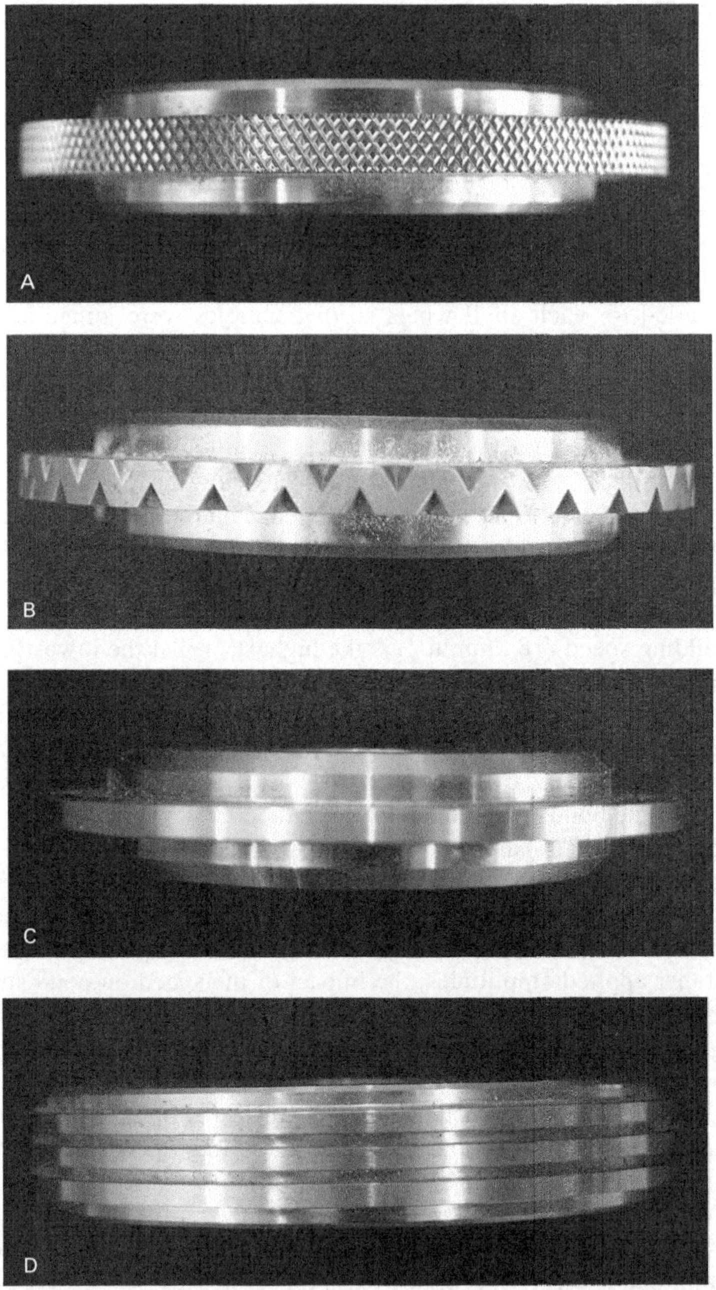

14.11 Geometry of the anvil wheels used: anvil wheel with point engraving (A), anvil wheel with zigzag engraving (B), anvil wheel with one-line engraving (C), and anvil wheel with three-line engraving (D).

Table 14.2 Resistance of the classic sail seam to tensile force

Sample K (classic sail seam)	Mean	CV (%)	PES Dacron® 3.7 us oz ($42.83\ gm^{-2}$)
Breaking force*, F_B (N)	629.59	0.95	360.00
Elongation, ε_B (%)	21.90	3.62	23.19
Strength, σ_B (Nmm^{-2})	13.66	0.87	165.06
Modulus of elasticity, E_0 (Nmm^{-2})	1.15	32.79	33.00

Note: PES sailcloth samples were joined on a special sewing machine Minerva-Boskovice class 525 especially designed for sail making (lengthened long-arm). The seam was made with a zigzag stitch, and a NM 100 needle and a polyester thread DABOND 2000 VVR (Heminway & Bartlett; V 92 Shade White) were used for sewing.
*EN ISO 13935-1:1999, ASTM D5035 - 06(2008)e1.

i.e. no clearly defined dependence of tensile strength on the amplitude and speed.

Figure 14.14 shows a parallel representation of the strengths for all the samples obtained by ultrasonic welding, using the anvil wheel with one-line engraving (C). The figure shows that the tensile strength of the ultrasonic bonds, obtained using the anvil wheel with one-line engraving, ranges from 0.71 to 6.47 Nmm^{-2}, whereby the tensile strength of the bond obtained at the lowest welding speed and amplitude (6 $dmmin^{-1}$ and 200 W; 50%) is highest, the lowest being the bond obtained at medium speed and amplitude (18 $dmmin^{-1}$ and 304 W; 76%). In the samples bonded at the lowest welding speed (6 $dmmin^{-1}$), tensile strength decreases with increased amplitude. If the values obtained for the tensile strength of the bond at the lowest amplitude of welding are analysed, it can be seen that a higher speed significantly reduces bond strength. These regularities are not valid for the other samples, i.e. at other welding speeds and amplitudes.

Figure 14.15 shows that the tensile strength of the ultrasonic bonds obtained by using the anvil wheel with three-line engraving (D) ranges from 0.52 to 17.34 Nmm^{-2}, which corresponds to the range of values recorded at different amplitudes at a speed of 30 $dmmin^{-1}$. For other speeds (6 and, 18 $dmmin^{-1}$), it is equally true that bond strength decreases with higher amplitude. If the values of the bond strength are observed at a constant amplitude, bond strength only increases with a higher speed of welding at 200 W (76%) amplitude.

The results shown in Fig. 14.16 indicate that ultrasonic welding, together with the selection of optimal parameters (speed and amplitude) can successfully replace conventional sewing in sail manufacture. Namely, the strength of the ultrasonic bond is, as a rule, higher than the strength of conventional sewn seams for sails (for the anvil wheel A up to 130%), which certainly

Table 14.3 Resistance of US-welded seam to tensile force (sample A*)

Sample A

A_1	A_2	A_3
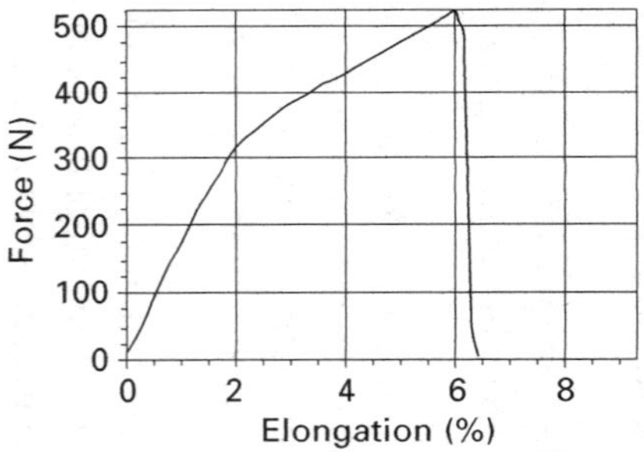	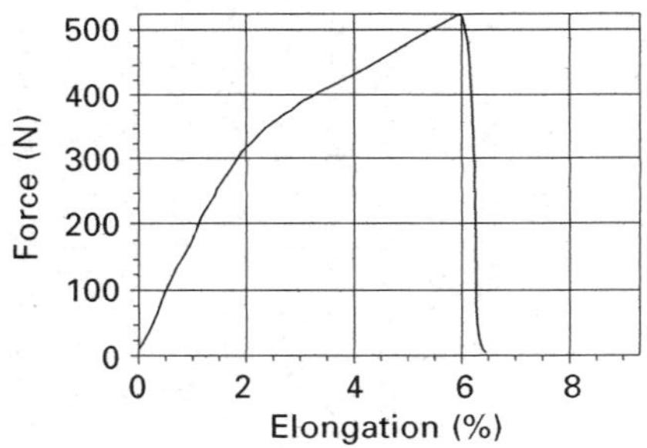	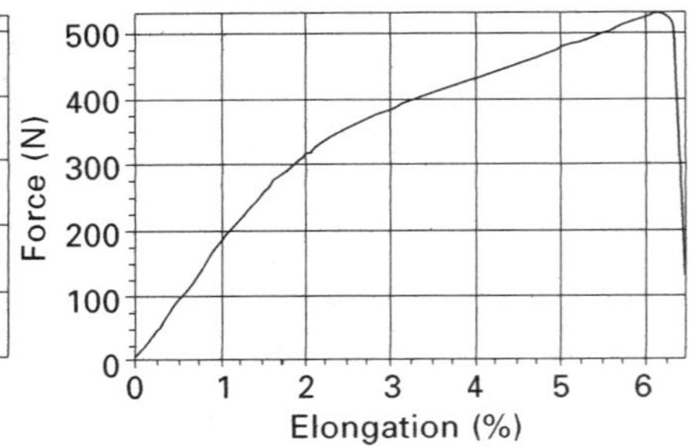

Breaking force; F_B				
x_m (N)	x_{min} (N)	x_{max} (N)	CV (%)	P_{le} (N)
532.17	525.64	539.37	1.29	± 0.24
Breaking elongation; ε_B				
x_m (%)	x_{min} (%)	x_{max} (%)	CV (%)	P_{le} (%)
6.04	5.98	6.14	1.48	± 0.00
Strength; σ_B				
x_m [Nmm^{-2})	x_{min} (Nmm^{-2})	x_{max} (Nmm^{-2})	CV (%)	P_{le} (Nmm^{-2})
30.13	29.78	30.51	1.22	± 0.01
Elasticity module; E_0				
x_m (Nmm^{-2})	x_{min} (Nmm^{-2})	x_{max} (Nmm^{-2})	CV (%)	P_{le} (Nmm^{-2})
10.36	10.16	10.49	1.68	± 0.01

* Sample A means ultrasonically welded seam with the use of anvil wheel with point engraving, designates velocity v = 6 $dmmin^{-1}$ and designates amplitude A = 200 W (50%).

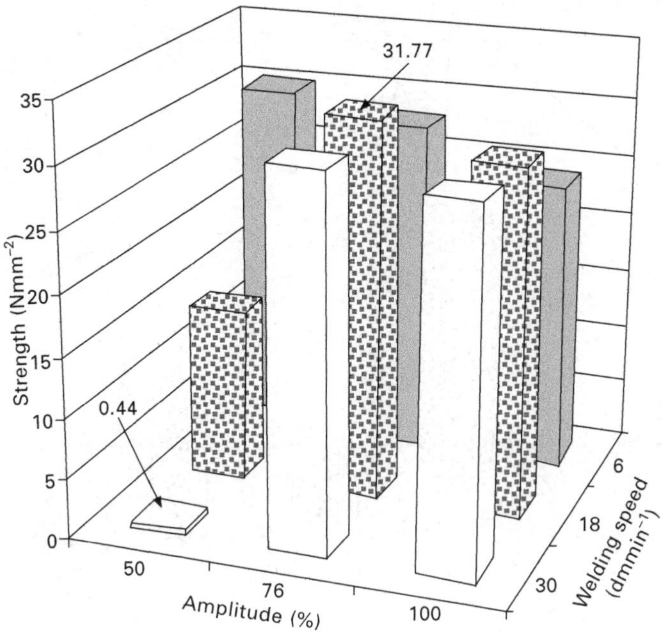

14.12 Tensile strength of the ultrasonic bonds obtained by the use of the anvil wheel with a point engraving (A).

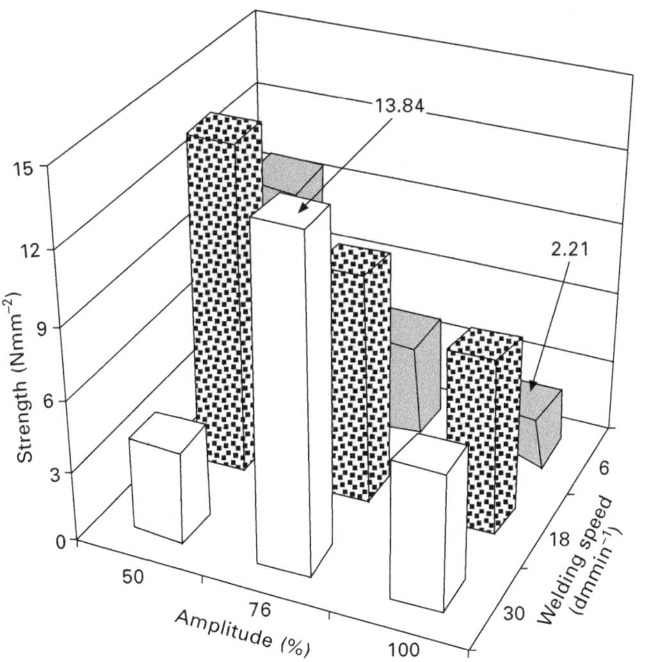

14.13 Tensile strength of the ultrasonic bonds obtained by use of the anvil wheel with a zigzag engraving (B).

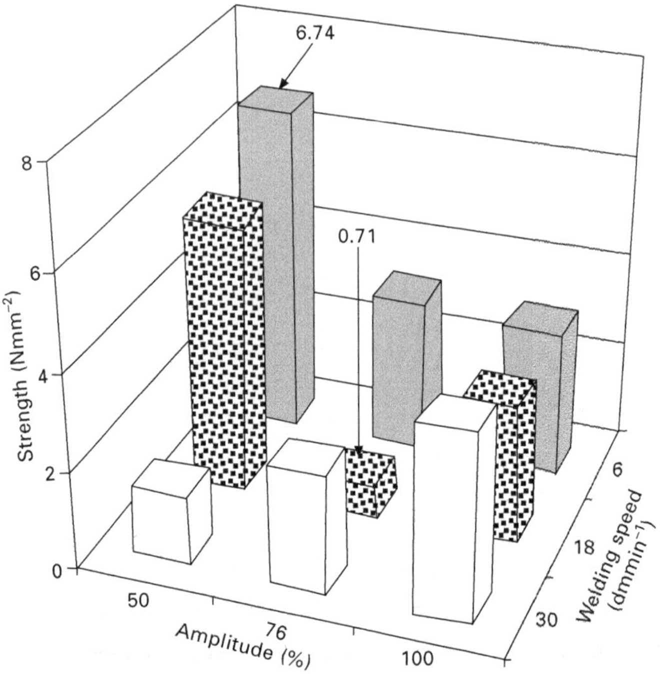

14.14 Tensile strength of the ultrasonic bonds obtained by the use of the anvil wheel with a one-line engraving (C).

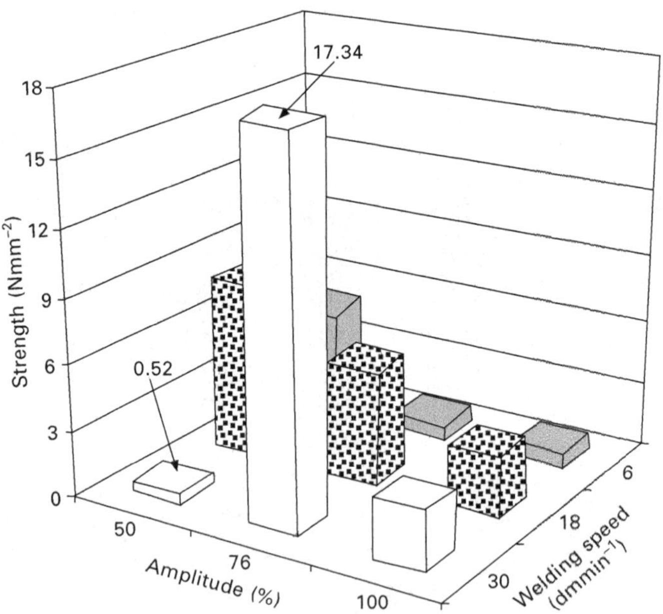

14.15 Tensile strength of the ultrasonic bonds obtained by the use of the anvil wheel with three-line engraving (D).

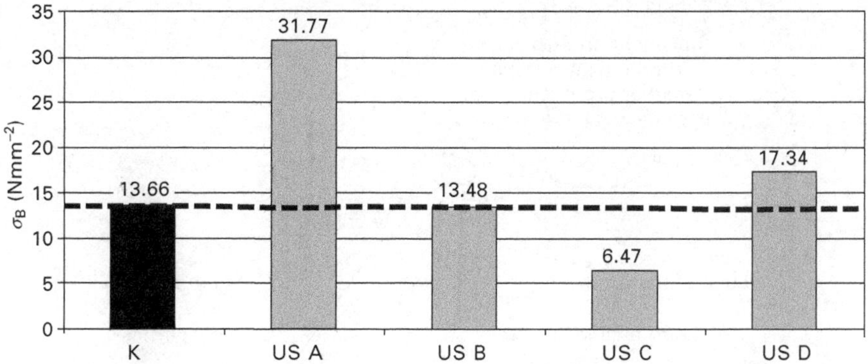

14.16 Parallel representation of the strengths for all the optimized sail samples (K – classical seam; US A – ultrasonically welded seam, anvil wheel with point engraving; US B – ultrasonically welded seam, anvil wheel with zigzag engraving; US C – ultrasonically welded seam, anvil wheel with one-line engraving; US D – ultrasonically welded seam, anvil wheel with three-line engraving).

contributes to safe sailing. Initial modulus is also higher, as well as the limit of aerodynamic efficiency of the designed sail shape, which is significantly augmented by a lower elongation at break.

An ultrasonic sail weld is approximately two-thirds thinner than a classic seam. It is less rough and, if welding conditions are optimally selected, it is completely air impermeable. Although sail strength represents a highly important element for the objective evaluation of sails from the point of view of safety and application (wind strength), the sail should also have certain aerodynamic properties to fulfil its basic function. The sail as a whole, as well as all of its parts, together with sailcloth weld (seam), should have its thickness and roughness reduced as much as possible.

14.3 Quality evaluation of welded or bonded seams

Quality evaluation includes performance, conditions and properties. If a seam is to be assessed in this light, a detailed requirement profile is needed. However, there are some indicators that are true for almost all seams (Fig. 14.17). Their evaluation is a prerequisite for assessing their quality, so every seam analysis should begin with the checking of these properties. Yet, different seams have different requirements for their end-use. Airbag seams, jeans seams, or upholstery seams – they all demand different properties of the seam joint and thus have their individual requirement profiles (Fig. 14.10). They must be worked out in accordance with the application and function.

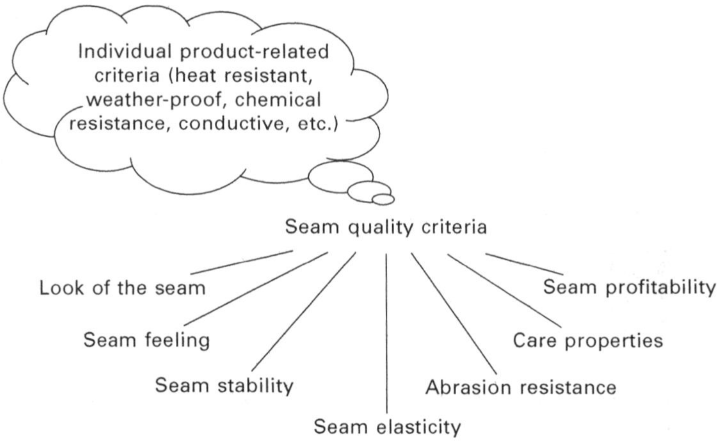

14.17 Seam quality descriptors.

14.4 Conclusion

The investigations performed on joining thermoinsulative chambers of intelligent clothing with adaptive technical protection (First Rogale *et al.*, 2009/10; Rogale *et al.*, 2007) and the investigations on joining sail-making fabrics (Vujasinovic *et al.*, 2006a; 2007b), both organized at the Faculty of Textile Technology, University of Zagreb, have concluded, without any doubt, that the seams made using the above mentioned and described joining techniques are most seriously influenced by the uneven energetic effect in treating the seam. All contemporary machines have a generator of appropriate energy as an energy source, which, in principle, has a fixed value. The power of the energy source is generally set before the operation of welding and stays such for the duration of making a number of seams. Stabilizers are often used to keep the set power constant, even in the case of significant variations in some other parameters (tension in the power grid, changes of material thickness, etc.), which can have quite detrimental effects.

Our investigations have established that using a uniform level of welding energy for the whole length of the welded part is of utmost importance. It is of extreme importance in order to achieve uniform energy distribution density at each segment of the seam made. This requirement is rather easy to meet when welding using HF energy, as the process takes place in a stationary state. The material to be joined and the HF electrode rest in relation to each other. Unfortunately, the other types of welding are of a dynamic nature, as the workpiece travels towards fixed machine parts. At the moment when joining begins, the workpiece rests, to be accelerated until it reaches the nominal shear speed. The workpiece then moves at the nominal shear speed for some time (depending on seam length), to be decelerated by the end of joining to rest again when the weld is completed. A constant amount of welding energy is brought to the workpiece and its weld, meaning that it is too high for the needs of the weld at the beginning, optimal when the workpiece moves at nominal shear speed, and then again too high in the deceleration stage up to the moment the workpiece comes to rest again. It can be concluded that linear density of the energy supplied, depending on the seam length, is not uniform. An uneven and exaggerated amount of energy in the weld at the initial and final stages of joining most often results in overheating and softening of the material that is joined, which is reflected as the weld is too thin in these areas. There are also visual imperfections and a considerably reduced weld strength in these areas, with possible vertical and horizontal porosity of the seam.

The technologists dealing with joining fabrics employing the above-mentioned contemporary methods are quite often faced with the only solution of finding an optimum of all the seam properties described in the predominant part of the seam, which includes the manner of machine work at the optimal shear speed. It is generally impossible to influence welding energy distribution in the stages of accelerating and decelerating the workpiece, which means that development in this area, of optimizing uniform welding energy supply to the seam in the accelerating and decelerating stages, is a clear direction of future research, as the aim is to keep the linear density of the energy supplied constant on each segment of the weld.

It is thus reasonable to expect that future development will be primarily aimed at studying the interactions of workpiece movement speed and the amount of the energy used, so as to create uniform linear density of the welding energy, to be attained at each segment of the weld produced. For this reason, the machines for welding using modern methods should be equipped with additional microcontroller systems for equalizing the welding energy, as well as with sensors to monitor workpiece movement speed. The control system of the microcontroller device should be supplied with the signal of the workpiece movement speed and the signal of the power generator output power (ultrasonic power, thermal power, the power of IR laser, etc.). The microcontroller should compare the speed and the power supplied. If the workpiece movement speed is low (e.g., in the workpiece acceleration and deceleration stages, at the beginning and the end of the weld), the output power should also be low. As the speed rises, so should the output welding power. In this way the power could be kept constant at each segment or part of the seam. The generator of the welding power should be equipped with an additional control input to be used to regulate the output power. The microcontroller device could, based on a quite simple algorithm, efficiently regulate uniform welding energy supply, which would result in a uniform and high-quality weld throughout its length, avoiding intensively molten parts at the beginning and end of the join, as well as on the spots of skipping the joining process in order to align the workpiece when manual handling of the workpiece is employed.

Future investigation in the area of high-temperature welding will be aimed at the thermodynamic phenomena of heat conduction dynamics in polymers, the ratio of contact heat transfer from metal wedge, from hot air stream or a similar tool onto the polymeric material, the rate of its movement, as well as the force parameters of the pressing roller. Investigations in the area of ultrasound welding will involve defining optimal ultrasound frequencies (the most often encountered range of application is between 18 and 42 kHz), power, or sonotrode tip vibration amplitude, force of sonotrode and counter roller pressure onto the welded material, material movement speed under the sonotrode and duration of welding, type and chemical composition of the fabrics used, as well as their thermal properties. Some of these investigations are already under way.

Investigations in the area of joining polymeric materials employing HF electromagnetic field technology will be also aimed at ensuring uniform welding energy supply. As already mentioned, the workpiece remains stationary in this type of welding, meaning the problem is not the same as in the other welding techniques. Uneven welding energy supply occurs when the material to be welded softens and the HF electrode impresses more deeply into the material, due to the constant pressure of pneumatic cylinders. All the key welding parameters are changed in this instance. The electrode, the material to be welded (possessing a particular dielectric constant) and the grounded counter-electrode initially constitute a plate condenser. The elevated temperature of the material changes its dielectric constant, while the electrode pressed into the material reduces the distance between the electrodes, which increases the capacity of the condenser. As a result of this, considerable change occurs in the impedance load of the high-frequency generator and the energy supplied to the polymeric material weld. The change is dynamically quite rapid, occurring in a fraction of a second, and is manifested as an enormous increase of high-frequency current, which is difficult to control and causes additional heat-up and possibly an avalanche effect.

Investigations are expected to continue in this area concerning a number of key parameters associated with optimal frequency of treating macromolecular welded polymer, with the level of change of polymeric material dielectric constant with temperature variations, with the type of treatment, its duration and the level of pressure exerted on the HF electrode, as well as with establishing a mathematical model of the changes. In the course of further development of the HF welding machine, it will be necessary to regulate additional welding power and HF electrode pressure, especially at the moment the material softens and the HF electrode starts sinking into it. Experimentally obtained parameters should be used at this moment and, depending upon the dynamics of the changes, the HF generator power and HF electrode pressure should be reduced on the treated weld.

Innovation and continued development of joining and surface technologies is key to achieving a competitive advantage in future manufacturing. Such new technologies can, in some cases, completely revolutionize manufacturing sectors. Other than this, improvements are always sought to meet the increasing industry demands for reduced costs and weight, better environmental performance and improved reliability, to name but a few.

Technologists need to ensure that their technologies reach the market as quickly as possible. Working with customers and developing mutually attractive intellectual property are essential steps which are necessary for this to happen. Some institutions of high reputation have offered good examples of this, and continue to be innovative, in order to supply new ideas to meet current and future market demands.

14.5 References

ASTM (2000), *ASTM D123-00b Standard Terminology Relating to Textiles*, Philadelphia, PA: American Society for Testing and Materials.

ASTM (2010), *ASTM F903-10 Standard Test Method for Resistance of Materials Used in Protective Clothing to Penetration by Liquids*, Philadelphia, PA: American Society for Testing and Materials.

Babic B, Jasarevic I, Kvasnicka P, Prager A, Schwabe Z and Smetin V (1995), *Geosintetici u graditeljstvu*, Zagreb: HDGI.

Berardinelli S P and Cottingham L (1986), *Evaluation of chemical protective garment seams and closures for resistance to liquid penetration*, Performance of Protective Clothing, ASTM STP 900, Philadelphia, PA: American Society for Testing and Materials.

Branson Ultrasonic Corp. (2009), *Ultrasonic Welding Characteristics of Textiles and Fibres*, Danbury: Branson Ultrasonic Corp.

Cubric G, Gersak J, Rogale D and Nikolic G (2010), Odredivanje mehanickih svojstava ultrazvucno spojenih poliuretanskih folija, *Tekstil*, 59, 64–70.

Fan J and Liu F (2000), Objective evaluation of garment seams using 3D laser scanning technology, *Textile Research Journal*, 70, 1025–1030.

First Rogale S, Rogale D, Nikolic G and Dragcevic Z (2009/10), *Controllable ribbed thermoinsulative chamber of continually adjustable thickness and its application*, Patent No. PCT/HR2009/000008 (WO2009115851); US 12/922, 761–2010.

Frank K K (1987), Seaming and joining methods, *Geotextiles and Geomembranes*, 6, 93–107.

Frankel E J and Wang K K (1980), Energy transfer and bond strength in ultrasonic welding of thermoplastics, *Polymer Engineering and Science*, 20, 396–401.

Humphries W and Jackson N (1996), Objective measurement of style, *Proceedings of Top-Tech'96 Symposium*, 11–14 November, Geelong, Australia: CSIRO, pp. 305–331.

Hustedt M, Stein J, Herzog D and Meier O (2008), Laser-based joining of technical textiles for airbag production, *Third World Automotive Congress Plastics-in-Motion*, 14–16 May, Prague, Czech Republic, pp. 1–11.

ISO (1999), *ISO 13935-1:1999 Textiles – Seam tensile properties of fabrics and made-up textile articles – Part 1: Determination of maximum force to seam rupture using the strip method.*

Jakubčionienė Ž and Masteikaitė V (2010), Investigation of textile bonded seams, *Materials Science (MEDŽIAGOTYRA)*, 16, 76–79.

Jones I (2005), Improving productivity and quality with laser seaming of fabrics, *Technical Textiles International*, 5, 35–38.

Jones I A and Wise R J (2005), Novel joining methods applicable to textiles and smart garments, Papers presented at Wearable Futures Conference, 14–16 September, Newport, Wales, University of Wales, Available from: http://www.twi.co.uk/content/spiajsept2005.html (accessed 5 July 2010).

Kang T J (2005), Fabric surface roughness evaluation using wavelet-fractal method – Part I: Wrinkle, smoothness and seam pucker, *Textile Research Journal*, 75, 751–760.

Luders G (2000), Quality assurance in hot wedge welding of HDPE geomembranes, *Proceedings of EuroGeo 2*, Bologna, Italy, 591–596.

Matsyuk L N and Bogdashevskii A V (1960), Ultrasonic welding of polymeric materials, *Soviet Plastics*, 2, 70–76.

McQuaid M (2005), *Extreme Textiles*, London: Thames & Hudson.

Mordvinteseva A V and Druzhinin N V (1964), Mechanism of bonding of plastics by ultrasonic welding, *Soviet Plastics*, 9, 34–39.

Mozgovoi I V, Konstantinopol'skaya M B, Berestneva Z Y, Kargin V A and Nikolaev G A (1968), Mechanism of bonding of plastics by ultrasonic welding, *Soviet Plastics*, 10, 51–54.

NASA (2009), *Space Age Swimsuit Reduces Drag, Breaks Records*, NASA Tech Briefs, Available from: http://www.sti.nasa.gov/tto/Spinoff2008/ch_4.html (accessed 28 April 2009).

O'Mahony M and Braddock S E (2002), *Sportstech – Revolutionary Fabric, Fashion & Design*, London: Thames & Hudson.

Racheel M (1995), *Modern Textile Characterization Methods*, New York: Marcel Dekker.

Rogale D and Dragcevic Z (2002), Tehnike konfekcioniranja tehnickog tekstila, *Tekstil*, 51, 64–77.

Rogale D, First Rogale S, Dragcevic Z and Nikolic G (2007), Inteligentni odjevni predmet s aktivnom termickom zastitom, Patent No. PK20030727.

Saville B P (2000), *Physical Testing of Textiles*, Boca Raton, FL: CRC Press.

Seymour S (2008), *Fashionable Technology – The Intersection of Design, Fashion, Science, and Technology*, Wien, Austria: Springer-Verlag/Wien.

Shi W and Little T (2000), Mechanisms of ultrasonic joining of textile materials, *International Journal of Clothing Science and Technology*, 12, 331–350.

Simmons T R, Runyan D, Liu K K Y, Paroli R M, Delgado A H and Irwin J D (1999), Effects of welding parameters on seam strength of thermoplastic polyolefin (TPO) roofing membranes, *The North American Conference on Roofing Technology*, 16–17 September, Toronto, Ontario, Canada, 56–65.

Sommerville P (1998), Objective Measurements – more than pretty numbers, *Proceedings of Australian Wool Testing Authority Seminar – Working with Wool*, 15 September, Canberra, Australia, 1–16.

Tao X (2001), *Smart Fibres, Fabrics and Clothing*, Cambridge: Woodhead Publishing.

Tucker R and Dugas J (2008), The lighting of the flame; but politics, and Speedo's 'Supersuit' are making waves instead, *The Science of Sport*, Available from: http://

www.sportsscientists.com/2008/03/beijing-olympic-torch-lit.html (accessed 3 June 2010).

Vujasinovic E, Jankovic Z, Dragcevic Z, Petrunic I and Rogale D (2006a), Investigation of Strength of Ultrasonically Welded Sails, *Book of Proceedings of 3rd International Textile, Clothing and Design Conference*, 8–11 October, Dubrovnik, Croatia, 510–516.

Vujasinovic E, Jankovic Z and Dragcevic Z (2006b), Ultrasonic welding – new method in sailmaking, *Annals of DAAAM for 2006 and Proceedings of The 17th International DAAAM Symposium Intelligent Manufacturing & Automation: Focus on Mechatronics & Robotics*, 8–11 November, Vienna, DAAAM International Vienna, 445–446.

Vujasinovic E, Dragcevic Z and Mucnjak N (2007a), Intensity of light as a measure of surface roughness, *Book of Proceedings of 5th IMCEP – Inovation and Modelling of Clothing Engineering Processes*, Moravske Toplice, Slovenia, 10–12 October, 169–173.

Vujasinovic E, Jankovic Z, Dragcevic Z, Pertunic I and Rogale D (2007b), Investigation of the strength of ultrasonically welded sails, *International Journal of Clothing Science and Technology*, 19, 204–214.

Zavec Pavlinic D, Gersak J, Demar J and Bratko I (2006), Predicting seam appearance quality, *Textile Research Journal*, 76, 235–242.

Part IV

Applications of joining textiles

15
The appearance of seams in non-iron shirts

G. K. STYLIOS, Heriot-Watt University, UK

DOI: 10.1533/9780857093967.4.467

Abstract: Non-iron shirts by definition should be iron free. Although wrinkle-free fabrics have become commonplace, flat seams in non-iron shirts still remain a challenge. This practical discussion generated by the author's own research and experience highlights how an armhole seam in a shirt is joined by a lap or French seam and how the insertion of a narrow bonding tape can alter the bending stiffness of the assembly, hence eliminating any seam puckering even in the most limp of fabric shirts.

Key words: non-iron shirts, seam pucker, lap seam, French seam, interlining, bonding tape, armhole seam, fabric stiffness, wrinkle-free fabric.

15.1 Introduction

This is a practical discussion on an important, perhaps the most important, problem associated with commercial garment manufacture; that of seam pucker. It is focussed on non-iron shirts because by definition seam pucker, creasing and buckling along the seams are regarded as faults of bad appearance, and are rejected by customers. This discussion is aimed at the very practical level of how to make non-iron shirts so that they do not possess creased, puckered or buckled seams, especially around the armhole sleeve area which is regarded as difficult to control because it is trying to take up the three-dimensional configuration of the arm and hence it is more demanding than other parts of a shirt. Therefore the question is how to eliminate seam pucker, ropiness, creasing or buckling in shirts during sewing and/or after laundering. The discussion will state some technical facts about wrinkle-free fabrics and interlinings leading to how these affect the selection of appropriate seams in the stitching process, underpinned by the mechanism of seam pucker and an understanding of the equilibrium state of a seam, as already discussed in Chapter 2.

15.2 Wrinkle-free fabrics

After many years of research and development, 'wrinkle-free' fabrics that do not need to be ironed, even after a garment is washed, are now commonplace. The earliest attempts to make these types of fabric started in the 1960s, by coating the garment with a synthetic resin based on urea-formaldehyde polymers, but they were not especially successful and fairly soon disappeared. One of the reasons was the fact that the synthetic resin used tended to depolymerise to a certain extent when the treated garment was cured in an oven, which gave rise to the release of formaldehyde, which is carcinogenic. Another problem was that the bonding of the resin to cotton was not good, so it tended to wear off after a short time, in addition to which the treatment weakened the cotton fabric. Moreover, the wrinkle-free fabrics made at that time were expensive.

A new range of wrinkle-free fabrics started to appear commercially in early 1994 after decades of research and development. The solution that was arrived at was the pre-treatment of the cloth with liquid ammonia to improve its strength and soften the handle, followed by impregnation of the cloth with a partially crosslinked resin which was cured by heat later. The first of these fabrics came from Japan, by the Japanese textile manufacturer Nisshinbo Textile Company. Hence a wrinkle-free fabric was now available, suitable for making shirts that did not possess the problems that previous wrinkle-free fabrics had in the past.

At first only one or two of these fabrics were available, but within a few years the number of them on the market mushroomed, and soon these fabrics were being widely used. The new wrinkle-free fabrics did still have some disadvantages compared to normal fabrics. They were more expensive, of course, and one of the major difficulties with them was that the resin coating had to be cured after the garment had been made up. This meant two things. First, stocks of the fabrics had to be kept refrigerated prior to use so that the curing process did not take place during storage, and secondly, the garments made with these fabrics had to be cured in an oven after they had been made and before they were packaged. The heat completed the crosslinking of the resin coating, thereby imparting the crispness that was required to keep the material smooth and flat, so that it looked as though it had been ironed.

15.3 Non-iron shirts and seam pucker

Once wrinkle-free fabrics became commercially viable and they started to become available, it meant that it became commercially worthwhile to address puckering because manufacturers could no longer rely upon the ironing process to remove any seam pucker. Well-known techniques for making pucker-free seams, include (a) balancing the seams and keeping the top and bottom thread tensions to a minimum, (b) the use of non-stretchable sewing threads, so that they do not shrink after sewing and laundering, (c) using pullers after the stitch line and/or differential feeding of the fabric so that the fabric was being stretched during sewing, and (d) the insertion of fusible interlinings to reduce the tendency of the fabric to buckle and for local stiffening. In relation to the technique of using fusible interlinings, the materials are inserted into various parts of garments, which are then pressed, causing the interlining (or the fusible part of it) to melt, and bond to one or both pieces of the garment fabric with which it is in contact. The effect of this is to reduce buckling of the fabric by bonding its fibres together flat and bonding the pieces of fabric to each other, thus stiffening it and overcoming pucker.

According to the mechanism of seam pucker discussed earlier in Chapter 2, the most obvious way to improve bending stiffness along the seam line of the fabric is by introducing stiffening along the seam line. This can be achieved by an interlining or by bonding the fabric layers locally with a chemical stiffening agent. Many trials using different linings (thermoforming, non-woven, adhesive) and incorporating them into seams during sewing of fabric samples with severe tendency to pucker have revealed their suitability and effectiveness. The effectiveness of the technique was investigated during the work of the author with Marks and Spencer and their suppliers between 1982 and 1994. Since then, fusible and non-fusible interlinings started to be used in all types of seams.

In most shirts, cuffs, collars and front plackets are stiffened by an interlining or by fusing, for improving their appearance. Around the armhole, however, there were still problems with the increasing demand of lightweight shirting fabric which, when laundered, with the contraction of the sewing thread made the pucker worse. Although armhole seams are not considered the most visible part of a shirt and despite the fact that fusing loses its stretching comfort, most non-iron shirts are now made by stiffening and fusing applied around the armhole seams of the shirt using folders (Fig. 15.1), and tape chutes for reducing standard minutes. Loose design fashion trends allow for the alleged loss of stretching comfort.

15.1 A lap seam fabric feeding folder.

15.4 Interlinings as sewing aids

Fusible materials, in the form of narrow strips or tapes are nowadays known and widely used to insert into many parts of numerous types of garments. Fusible materials, such as fusible adhesive plastics and interlinings coated with adhesive, are being used in mass produced garment construction. They come in many different forms such as webs of randomly oriented fibres (non-woven), meshes or tapes of thermoplastic polymer, or fabric tapes with polymer adhesive applied to one or both sides (either plain or in patterns). In general, a lighter weight fusible is used if the fabric from which the garment is made is lighter weight and hence prone to deformation, and for very light materials, such as cotton poplin or satin, a thin fusible polymer film is known to offer benefits over a woven or non-woven interlining that had been coated with a fusible adhesive. These fusible polymer webs or films are thermoplastics and act as adhesives and, once sewn into a seam, they are ironed or pressed under heat and pressure to bond together the fibres of one or more pieces of fabric, and to possibly bond together the two pieces of fabric in the seam between which they are placed. Consequently, this makes the fabric less likely to buckle, it increases its stiffness locally and prevents puckering of seams.

The weight of fabrics and fusible interlinings is measured in grams per square metre (gm^2) throughout the industry. The polymer fusibles are provided in a range of densities by their suppliers. Typical weights of fusibles are in the range of about 45–100 gm^2. Better still, a technologist should choose from a range of densities and, with a little experimentation, identify the one with the density which is most suitable for the fabric type used in the garment and on the design of the garment. A technologist should decide by measuring the fabric mechanics, particularly fabric bending stiffness and use compliance charts. This applies whether you are using a web, mesh or tape of thermoplastic polymer.

The fusible materials used are often polyester or polyolefin which have fusing temperatures ranging between 80 and 120°C. Polyamides are also used for certain purposes, but less often because their melting points may be considerably higher. The fusing is done using a heated press, which has the same effect as a hot iron, but is of course faster and more convenient for a mass production process. The temperature, pressure and time of pressing are adjusted according to the melting point of the particular fusible that is being used, and the type of fabric from which the garment is made. The technique of using fusible interlinings is also used in collars, cuffs and front pockets of shirts because these are even more visible than armhole sleeves when shirts are sold in transparent packaging and during wear.

The Vilene Company, now part of the Freudenberg Group, is one of the main suppliers of interlinings to the industry. In the early 1980s Vilene manufactured fusible interlinings which were web-like materials made of a thermoplastic cut to various widths that were entirely made of randomly oriented thermoplastic fibres. I experimented with these materials from Vilene during the 1980s (in order to investigate the mechanism of seam pucker and its prevention). Such materials come in at least three qualities depending on their stiffness/weight (light, medium and heavy) for stiffening lightweight fabrics prior to stitching them. A sample of the Vilene that I used in 1982 in investigating seam pucker in microfibre fabrics is shown in Fig. 15.2. The photograph shows one continuous seam which appears partly buckled when no interlining is between the two fabrics and partly flat with the interlining inserted between the two fabrics. This photograph became like a 'mascot' across Marks and Spencer Plc and their suppliers in the 1980s and early 1990s, when explaining in many technical seminars the mechanism of seam pucker that I put forward, and the importance of fabric stiffness.

15.5 The stitching of non-iron shirts

Both the lap seam and the French seam can be used for the stitching of the armholes of sleeves to the garment body of a shirt (Fig. 15.3). The French seam is really a particular type of lap seam. In both types of seam the edges of both pieces of fabric that are joined by the seam are folded inside the seam for neatness, but the difference is that in a true lap seam the folded-over edges overlap each other (the folded-over part of the lower fabric going over the folded-over part of the upper fabric), whereas in a French seam the two folded-over edges are shorter, and just meet each other in the middle of the

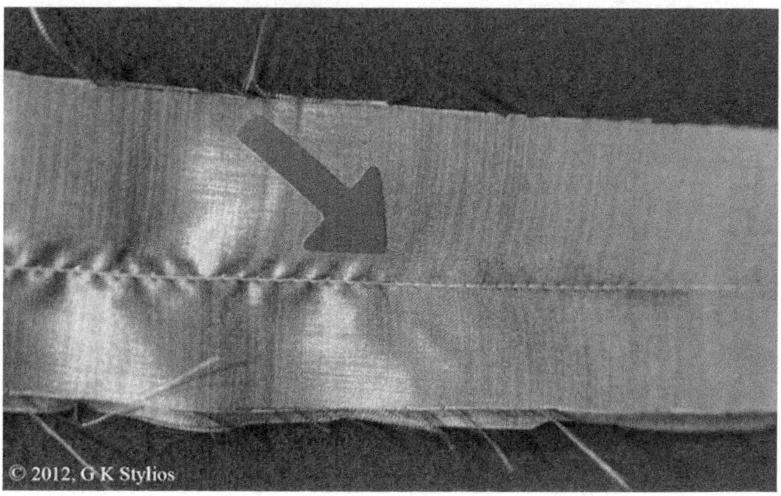

15.2 Elimination of seam pucker by attaching an interlining along the seam between the upper and lower fabric.

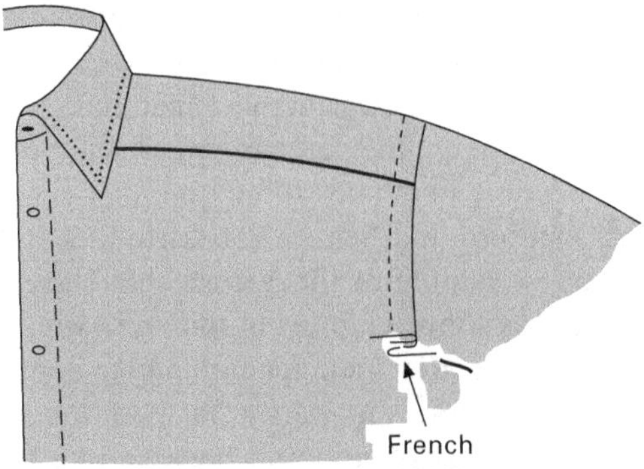

15.3 A French seam of the armhole of a shirt.

seam. The French seam is now more commonly used for armhole sleeves, because it is flatter, despite not being as strong as the lap seam. The most common way of making an armhole seam on a good quality shirt is to lay the sleeve fabric flat on the sewing machine table, feed it into a 'folder', lay the body fabric (which was made up of a front panel, yoke and rear panel) on top and use the folder to fold both fabrics into the configuration shown in Fig. 15.4. This configuration assembly is then fed through a twin needle sewing machine making a chain stitch. It is more common to use a twin needle machine, because it allows the stitching to be carried out faster, and it makes a more stable seam by holding the stitched components together.

A tape chute is a guide that is used for inserting a stiffening tape into a seam. The tape is usually held on a reel, and the end of it is fed through the chute to the seam into which it is to be inserted. The tape is pulled off the reel and through the chute by the operation of the sewing machine, in particular by the presser foot, pulling the pieces of fabric along. A tape chute, when it is needed, could be fastened onto any suitable part of the sewing machine, but a folder or the back of the presser foot are obvious places at which to attach it for that purpose. This will allow the insertion of the bonding tape in the seam, shown in Fig. 15.4, locally stiffening and/or bonding the seam and hence overcoming seam pucker. Figure 15.5 shows an unpicked French seam showing the insertion of the bonding tape.

As a general rule, during the construction of the armhole seam of a shirt, either the lap seam or the French seam, shown in Figs 15.6 and 15.7 respectively, incorporates a bonding tape which eliminates any deformation during sewing or after laundering, and bonding is performed during heat setting which is regarded as an additional operation.

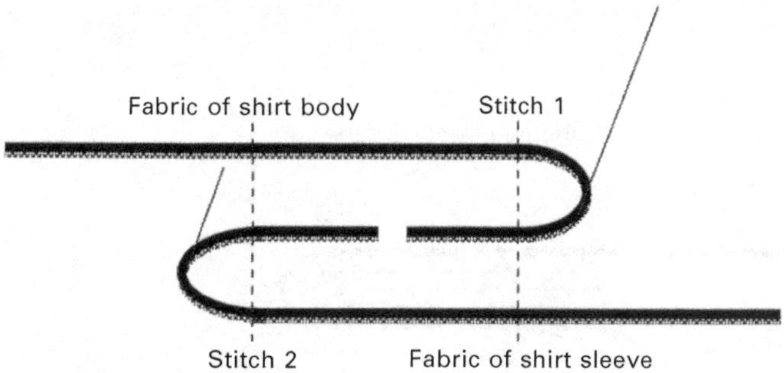

15.4 The configuration of a French seam.

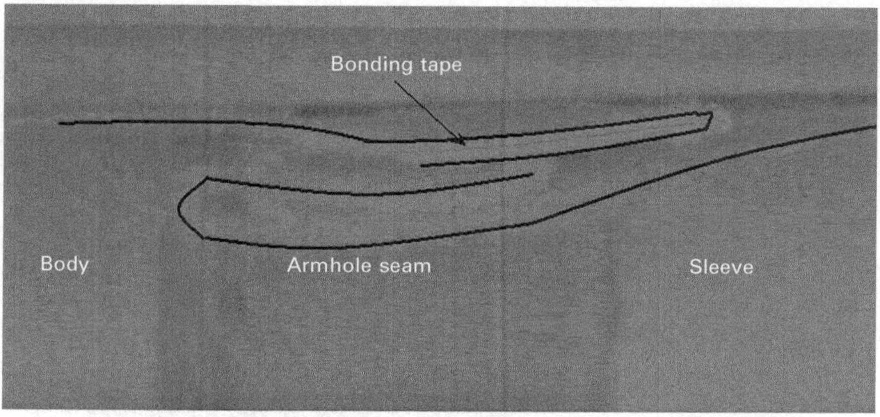

15.5 An unpicked French seam showing the insertion of the bonding tape.

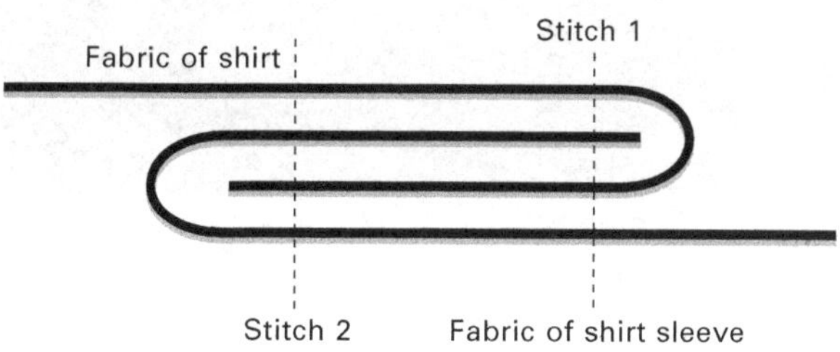

15.6 A lap seam joining the sleeve to the body of a shirt.

15.6 Discussion and conclusion

During my investigations, as early as the 1980s Marks and Spencer garment suppliers started to improve the appearance of their shirts and other microfibre and shingosen garments by stiffening and fusing various parts to prevent pucker. This was principally done on the parts of the shirts which were most obvious when they were sold in see-through plastic packaging, i.e., on the

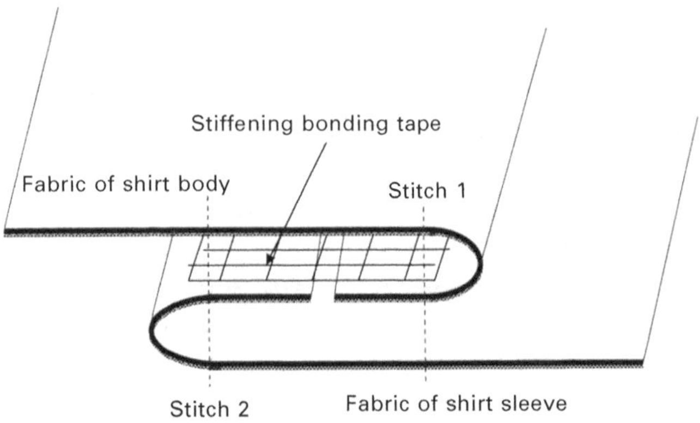

15.7 A French seam with a bonding tape inserted to join the sleeve to the body of the shirt and at the same time curing seam pucker.

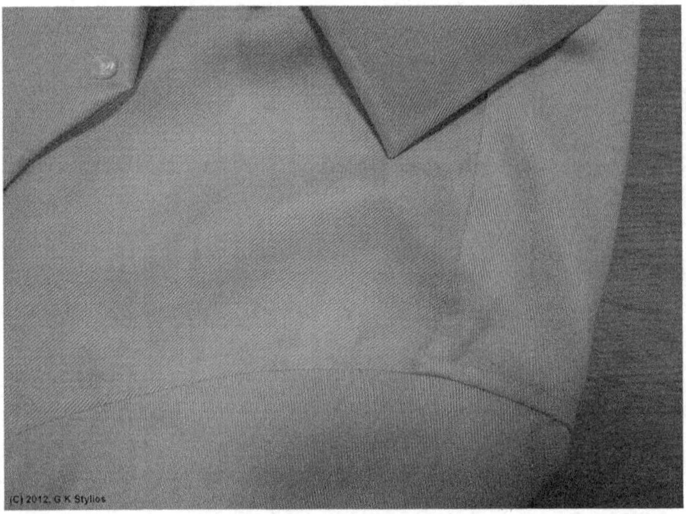

15.8 A French seam with a bonding tape fusing the arm to the sleeve and eliminating seam pucker.

collars, cuffs, front placket and pockets, unless a particular garment, fabric or customer had a problem in other seam parts. When successful wrinkle-free fabrics came on the market in 1994 and later on, when most of the shirts started to be made in the Far East and lower production costs allowed, the bonding of armhole sleeves became commonplace. Nowadays, more than 80% of non-iron shirts are made by bonding of the armhole seams made either by a French seam or lap seam construction. Figures 15.8 and 15.9 show an example of a currently produced formal shirt with flat seams around the armhole made by inserting a bonding tape during seam joining.

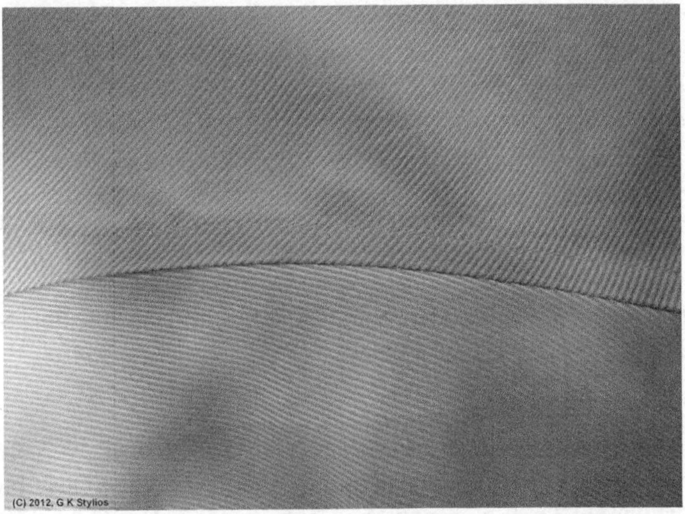

15.9 A close-up of the French seam in the shirt shown in Fig. 15.8.

The method chosen depends on the garment and the general rule may be to try to preserve seam properties such as stretching and shearability, and if the material is carefully selected, it may also not affect garment drape significantly. Although fusing somewhat restricts the seam extensibility and hence its stretching comfort as the two fabrics are bonded together, it seems to be preferred in the case of non-iron shirts, especially since loose shirt styling is preferred.

As the stiffening material needs to have stability with normal wear, neutrality in colour, ease of handle, and it must be relatively inexpensive, widely available and easy to insert in the seam, local and narrow bonding around the arm sleeve has been found to be the most appropriate.

In conclusion, we have seen how local stiffening can overcome seam pucker in commercial garment manufacture and in the case of non-iron shirts how armhole seams may be joined using either a French seam or a lap seam with a folder and tape for the automatic folding and insertion of a fusible interlining. Consequently, any seam pucker is eliminated, after sewing and even after shirt laundering (by the recovery of the sewing threads), hence always ensuring a flat seam appearance necessary for non-iron shirts.

15.7 Acknowledgements

The author acknowledges the following EPSRC government grants supporting this work:

1. *Investigation of Sewability Problems in Commercial Garment Manufacture Using Objective Measurement Technology*, in collaboration with Abbey

Textiles Ltd and South Knighton Dyeworks Ltd from 1990 to 1993 and continuation of that work in the following projects.

2. *Automation of Sewing Machine Settings in Difficult to Sew Fabrics Using Objective Measurement Technology*, in collaboration with Bellow Machine Co., Claremont Garments Plc and Burberrys of London from 1993 to 1995. A substantial part of this work was done at the University of Leeds and at Bradford University by the author and his team.
3. Engineering the performance and functional properties of technical textiles 2004–06, at Heriot-Watt University.
4. Multi-scale Integrated Modelling for High Performance Flexible Materials 2009–11, at Heriot-Watt University.

Other contributions:

- Science and Technology Agency, Japan Fellowship 1995–96 at the Institute of Polymers and Textiles in Tsukuba Science City, Japan.
- Department of Trade and Industry, Centre of Objective Measurement and Innovation Technologies COMIT; 1995–99, at Bradford University.

15.8 Appendix: key terms and definitions

Adhesive is any material that promotes adhesion (i.e., sticking to one surface or sticking two surfaces together) which would include some thermoplastic materials.

Centre placket is the front strip of the shirt which carries the buttons and button holes, and which forms the most conspicuous part in a formal shirt.

Folder is a fabric guide which is used for folding the fabric to a configuration dictated by the shape of the folder, prior to stitching, e.g. for making a lap seam or a French seam.

French seam is a seam in which the ends of the cut panels of fabric are folded in towards each other so that they just meet each other in the middle of the seams. The seam is flat and neat because no edges are exposed and only one line of stitching can be seen.

French seam construction – a French seam is made in two sewing operations: in the first operation one or two lines of stitching are made across the folded fabric part (e.g., the sleeve of a shirt) and the flat fabric (e.g., the body of a shirt) meeting each other in the middle, and then one of the pieces of fabric (in the case of an armhole seam this is the body fabric) is folded over the one or two lines of stitching and then stitched down, thus completing the second operation.

Fusible interlinings are materials heated so as to fuse to one or both of the layers of fabric between which they are placed.

Interlining refers to a material that is inserted between the outer fabric and regular lining of a garment, or between two fabrics. The main purpose of using interlinings in garments is to stiffen the part of the garment that is being applied. In shirts, collars and cuffs are stiffened using interlinings, as are lapels and shoulders in jackets.

Iron-free, **non-iron** or **easy care shirts** refer to shirts that do not need to be ironed even after the shirt is washed.

Lap seam is a seam in which the ends of the cut panels of fabric overlap each other so that the folded-over part of the lower fabric is going over the folded-over part of the upper fabric.

Microfibre fabric is a fabric made of yarns consisting of microfibres. A microfibre is a fibre with less than 1 decitex per filament with the most common types being from polyesters and polyamides (1 dtex = 1 g/10,000 m). These fabrics are very fine and limp.

Olefinic refers to any of a class of unsaturated open chain hydrocarbons such as ethylene, having the general formula C_nH_{2n+2}. The term in this context refers to a class of thermoplastics, which are polymers made from olefins, or, in other words, polyolefins, such as polyethylene.

Polyamide refers to a polymer containing repeated amide groups, as in various kinds of nylon.

Polyester refers to a large class of synthetic materials that are polymers containing recurring ester groups.

Seam is the joining of at least two pieces of fabric by at least a sewing needle and thread. Seams in garments production are made using a sewing machine.

Seam pucker is understood as a quality problem of appearance, in which a stitch, instead of being flat, is deformed showing creases and buckles along its stitch line.

Shingosen are fabrics made from ultra-fine polyester filament yarns that undergo modern fibre and fabric processing techniques to mimic various natural textures, such as peach skin, silk, butterfly wing, wool, leather and others.

Shirt yoke is the part of a shirt that stretches over the shoulders, usually made out of a doubled piece of fabric.

Tape chute is a guide that is used for inserting a tape into a seam.

Thermoplastic refers to a synthetic plastic or resin material that becomes soft when heated and solid when cooled.

16
Seams in car seat coverings: properties and performance

S. KOVAČEVIĆ and D. UJEVIĆ,
University of Zagreb, Croatia

DOI: 10.1533/9780857093967.4.478

Abstract: This chapter first provides an overview of the basic role played by textiles in the production of car seat covers, along with a description of the types of fabrics used and their properties and performance. Particular attention is given to the properties of a variety of car seat coverings and the possible problems associated with them, as these determine the durability, comfort and aesthetic appearance of the car interior. The chapter then discusses the importance of correct seam selection and its impact on the properties and appearance of car seat covers: as the cover should not lose strength in the seam, the correct sewing thread, seam type and sewing machine, with corresponding sewing needle, must be employed. The mechanical properties of composite materials are examined, along with the consequences of reducing the elasticity and strength, while increasing the comfort of the car seat cover in the seam area. The chapter offers possible means of improving the properties of seat cover seams, followed finally by an overview of future trends in the field, on the basis of previous studies into the use of certain composite materials for car seat covers and the recent advances in the development of materials.

Key words: car seat covers, materials for making car seat covers, sewing car seat covers, mechanical seam properties, seam types for sewing car seat covers.

16.1 Introduction

Car seats should provide optimum body posture and, therefore, a comfortable sitting position. Optimum posture contributes to a reduction in energy needed, as well as facilitating circulation, which is particularly important for a vehicle driver. The interior of the vehicle should be designed in such a way as to provide optimum comfort and safety during driving. The seat dimensions and the entire interior surrounding the seat should match the anthropometric measurements of each passenger. The materials used to improve the appearance of the vehicle interior and those used for passenger safety functions are made of different materials and are produced by different manufacturing methods. These materials are classified as technical textiles, and account for approximately 15% of all manufactured technical textiles, and about 4% of all manufactured textiles worldwide.[1]

The shape of car seats is becoming increasingly complex and ever more ergonomic shapes are being developed. Material selection, cutting precision and qualitative joining of cut parts by sewing are all employed in attempts to produce car seat covers that last at least as long as the average life span of the car. The life of the cover is not the only consideration, however: initial properties such as comfort, dimensional stability, colourfastness to light, abrasion resistance and seam stability must also be taken into account. There are a number of high performance requirements for car seat covers; consequently, multi-layered materials are used, which contain different layers composed of materials produced by a variety of methods, with each layer contributing to the fulfilment of one or more of these requirements. One extremely important requirement for car seat covers is that the cut parts, which are often made of different materials and have a different thickness and penetration hardness, should be qualitatively joined. The adaptation of the car seat cover to an increasingly complex seat shape requires an increasing number and complexity of cut parts, and thus more seams and increasingly complex stitching. On average there are 15 cut parts with a seam of 25 m per car seat cover, while per car there is on average a total of 120 cut parts with a total seam length of 175 m. This explains the growing importance of seam quality in manufacturing car seat covers. Sewing is currently the only possible method of joining the cut parts, due to the need to join composite materials of different thicknesses. Ongoing research is required to develop new materials that will allow the optimisation of the seam quality of car seat covers.

16.2 Materials and machines for sewing car seat covers

The interior design of vehicles is determined mostly by the appearance and properties of principally woven and non-woven fabrics. Besides textile products, polyurethane sponge, filling materials, and artificial and genuine leathers are also used on the reverse of the material. Woven fabric is mostly used for elements such as car seat covers, air bags and seat belts because of its good physical and mechanical properties, and because a variety of designs are available at acceptable prices. Non-wovens are mostly used for the places prone to lower stresses, such as rear seats, door lining and other visible metal parts. Moreover, the price of non-wovens and their ease of manufacture mean that they are preferred over other surface materials. Artificial leather is mostly used for the side parts of car seat covers and the front interior, whereas genuine leather is less frequently used because of its high cost and because it has several properties that do not satisfy the requirements for materials used for car seat covers. The objective of using these materials is to provide comfort for a car passenger since he/

she is rather inactive, spending a relatively long time in a seated position. Moreover, it is desirable that the materials used retain their appearance and comfort during their entire lifecycle; composite materials, whereby several different materials are thermally bonded, are more durable and are therefore used for this purpose today. In most cases, artificial or genuine leather is used on the front side, polyurethane sponge in the centre and knitted fabric on the back of the composite material (Fig. 16.1). This results in maximum comfort, durability and stability airiness in car seat covers made with these materials.[2,3] Moreover, in manufacturing car seat covers, different composite materials are used for the various cut parts: the seat section and backrest are made of woven fabric, the side parts from artificial leather, and the rear sections from non-woven materials. In this way, maximum flexibility and durability of the car seat cover are achieved at an optimum price.

Materials used in the automotive industry require high elasticity: this is difficult to achieve, because a surface material must also be resistant to abrasion and have relatively high strength and stability. The combination of polyurethane foamy mass or expanded sponge on the back of the fabric with a very thin and airy knitted fabric achieves the required stability, softness, elasticity and strength. In summary, to cover car seats, woven fabrics, genuine and artificial leather are used on the face of the material, with thermally

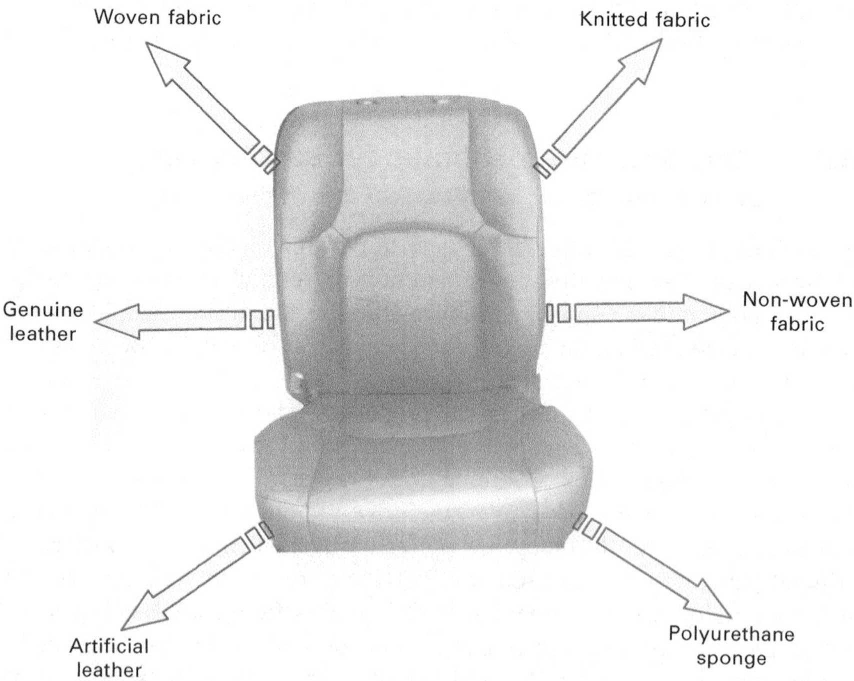

16.1 Different materials are used for car seat covers.

bonded polyurethane sponge and knitted fabric on the reverse. Non-woven materials of varying thickness and with differing properties are used to cover the back seats and other parts of the interior.[4,5]

Due to the complex nature of the materials used in vehicles, and the changes in their properties at the seam, this chapter will describe only the materials used to make car seat covers. The basic properties of single materials in composites will be described, as will composite materials as final units of the surface material for making car seat covers. Table 16.1 lists the mechanical properties of three types of material that are most commonly used for manufacturing car seat covers.[6]

16.2.1 Types of materials

This section describes the most important properties of the various materials used for making car seat covers.

Table 16.1 Mechanical properties of several materials used for making car seat covers

Multi-layered materials	Standards	Woven fabric/ polyurethane sponge/knitted fabric	Artificial leather/ polyurethane sponge/knitted fabric	Non-woven material for the back of the seat
Weight per unit area of fabric (g/m^2)	ISO 3801	300–600	700–1,200	150–250
Thickness (mm)	ISO 5084	1–3	1–4	1–1.5
Breaking force in the warp direction F_1 (N)	ISO 13934–1	800–2,500	400–2,000	250–350
Elongation at break ε_{F1} (%)		30–35	35–60	45–55
Breaking force in the weft direction F_2 (N)		500–1,300	350–1,500	200–300
Elongation at break ε_{F2} (%)		20–25	25–50	30–50
Breaking force to bursting F_3 (N)	ASTM D3787–89	1,500–2,000	600–1,800	20–50
Elongation at break ε_{F3} (%)		8–15	5–15	5–15
Abrasion resistance (mass loss after 10,000 cycles, %)	ISO 12947–3	1–3	0.4–1	40–60

Woven fabrics

It is impossible to imagine our lives without woven fabrics, which have good properties, offer excellent comfort, and are widely used for different purposes in everyday life. They can be designed and produced with a variety of desirable properties. The woven fabric used for making car seat covers is irreplaceable: it has several properties that cannot be achieved with the use of any other product. The greatest portion (approximately 85%) of the woven fabrics found in vehicles is used in seat covers and upholstering the car interior. Their other uses are in airbags (approximately 11%) and safety belts (4%).[7]

The woven fabrics used in car seat covers are mostly of synthetic origin, as these have superior properties compared to their natural counterparts: they have better strength, resistance to abrasion and sunlight and higher stability; they offer improved comfort and easy care; they do not generate static electricity and pilling; and they have good physical and mechanical properties that could not have been imagined a few decades ago. They are also usually cheaper than fabrics made of natural fibres. Synthetic fabrics can also be non-flammable, a requirement in some vehicles. This means that they do not burn when the oxygen concentration is as high as 25% or when the material is heated to a very high temperature, or that they stop burning when the flame is removed.[3]

Natural materials are also used for making car seat covers, but to a lesser extent than synthetic fabrics. Natural fibres have several properties that are less desirable than those of synthetic fibres. The principal problem is that pesticides and other pollutants can affect natural fibres at the source, and these can be transmitted by a number of different means not only to plant fibres, but also to animal fibres through the food chain.

Woven fabric is mostly used as a surface component of a multi-layered material in car seat covers, to provide comfort and softness in contact with the body. This type of composite material allows ergonomic design of the seat and other interior products. In most cases the surface fabric is woven in standard weaves or in derived weaves with smaller weave units. Velours and velvets woven with multi-coloured warps and wefts in jacquard patterns can also be used; however, these surface materials consist of a fibrous structure on the front side, leading to lower abrasion resistance than in the other types of fabrics mentioned. Due to their softness and comfort, they are used to a certain extent, but a higher weave density and an additional treatment are desirable.

Knitted fabrics

The advantage of knitted fabric lies not in its strength, but in its elasticity and softness, making it more comfortable than woven fabrics. However, knitted fabrics have lower abrasion resistance, are less stable and durable, and have a lower breaking force compared to woven fabric and artificial and genuine leather; they are therefore mostly used as the third layer of a composite multi-layered material for car seat covers. The fabric is bonded to the back side, protecting the polyurethane sponge and contributing to the elasticity, strength and softness of the car seat cover. Recently, knitted fabric has been used in the part of the car seat cover which is in contact with the passenger's head, as a two-dimensional fillet warp knitted fabric. In this case it is not necessary to add other materials because this type of knitted fabric is very soft, airy and thick enough to impart special comfort in contact with the head. Since it lacks sufficient strength and stability, it is sewn on in the head region where stresses are lower (Fig. 16.2).

Non-woven fabric

Non-woven fabric is classified as a technical textile and ranks third in the manufacture of textile surface materials after woven and knitted fabrics. Non-woven fabric has a number of advantages over woven and knitted

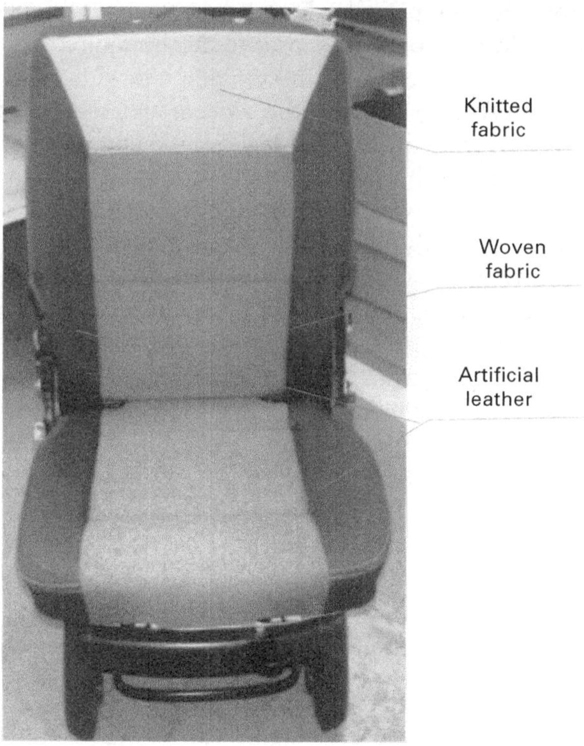

16.2 Car seat cover made of different materials.

fabric: it can be designed with specific targeted properties; can be produced with substantial variations in thickness, mass, voluminosity, elasticity and stiffness; and is comparatively quick and cheap to manufacture. Data from previous studies on the production of non-wovens shows a continued growth and an increasing usage. Non-wovens are the most widely used textile product in the automotive industry and are unrivalled in their sound and vibration insulation properties. The non-woven fabric used in the automotive industry is in most cases of synthetic origin. Thus, the most demanding requirements for automotive textiles, such as non-flammability, can be met through the correct selection of fibres and subsequent treatments in the production process.[8–10] Non-woven fabric is used as the back side of the car seat and is part of the car seat cover. One of the problems associated with this is the method of joining the non-woven fabric to another material by sewing, which can result in seam instability. Therefore, a non-woven fabric with optimal fibre density, compactness and softness is used.

Artificial leather

Artificial leather offers high abrasion resistance, but can cause discomfort due to coldness over longer periods. Developments in artificial leather production have brought its properties closer to those required for car seat covers. Artificial leather is often used in combination with other surface materials, and is invaluable in keeping the side of the seats stable (Fig. 16.2). Due to the higher price of artificial leather, the automotive industry generally opts for a higher proportion of woven, knitted and non-woven fabric in the car seat cover. However, the excellent reputation of some car manufacturers is based not only on the quality of the car, but also on the quality of the design of the car interior, in which the seat covers play a significant role. These manufacturers do not economise on interior design and use mostly expensive materials for seat covers, such as artificial leather with specifically tailored properties, in order to offer safety, comfort and flawless interior design. New technologies for manufacturing artificial leather can offer perfect emerised materials with very comfortable handle, which are also environmentally friendly.[11]

Genuine leather

Genuine leather is one of the most expensive materials used for car seat covers. The leathers most commonly used for making car seat covers are those made from the hides of cows, calves, horses, pigs, goats and sheep. They are used for high strength, elasticity and abrasion resistance, and are also soft and smooth with good air permeability. It is believed that genuine leather is both the most comfortable material in contact with the human body as well as the healthiest, especially if it is vegetable-tanned and naturally coloured. Vachetta leather, either vegetable-tanned or chrome-tanned, is often used for car seat covers and belongs to the category of thin leathers used over large areas. These leathers offer softness, gentle handle and easy care. High braking forces, elongation, abrasion resistance and stability ensure that leather is a very durable and resistant product. As a consequence of its high price, genuine leather, like artificial leather, is mainly incorporated into the seat covers of more expensive cars produced by renowned car manufacturers, and is a highly coveted feature for many customers. Genuine leather will therefore continue to be used for car seat covers in the future, irrespective of developments in synthetics materials. By selecting the leather direction in cutting, the length of the stitch, and the sewing thread and needle type, it is also possible to achieve good properties in the seam.[12] The various stages of leather production can often involve substantial distances; for example, cattle breeding in South America; processing, tanning and dyeing in Europe; making up in Central and Eastern Europe and Asia. This should be taken into consideration in order to ensure a high quality final product.

Polyurethane sponge

The use of polyurethane sponge as the inner component in car seat covers provides a particularly high level of softness and comfort, and offers excellent cover stiffness, preventing bending, puckering, wrinkling and stretching on the seat and back rest. However, the comfort level is not maintained over long periods, due to its relatively high water and air impermeability. Polyurethane sponge also has low strength and weak abrasion resistance. Substitute materials should therefore be used in future; alternatively, polyurethane sponge with higher hydrophilicity, but with equal softness and durability, will need to be developed, without a concomitant price increase. Materials of different thickness ranging from 1 to 11 mm are used for making car seat covers. In the parts of the car seat cover which are exposed to higher pressure, a material with a thicker polyurethane sponge is sewn in. The seat and back rest have the thickest sponge, while the side and rear parts of the cover are made of a material with a thinner sponge. This results in improved cover stability with no decrease in comfort.[11]

Sewing threads

The significance of the sewing thread in contemporary manufacturing processes for making car seat covers, which are often aimed primarily at reducing costs, is often ignored. Quality, optimal colour selection and easy availability are taken for granted, and manufacturing problems arise if one of these parameters is not fulfilled. The fineness, raw material composition, colour and strength of the sewing thread should be carefully selected to ensure that the joins satisfy quality and durability requirements. A higher or lower thread tension entails an irregular and poor seam appearance and a risk of seam weakening and separation of cut parts (Fig. 16.3).

The properties of the sewing threads for sewing car seat covers are as follows:

- count: 50–100 tex
- breaking force: 40–80 $cNtex^{-1}$
- elongation at break: 20–30%
- modulus of elasticity: 0.5–4 $cNtex^{-1}$
- raw material composition: polyester, polyamide, cotton – singed, mercerised and lubricated.[13]

Good thread properties are defined as strength, elasticity, elongation, softness, regularity, lustre and smoothness. It may be 'bonded', meaning that the thread components are glued together. This prevents 'opening' of the thread when it moves in different directions during sewing and is exposed to relatively high abrasion.[14]

The sewing threads used for car seat covers should be extremely strong. If the stitch type, seam and sewing thread are correctly selected, seams of precisely specified strength and elasticity can be produced; this is particularly important for car seat covers. Furthermore, the sewing thread should be resistant to abrasion, weathering and extremes of temperature. The aim is to achieve a sewing thread and seam with durability equal to that of the materials used in the car seat covers.

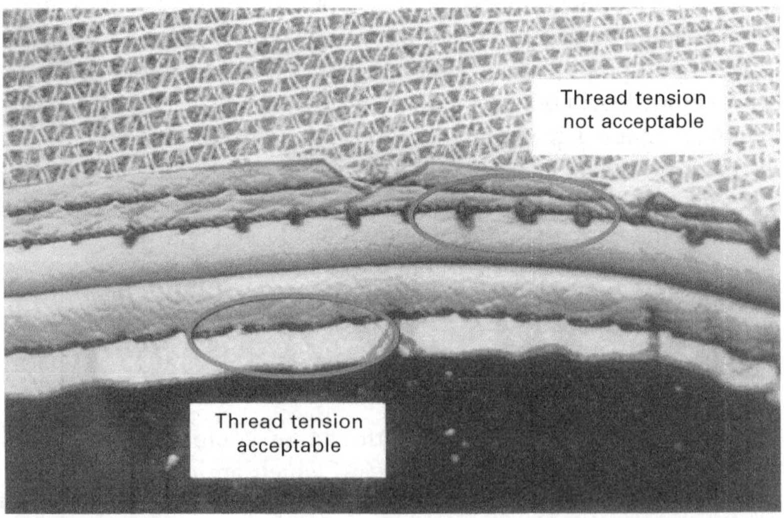

16.3 Appearance of the correct and incorrect seam caused by a difference in thread tension.

During sewing, the thread is dynamically loaded according to the stitch speed, causing changes in the mechanical properties of the thread. During stitch formation the thread is subjected to tensional, frictional and bending loads. The highest stresses occur at folds, near the eye of the needle, and where the upper and lower threads are interlaced. In addition to dynamic loading, the thread is also subjected to friction and bending to a small extent, pressure, elongation and abrasion, as well as to high speeds and accelerations. Due to cyclic stress at various speeds, thermal stress of the thread occurs. The mechanical properties of the thread for sewing car seat covers depend on the properties of the material as well as on the construction parameters of the thread. Stitching at higher speeds causes a higher tensile strength and higher friction of the thread with the thread guiding elements, resulting in changes in the thread strength after sewing. Multi-layered materials used for making car seat covers are made of different raw materials, with differing burst hardness, and therefore require an especially strong sewing thread that is resistant to the stresses mentioned above. Most forces acting on the thread are cyclic, which results in thread fatigue related to the fatigue of the single fibres in the thread. An optimum thread twist level and number of components of the thread, combined with regular step twisting, create a circular form of the cross-section of the thread, resulting in a minimum area of contact between the thread and the sewing needle. Optimum twist level and regular thread twisting also provide the sewing thread with the viscoelastic properties required to achieve resistance to dynamic loads in the sewing process. If dynamic loading of the sewing thread exceeds the yield point, undesirable plastic deformation of the thread takes place.[13]

16.2.2 Cutting machines for car seat covers

The materials used for making car seat covers must be cut with great accuracy, so that the cover fits the seat exactly. Cutting requires a high level of precision, which can be provided by computer-controlled digitised cutting patterns. Since car seat covers are made from materials with multiple different components, care should be taken when cutting layers of material. To ensure that the layers are cut accurately, they are covered with foil before cutting and then air is squeezed out of the material. This method creates compactness and hardness in the material to be cut, facilitating a high level of precision in cutting (Fig. 16.4). The main advantage of contemporary cutting technology is that the cutting process is automatic, with an operator monitoring and controlling the operation of the machine. This operator must undergo special training to correctly monitor the cutting process.

16.4 Cutting machine for car seat covers (Bullmer, model: Product XL 7500 LV).

16.2.3 Sewing machines for car seat covers

The selection of the correct seam type and length is particularly important in making car seat covers, as the seam must meet regulatory standards (ASTM D6193). Sewing car seat covers requires special sewing machines adjusted for sewing multi-layered thick materials (Fig. 16.5) without damaging and tightening single components. The sewing machines used for car seat covers are sturdier than conventional sewing machines. They can be made with an extended head and free arm to facilitate the handling of the material during the sewing process. Figure 16.5 shows a plant for sewing car seat covers with Dürkopp Adler 367 sewing machines. These machines sew with double lock stitch, and have automatic thread cutting off and bartacking. They are equipped with bottom feed, needle feed and differential top feed. Their nominal stitching speed is about 2,800 min^{-1}. In the development of sewing machines for multi-layered materials, particular attention has recently been paid to ergonomic design, efficiency increase, simpler and easier handling of larger cut parts, higher pressure foot lift, even stitch length and flexibility.

Sewing needles

All the technological sewing parameters that influence the quality of the seam should be matched with each other, from the design of the machine

16.5 Sewing car seat covers on the Dürkopp Adler sewing machine.

to the guidance of the thread and tension on the needle. During sewing, the material can become damaged if there is a lack of coordination between the sewing needle size and point style and the material to be sewn. The material, needle and sewing thread form a trio with a direct influence on the seam quality. The size of the eye of the needle and the thickness of the thread should be precisely matched so that the sewing thread can pass through the eye with as little friction as possible. The importance of this becomes particularly clear if we imagine a scenario in which the upper thread, using a double lock stitch, is passed through the eye 25–60 times in alternating directions until it is joined in the middle of the material as a seam with the bottom material.

The correct selection of needle size to avoid damage depends on the specific material used. The multi-layered materials used for making car seat covers add complexity to the question of needle selection. There are some rules that can be followed: the finer the structure of the material, the finer the needle; and the smaller the needle diameter, the less the stitch hole in the material is pressed or expanded. Moreover, if the thread is more elastic and the needle thinner, the damage to the material will be less if the other parameters are optimally set; on the other hand, if the thread is non-elastic and thicker, the needle increases friction during stitching and causes the material to be damaged around the stitch area. Besides needle size, the point of the needle also plays an important role in the quality of the seam and the level of material damage. Since the needle penetrates several different materials of different thickness and sensitivity, the selection of needle point style and needle thickness is more challenging for car seat covers than for other materials.[15]

Special sewing needles are necessary to join cut parts without damaging the cover and the thread (Fig. 16.6). The needle point depends on the type of material and the type of seam used. The needle should be especially hard and heat-conductive, so that the cover is damaged as little as possible at the stitches while still forming a correct seam. The shape of the needle point is essential in achieving these aims. In general, both ball point and round point needles are used for sewing car seat covers.

16.3 Influence of the seam on materials for making car seat covers

During sewing, the material is stressed. After sewing is complete, external forces stop acting, and the relaxation phase commences. The forces that act on the material during sewing are not constant. The feed dog of the sewing machine follows an elliptical path, and as a consequence the pressure and friction forces within the material change. This means that different forces affect the material during sewing, leading to consequent differences in the possible deformations. Since these forces are relatively small in the sewing process and are in the elastic and visco-elastic range, it is not possible to analyse them after sewing. Thus, they are often neglected, although they can partly affect seam length. They are determined by the type of the material and the construction parameters of the single material being sewn. One of the most important factors for the quality of the car seat cover is the appearance of the seam and its efficiency during sewing. It is extremely important to optimise the seam length and type, thread strength and fineness and needle type, which are all dependent on the type and thickness of the material. To

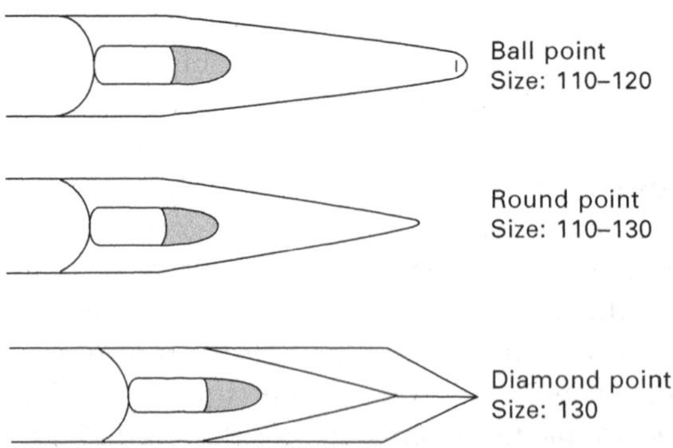

16.6 Sewing needles for sewing car seat covers.

obtain a correct seam, the tension of the lower and upper thread and the two-sided shift of the material must also be optimised; this is very complex due to the different thicknesses of the materials involved (Fig. 16.7).

Due to the exceptionally high stress exerted on the car seat cover in the sitting and backrest sections, particularly at the seams, threads and fibres are pulled out, especially where the stitches have been made. This impairs the construction of the material at the stitches, especially when woven fabrics are used. As a consequence, gaps or holes can be formed, followed by breaking of both the sewn material and the sewing thread (Fig. 16.8).[16]

A skipped stitch also creates a poor appearance, weakening the seam and leading to the separation of cut parts; this is a significant problem if the fault is not recognised in due time. To re-sew the material in the same location is not advisable, as this leads to marks on the material, such as holes caused by sewing. As well as causing poor aesthetic appearance, this also results in damage to the material and a lower quality product. Since choice of different seams for car seat covers is based mostly on aesthetic considerations, the

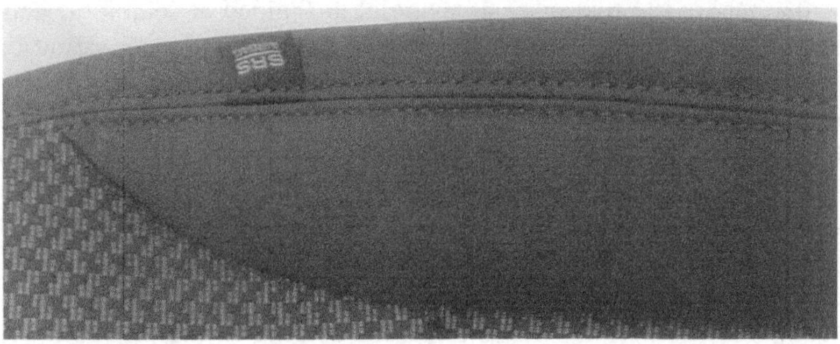

16.7 Sewing variously thick materials.

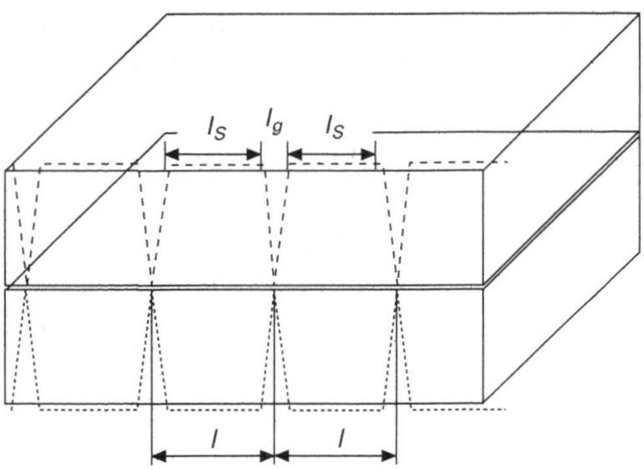

16.8 Visual division of stitch length l to stitch l_s and the gap formed between stitch l_g.

strength and elongation of the stitched material is different from that of material without a seam. Figure 16.9 shows the seams most commonly used for sewing car seat covers.

Table 16.2 shows the unit values of breaking force and elongation at break of two types of the multi-layered material used for making car seat covers.[17] Ten samples with woven fabric on the front side and ten with artificial leather on the front side were investigated. The polyurethane sponge used was of varying thickness and different knitted fabrics were used on the reverse of the materials. The investigations were carried out by the authors of this chapter as part of a joint project with the textile company Prevent d.o.o. of Zlatar, Croatia, where car seat covers are sewn for European cars. The data obtained show that breaking strengths differ according to seam types, and in most cases they are lower than the breaking strengths of the same seamless material. The breaking force of the material (woven fabric, sponge, knitted fabric) with seams (a) and (c) decreases to 30% of the average breaking force without seam, meaning that the seam is a weak point; the car seat cover is therefore at risk of breaking just in this location. The reason for this effect is that where the two parts of material are joined at seams (a) and (c), the seam takes over the entire stress or the sewing thread of a count of 80 tex, raw material composition of 100% polyester and a stitch length of 5 mm. In order to increase seam strength and material stability under loading, a special tape, shown in Fig. 16.9(a) and (c), is sewn onto the reverse of the material. The tape also plays a role in protecting the edge of the material on the back of the seam which is prone to weakening and coming apart (Fig. 16.9(a)). The elongation at break of the material in the seams generally decreases to 50% of that of the seamless material, but higher elongation at break in the seams may sometimes be obtained. According to the results of the study, seams (a) and (c) (Fig. 16.9 and Table 16.2) are not recommended for use in areas subject to higher stresses.

In the case of multi-layered materials composed of artificial leather, sponge, and knitted fabric, the difference in breaking force between seam and seamless locations is not as high as when woven fabric is used as the upper layer. The breaking force at the seam usually decreases to 75% in seam (c), but in the case of seam (b), the breaking force at the seam may be even higher than that of the seamless material.

Seam (b) is therefore the strongest type of seam, with breaking force value close to those of seamless materials, particularly when artificial leather is used for the upper side of the material. Elongation at break is also the highest in seam (b): this is the case in both investigated materials. Thus, seam (b) is recommended for use at the points of greatest stress, namely the sitting section and the front passenger backrest.

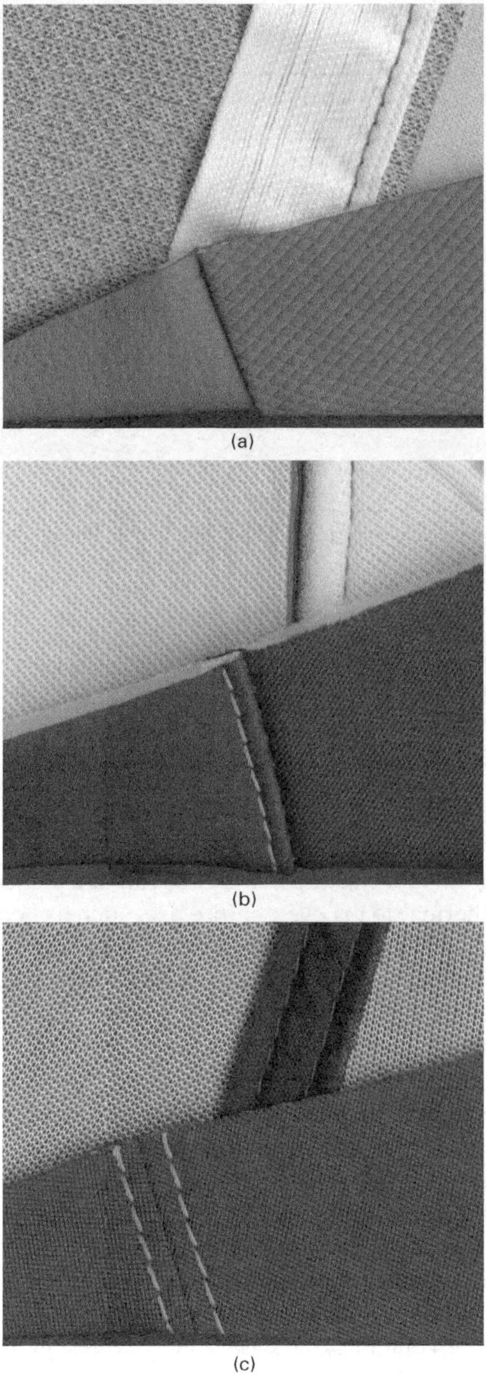

16.9 Seam types for sewing car seat covers: (a) seam variety: 1.01.01, (b) seam variety: 2.02.03, (c) seam variety: 4.03.03.

Table 16.2 Breaking force and elongation at break over the seams if the unit value for seamless samples is 100

Seams according to Fig. 16.9			Breaking force (unit values)		Elongation at break (unit values)	
			In the warp direction	In the weft direction	In the warp direction	In the weft direction
Seamless			100	100	100	100
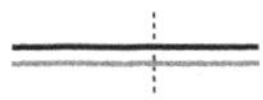	(a)	Multi-layered material: woven fabric, sponge and knitted fabric	30–60	40–70	40–50	50–80
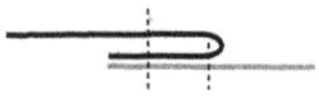	(b)		50–70	60–100	60–80	80–110
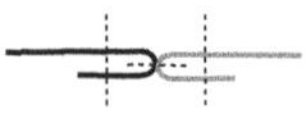	(c)		30–50	30–70	40–60	50–80
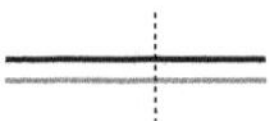	(a)	Multi-layered material: artificial leather, sponge, and knitted fabric	80–95	85–95	70–100	80–100
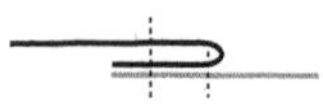	(b)		85–100	90–120	90–120	90–120
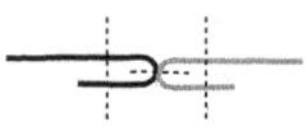	(c)		75–90	80–95	80–90	85–95

16.3.1 Seam quality

Textile multi-layered materials for making car seat covers are subjected to a series of mechanical and thermal loads even before use, such as heat setting of single components into a surface material, followed by sewing and fitting to the car seat. The requirements for consistently high quality seams and fit of the cover on the seat cannot be met if the shrinking values during cutting and sewing are not accurately predicted. The shape of the car seat cover changes when it is loaded in use. If the passenger is heavier, the loading and deformations of the material will be greater, especially on the seams of the cover where the passenger sits. This change in the shape is partly temporary due to the elasticity of the material. A change in the shape of the car seat cover is mostly visible at the seams. Permanent deformations occur in the case of a longer and greater deformation. It is important to choose an appropriate material to reduce the permanent deformations of the material that occur as a result of varying loads on individual parts of the car seat cover. Thus, materials with greater strength and thickness are used for those parts that are subjected to the greatest loading over the longest period of time.

When assessing durability, it is important to take account of material abrasion against the passenger's body. Any protruding parts, along with corners, are most prone to abrasion in the seams. Moreover, the car seat cover is thicker at the seams, meaning that the seam is subjected to abrasion to an even greater extent. According to the investigations performed by the authors and the industry, these parts are up to 30% less durable than seams not subjected to abrasion.

Seam quality is an important parameter which determines the performance characteristics of car seat covers. An incorrect seam reduces the life cycle of the car seat cover and negatively affects its appearance. The mechanical and dimensional properties of the sewing thread and the material affect the seam quality, and should be taken into consideration when analysing the material. Two values used to describe the sewing quality of the material are efficiency and seam slippage. Seam efficiency depends on a number of factors: the breaking behaviour of the material and sewing thread; the combination of the material and sewing thread; the type and shape of the needle; and the parameters of the sewing machine and sewing process. Sewing conditions, such as seam types, stitch density and selection of thread and needles, can therefore be optimised to maximise seam efficiency. As is well known, during sitting, different static and dynamic forces affect car seat covers in different directions. As already stated, these forces may cause deformation and eventual breakage of the car seat cover, in most cases along the seam. Sewn joints with higher seam efficiency act in opposition to these forces. If the sewing thread has not been properly chosen with regard to the construction of the material, the seam efficiency will not be acceptable.[18]

16.3.2 Parameters influencing the stable length of sewing stitch and seam puckering

Besides strength, one of the most important parameters in assessing the quality of sewn car seat covers is stable stitch length. Stability of stitch length depends on the mechanical properties of the materials used, the sewing conditions and the sewing machine chosen. Since the car seat cover often consists of cut parts of different materials, the frictional forces between machine parts and the material, and between materials, are not always equal. This may cause differences in the feeding of material even when the feed dog is running correctly. The inertial movement of the material depends on the stitching speed, which is obtained from the stitch plate and affects the stitch length. Another parameter that affects the feeding of the material is the vibration of the sewing machine caused by the driving belt. Investigations into this phenomenon have been closely related to the study of changes in the geometric shape of the V-belt and its dimensional change during use.[17,19] Differences in tension between the upper and lower thread and changes in the stitching speed can also cause uneven stitch length, stitch skipping, and interlacing of the upper and lower thread, resulting in a low quality seam.

Seam puckering is a major problem in sewing car seat covers. A study of the impact of the physical and mechanical properties of the sewing thread on the appearance of the seam revealed the following: a finer thread is better because there is less irregularity, less elongation in case of low loads, a reduced surface load, lower surface friction and less recovery deformation. Material slippage is the main cause of seam puckering when sewing car seat covers; this is confirmed by a wide range of different multi-layered materials. The various devices and mechanisms of the sewing machine (tension regulator of the upper thread, needle thread mechanism, thread looper and feeding device and needle properties) also affect seam puckering. The different mechanical properties of the individual components of the composite material used to make the covers cause differences in the pushing movement of the material within the zone of the seam, resulting in a wavy or puckered seam. During sewing, the feeding device pulls the contact layer faster than the inner layers, and this phenomenon can also cause seam puckering. The inertial forces acting on the material during feeding are dependent on the stitching speed, meaning that the stitching speed also has a significant impact on seam puckering.[20]

16.3.3 Automation and robotisation in sewing car seat covers

Materials for car seat covers are mostly made of three layers which are doubled at least once where they are sewn. The main problem faced when sewing such complicated and thick materials is the occurrence of material slippage as a result of non-uniform material feeding, especially when smooth materials are used. Uniform feeding of such materials is difficult to achieve even on state-of-the-art sewing machines, where, in addition to material complexity and thickness, a number of other factors play a significant role, including speed of the sewing machine; its possible acceleration and deceleration; and the behaviour of the operator who adjusts the material and supports the feeder to pull the material. Greater feeding of one or of both material layers causes non-uniform material and sewing thread tension as well as non-uniform seam length, resulting in lower material strength, and a tightening of the material or sewing thread at the seams. This part of the cover then becomes weak, resulting in premature breakage or impairing the aesthetic appearance. Since one layer of the material is polyurethane sponge, which becomes thinner under pressure, it is necessary to optimise the pressure of the presser foot on the material. A higher or lower pressure causes the sewing thread to be too taut or not taut enough, resulting in lower material strength of the seam and an unsightly aesthetic appearance. High thread tension tightens the material and creates larger holes in the material (Fig. 16.8), while low thread tension does not create a strong enough join between the two materials.

Semi-automated and automated sewing machines and robots are increasingly replacing conventional sewing machines for sewing car seat covers. Their main advantages are as follows: the material is equally taut through the whole sewing process with minimum bending and slippage; the seam is of high quality; the material ends are neat and equally long; the operator's role is minimal; and all the parameters related to the sewing thread and the material are optimised according to the given values. Their efficiency and seam quality compared to that offered by conventional machines are particularly advantageous for short and complicated seams, such as short rectangular hemming and topstitching, sewing on of plastic parts and other similar operations that require a high level of complexity and precision.

Adjustment of the seat to the body begins with the ergonomic design of the solid seat base. Despite the ergonomic shape of the seat, additional adjustment is possible using foam designed in segments so that the load at the point of sitting is evenly distributed with less vibration while driving, with greater comfort and body stability on the seat. The seat is thus fully adapted to the body shape so that the person can continue to feel comfortable over time. However, the subsequent adjustment of the seat to the body in some places (contact points of the body skeleton and the seat) causes greater multidirectional stresses and strains, and thus greater material deformations. By robotic simulation, it is possible to monitor surface stresses on the seat created by body pressure and thus adjust design and foam type.[21]

Multilayered materials and various composites used to cover the seat should be, among other things, comfortable to the touch, soft, elastic and resistant to force and friction. Softness is achieved with polyurethane sponge in the middle layer, and elasticity is achieved by use of warp knitted fabric on the back side and elastic fabrics on the face side with a share of elastane fibres or yarn selection and fabric structures. Polyurethane foam is not elastic enough to withstand greater strain and there is a danger of tearing. Also, due to higher elongation at the time of seat adjustment to the body, textile surface materials (woven and knitted fabrics) suffer multiple stresses and higher deformations, which indicate additional problems of durability of covers and their aesthetic appearance. Thus, the ability to further adapt the seat to the body leads to new requirements for cover materials, where all layers of the material should have sufficient flexibility, strength and wear resistance so that durability and aesthetic appearance are satisfactory.

16.4 Seaming problems in car seat covers and their solution

A variety of materials are currently used for making car seat covers. Surface materials, such as woven, knitted and non-woven fabrics and artificial and genuine leather, are predominant. The most frequently used of these are synthetic polymers such as polyester, polyamide, polycyclic and viscose, and natural fibres such as wool, cotton, linen and animal leather. More resilient and softer surface materials are used to provide comfort, but these should also be durable and abrasion-resistant.

As a consequence of the development of new materials, a further investigation of seams has become necessary, with regard to their strength, elasticity, aesthetic appearance and resistance to long-term use (principally abrasion resistance). The seam has been shown to be the weakest area, and represents a major risk area in car seat covers, as it is thicker and more prominent than the rest of the cover, and is often located at the folds, where the greatest loads and abrasion occur and where exposure to light and soiling is highest.[22]

Since the materials used for car seat covers are multilayered composite materials, the joining of cut parts by sewing is more complicated. Each material layer has completely different properties in relation to the other layers. To achieve a strong seam that is also both functionally and aesthetically suitable, each material must be examined separately. It is crucial to find the most suitable sewing thread with respect to fineness, raw material composition, number of strands and twists, if necessary with step twisting and surface treatment. Furthermore, the thread should have a normally cylindrical and regular cross section that is resistant to abrasion and thermally stable. At most joins, one seam takes over the stress forces in this area of the cover (Figs 16.9(a) and (c)). The correct choice of sewing thread, seam length, needle type and other essential parameters of the machine and the material is therefore of utmost importance. All components of a multilayered material affect the material's properties. In terms of their mechanical properties, multilayered materials are mostly anisotropic, displaying different types of anisotropy. Materials are said to have anisotropic properties when their mechanical values differ depending on the direction in which they are measured. Since the materials for car seat covers are stressed in different directions, as well as at the joints, seamed material exhibits a higher degree of anisotropy. After a period of use, car seat covers display different deformations that directly affect their life span and aesthetic appearance. The higher the quality of the material and of the seam, the longer the life span and the better aesthetic appearance will be. The directions with the lowest material strength, excluding the weft or warp direction, are at risk (Fig. 16.10). The multi-axial properties of the material should ideally be as

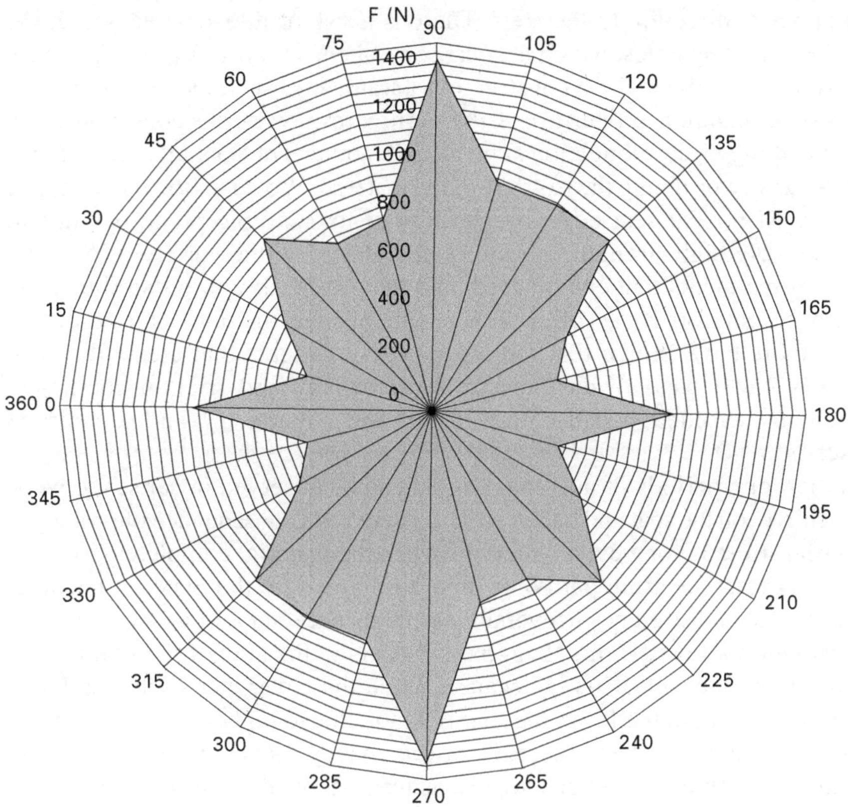

16.10 Breaking force in degrees of a composite material for car seat covers with woven fabric on the front side (2/2 twill right direction), polyurethane sponge and knitted fabric on the back side. Tests were carried out according to the standard ISO 5081 F – Breaking force of the material (N).

balanced as possible to ensure that the material is suitable for making car seat covers.[21,23,24]

Multilayered textile materials that contain a woven fabric as one layer are markedly anisotropic. Anisotropy decreases if the woven fabric is combined with polyurethane sponge and knitted fabric. Previous investigations have shown that several materials used for making car seat covers, including non-wovens, have almost orthotropic mechanical properties. Orthotropism of a multilayered material cannot be achieved if one layer of the material is a woven fabric; woven fabric impairs orthotropism in multilayered materials, especially in the direction opposite to that of the warp and weft.[25–28]

Isotropy of a multilayered material with woven fabric depends on the construction of the fabric: the direction of the warp and weft plays a decisive role as do the parts around the sewn joint, particularly if the joint runs in the opposite direction to the weft. There is a risk of threads creeping at the stitches, creating holes, which may cause weakened seams. A lesser influence is exerted by other fabric construction parameters, such as weave pattern, density, elongation properties of the yarn, stress, weaving conditions, etc. The breaking force, elongation at break and modulus of elasticity are decisive parameters determining the isotropy of the material. Woven fabric will continue to be used for car seat covers regardless of its isotropic properties due to its relatively high strength, durability and elasticity, the variety of possible surface treatments that can be used, and the greater design possibilities and improved comfort it can offer. The results of investigations carried out by the authors have shown that fabrics in weave patterns that create diagonal outlines (twill weave) and have a lower density and friction coefficient should be avoided because they affect the asymmetry and anisotropy observed.

After long periods of use, 'bagginess' can be observed in car seat covers, particularly on the seat and backrest. This occurs as a consequence of prolonged material stress in long cyclic intervals when the material completely loses its elasticity after a certain period of time. In these areas the material becomes viscoelastic and stretches irreversibly as far as necessary, or as far as the seat construction allows. If this 'bagginess' occurs in the area of the seam, seam breakage may occur. This is because, while the cover can stretch as far as the seat construction allows, the seam cannot; due to higher stress the sewing thread breaks first. The occurrence of this 'baggy' appearance is dependent on the strength and elasticity of the material as well as on its isotropic or orthotropic nature. It further depends on the mass of the passenger or driver, the timing cycles of material stress and the application conditions.

Special attention is paid to the placement of the sewn seam in woven and knitted fabric, and especially to the direction of the weft and warp. This is to ensure that the light refraction does not differ at the seam, as this would impair the appearance of the whole product. The join in woven fabrics should follow the same direction as the warp or weft, while in knitted fabrics should be joined in the direction as the rows and wales. Fibrous fabrics should similarly be joined in the same direction of the fibres; sewing must be carried out in the direction of the fibres, because otherwise the upper layer of the material would slip. Since the surface materials used in the automotive industry must satisfy minimum strength requirements, the seams should have similar or almost equal properties. The properties of the join are dependent on the tension of the sewing thread, the seam type and the joining method. In most cases, if a specific level of strength, elasticity and abrasion resistance cannot be achieved in a seam, the seam must be reinforced with additional tapes and treated in a special way so that warp or weft threads are not pulled out. In the case of weft knitted fabric, it is also important to ensure that no unravelling should take place. The use of special machines to sew the woven or knitted fabrics used for making car seat covers makes the work of the sewer easier and reduces the number of possible faults.[29]

Covers used in vehicles such as powerboats, yachts and the like require a waterproof material; the places where the fabric is joined therefore also need to be waterproof. Sealing and cutting the material by ultrasonic, thermal or laser welding provides accurately made high quality seat covers. To achieve a water and air-impermeable seam, the material must be of synthetic origin and thermally suitable for sealing. Materials for car seat covers, being mostly composite materials consisting of several different layers, can develop different ultrasonic vibrations that cannot be harmonised. If the melting point of the different materials is not approximately equal, the materials cannot be effectively sealed by ultrasonic welding. In this case new synthetic materials with similar thermal and ultrasonic properties are chosen, so that the materials can be effectively sealed, producing a high quality seam. The sewing machines for this type of sealing allow continuous sealing and cutting of material, although point bonding or seam reinforcement through the welding of an additional tape is also possible. Thermal material welding with hot air or a hot wedge is carried out by welding the tape at the sealing point, preventing the passage of water or air through the seam. In this case the material is first joined by sewing, before the application of the thermally bonded tape makes the seam impermeable. Laser welding is a relatively new way of sealing car seat covers using new materials that effectively absorb infrared beams. The advantages of this kind of sealing are a relatively high sealing speed, limited area of heating, radiation access from one side, retention of seam flexibility and complete water and air impermeability through the seam.[1]

16.5 Future trends

Developments and improvements in the properties of car seat covers are aimed at ensuring driver and passenger comfort even after sitting for long periods. Firstly, this requires an ergonomically designed seat, covered with a material that will improve comfort with softness, airiness, heat regulation and even massage. Ongoing developments and improvements in the field of vehicle and interior design are contributing to increasingly intensive investigations into new materials offered on the market that could be used for making car seat covers. Materials should have certain properties required for this purpose: they should be hypoallergenic, non-flammable, easy-care, fast to light and UV radiation; their durability should be as long as the life cycle of the car; and their mechanical properties should be satisfactory.

The use of the new generation of synthetic fibres is growing every day; for example, the use of hollow polyester fibres for non-wovens improves their resistance to thermal conductivity. The car interior and parts of the car seat cover made of the non-woven material would retain warmth and would be comfortable in contact with the body.[30] The surface application of nanotechnological processes and the development of car seat covers with new properties are some of the most interesting challenges for the technological progress of materials. Abrasion resistance and fire protection are two properties that can be improved through the use of nanotechnological sol-gel methods on the surface of car seat covers.

Future materials for car seat covers will offer a number of advantages. The first of these will be the ability to recycle the materials when the car is destroyed. New materials can also be expected to provide thermal insulation and breathability; and there is potential for the development of new intelligent materials that will automatically identify and react to varying environmental conditions. These will be materials that can change their properties independently, for example, by changing from hydrophilic to hydrophobic when required.

In the near future car seat covers will probably be developed that can check the vital functions of the car driver such as blood pressure and heart rate, providing warning if these values reach a critical point. The installation of transponder systems into car seat covers will provide information about any irregularities in the operation of the car. The car seat cover will also prove to be the most suitable medium for the installation of a number of high-tech electronic systems. Textile materials with incorporated microcapsules for the measurement and regulation of temperature and moisture have been used for some time in the aerospace industry and for sports; these will also soon be used in vehicle interiors. The integration of thin, flexible cushions that can vibrate into covers in the seat and backrest will provide soft massage and body relaxation during travel. These soft vibrations can help renew energy and improve blood circulation; this is especially important for the driver during trips of long duration when there is a risk of loss of concentration and driving fatigue. Car seat covers have also recently been developed with integrated pulsating air chambers to massage the lower parts of the body. Ventilated car seat covers, which allow refreshing air flow, and help to prevent sweating of body parts in contact with the seat, use low-voltage power (12 V) and surely represent the beginning of the development of intelligent car seat covers.

In parallel with the development of microchips, fibres with numerous interesting properties adapted for use in car seat covers are being developed. These include fibres with health benefits, fibres made of renewable resources, environmentally friendly fibres with antibacterial properties, fibres releasing fragrant substances, and fibres with a particularly soft touch. Car seat covers should provide easy care and good weathering resistance, which can be obtained through the selection of the correct raw materials and through the use of after-treatments. The materials used tend to be composed of man-made fibres with predetermined properties; however, natural fibres should not be ignored, as many of these have numerous good properties; for example, the seed fibres of the poplar tree provide exceptionally good thermal and sound insulation and have hypoallergenic properties.

Woven fabric made of metallised fibres coated with a layer of silver is used for a number of purposes in the clothing industry, and is especially suitable for medical use because of its antibacterial action and warmth retention properties. It is only a matter of time before these fabrics will also be used for car seat covers, thanks to the following properties: long-term life cycle; optimum air permeability and weathering resistance; benefits to human health; and easy-care characteristics, among others.

It is difficult to predict future developments in the materials and machinery used for making car seat covers, particularly in light of previous developments of new materials from new raw materials, and the ongoing developments in surface treatments. Market globalisation of the automotive and textile industry will also have an impact on further development and fashions and customer preferences need to be taken into account. In order to determine development trends, the technique of sewing car seat covers must be considered as a complex process, determined by the level of development of the sewing machines, the materials, and the technology used to create multilayered composite materials. Textile materials have nonlinear mechanical properties that should be adapted to meet the requirements of their end-use. To achieve these goals, new technological approaches are necessary, including improvements in the methods of measuring process parameters and their optimisation, the design and monitoring of the process of sewing car seat covers, the selection of suitable materials and equipment, and the adaptation of the material and equipment to meet new requirements.[31]

16.6 Conclusions

Historically, the basic raw materials used for making car seat covers were wool, flax, cotton, leather and other natural materials. Nowadays these have largely been replaced by synthetic materials, which have even better properties.[3] In order to achieve comfort over long periods of sitting, as well as appropriate body posture, car seats and their covers must meet several construction standards:

- The seat structure must be strong and ergonomically designed according to proper body posture in a seated position.
- Each segment of the car seat cover must be properly designed from the most appropriate material in terms of durability and cost-effectiveness, and it must make a contribution to the safety and comfort of the seated passenger or driver.

- The car seat cover should have a thicker polyurethane sponge in sitting regions and the backrest.
- The outer layer of the material must have appropriate strength, elasticity, resistance to abrasion, pilling and ignition, good hygroscopic and hypoallergenic properties, as well as many other characteristics that are essential if the car seat cover is to meet high performance and aesthetic requirements.

Check fabrics sometimes have a certain level of weft distortion which can create a problem in cutting larger parts for car rear seats where the distortion can be most clearly observed. This problem can be solved if the cut parts are made smaller, thereby avoiding the issue of weft distortion. However, one consequence of this method is that the total length of the seam becomes longer over the car seat cover; furthermore, the risk of material deformation and tearing at the seams becomes greater.

Caution is necessary in the construction of several types of seam because their material strength is considerably lower than the average strength of seamless materials. Consequently, the strength and durability of the car seat cover as a whole depends on the strength of these areas. The goal of sewing cut parts is not only to create a secure join, but also to achieve a material whose properties at the seam are approximately equal (as far as is possible) to those away from the seam. Where the seam is stitched, the needle should not damage the material, as this causes a larger displacement of threads and fibres in the material. The optimisation of the upper and lower sewing thread, uniform material guiding, and selection of a suitable needle and sewing thread contribute to minimum material damage, uniform stitch length, and create a high quality seam with good aesthetic appearance.

16.7 Acknowledgement

The authors would like to thank the company Prevent d.o.o. for their assistance in writing this chapter.

16.8 References

1. Nikolić G, Rogale D, Ujevic D and First Rogale S., 'Textile material joining and cutting machines with new technologies – IMB 2000', *Tekstil*, 2007, 56, 308–317.
2. El Mogahzy Y E, *Engineering Textiles: Integrating the Design and Manufacture of Textile Products*, Cambridge: Woodhead, 254–322, 2009.
3. Fung W and Hardcastle M, *Textiles in Automotive Engineering*, Cambridge: Woodhead, 455–474, 2001.
4. Hardt K, Fischer T and Horstmann G, 'Gewebesimulation und Datenmodell: Grundpfeiler eines effizienten Informationsmanagements im Textilbetrieb', *Melliand Textilberichte*, 1990, 10, 748–750.

5. Schepers J, 'New generation of monitoring systems for the textile industry', *Melliand International*, 1998, 3, 174–175.
6. Zavec D and Geršak J, 'Influence of mechanical and physical properties of fabric on their behaviour in garment manufacturing processes', *International Conference Innovation and Modeling of Garment Manufacturing Processes, IMCEP 2000*, University of Maribor, Slovenia, 249–257, 2000.
7. Čunko R, 'Textiles – an important component in the contemporary car production', *Tekstil*, 2006, 55, 279–290.
8. Albrecht W, 'Nonwovens – outdated but modern anyway and outlook', *Tekstil*, 2000, 49, 568–570.
9. Höffer D, *Nonwovens*, Croatian Association of Textile Engineers, Zagreb, Croatia, 1976.
10. Anon. (2001), 'Nonwovens in West Europe', *Tekstil*, 2001, 50, 99–106.
11. Kovačević S, Ujevic D, Schwarz I and Brlobasic Sajatovic B, 'Analysis of motor vehicle fabrics', *Fibres & Textiles in Eastern Europe*, 2009, 71, 32–38.
12. Ujević D, Kovačevic S, Ha□tina J and Karabegovic I, 'Impact of Joined place on the fabric intended for manufacturing car seat covers', *2nd AUTEX Conference*, Bruges, Belgium, 1–3 July 2002.
13. Rudolf A and Geršak J, 'Influence of sewing speed on the changes of mechanical properites of differently twisted and lubricated threads during the process of sewing', *Tekstil*, 2007, 56, 271–277.
14. IMB Köln, Technical information firm Schmetz and Dürkopp Adler, 2006.
15. Rogale D, Ujevic D, First Rogale S and Hrastinski M, *Technology of Clothing Manufacturing with Work Study*, 2nd edn, Technical Faculty of the University Bihać, Bosnia and Herzegovina, 2005.
16. Juciené M and Vobolis J, 'Analysis of seam parameters in upholstery of upholstered furniture', *Tekstil*, 2008, 57, 352–356.
17. Tartilaité M and Vobolis J, 'Effect of fabric tensile stiffness and of external friction to the sewing stitch length materials science', *Medžiagotyra*, 2001, 7, 57–61.
18. Kawabata S M and Niwa M, 'Fabric performance in clothing and clothing manufacture', *Journal of Textile Institute*, 1989, 80, 19–50.
19. Vobolis J, 'Quality estimation of sewing machine V-belt drive', *Mechanika*, 2000, 23, 32–36.
20. Dobilaité V and Juciené M, 'Influence of sewing machine parameters on seam pucker', *Tekstil*, 2007, 56, 286–292.
21. Schröer W, 'Polyurethane coating on textiles', *Tekstil*, 1989, 38, 147–154.
22. The experimental part of this chapter results from the scientific programs (Advanced Technical Textiles and Processes, code: 117-0000000-1376; and Anthropometric Measurements and Adaptation of Garment Size System, code: 117-1171879-1887), conducted with the support of the Ministry of Science, Education and Sports of the Republic of Croatia.
23. Sengupta A K, De D and Sarkar B P, 'Anisotropy in some mechanical properties of woven fabrics', *Textile Research Journal*, 1972, 42, 268–271.
24. Sengupta A K, De D and Sarkar B P, 'Anisotropy of breaking load of woven fabric', *Textile Research Journal*, 1971, 41, 277–278.
25. Frontczak-Wasiak I, 'Measuring method of multidirectional force distribution in a woven fabric', *Fibres & Textiles in Eastern Europe*, 2004, 12, 48–51.
26. Frontczak-Wasiak I, Snycerski M, Stempleń Z and Suszek H, 'Isotropy of mechanical properties of multiaxial woven fabrics', *5th World Textile Conference Autex*, 27–29 June, Portorož, Slovenia, 2005.

27. Skoko M, 'Investigation of the properties with multiaxial strengths and deformations of coated fabrics', *Tekstil*, 1998, 47, 345–349.
28. Skoko M, 'Contribution to investigations of stresses and deformations of particularly loaded textiles for particular purposes', *Tekstil*, 1986, 35, 403–410.
29. Ferd. Schmetz GmbH, *Taschenbuch der Nähtechnic*, Schmetz, Herzogenrath, 1983.
30. Murárová A and Jambrich M, 'Profiled and hollow polyester fibers in nonwoven fabrics and their heat transfer', *Tekstil*, 2006, 55, 451–457.
31. Geršak J, 'Development trends in sewing techniques and garment engineering', *Tekstil*, 2001, 50, 221–229.

17

Joining of wearable electronic components

D. J. TYLER, Manchester Metropolitan University, UK

DOI: 10.1533/9780857093967.4.507

Abstract: Wearable electronics bring new challenges for joining technologies, as the electrical connections must fully support the functionality of the product and, at the same time, deliver comfort to the user. The connections should have the durability expected of textile materials: the ability to withstand repeated flexure in wear and the rigours of laundering. However, a decade after the launch of the first commercial product, despite numerous advances in joining technologies, the field should be perceived still as under development.

Key words: wearable electronics, e-textiles, apparel design, new product development, interconnects.

17.1 Introduction

The context for this chapter is wearable electronics. The vision is for people to wear clothing with embedded intelligence that imparts new capabilities. These garments are able to sense and be sensed; able to receive data and transmit it; able to interact meaningfully with the environments around us. When these ideas were first discussed, people referred to them as 'wearable computers'. The MIT Wearable Computing Lab has this manifesto statement on its home page:

> To date, personal computers have not lived up to their name. Most machines sit on the desk and interact with their owners for only a small fraction of the day. Smaller and faster notebook computers have made mobility less of an issue, but the same staid user paradigm persists. Wearable computing hopes to shatter this myth of how a computer should be used. A person's computer should be worn, much as eyeglasses or clothing are worn, and interact with the user based on the context of the situation.

To achieve these goals, the resource designed for the desktop has to be deconstructed and the various elements repackaged in ways that are compatible with clothing. The design brief is not just to provide wearers with computing power, but to incorporate a variety of sensors and communication tools to create new functionality and user benefits.

Whilst prototype garments have been with us for some time, there is still a significant gap between concept and commercial reality. Berzowska (2005a) writes: 'It is ironic that although they are powerful, these wearable computers are not very wearable'. Often, the materials used are hard and uncomfortable. In some applications, users find the clothing heavy or lacking in flexibility. Some materials do not drape well, putting constraints on design. Durability and performance in laundering may be an issue.

This provides the rationale for electronics to be embedded in clothing as conducting textile materials. Conductive yarns and textile-based sensors should be, in principle, as robust as conventional yarns and textile components of garments. Research (mostly proprietary) into these aspects has been extensive.

This is an emerging field and many avenues are being explored. Not all of them will survive the test of time. This review reflects the current state of the technologies on offer. Initial attempts to introduce electronic functionality to textiles made use of wires, but there are now a variety of conductive yarns available. Many of these yarns have been developed for anti-static applications and electromagnetic shielding and are not being marketed for smart textiles applications.

The first commercial product that can be classed as 'wearable electronics' was the Philips/Levi ICD+ jacket, launched in September 2000. This provided users with remote-control of a mobile phone and an MP3 player.

> Levi's ICD+ is an unprecedented collaboration between Philips, Europe's largest electronics company, Levi's, one of the world's largest clothing manufacturers, and designer Massimo Osti, who was last year voted most influential designer of the 90s by menswear magazine Arena Homme Plus. Between them, the trio hope to produce the first really credible, successful range of technical clothing. The first ICD+ range incorporates Philips mobile phone and MP3 technology into a range of jackets designed by Osti and made by Levi's. (O'Connell, 2000)

The project was ambitious. The jacket was intended to set the standard for a new generation of apparel.

> This autumn's jackets feature phones which can be dialled using voice recognition technology, and a microphone and earphones built into the hood or collar. The MP3 player automatically cuts out when the phone rings, and the whole lot is machine washable. Everything is controlled via a keypad hidden beneath a pocket flap. The most challenging aspect of the project, according to Osti, has been to soften and humanise the technology. 'The challenge is not miniaturisation as we currently think of it – making something smaller – but to make it flat and flexible so it becomes a second skin, completely invisible. We aim to massage the technology into the garment,' say Osti. (O'Connell, 2000)

According to reports, demand for the limited number of garments on offer was strong. After the autumn range of products, Levi produced designs for the spring market but, by June 2001, they withdrew the product from sale. According to a spokesperson, 'although Levi's brought out a spring range, we decided that it would be bad business simply to redesign the jackets without offering updated technology to go with the new designs' (Mallin, 2001).

The decision to discontinue was linked, at least in part, with technology factors. The electronic circuitry was of a traditional nature: a wire harness was used to connect the mobile phone, the MP3 player, the control panel and the audio devices. There were pockets for the devices and channels to carry the wiring. Connectors were also traditional (see Fig. 17.1). With this architecture, it was unavoidable that obsolescence was in-built.

Levi's ICD+ faded from view and was not replaced. Numerous other branded products appeared on the market, mostly using improved electronic circuitry and joining technologies. Many journals carried feature articles documenting progress, such as Service (2003) in the journal *Science*. The field is still emerging, and further comments follow in this chapter.

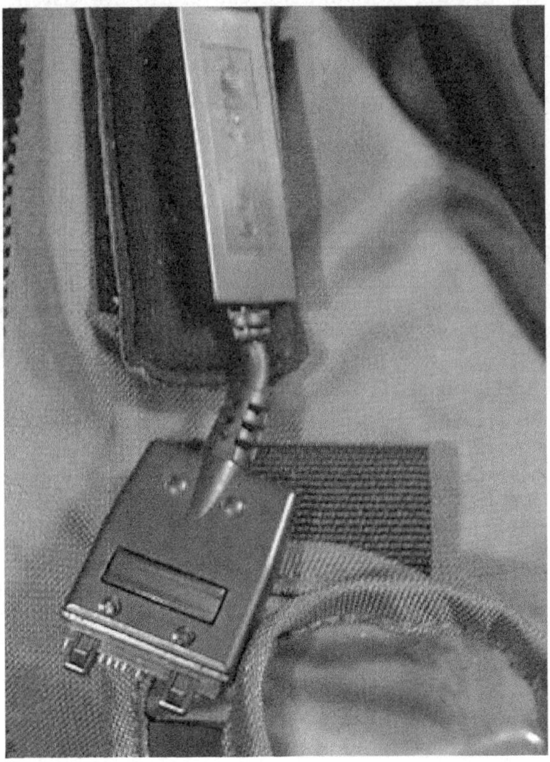

17.1 The mobile phone connector used in Levi's ICD+ jacket (source unknown).

17.2 Conducting fibres

The first step in developing wearable electronics is to make use of fibres and yarns that can conduct electricity. A range of technologies for achieving conductivity are employed. Suppliers of these yarns target application areas where there are financial benefits. Wearable electronic products are, at present, in the early stages of commercial development. Consequently, the technical data provided for the available fibres and yarns are selective and only limited comparisons can be made. Anderson and Seyam (2002) provided an overview of the different ways of imparting conductivity to textile materials, and further information can be found in Lam Po Tang and Stylios (2006) and Harlin and Ferenets (2006).

17.2.1 Metal fibres

These are made from stainless steel, silver, copper, aluminium and a variety of alloys. They have been produced as continuous filament and staple yarns. Blends of metal fibre and polyester fibre are also available. These yarns have high conductivity and are resistant to heat and to corrosion. Most of the applications are for products requiring anti-static properties (especially in carpets and industrial textiles) and the ability to provide electromagnetic shielding. Examples are from Epitropic Fibres Ltd, Bekaert and R.STAT (see Table 17.1).

Table 17.1 Selected metal fibres

Brand	Notes	Web link
Epitropic Fibres Ltd	Resistivities quoted for stainless steel fibres are 70 ohm/cm, for 20% steel/80% polyester yarns 1000 ohm/cm	http://www.epitropicfibres.co.uk/steel-fibres-02.htm
Bekaert Bekinox VS, LT and VN	VS (100% stainless steel staple), LT (blend of steel and polyamide fibres) and VN (100% stainless steel continuous filament fibres)	http://www.swicofil.com/bekintex.html
R.STAT/S	The range is composed of stainless steel fibres with diameters of 8 μ, 12 μ and 22 μ. They are available as spun or continuous filaments. Spun blends are with PES, nylon or aramid fibres. Resistivities are for 8 μ: 150–170 ohm/cm, 12 μ: 60–80 ohm/cm, and 22 μ: 10–30 ohm/cm.	http://www.r-stat.com/GB/gbconducteur.html

17.2.2 Carbon fibres

These conduct electricity and can be made into continuous filament and staple yarns. These yarns have high conductivity and are reputed to have good abrasion and wear resistance. An example is Hexcel carbon fibres. Epitropic Fibres Ltd and Swicofil AG produce carbon-impregnated polyester fibres designed to dissipate static. These have electrical resistance properties in the range 10–100 megaohms/cm. Resistat produce their Resistat® fibre collection and Sanstat® monofilaments by a carbon suffusion process which chemically saturates the outer skin of nylon fibres with electrically conductive carbon particles (see Table 17.2).

17.2.3 Conducting polymers

These have been developed to act as semiconductors. The most prominent of these are polyaniline (Pan) and polypyrole (Ppy). Whilst much research has been carried out, commercial products are still under development. Further information can be found in Harlin and Ferenets (2006).

17.3 Conducting yarns

This section is indicative of materials that need to be joined. Berzowska and Bromley (2007) discuss a number of conductive threads considered prior to constructing an 'animated quilt'. The web links provided in Tables 17.1 and 17.2 also carry information about yarns made from conducting fibres.

17.3.1 Metallic or metallised yarns

Yarns incorporating metal fibres as a component are known as metallic yarns. In multi-end yarns, one end might be completely metal, and this has the best potential to give good, reproducible conductivity.

Ohmatex (see Section 17.9) has found that the best results are achieved

Table 17.2 Selected carbon and carbon-impregnated fibres

Brand	Notes	Web link
Hexcel HexTow™ fibres	Continuous and staple	http://www.hexcel.com/Products/Carbon+Fibers/
Epitropic fibre	Electrical resistance quoted as 10–100 megaohms/cm	http://www.epitropicfibres.co.uk/products.htm
Resistat® fiber collection and Sanstat® monofilaments	Primarily used in antistatic applications	http://www.resistat.com/

by combining natural or artificial textile fibres with silver-coated copper filaments – insulated or uninsulated depending on the application. For example, uninsulated filaments are used in embroidery yarns designed to establish an electrical connection. Electrical resistance is said to be unaffected, even though some conductive fibres break.

17.3.2 Coated yarns

Some yarns have a conducting layer deposited on the fibres by chemical means, by electrical activity or by using adhesives. Examples of coatings are metals, titanium dioxide and tin dioxide, antimony dioxide.

17.3.3 Yarns with a conductive filler

Fillers may be carbon black, fragmented carbon fibre or graphite. The filler is engineered to match the needs of the application, but these yarns typically have a low conductivity. Although carbon is electrically conducting, particles embedded in a polymer do not provide a good pathway for current flows.

17.4 Fabrics and composites

For a robust product, the electronic elements should have the same ability to extend as the fabric, without any adverse effects on performance. In the past, this has been a problem for wiring contained within the garment, which has led to failure in use. Hence, the interest now is to embed the electronic elements within the fabric, so that the integrity of the product can be maintained. As well as the textile compatibility issues, two or more adjacent conducting yarns introduce challenges for maintaining signal integrity. Crosstalk is a significant issue to address, and some relevant research is reported by Dhawan *et al.* (2004b).

17.4.1 Ribbon cable connectors

Conductive ribbons are narrow woven fabrics, with non-conducting weft yarns and an engineered mix of non-conducting and conducting warp yarns. The multi-channel cables are suited to connecting sensors and displays in clothing with electronic control devices. They are designed to be 100% washable and robust in use.

The ribbon illustrated in Fig 17.2 has conductive yarns made from a mixture of polyester and varnish-covered metal filaments. The varnish provides insulation and protection. It is removed for the purpose of making connections. Ohmatex suggest the use of laser stripping methods, a soldering bath (temperature above 450°C) or chemically, using sulphuric acid. The

17.2 Conductive ribbon with metallic conducting filaments. The end of this four-cable ribbon is being prepared for making connections (courtesy of Ohmatex ApS).

connections can make use of standard connecting technologies discussed in Section 17.5.

17.4.2 Woven and knitted fabrics incorporating conducting yarns

Much of the research effort into wearable electronics has been concerned with fabrics: embedding sensors and circuitry and incorporating functionality. A review by Seyam (2003) showed the potential. Much of this work has adopted a multilayer design approach: two or three layers are constructed. One layer is internal and is designed to enhance wearer comfort. The next layer carries the electronics and provides functionality. The third layer, if present, acts as the skin, enclosing the electronics and, if appropriate, supplying additional protection. There have been numerous application areas in medicine and sport, but these lie outside the scope of this chapter. A listing of commercially available conductive fabrics is given by Ghosh *et al.* (2006). Further comments can be found in Lam Po Tang (2007).

17.4.3 Quantum tunnelling composite

Wearable electronics applications have made significant use of a new class of flexible polymer: the quantum tunnelling composite (QTC). This is patented technology, owned by Peratech. The base polymer is usually silicone rubber and it contains microscopic metal particles with engineered irregular surfaces. When uncompressed, the material is an electrical insulator with a resistance of 10^{12} ohms. Under pressure, the QTC material becomes a conductor, with a resistance that can be less than 1 ohm. This behaviour is quite different from a carbon composite, whose resistance can change under pressure from several thousand ohms to several hundred. The reason for the difference is quantum tunnelling.

In classical physics, the electron is considered a particle, and the barrier between conducting particles can only be jumped if the electrons have enough energy. Thus, a carbon composite is conducting when particles touch, and pressure increases the number of particles touching. Quantum theory considers electrons as waves, and probability considerations govern whether they can pass through a potential barrier. The metal filler particles in a QTC material are engineered so they do not physically touch, but rather they provide the opportunity to vary the potential barriers faced by the electrons. All current flow occurs through electrons tunnelling through these barriers. Further information on this can be found on the Peratech web site: see Section 17.9. These materials have been used as switches and sensors in wearable electronics applications. For further comment, see Section 17.7.1.

17.5 Connecting technologies

Connections are needed between conductive yarns and sensors (and other devices) located in the product, and between conductive yarns and the terminal zone where signals are transferred away from the product. These connections must be robust in use. For example, if they are not detachable for laundering, the electronic components need to be durable to withstand washing. The literature base on these connecting technologies is limited. Lehn *et al.* (2004) identified nine factors to consider when choosing a connection method: physical strength, electrical reliability, ease of attachment, repeatable re-attachment, aesthetics, size, comfort, cost and availability. They discussed soldered connections, buttons or snaps, ribbon cable connectors and something they called raised wire options. This led to a table of advantages and disadvantages, also presented and summarised by Ghosh *et al.* (2006). A craft-orientated perspective is provided by Buechley and Eisenberg (2009), focusing on three novel techniques for making e-textile technology available to 'crafters, students and hobbyists'. These are: iron-on fabric printed circuit boards, the use of electronic sequins and the use of socket buttons. Their techniques for achieving electrical connectivity are, however, the same as those used by others working in this field.

An indication that 'wearable electronics' is still in its infancy came from a workshop to address design issues. One presentation was by Papadopoulos (2007) who reflected on ten years of working in this field:

> Still, lack of adequate connectors between conductive threads that have the tendency to fray, and attractive connectors for removable, rechargeable

batteries, make the development process so much more time consuming and frustrating. It also makes it harder for the unseasoned practitioner to innovate and focus on the aesthetics and 'content' of the work. At this stage we have experimented with all sorts of connectors: magnets for our custom made battery boards, snaps, pins, sam browne buttons … we have tried it all and we have yet to come up with a satisfying solution and one that would make manufacturing easy. At this point our garment construction is more akin to couture than to ready-to-wear. When we have the right tools and connectors will wearable technologies take off?

The clear implication is that making connections that meet the expectations of product designers requires significant development work.

17.5.1 Soldered connections

Solder is an option with metal yarns that can withstand the high temperatures. It is essential that the solder flows, wets and results in a good electrical connection. Traditionally, tin-lead alloys solders have been used in the electronics industry, but the toxicity hazard combined with wider environmental concerns have restricted their use. Japan banned solders containing lead in January 2005 and the European Union prohibited their use in July 2006. The acronym RoHS is used within the EU. This stands for Restriction of Hazardous Substances. The adopted legislation bans six hazardous substances from manufacturing processes: cadmium (Cd), mercury (Hg), hexavalent chromium (Cr (VI)), polybrominated biphenyls (PBBs), polybrominated diphenyl ethers (PBDEs) and lead (Pb). In response to these developments, two alternatives have been developed: lead-free alloys (considered below) and electrically conductive adhesives (considered in Section 17.5.3).

According to Li *et al.* (2005), tin/silver/copper alloys yield the most promising lead-free solders. They report that a mixture of 95.4% tin, 3.1% silver and 1.5% copper provides optimal strength, fatigue resistance, plasticity and reliability. The main drawbacks concern their higher processing temperatures and reduced reliability of the solder joints.

The processing temperature problem can be gauged by looking at minimum melting points. The lowest melting point for a tin-lead alloy is 183°C. Tin melts at 232°C and the optimal strength tin/silver/copper alloy melts at 217°C. Consequently, the processing temperatures in assembly are 30–40°C higher with lead-free solders. The practical effects of this are to impair the reliability and functionality of the various components.

Fatigue resistance is an issue with soldered connections, and this problem has been noted with textile and clothing products. The durability of the join and the mechanical properties of the solder must be compatible with the intended use, especially when textile products experience repeated flexure.

However, engineered solutions are feasible and soldered connections remain a major means of achieving joins (Figs 17.3 and 17.4).

17.5.2 Welded connections

Welding is a general term covering joining or sealing operations. Most uses in textiles/clothing relate to joining two layers of material with sufficient thermoplastic content to obtain a flow of molten polymer with subsequent bonding. The most frequently used technologies are as follows:

17.3 Conductive yarns in a woven textile ribbon are connected using lead-free solder to a small printed circuit board (courtesy of Ohmatex ApS).

17.4 A laser has been used to expose metal filaments in a conductive ribbon, which are then soldered to a printed circuit board. These components should be subsequently encapsulated for protection against strains and moisture (courtesy of Ohmatex ApS).

- ultrasonic welding (the use of a patterned horn or anvil which vibrates at ultrasonic frequencies to impart energy to a polymeric substrate sufficient to melt it)
- high frequency welding (where the substrate materials are clamped between two electrodes between which a radio frequency electric field is applied; the materials must contain suitable polymers which are stressed by the applied field, thereby increasing their temperature to achieve melting and flow)
- hot air welding (where a nozzle is used to direct hot air onto a thermoplastic polymer that melts, flows, and forms a bond when it solidifies)
- laser welding (where the radiation energy is transmitted through the materials to be joined but absorbed by a polymeric film in the joining zone; the film melts, flows, and forms the bond)
- resistance welding (heat energy is generated by the flow of a welding current because of the contact resistance between two conducting materials; the best known example of resistance welding is spot welding).

These technologies have been considered for their potential to join conductive textile materials. Dhawan *et al.* (2004a) considered resistive welding to have 'the greatest potential for applications in an automated manufacturing environment' and reported measurements of electrical resistance of interconnects using this technology. Resistance welding technologies (microspot welding) feature in Post *et al.*'s (2000) study of e-broidery. Applications to date all appear to be in experimental work and prototypes: the major issue for commercialisation is durability.

17.5.3 Conductive adhesives

Electrically conductive adhesives (ECAs) have been developed as a way of replacing tin-lead solder in the assembly of electronic circuits. ECAs have two components: a polymeric resin and a metal filler. The resin may be thermosetting (e.g., epoxy or silicone) or thermoplastic (e.g., polyimide). This component has to provide the necessary physical and mechanical properties of adhesion, mechanical strength and impact strength. The metal filler can be particles or flakes of silver, gold, nickel, copper or carbon, in sufficient quantities to ensure good electrical conductivity. Silver is most commonly used because it has a conductive oxide and is cost-effective. There are numerous categories and variants that have been explored by developers of ECAs and a useful review is by Li and Wong (2006).

ECAs bring both advantages and disadvantages when compared with conventional solder technology. The recognised advantages include improved environmental impact, gentler processing conditions (allowing the use of heat-sensitive and low-cost components and substrates), and fewer processing steps (reducing processing cost). However, their disadvantages have restricted commercial applications: electrical conductivity is lower than metal alloy solders, conductivity fatigue is a problem (elevated temperatures, high humidity and wear-and-tear in use can impair conductivity), the connections can fail reliability tests, there is limited current-carrying capability, and the impact strength is poor. As can be expected, such performance issues are the subject of research and the situation is not static. However, the research has a focus on the needs of the electronics industry and the specific interests of companies producing wearable electronics do not attract the same resources.

Conductive adhesives are a promising way for bonding yarns to other yarns and substrates. They are non-toxic, conductive and can be engineered to have properties compatible with the intended end use. Examples of their use are provided by Berzowska (2005b) and Berzowska and Bromley (2007).

17.5.4 Pressure-based connectors (grippers, staples, stitching, crimping, snap fastener or press stud)

The principle with all these techniques is to bring wires or conductive yarns firmly together and to secure them so that a good electrical connection is maintained. Grippers can be spring-loaded or screwed down to ensure that there is a connection between the conducting wires/yarns. In most cases, if needed, the pressure can be released (i.e. the connection is reversible). Staples can be used to perform the same task as grippers: a metal staple effectively clamps the materials in place and achieves a conducting pathway. Stapling is non-reversible.

Stitching uses a conductive yarn as a sewing thread. The seam should be of short length but with a high stitch density. Post *et al.* (2000) discuss ways of creating circuitry by embroidery using conductive threads. Bartacking machines provide a route for joining one conductive textile to another: a conductive thread is sewn to ensure it has electrical contact with the conducting materials in the fabrics being joined. Crimping makes use of suitable metal holders into which the wires/yarns are inserted and then the holder is permanently deformed or crushed to ensure an electrical connection. Crimping is non-reversible. A typical application is for crimping to join the conducting materials to a metal connector which can then be used in conventional ways.

The snap fastener or press stud is a particular type of gripper made up from four components. They are generally used for attaching one fabric to another and, given that the components are conducting, they can establish an electrical connection between two conducting fabrics. Part of the attachment process involved metallic teeth penetrating the substrate, and these teeth can be designed to maximise contact with conductive yarns. For an example, see Fig. 17.11.

As a general comment with all these means of joining, any electrical contact has the potential of being degraded with wear and tear. The conducting materials may be affected by mechanical and electrical fatigue, and there may be chemical changes (such as oxidation). Prototyping may give good results, but the performance in use may become impaired.

17.5.5 Connectors for use with ribbon cables

ProeTex connector from Ohmatex

The ProeTex project was to produce a jacket with an integral communications system. The connector (Fig. 17.5) was designed to provide a connection between the sensor arrays in the inner garment and the electronic controls in the jacket itself. It was designed as a 'press fastener' type solution with strain alleviation on the soldered connections from 'cable grips' in the plastic housing. The connector is still at the prototype stage. According to Ohmatex, the concept appears to be viable and will be further developed in future projects.

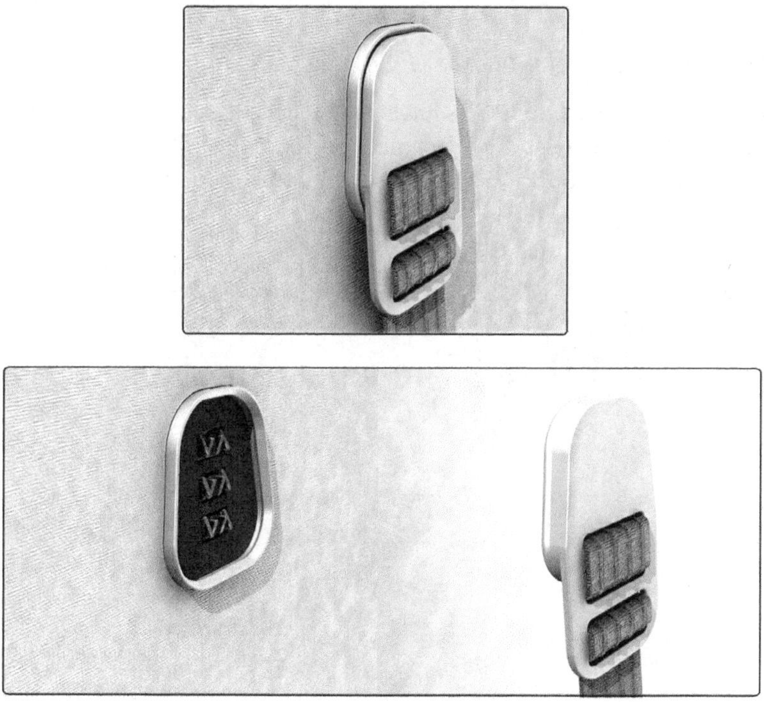

17.5 The ProeTex connector for strain alleviation (courtesy of Ohmatex ApS).

Stitching (bartacking)

Another example of a ribbon cable but with different connections comes from the Inteltex project (an FP6 EU project) and is illustrated in Fig. 17.6. The project is developing the concept of embedding intelligent materials within yarns and using them to sense the environment: detecting strain, shear, temperature or liquids. Textile sensors are connected to textile cables and then to an external (removable) electronic pack.

The sensor illustrated is a rectangular textile sample with a cotton weft and black warp sensing threads. The sensing utilises carbon nanotubes carried by polymeric fibres. The connector ribbon has three conducting pathways (silver-loaded), although in this example only two are active. One end of each conductive pathway is sewn onto the sensor using silver coated threads using bartack stitching. The other end of the connector ribbon has gold-plated crimp connectors to attach the assembly to the external electronic pack.

According to Peratech, this configuration leads to a very tough sensor with connectors that will withstand flexing, pulling and tugging as well as wetting or even washing. A similar assembly is used to make the bank of textile switches which are laid into the arms of ski jackets to control iPods within the pocket of the jacket.

Fabric printed circuit boards (PCBs)

Although conventional PCBs have been incorporated into wearable products, and textile-based PCBs are being made at a craft level (Buechley and Eisenberg, 2009), textile-based PCBs for commercial use are still at the development

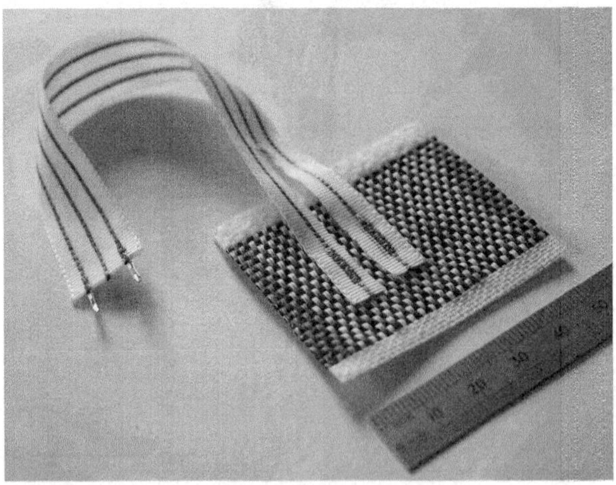

17.6 Textile sensor bartacked to a conductive ribbon with crimp connectors to link to external circuitry (courtesy of Peratech).

stage. Sefar (2009) has announced its PowerMatrix fabric, saying that it can be used as entirely flexible circuit board in order to connect electronic components. The fabric itself is a mixture of PET monofilaments and insulated copper monofilaments in both warp and weft. Resistors, capacitors and integrated circuits can all be assembled with soldered or conductive adhesive connections and any desired wiring structure can be established. Comments on earlier stages of development appear in Kirstein *et al.* (2006) and Locher and Tröster (2008). An illustration to show contemporary use is shown in Fig. 17.7.

Four manufacturing steps for creating an interconnect between copper wires are described by Locher and Tröster (2008):

1. Coating removal on copper wires at defined intersections using laser ablation.

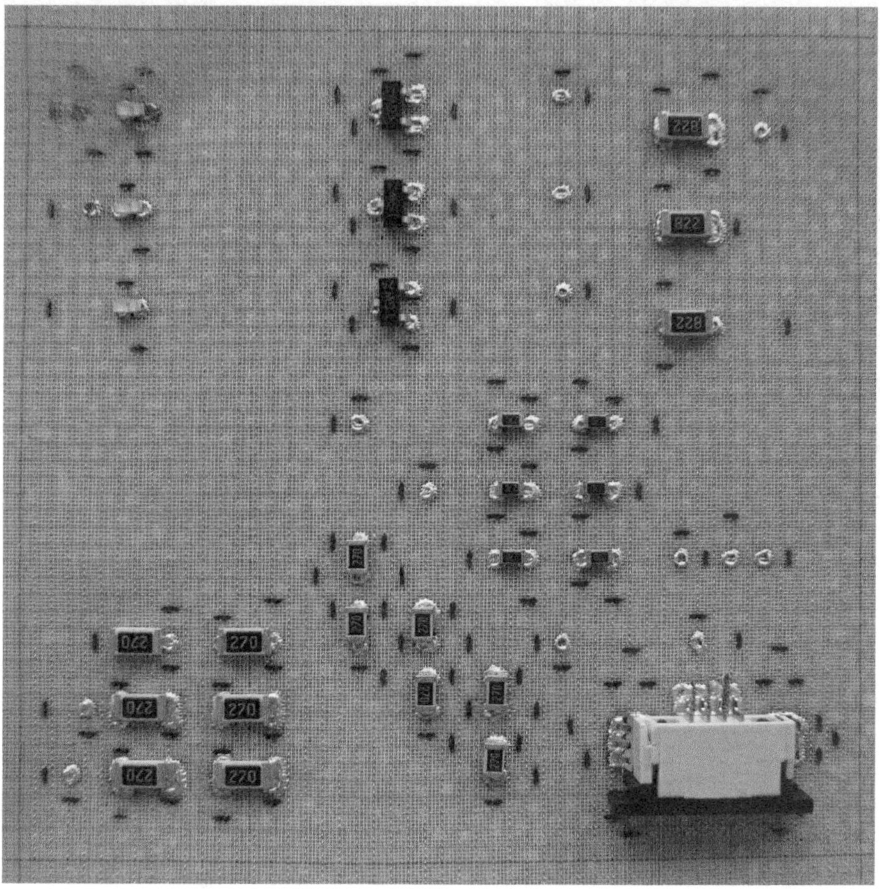

17.7 The Sefar PowerMatrix fabric with electronic components soldered in place (courtesy of Sefar AG).

2. Cutting of certain wires in order to avoid short-circuits with the rest of the routing. This uses a high energy laser.
3. Interconnection of the skinned wire sections. A drop of conductive adhesive is dispensed on the ablated spot connecting the two crossing wires.
4. Adding mechanical and electrical protection to interconnection. Epoxy resin is deposited to provide an insulating covering.

The documentation provided refers to Sefar supporting customers by specifying and defining the product according to stated needs. As an example of this, different cables/connector solutions may be needed depending on the electrical current drawn. This is particularly the case with heating applications as high currents are needed and, therefore, the cross section of the cable and connector type must be suitable.

17.5.6 Summary of techniques

A brief overview of the different techniques is presented in Table 17.3.

17.6 Requirements of electronic interconnects

It must be emphasised that connector design and selection is but a small part of a design framework for wearable electronics. Thinking at this level became more prominent after 2003. Prior to this time, it could be said that: 'E-textiles to date have been created in a trial and error fashion (prototype,

Table 17.3 Summary of joining technologies for conductive yarns

Technique	Application areas	Fixed or detached	Performance aspects
Soldering	Metal yarns	Fixed	Localised temperatures exceeding 200°C Fatigue and durability issues
Welding	Metal yarns	Fixed	Fatigue and durability issues
ECAs	Most conductive yarns	Fixed	Gentle processing conditions Lower conductivity Reliability and poor impact strength
Pressure-based connectors	Most conductive yarns	Detachable or fixed	Potential for degradation in wear
Ribbon cable connectors	Ribbons	Detachable	Prototype stage of development
Fabric PCBs	Most conductive yarns	Fixed	Prototype stage of development

evaluate, prototype, evaluate)' (Martin *et al.*, 2003). The problem is that the design-state space for e-textiles is very large and the challenge is to find a structured way of moving towards an effective and potentially commercial solution. The diversity of design issues was illustrated by Martin *et al.* (2003) using eight categories:

- physical environment
- sensor behaviour
- human body and motion
- motion/draping of clothing
- manufacturability
- networking
- power consumption
- software execution.

Most, if not all, of these have implications for interconnections. Notably, the means of joining conducting materials must be compatible with the physical environment (whether it be a firesuit or a sports shirt). Furthermore, the stresses and strains experienced during use must not lead to premature failure of the joins. These design considerations can be expressed in terms of three broad areas for assessing the appropriateness of a specific connection technology. These are discussed in the sections below.

17.6.1 Functionality issues

Connectivity should not be a dominant consideration affecting product design. Making connections is a means to an end. The goal is to achieve a product that has the functionality intended by the designers. This puts the spotlight on the sensors and other active elements of the e-textile system. Connectivity should then ensure that this functionality is not eroded by compromises in the way the electrical connections are made.

17.6.2 Comfort, ease of use and well-being issues

Considerations of comfort relate to the finished product, of which connectivity is but a small part. The NuMetrex sports garments (Section 17.7.3) have received praise because the sensors and the transmitter are unobtrusive. It can be noted that the means of connecting the sensors to the transmitter helps to achieve this level of consumer satisfaction.

In a study of consumer attitudes to iPod jackets, Anderson and Lee (2008) identified six adoption factors from various sources. These were: trialability (before purchase), convenience (ease of use), perceived social prestige, compatibility (meeting consumer needs), complexity (difficulties of understanding in use) and observability (visibility to peers). The respondents were students considered to be likely adopters of iPod jackets. Of these factors, convenience and compatibility emerged as the most significant. This led the authors to investigate in more detail the convenience factor – analysed as follows (with the responses in brackets):

- less hassle untangling headphone wires (87%)
- no loose headphone wires catching on objects (57%)
- less time searching for the iPod in bags and pockets (39%)
- no software or assembly required, making it immediately beneficial (35%).

The authors draw the conclusion that convenience and compatibility issues should be more prominent in the thinking of product developers:

> The most important adoption factors are convenience and compatibility, and the least important are perceived social prestige and observability. This finding might not be intuitive, considering that potential consumers of wearable electronics, especially technology-savvy young adults, are thought to be greatly influenced by external forces such as peer pressure, trends, and perceived social prestige. Our research suggests to the contrary, that consumer adoption of an iPod jacket is likely to be driven primarily by convenience, compatibility, complexity, and usefulness. We suspect that the struggle for mass consumer adoption of smart wearable electronics is partly due to companies' misguided overemphasis on products' social prestige and visibility instead of product convenience and compatibility.

Medical applications have greater demands on functionality and also on wearer comfort. A questionnaire survey by Anliker *et al.* (2004) was concerned with a wearable medical monitoring and alert system. They found that user feedback 'underline[d] the importance of the visual and comfort aspects of the device. It is required that the functioning of the device does not hamper the daily life of the patient.'

Comfort issues affecting wearable computing products have been reviewed by Knight *et al.* (2002). Their analysis has six dimensions: emotion, attachment, harm, perceived change, movement and anxiety. The objective was to aid designers to identify the modifications needed to make wearable computers more comfortable. The topic has subsequently been discussed by Bodine and Gemperle (2003) and Duval and Hashizume (2005), showing that there is a strong interconnection between functionality and perceived comfort.

17.6.3 Performance issues

Numerous wearable electronics prototypes proceed no further. This is because the prototype is just not fit for regular use. The concept may be feasible, but the reality is that consumers will not purchase products that degrade quickly or fail. The major concerns here are robustness in wear and durability in laundering. The product will experience numerous stresses and strains in use, and the connections may start to weaken leading to impaired functionality. Combine this with the rigours of laundering, and the challenge for product developers is substantial. An example of this problem affecting connectivity is given in Section 17.7.2.

17.6.4 Manufacturability issues

In 2002, I had the opportunity to talk with someone who was involved with the launch of the Philips/Levi ICD+ jacket. I asked what had been learned from the experience. The answers were informative:

1. Wires were considered by the technologists to be a necessary part of the product, but consumer resistance was high. Wires were not welcome.
2. Communication problems were numerous. Levi's worked at a different pace from Philips. The Philips engineers used a different language to the designers: instead of 'warp and weft', the engineers responded to 'matrix'.
3. Grading of the wire harness. This was an unexpected problem. It all had to be done by Philips, whereas the garment patterns were graded by Levi's.
4. Making up was in the Far East. The Philips staff spent two weeks with the manufacturer and thought all was going well. However, after they left, hundreds of garments were ruined because the machinists sewed over the wires. There was no real understanding of the damage this would cause.

This experience shows not only that multi-disciplinary teams are crucial, but also how important it is to manage the process (Tyler, 2008a). It appears that none of the team members had expertise in branding and market research and the original Levi product did not fit well with market expectations/needs. Philips had some significant learning to do about the brand awareness of the clothing market. Levi's appear not to have appreciated the importance of 'production' involvement in the product development team – as the products were made up in the Far East and production people were deemed to need 'support' (rather than contributing to product development). As is typically the case, manufacturing is where expensive mistakes can be made and where late stage changes become very costly.

Lu and Wood (2006) point out that product design, process design and manufacturing are three distinct stages in bringing new products into existence. This means that there are two interfaces (product design – process design) and (process design – manufacture) which need to be managed effectively if design-for-manufacture is to be achieved.

Many wearable electronic products bear the marks of inadequate new product development and the 'limited editions' should be regarded as prototypes rather than commercial products. There are problems when electronic engineers work with garment technologists and garment designers: the three groups have different cultures, operate on different timescales and often have a different professional language.

An additional factor to consider with wearable electronics is the end-of-use phase. Will the garment be recycled or reused? The landfill option is increasingly unrealistic. Since 2007, there have been EU regulations on the disposal of waste electrical and electronic equipment (WEEE directive). If the wearable electronic product falls into any of the categories covered by the directive, then there is producer responsibility for disposal. There are inevitably ambiguities, and product development teams need to be alerted to aspects that may affect design decisions. It seems clear that medical products are covered by the category: 'other appliances for detecting, preventing, monitoring, treating, alleviating illness, injury or disability'. In other industries, formal tools permitting Design for Environment have been used to ensure products can be disassembled for reuse, recycling or disposal as appropriate. To achieve this in textile/clothing products requires the adoption of concurrent product development approaches (Tyler, 2008a, 2008b).

17.7 Applications

The market for wearable electronics is barely a decade old, and there is limited expertise in design, marketing and garment technologies. In this section, applications are presented as case studies to give some insight into products offered to consumers.

17.7.1 Softswitch/Eleksen/Qio controls for jackets

Subsequent to the Philips/Levi ICD+ jacket, other brand-owners have tested the market by producing limited edition versions of this garment. These garments have provided companies with first-hand experience of the technologies, and have given consumers opportunities to explore the potential of wearable electronics. The technology lead during the past decade has been with two companies: Softswitch and Eleksen. Softswitch was acquired by Peratech in 2006, and Eleksen in 2008. The textile-related applications of these technologies are now handled by a newly formed independent company called Qio. This company has developed an affordable system to promote wearable electronics: PANiQ (2009). Concept images to show potential for garment applications are presented in Fig. 17.8.

As examples of products using these technologies, Gul International has

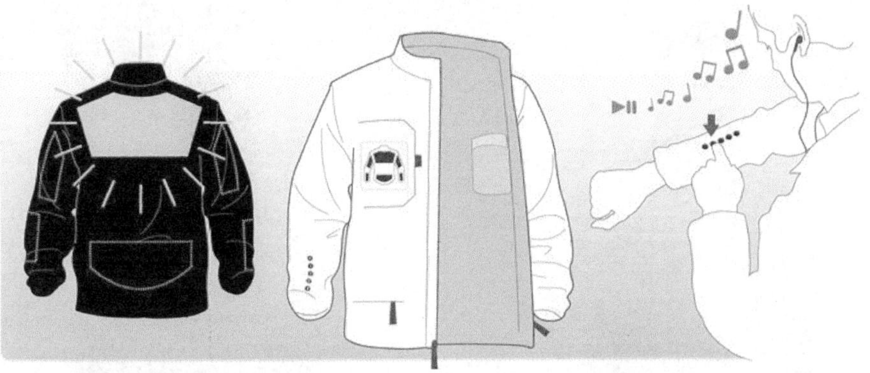

17.8 Wearable applications for military, industrial, sports or consumer products. QTC sensor provides a flexible control interface for high-visibility garment, allowing the wearer to select which part of the garment to illuminate for optimum safety. Playback buttons on sleeve to control music/mobile device without any contact with device (image courtesy of Peratech).

introduced the Vigo Mens and Ladies Coastal Jacket to its 2009 range. The iPod control version is only available to consumers in North America. Also in 2009, Bailo introduced the BTR Rocket jacket for men and women, with integral iPod control. The Qio eSystem solution is supported, so customers have the option of enhancing the functionality to include a Bluetooth module, AM/FM radio or a Walkie-Talkie system. Further information about these two products can be found from the relevant websites listed in Section 17.9.

It is possible for consumers to enhance existing jackets: by acquiring the My Sew EZ Touch Kit. The sales blurb says: 'The kit has everything you need to upgrade your jacket to control your iPod or Cell Phone right from the sleeve.' The main components are: a standard six-button PANiQ Wearable User Interface (ElekTex Textile Touchpad) and a PANiQ Controller for iPod. Further information is available on the relevant websites listed in Section 17.9.

17.7.2 VIKING fire-fighters jacket

VIKING Life-Saving Equipment is a global market leader in maritime and fire safety. In 2009, it introduced the first fire suit to have built-in thermal sensor technology (TST). Thermal sensors built into the clothing provide a warning to fire-fighters of critical temperatures that will cause heat stress and burn. Two sensors are used to monitor temperatures near the fire-fighter: one on the inside and the other on thc outside of the suit. LED displays are provided in two places: the left sleeve display provides a warning to the wearer and the upper left shoulder alerts other fire fighters in the team (see Fig. 17.9). The LED display design has a long line and an outer circle. The

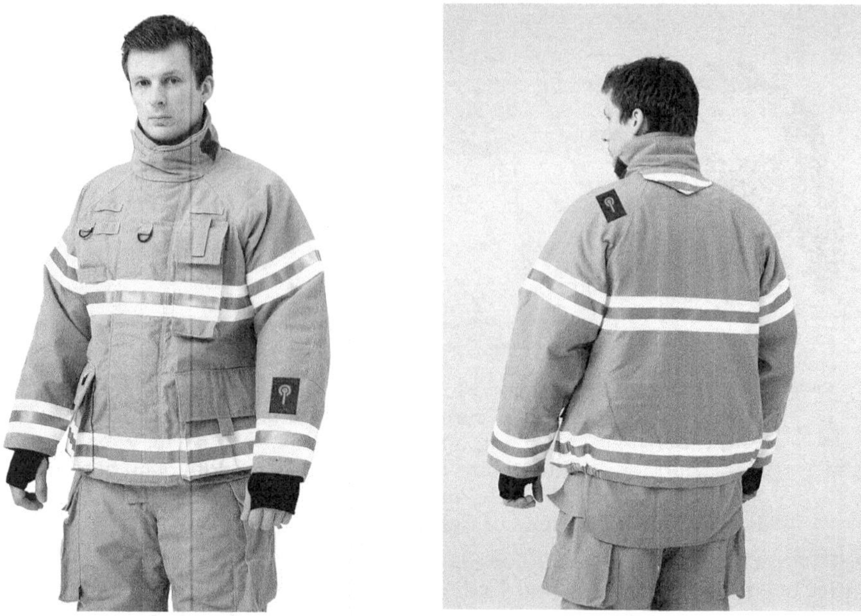

(a) (b)

17.9 Viking TST suit (images courtesy of Viking Life-Saving Equipment).

long line conveys information about the temperature inside the suit and the outer circle does the same for the outside. The details are as follows:

- outer circle flashes slowly when external temperatures reach about 250°C
- outer circle flashes rapidly when external temperatures reach about 350°C
- inner circle flashes slowly when internal temperatures reach about 50°C
- inner circle flashes rapidly when internal temperatures reach about 68°C.

The electronic circuit is powered by a removable battery that is stored in the lining of the suit along with a control chip that calculates the temperature and activates the LED displays. The sensors are embedded in the suit and covered with flexible waterproof plastic to protect against fluids and heat. Initially, the connections between the various components were made using conductive ribbon sewn into the fabric layers. However, a performance requirement for the microelectronic circuitry is that it must withstand at least 25 wash cycles – after the battery has been removed from its pocket. Unfortunately, durability problems were experienced with the conductive ribbon and its connections with the sensors/displays. As a consequence, more traditional wiring (metal wires and soldered connections) was used in the product released to the market.

17.7.3 NuMetrex sports bra and sports shirt

Research efforts in many countries have addressed medical and sporting applications (e.g., Catrysse *et al.*, 2004; Van Langenhove and Hertleer, 2004; Paradiso *et al.*, 2005). Both sectors have an interest in monitoring and this extends, with medical applications, to diagnosis. A review of this field was conducted by Solaz *et al.* (2006). The authors point out the advantages of intelligent textiles being in contact with the skin: they are able to capture biometric data and they can be designed to follow the user's movements. Two drawbacks are identified: the high impedance of skin contact for recording physiological signals and the draping movements of garments. These are the major design issues faced by product developers.

The application considered here concerns the monitoring of heart rate during sport by sensing its electrical activity. Medically, this approach results in an ECG (electrocardiogram), but for sporting applications, data gathering and analysis can be simplified. Wearable electronic solutions make use of textile sensors in direct contact with the skin. Most of these sensors make use of stainless steel fibres: suitable conductivity, low toxicity, no allergy problems, easily laundered, easily woven or knitted and comfortable in wear (Solaz *et al.*, 2006). The garments are designed to bring the sensors in contact with the skin and to maintain contact as the wearer moves around. The electrical signals recorded by the sensors are then processed and sent to an output device. Some of the early research utilised a skin gel to obtain a good contact with the electrodes, but more recent initiatives gather the signal without such facilitators.

The NuMetrex Heart Sensing Sports Bra was launched by Textronics, Inc. in December 2005. The following year, it was named 2006 Sports Product of the Year by the Sporting Goods Manufacturers Association. The Cardio Shirt for Men was released in 2007 followed by several other variants. The garments are made from nylon/Lycra® fabric with electrodes knitted into the garment forming a strip positioned at heart level, as shown in Fig. 17.10. The signal is processed by the control electronics and transmitted to a wrist watch that displays the heart rate.

The electrodes are knitted using uncoated metal filaments which are in direct contact with the skin. The connections between the transmitter and the metal filaments forming the electrodes are achieved using press studs, as illustrated in Fig. 17.11. As the metal filaments are uncoated, the press stud mechanism results in a robust low resistance connection. This arrangement can withstand repeated washing.

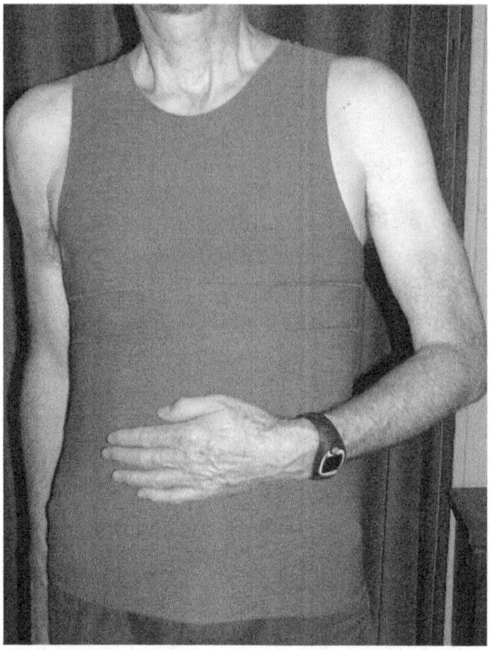

17.10 The NuMetrex Cardio Shirt for Men (source: D. Tyler).

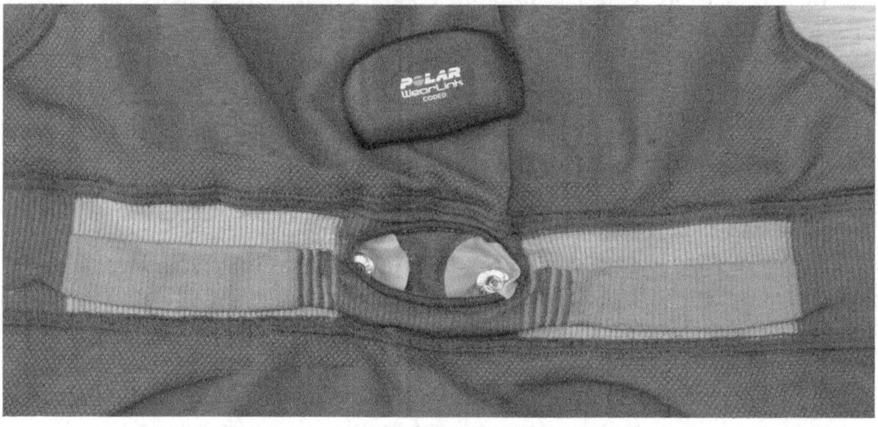

17.11 Cardiac Shirt electrodes, transmitter pocket and transmitter, showing press stud connectivity (source: D. Tyler).

17.8 Future trends

The market for wearable electronics continues to be uncertain, and most products that are launched are limited editions. Product developers, brand owners and consumers are all changing with time. The two things that all are agreed on are that the potential market is very large and that the outlook is very promising (McCann and Bryson, 2009). One of the major challenges is to build product development teams that communicate well and can develop common goals. However, in the past, it has not been easy to get textile and garment specialists to work together, and, with wearable electronics, the need is to bring in electronics expertise also. Team-based approaches, committed to the principles of concurrent product development, would appear to be essential (Tyler, 2008a). Sensor development must be a key issue for gaining a market for these technologies. From the discussion above, such developments must always consider both electrical and textile interfaces.

Applications for nanotechnology are numerous, as shown by a review by Coyle *et al.* (2007). Specifically, these authors identify new applications in the areas of self-cleaning, sensing, actuating, and communicating. Peratech has actively researched switching and sensing applications that utilise QTC. Wearable electronics is just one of many applications that are to be found in a wide range of industries. Current research involves the exploitation of nanotechnologies: the INTELTEX project (see Section 17.9) is concerned with creating smart, multifunctional materials. Adding nanotubes to a base polymer can give fibres and yarns the same properties as QTC materials. Temperature and pressure change affects the configuration of these nanotubes and the conductivity of the yarn changes. This change in conductivity can be picked up by electronic monitoring circuitry. Applications are being designed to monitor near skin temperature, exterior temperature and mechanical stress as well as the presence of toxic substances.

Ink-jet printing provides another avenue for imparting conductivity to textile substrates. Information on Xennia's conductive technology may be found on their website (see Section 17.9). Results of depositing silver on textile substrates are reported by Bidoki *et al.* (2007) who conclude:

> Although printing a high quality pattern, appropriate for use in the electronics industry, needs more research, the present findings in conjunction with the advanced capabilities of recently available ink-jet printers mark a starting point for a safer, low cost and digitally controllable method for the build-up of metallic patterns on hydrophilic flexible substrates. (p. 974)

Blum (2007) includes a paragraph on ink-jet printed circuitry on textile substrates, but this indicates potential and prototype applications only. A Nanomarkets report (2008) considers that ink-jet technologies will become a key fabrication approach for wearable electronics, but much more R&D needs to be done. There appears to be greater progress with printing on flexible polymer substrates, as appears from research papers (e.g., Kim *et al.*, 2009) and commercial ventures (e.g., Xennia's Conductive Inkjet Technology). Screen printing offers a way of achieving the same goals.

Karaguzel *et al.* (2009) used flexible non-woven materials for the substrate and applied conductive silver inks with elastic properties matching that of the substrate. The printed material was then laminated using a meltdown layer of thermoplastic urethane that provided a 40–50 micron protective film with microporous properties. The authors report improved results with higher viscosity inks, with substrate-ink combinations capable of passing a test involving 25 washes.

17.9 Sources of further information and advice

- Bailo Trail Running (BTR) Rocket jacket
  http://www.qiosystems.com/?page=solutions/sportswear.php
  http://www.talk2myshirt.com/blog/archives/801
- Conductive Inkjet Technology
  http://www.conductiveinkjet.com/
- Gul's Limited Edition iPod Coastal Vigo jacket
  http://www.gulusa.com/sailblog/?p=117
  http://www.talk2myshirt.com/blog/archives/2891
- Inteltex project (*Intelligent multi-reactive textiles)*
  http://www.inteltex.eu/
- MIT Wearable Computing Lab
  http://www.media.mit.edu/wearables/
- MY SEW EZ TOUCH
  http://www.rnkdistributing.com/cid-80-1/My_Sew_EZ_Touch.html
- Numetrex
  http://www.numetrex.com/
- Peratech
  http://www.peratech.com/
- Ohmatex Aps
  http:/www.ohmatex.dk/
- QIO Systems
  www.qiosystems.com
- QTC (Quantum Tunnelling Composite)
  http://www.peratech.com/qtcscience.php
  http://www.mutr.co.uk/images/QTC.pdf
- Sefar AG
  http://www.sefar.com/
- talk2myshirt ('everything you wanted to know about wearable electronics')
  http://www.talk2myshirt.com/blog/
- *Role of Design in Wearable Computing*, Workshop in Boston, Massachusetts, October 13, 2007. Available at: http://www.ece.vt.edu/tlmartin/iswc07_design_workshop/iswc2007-design-workshop-proceedings.pdf

- Viking TST suit
  http://www.viking-life.com/viking.nsf/public/newspress-built-inthermalsensorsrevolutionisenfpaturnoutgear.html
  http://www.viking-life.com/viking.nsf/public/downloads-downloadbrochures.html/$file/VIKING_TST.pdf
- Xennia Technology
  http://www.xennia.com/conductive_inkjet/
- XS Labs: Design Research for Soft Computation
  http://xslabs.blogspot.com/

17.10 Acknowledgements

The author is very grateful for feedback and images provided by industrial developers of these technologies: David Lussey of Peratech, Christian Dalsgaard and Anne Jensen of Ohmatex, Jens Peter Kruse of Viking Life-Saving Equipment A/S, and Ivo Locher of Sefar AG.

17.11 References

Anderson, G. and Lee, G. 2008, Wearable computing – why consumers (don't) adopt smart wearable electronics, *IEEE Pervasive Computing*, 7(3), 10–12.

Anderson, K. and Seyam, A.M. 2002, The road to true wearable electronics, *Proceedings of the Textile Institute 82nd World Conference, Cairo*, 23–27 March.

Anliker, U., Ward, J., Luckowicz, P., Tröster, G., Dolveck, F., Baer, M., Keita, F., Schenker, E., Catarsi, F., Coluccini, L., Belardinelli, A., Shklarski, D., Alon, M., Hirt, E. and Vuskovic, M. 2004, AMON: A wearable multiparameter medical monitoring and alert system, *IEEE Transactions on Information Technology in Biomedicine*, 8(4), 415–427.

Berzowska, J. 2005a, Electronic textiles: wearable computers, reactive fashion, and soft computation, *Textile*, 3(1), 2–19.

Berzowska, J. 2005b, Memory rich clothing: second skins that communicate physical memory, *Proceedings of the 5th Conference on Creativity & Cognition* (C&C '05), available at: http://www.xslabs.net/papers/cc05-berzowska.pdf

Berzowska, J. and Bromley, M. 2007, Soft computation through conductive textiles, *Proceedings of the International Foundation of Fashion Technology Institutes Conference*, (IFFTI '07), available at: http://www.xslabs.net/papers/iffti07-berzowska-AQ.pdf

Bidoki, S.M., Lewis, D.M., Clark, M., Vakorov, A., Millner, P.A. and McGorman, D. 2007, Ink-jet fabrication of electronic components, *Journal of Micromechanics and Microengineering*, 17, 967–974.

Blum, J.B. 2007, Ink jet printing for high-frequency electronic applications: nanoparticle inks and drop-on-demand ink jet printers offer a unique opportunity to generate fine-line additive circuits on flexible, three-dimensional substrates. *Printed Circuit Design & Fabrication*, 24(10), 38-42.

Bodine, K. and Gemperle, F. 2003, Effects of functionality on perceived comfort of wearables, *International Symposium on Wearable Computers*. White Plains: IEEE Computer Society, pp. 57–60.

Buechley, L. and Eisenberg, M. 2009, Fabric PCBs, electronic sequins, and socket buttons: techniques for e-textile craft, *Personal and Ubiquitous Computing*, 13(2), 133–150.

Catrysse, M., Puers, R., Hertleer, C., Van Langenhove, L., van Egmond, H. and Matthys, D. 2004, Towards the integration of textile sensors in a wireless monitoring suit, *Sensors and Actuators A: Physical*, 114(2–3), 302–311.

Coyle, C., Wu, Y., Lau, K-T., De Rossi, D., Wallace, G. and Diamond, D. 2007, Smart nanotextiles: a review of materials and applications, *MRS Bulletin*, 32, 434–442.

Dhawan, A., Seyam, A.M., Ghosh, T.K. and Muth, J.F. 2004a, Woven fabric-based electrical circuits: Part I: Evaluating interconnect methods, *Textile Research Journal*, 74(10), 913–919.

Dhawan, A., Ghosh, T.K., Seyam, A.M. and Muth, J.F. 2004b, Woven fabric-based electrical circuits: Part II: Yarn and fabric structures to reduce crosstalk noise in woven fabric-based circuits, *Textile Research Journal*, 74(11), 955–960.

Duval, S. and Hashizume, H. 2005, Perception of wearable computers for everyday life by the general public: impact of culture and gender on technology, *Lecture Notes in Computer Science*, 3824, 826–835.

Ghosh, T.K., Dhawan, A. and Muth, J.F. 2006, Formation of electrical circuits in textile structures, in: Mattila, H.R. (ed.), *Intelligent Textiles and Clothing*, Woodhead Publishing, Cambridge, 239–282.

Harlin, A. and Ferenets, M. 2006, Introduction to conductive materials, in: Mattila, H.R. (ed.), *Intelligent Textiles and Clothing*, Woodhead Publishing, Cambridge, 217–238.

Karaguzel, B., Merritt, C.R., Kang, T., Wilson, J.M., Nagle, H.T., Grant, E. and Pourdeyhimi, B. 2009, Flexible, durable printed electrical circuits, *The Journal of The Textile Institute*, 100(1), 1–9.

Kim, H-S., Kang, J-S., Park, J-S., Hahn, H.T., Jung, H-C. and Joung, J-W. 2009, Inkjet printed electronics for multifunctional composite structure, *Composites Science and Technology*, 69(7–8), 1256–1264.

Kirstein, T., Tröster, G., Locher, I. and Küng, C. 2006, Context aware textiles for wearable health assistants, in: Mattila, H.R. (ed.), *Intelligent Textiles and Clothing*, Woodhead Publishing, Cambridge, 399–433.

Knight, J.F., Baber, C., Schwirtz, A. and Bristow, H.W. 2002, The comfort assessment of wearable computers, *6th International Symposium on Wearable Computers*, Seattle, WA, 7–10 October, 65–72.

Lam Po Tang, S. 2007, Recent developments in flexible wearable electronics for monitoring applications, *Transactions of the Institute of Measurement and Control,* 29, 283–300.

Lam Po Tang, S. and Stylios, G.K. 2006, An overview of smart technologies for clothing design and engineering, *International Journal of Clothing Science and Technology*, 18(2), 108–128.

Lehn, D.I., Neely, C.W., Schoonover, K., Martin, T.L. and Jones, M.T., 2004, e-TAGs: e-textile attached gadgets, *Proceedings of Communication Networks and Distributed Systems*, San Diego, January.

Li, Y. and Wong, C.P. 2006, Recent advances of conductive adhesives as a lead-free alternative in electronic packaging: materials, processing, reliability and applications, *Materials Science and Engineering: R: Reports,* 51(1–3), 1–35.

Li, Y., Moon, K-S. and Wong, C.P. 2005, Electronics without lead, *Science,* 308(3 June), 1419–1420.

Locher, I. and Tröster, G. 2008, Enabling technologies for electrical circuits on a woven monofilament hybrid fabric, *Textile Research Journal*, 78(7), 583–594.

Lu, O. and Wood, L. 2006, The refinement of design for manufacture: inclusion of process design, *International Journal of Operations & Production Management*, 26(10), 1123–1145.

Mallin, J. 2001, Fashion and technology – an ideal match? *just-style.com*, 12 June, available at: http://www.just-style.com/article.aspx?id=92826

Martin, T., Jones, M., Edmison, J. and Shenoy, R. 2003, Towards a design framework for wearable electronic textiles, *Seventh International Symposium on Wearable Computers*, October 2003, 190–199.

McCann, J. and Bryson, D. (eds) 2009, *Smart Clothes and Wearable Technology*, Woodhead Publishing, Cambridge.

Nanomarkets, 2008, Materials for Functional Inkjet Printing: A Market Forecast, 2009–2016, Nano-062 (November).

O'Connell, S. 2000, Fibre options, *The Guardian*, 14 August, available at: http://www.guardian.co.uk/Archive/Article/0,4273,4051264,00.html

PANiQ – A new wearable consumer electronics brand, *Future Materials*, 2009, March, 21, availavle at: http://www.webmags.co.uk/mag.aspx?magcode=Future_Materials_Issue_3_2009_9835

Papadopoulos, D. 2007, Natural language, connectors and fraying threads, in: *The Role of Design in Wearable Computing,* Workshop in Boston, MA, 13 October, pp. 21–27. http://www.ece.vt.edu/tlmartin/iswc07_design_workshop/iswc2007-design-workshop-proceedings.pdf

Paradiso, R., Loriga, G., Taccini, N., Gemignani, A. and Ghelarducci, B. 2005, WEALTHY – a wearable healthcare system: new frontier on e-textile, *Journal of Telecommunications and Information Technology*, 4, 105–113.

Post, E.R., Orth, M., Russo, P.R. and Gershenfeld, N. 2000, E-broidery: design and fabrication of textile-based computing, *IBM Systems Journal*, 39(3–4), 840–60.

Sefar, 2009, SEFAR PowerGlow – fabric as circuit board, available at: http://www.sefar.com/htm/609/en/SEFAR-PowerGlow.htm?Folder=29396 and http://www.sefar.com/download/htm/1229/en/SF-PDF-Smart-Fabrics-CI-13-PowerMatrix-EN.pdf?Article=&Page=&ArticleReturn=609

Service, R.F. 2003, Electronic textiles charge ahead, *Science*, 301(15 August), 909–911.

Seyam, A.M. 2003, Electrifying opportunities, *Textile World*, February.

Solaz, J.-S., Belda-Lois, J.-M., García, A.-C., Barberà, R., Durá, J.-V., Gómez, J.-A., Soler, C. and Prat, J.-M. 2006, Intelligent textiles for medical and monitoring applications, in: Mattila, H.R. (ed.), *Intelligent Textiles and Clothing*, Woodhead Publishing, Cambridge, 369–398.

Tyler, D. 2008a, Advances in apparel product development, in Fairhurst, C. (ed.), *Advances in Apparel Production*, Woodhead Publishing, Cambridge, 157–176.

Tyler, D.J. (ed.). 2008b, *Carr & Latham's Technology of Clothing Manufacture*, 4th edn, Blackwell Publishing, Oxford.

Van Langenhove, L. and Hertleer, C. 2004, Smart clothing: a new life, *International Journal of Clothing Science and Technology*, 16(1/2), 63–72.

18
Joining of technical textiles with stringent seam demands

S. KOVAČEVIĆ and D. UJEVIĆ, University of Zagreb, Croatia

DOI: 10.1533/9780857093967.4.536

Abstract: Sewn seam is usually the most critical position for the permeability of the material, but not in case of strength, so its use is indispensable where seam tightness is not a requirement. Seam rigidity, especially in folded places, results in weakening the joint and it is important where possible to avoid it in folds. Special prominence is given to the essential properties of multi-layer fabrics with woven fabric as a basic component for use in tarpaulins, tents, awnings, sailboats, aircraft, materials for protective clothing, etc. Basic properties of materials with limited values are emphasized, and an experimental study of some selected materials is given. In this chapter, performance of individual seams on various materials is studied.

Key words: sewn seam, high-frequency welding, hot air welding, technical protective fabrics, fabrics for tarpaulins, tents, awnings, sailboats, aircraft, protective clothing.

18.1 Introduction

Textile surfaces such as woven fabric, knitted fabric and non-wovens play a very significant role in many parts of our lives. According to historical records and archaeological findings, it is assumed that woven fabric is the oldest textile material used by humans, primarily for body protection. It is possible that animal skin was used prior to woven fabric; however, there is no reliable evidence in that respect. Degradability of organic substances has prevented the discovery of leather and woollen cloth from prehistoric times. According to the oldest discovered textile samples (5,000 years BC) and sewing tools (20,000 years BC), it can be claimed that clothes were made much earlier than the samples were found.[1] In Egypt, coarse woven fabrics were made about 4,000 years ago, which are today classified as technical textiles such as sails, tents and various covers for protection from sun, rain and other natural disasters. The first fabrics were used for making simple garments, and for other purposes such as dwelling covers, walls made of twigs which were coated with mud on both sides. According to records from the Bible, Apostle Paul was an excellent weaver who learned the skill of weaving a tent from his father. Therefore, it can be assumed that the fabric was used very early not only for clothing but also for other purposes. Sewing seams was probably known already from the use of animal skins for covering the body, dwellings and other items. Initially yarns were rough, and needles were made from the bones of dead animals. By improving the production of textile fabrics they have become increasingly finer, cheaper and more available. Sewing also developed, especially due to the advent of sewing machines. Due to the increase in the production of fabrics, knitted fabrics, non-wovens and technical textiles, the need for sewing also increased. One of the greatest requirements is to obtain a target product that will correspond to the purpose which is required by the properties of materials and seams. Technical textiles are often cheaper, more durable and stronger than conventional textiles, and due to their good qualities they supplement and replace some materials in civil construction such as steel, concrete, etc. Technical textiles can be found in almost all parts of human life, and there are virtually no industries that do not use textiles in some form. The development of technical textiles has been on the rise in the past few decades, as illustrated by the following data: 10% of the total textile production in the world in 1982, 15% in 1988, 20% in 2001 and even 40% in 2010.[2]

Multilayer technical textiles with woven fabric as the base components that are used for architecture, parachutes, planes, boats, awnings, NCB clothing (nuclear, chemical, biological), protection, tents, sails, etc., differ from one another, and their production requires target raw materials, construction parameters, special weaving and finishing machines. Common to all these materials is the need to protect people effectively from environmental threats, harmful substances and to provide safe transport (Fig. 18.1).

A high-quality and safe seam affects the quality of these products. The main function of the seam is to enable a uniform load transfer from one part of the fabric to another and so preserve the overall integrity of the joined fabrics. Seams are mostly the weakest link regarding strength and media permeability in the finished product. Therefore, the quality of joining cut parts is extremely important. It would be most ideal to produce target products without the use of seams or with a minimum portion of seams. Incorrect joints with insufficient strength or higher permeability of air, water, gas, x-rays or hazardous fluids may be detrimental to human lives. This is the reason why it is necessary to find optimal parameters in the sewing process or to join seams using either thermal or high-frequency or any other method. To perform the joining process satisfactorily, the seam must have such properties that very closely match the characteristics of the selected base material with compatible conditions and methods of seam joining. From the aspect of utility value, breaking force and elongation at break in seamed and seamless places can be considered as imortant mechanical properties of technical fabrics. Increasingly diverse and demanding materials and their versatile properties require the development of more complex joining methods.[3]

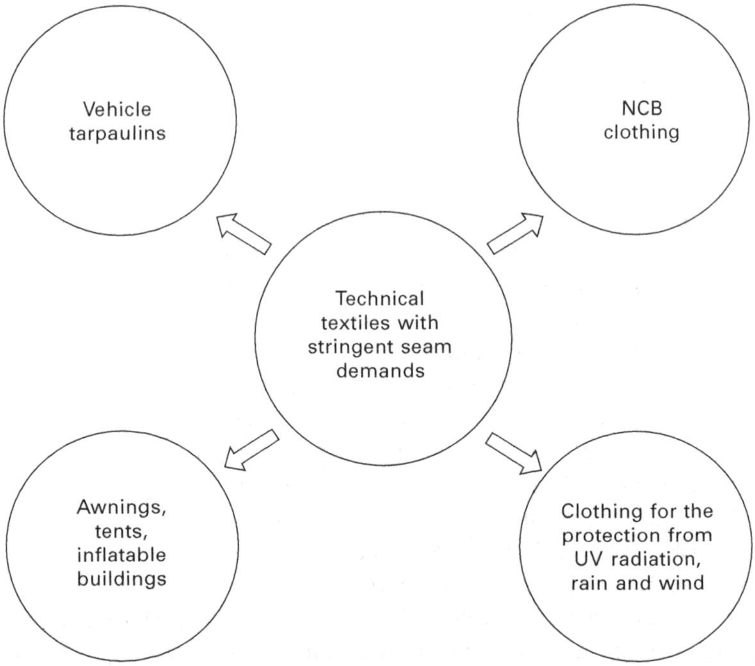

18.1 Division of technical textiles with stringent seam demands.

18.2 Joining techniques for textiles with stringent seam demands

Durability and safety of the seam depend on the quality, type of seam and material to be joined. The welded seam completely closes the joint, and its strength can be the same or even better than that of the seamless material. However, the place joined using the high-frequency, ultrasound, laser or thermal technique with hot air creates a joint having higher stiffness and harder bending in comparison to the sewn seam. Stiffness of the material results in a more difficult adaptation of the product and the material loses its elasticity. The joint is usually found in folds, which increases abrasion and causes greater stress, and thus greater deformations. The correct selection of the seam for a specific application of technical textiles for interior decoration represents the most important criterion for the finished product.

The basic properties that are essential for joining critical seams include strength, abrasion resistance, durability, yield resistance, low weight, flexibility and resistance to extreme external conditions, chemicals and UV radiation.

Joining large-area materials such as tarpaulins, tents, awnings, etc., requires a sufficient handling space. With regard to the machines for joining these materials, it is very important that a good workplace layout is available which allows materials to be handled without excessive bending, twisting and stretching, and to allow the optimization of the needle feed. High frequency and thermal welding has certain advantages over traditional sewn seams. Welding is faster and easier, and a higher-quality seam with a lower risk of tightening the material is produced. High frequency and thermal welding is carried out on the principle of the press where the material is stationary during welding. It is also possible to reverse the procedure, i.e. the material is moveable, while the welding machine is stationary. Despite the advantages in quality and appearance of the seam obtained by high-frequency or thermal welding, sewn seams are still in use, but only when the tightness of the material is not a requirement.

Since high-frequency and hot air welding machines can make flat seams with limited curvatures and material thicknesses in most cases, cutting parts with a variety of short and irregular curves and different thicknesses are joined with hot air, while the machine is stationary, and the material moves or it must be joined with a sewn seam. However, in hot air welding there is a risk of a poor joint at different joining speeds and different material thicknesses. Sewing machines for sewing heavy, large-area materials have a longer arm for easier material handling. A properly made seam must have a certain strength, elasticity, durability, safety and appearance. Despite specialized sewing machines with a double feed, the surface smoothness of multilayer heavy materials makes the same transport of the upper and lower material in the sewing process more difficult. The materials used for awnings and tents include laminates made from fabrics as the base material and a specific polymer film on both sides. When designing tarpaulins, tents or other similar items, these materials are subjected to various deformations such as bending, puckering and stretching in all directions. High levels of stiffness can negatively affect the quality of the material. Bending load is in fact a combination of tensile, compressive and and shear stress.[4] Bend strength may be different when the material is bent on the face or on the back. The material will weaken faster at bending points where the coated film breaks and leaks water and air after some time. If the bending point is also a joint, whose stiffness increased after joining, the material can weaken earlier which may also be caused by extreme conditions (heat, cold, rain, snow, wind, etc.). Therefore, it is beneficial to avoid bending points in the folds of the material by moving the material so that the joints are where the material is less stressed. In folds it is desirable to weld or glue a special reinforcement tape on the back of the material.

18.2.1 Thermal and high-frequency welding

Coating a polyurethane or polyvinylchloride layer on the fabric or fabric lamination enables high-quality thermal or highfrequency welding. The use of multilayered and composite materials with fabric as the basic component is desirable for architecture, parachutes, belts, airplanes, boats, inflatable buildings, roof coverings, tents, sails, etc. For these purposes tight seams with high-frequency welding or hot air are often used. Thermal welding requires a specific temperature or frequency if it is joined with high-frequency welding. The aesthetic appearance of the seam is also important, especially if it is located in a visible and folded place. Joining the laminate can be performed with different joint widths (about 2 cm). Joining materials in accordance with Fig. 18.2(a) makes a very strong joint with the lowest increase in stiffness and thickening. The disadvantage of these joints is a low strength of radial loads (failure location is shown in Fig. 18.2b, c and d). This is because the laminate or multilayered material is made of multiple layers of different materials that are easily separated at the joint and represent the greatest problem. This problem could be solved only through a combination of the sewn seam and then the high-frequency or thermal weld with or without added tape for more efficient joining. The problem of the sewn seam is needle penetrations which ought to be completely closed to provide a tight joint.

High-frequency welding (Fig. 18.3) and hot air welding (Fig. 18.4) are mostly performed on the flat joints of the material using a press. High-frequency oscillators oscillate at a frequnecy of 27.12 MHz with a power of 4 kW. High-frequency welding is based on the electrically generated high frequency vibrations which are mechanically transferred to the material.

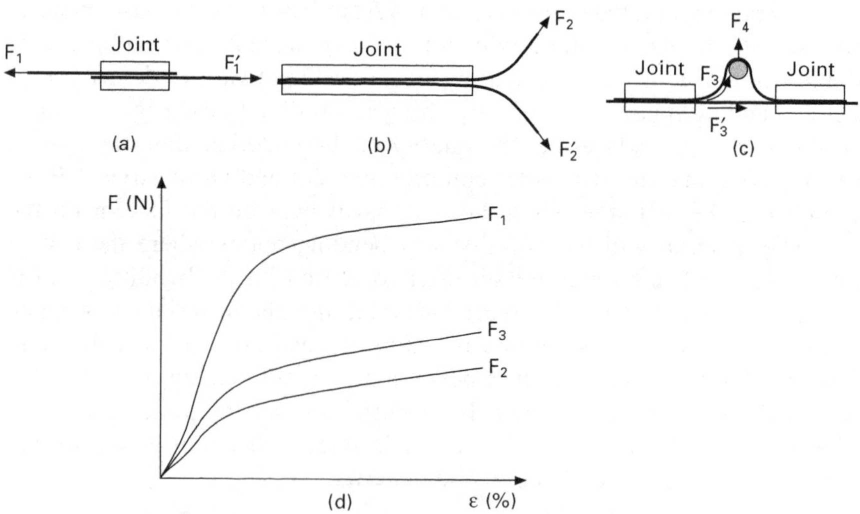

18.2 Forces that can occur in the joints of the material: (a) axial forces, (b) radial forces through component separation, (c) radial force by dragging one component, (d) force diagram in accordance with the design of the joint; F – breaking force (N), ε – elongation at break (%).

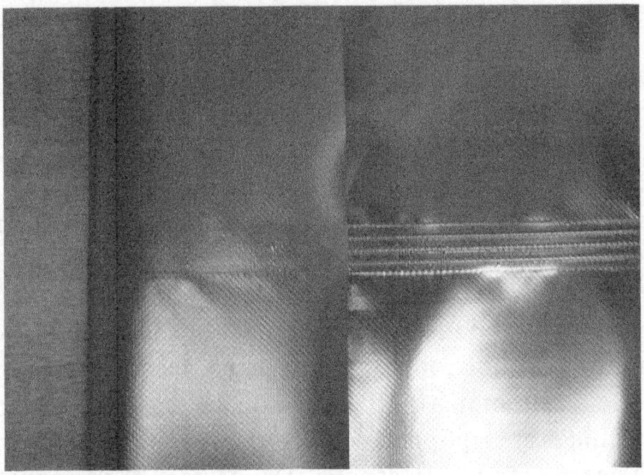

18.3 High-frequency welded seam.

18.4 Hot air welded seam.

This leads to the molecular motion of particles that through mutual friction create heat, which in turn converts the material into the plastic state, and these two materials are joined under pressure. The most important element of welding in some machines is the lower vibrating wheel, the so-called sonotrode with an axial parallel synchronous counter roller. Materials to be welded are placed between the rollers, and through ultrasonic vibration they are melted and pressed. The rotating sonotrode allows a very precise adjustment of welding parameters. There is no friction and resistance as in systems with a fixed probe. By special regulation and moving the sonotrode, welded seams of different shapes and requirements can be made.[5]

18.2.2 Ultrasonic and laser welding

Ultrasonic welding machines make a secure, strong and durable joint in different dimensional material forms. In these machines the ultrasonic transducer converts the alternating voltage with frequencies from 25 to 40 kHz into ultrasonic mechanical oscillations transferred by the sonotrode onto the material. At the mentioned frequencies macromolecules of artificial materials begin to move through the material thickness and mix from one material into the other one, making new bonds. Mutual friction of macromolecules, breaking and creating new bonds can create heat at the material joint and softening (as in high-frequency joining) which causes less material pressure. The joint may have a different mark depending on the engraved pattern on the cylinder. These machines are equipped with a process computer that optimizes ultrasonic joining energy at uneven speeds. The design of ultrasonic joining machines is very similar to conventional sewing machines. This procedure facilitates the joining of curved seams, such as seam joining. Mostly on the bottom side where the lower thread looper is located, there is an ultrasonic device with fixed sonotrode, and on the upper side, instead of the needle, there is a counter electrode oscillating in the rhythm of the upper shaft. The work process is similar to the sewing process.[5,6]

Laser welding is performed using the exact quantities of heat only on the contact surface of the fabric. Such selective heating can be achieved by applying the liquid that has the capacity to absorb, and is applied by spraying. Laser energy passes through the fabric and creates a strong place on the contact points of both fabrics. The advantages of laser joining are high speed joining, limited space of heating, retaining the flexibility of the seam, impermeability to gas and water, control of the volume of the melted material, different material thickness, possibility of joining multiple material layers and the regulation of shape and curvature of the seam. Joints are flexible and strong, and they are made using well-controlled laser power and heat through material layers.[5,6] The author of the project[7] for robotic laser welding in the 2D and 3D process built an integrated simulation model that allows the joining of the curved shape and varying material thicknesses. Robotic laser welding of the material provides many joining opportunities as well as joining several garments simultaneously.

18.2.3 Sewn seam

For technical outdoor fabrics that protect people, sewn seams are still used, but to a smaller extent. To make a safe sewn seam, it is necessary to adapt it to the materials and use. The forces acting on the lower and upper thread in the process of forming lock stitches occur as a consequence of maintaining the thread tension to obtain a regular seam. The needle and the lower thread interlace or wrap, respectively, and that place is, in the case of a regular seam, in the middle of the fabric, observing its cross section. For thicker, multilayer or composite materials, the maintenance and equalization of tension of the needle and lower thread is more difficult than in the classical and homogeneous layers. For superficially smooth materials, which are used for various protective covers, there is a risk of slipping in sewing as a result of unevenly long stitches and puckering on one side of the material. This problem causes more difficult material handling and uneven material feed to the sewing machine. Thanks to the influence of different factors in the sewing process, one of the greatest faults occurs, namely uneven stitch length. Conditions of moving the material greatly affect the stability of the stitch length. The movement of the material after it moves away from the feed dog is particularly influential.

The inertial motion of the material virtually depends on the speed coming from the throat plate. It is a complex task to maintain a constant sewing speed in order to obtain stitches of the same length. Sometimes it is necessary to apply the force of the pressure foot, and in that case material stoppage is additionally adjusted. Despite the growing automation and robotization in the sewing process, it is still a problem to handle the material from cutting layers to feeding into the sewing machine, especially for heavy materials with larger surfaces of cut parts. Uniformity and fineness of the thread with a correct circular cross section and its slipping properties greatly contribute to the seam quality. Composite and multilayered materials of the group of technical outdoor fabrics made of several different materials are generally not recommended to be joined with a sewn seam. The needle penetration through the layers of the material damages the material and creates holes that are partially not closed by the thread, and there is a risk of media penetration through the holes (Figs 18.5 and 18.6).

Multilayered materials usually have a woven fabric or a knitted fabric in the inner or outer layer which is fixed with a plastic coating, and thread spacing at the place of the needle penetration is not possible. As a result, material and thread damage is inevitable at the moment of needle penetration. Skipped stitches or unwanted loops weaken the seam and increase the risk of creating holes and finally the separation of cut parts. Therefore, sewing multilayered materials is not recommended for the protection from media such as air, water, chemicals, microorganisms and UV rays. The consequence of shear deformation increases the size of the holes and material deformation becomes increasingly greater, particularly in the tense and wet state. Weakening and deformation of the sewn seam will be greater if the length of the stitch is uneven or if it has skipped stitches where the needle made a hole, but without the thread. Microfiber fabrics for protective clothing are light-weight, dense, smooth, soft and extremely sensitive in the sewing process. Sewing difficulties occur because of uneven fabric transport using one-sided feed of

18.5 Sewn seam for tarpaulins and tents. Seam designation: 2.02.03/301.

18.6 Sewn seam for awnings and sunshades. Seam designation: 2.02.03/301.

the material, resulting in seam puckering. Therefore, sewing these materials is more suitable by multiple feed and differential top feed. Besides that, it is necessary: to use a spun thread with greater surface smoothness, to wind lower thread with lower tension, to use a needle with thin needle point, to use a thread tension as low as possible, to limit stitch length, to move the feed dog forward as far as possible, to use a throat plate with a cross element, to use a sewing machine with an intermittently movable puller, to use double chain stitch requiring up to 50% lower tension than double lock stitch. Basic thread parameters are: balanced twist level, frequency of weak, thick and thin places as low as possible, uniformly and qualitatively lubricated, satisfactory mechanical properties and high resistance to abrasion, bending and torsion and low friction level.[8]

The material breaking force at the seam place causes shear or shift of threads which are parallel to the sewn seam. This is opposed on a thread forces at fabric joints or at stitch points, respectively. Seam strength can be defined by the theoretical basis and that a break will not occur until the force acting on the seam through the warp and weft system is greater than the joining force created by the thread in stitch points.[9–11] It is anticipated that soon new intelligent machines will replace traditional sewing machines. On these machines the thread guidance is tracked by an intelligent UV camera detecting the loop and optimizing the position of the thread and the needle during sewing. These intelligent joining machines are actually easy to use. Sewing threads of different fineness can be used without additional regulations.[12]

18.2.4 Combination of sewn and welded seam

Regardless of its method of joining, a joint poses a certain risk that after some time it will be weakened. In technical fabrics used for protective clothing, the sewn seam poses a risk of the penetration of certain media from outside to the body, especially at places of needle penetration. One part of the hole closes the thread that remains in the penetrated places, but the other part of the hole is free and permeable. By stress of the material at the seam, a deformation of the sewn seam is caused, which enlarges the holes of the stitch points, resulting in increasingly permeable holes. Considering the irregular and short curves of the parts of protective clothing to be joined, they cannot be joined by welding, but only by sewing. Therefore, the inner part of the sewn seam is reinforced by thermal application of a plastic tape, and in some cases using a liquid coating as in the case of airbags (Fig. 18.7). It may be worth mentioning that seams can also be sealed using a liquid coating. Some automotive airbags are sealed internally with a coating after stitching.

18.3 Applications of stringent seams in technical textiles

For technical textiles, seam tightness and strength are extremely important. Materials used for covering products such as tarpaulins, tents and awnings must be tight as well. Seam quality is particularly significant due to the complexity of material, large area and thickness, as well as demands on the resistance to weather. Likewise, protective clothing must meet many standards to achieve effective protection. Quality joining the cut parts must guarantee protective clothing safety.

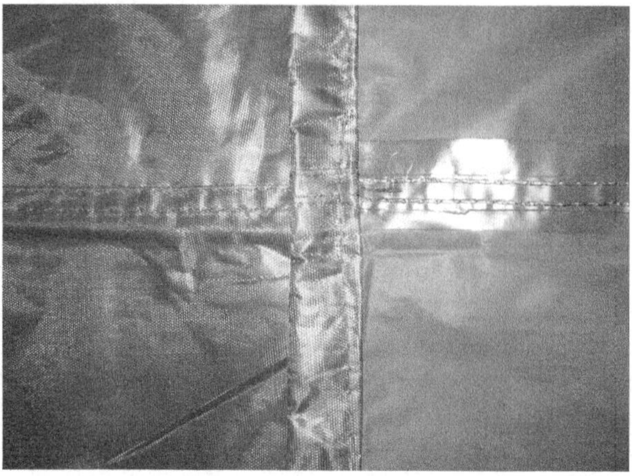

18.7 Combination of the sewn and welded seam.

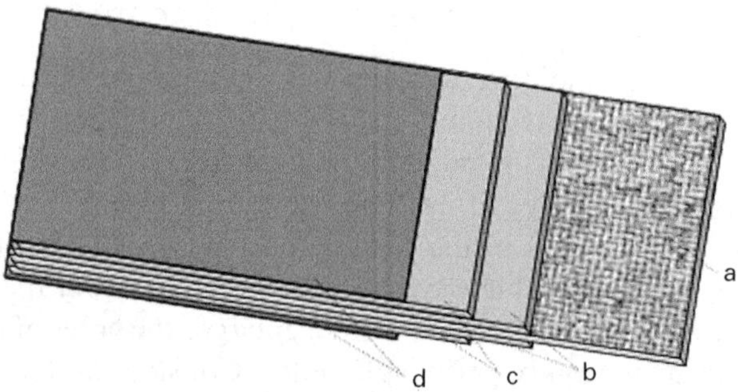

18.8 Manufacture of tarpaulin material: (a) woven fabric, (b) first polyvinylchloride layer, (c) second polyvinylchloride layer, (d) third layer – varnish.

18.3.1 Lorry tarpaulins

Tarpaulins are technical textiles used for motor vehicles. Since tarpaulins are subjected to weather through all four seasons, including severe weather conditions such as severe wind and bright sun rays, resistant and very strong materials are therefore required. These materials consist of a high-strength polyester fabric with two-sided lamination of two polyvinylchloride layers on each fabric side and surface polyethylene coating on one or on both sides (Fig. 18.8).[13] Joints are mostly high-frequency or thermal welded, which guarantees complete seam tightness and strength. The iron structure, over which the tarpaulin is stretched, damages the material over time, and it is desirable to strengthen these places with additional materials. Lorry tarpaulins with welded belts reinforce folding seats and increase their performance (Fig. 18.9).

18.9 Lorry tarpaulin (source: Krika).

18.3.2 Tents

The production of the first tents on hand weaving looms dates back into prehistoric times. Records testify that the tents and various covers for dwellings were produced 4,000 years ago for protection against rain, wind and sun. At that time the fabric was coarse and made from natural fibres, so that after some time it leaked and lost strength, and durability was relatively short. Cut parts were sewn by hand using a relatively coarse needle and coarse thread, resulting in leaking at the seams. Therefore, seams were covered with various mixtures of soil, clay, ash and tiny plant particles to achieve a better adhesiveness and strength as well as tightness. Today's tents are mounted on an iron structure with horizontally placed joints (Fig. 18.10). Larger versions that are used in all seasons, mostly as sports halls, usually consist of two layers, an outer and inner material with a certain gap. By circulation of hot or cold air between the interstices of the layers, optimal conditions in the space can be achieved. The outer fabric is usually made of aluminised polyamide or polyester. This material protects against UV radiation and is waterproof. Seams are joined by high-frequency or thermal welding. There are almost no sewn seams because of possible leaking at seams where the

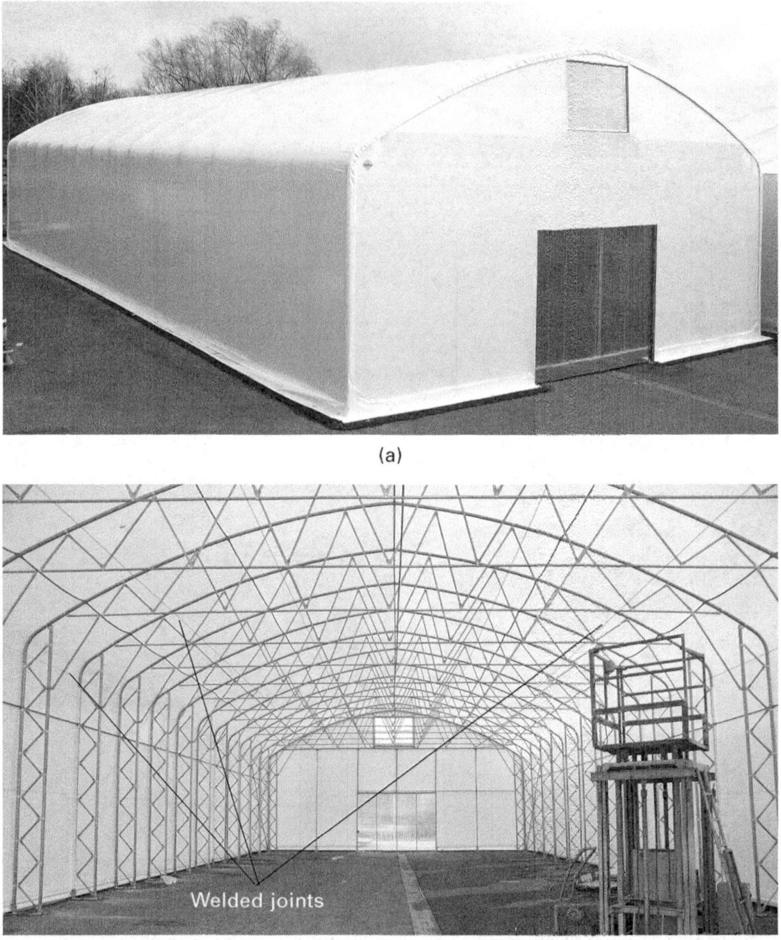

18.10 Tents: (a) external side-view, (b) internal side-view with visible joints and iron construction (source: Krika).

needle would penetrate the material, thus enabling easy water leakage after some time. The inner material is also made of synthetic lightweight materials, with or without a carrier material. Seams are mostly joined with sewing because this layer is not in contact with the external conditions. These joints are often in combination with sewn and welded seams that have a relatively high mechanical resistance to mechanical shocks.

18.3.3 Awnings and sunshades

Awnings are heavy technical fabrics densely woven in plain weave. These fabrics are typically woven polyester yarn, impregnated and additionally treated with polyvinylchloride and polyurethane. Fabrics are usually finished with dyes and treated with agents for providing water-tightness, protection from UV radiation and colourfastness. Despite the development of new materials which provide many advantages, there are disadvantages as well, for example after some time leakage occurs at seams and at seamless places. Another disadvantage is that after a certain period of time, so-called pockets are created in places that are not tightened enough or the iron structure is not in contact with the fabric, resulting in water gathering, deformation of materials and leakage. The material in the wet state stretches and deforms, becoming looser and more sensitive, which has the effect of creating bagginess. In the winter time, the retained water on the material freezes and the material expands and weakens further. Joining the awning is performed mostly with sewn seams (Fig. 18.11), which leads to a risk of water leakage through needle penetration points that will expand through time because of thread and material tension. Since the position of the seam is along the awning length, the seam does not require higher strength, so that the seam length is about 7 mm. The material is twice as thick in joints, and there is a risk of material waveness and tightening in the seam area. Solar radiation causes a surface effect (pigment change) which is visible as a stain. Longer exposure of this material to UV radiation, wind, high and low temperatures and mechanical stresses causes ageing and thus the loss of physical-mechanical properties.[14]

Parasols for the protection of people from the sun have an unusually

18.11 Awning (source: Krika).

long history. From the beginning it surpassed its primary function, being a symbol of higher status in society. The parasol symbolized the sun, the bars represented the sun's rays, and the rod the cosmic axis. Their disadvantage was their great mass, and after some time they would leak. Over time it has developed into a fixed item in the form of today's awnings or as a carrying umbrella or a smaller parasol; along with the protection purpose, they were also used as a fashion accessory. Since we are increasingly exposed to UV radiation in the summer, we need protection from the sun wherever we are. Most often these are awnings stretched over a solid structure or as a stand-alone facility. In evaluating the performance of awnings and parasols, it is necessary to consider two factors, namely, level of shading and protective efficiency. The physico-mechanical properties of the material as a core component of the materials for awnings, boat covers, sunshades, tents, etc., greatly influence the utility value of the final material and product. In order to produce textile fabrics of desired properties to protect people from weather, multilayered composite materials or only surface-treated fabrics are most commonly used. Most often these are very strong fabrics woven in plain weave. The basic parameters that affect the fabric properties of the finished materials are raw materials, yarn fineness, density of warp and weft, mass and weave. This chapter will show the experimental testing of different protective materials of different manufacturers and thus contribute to some conclusions about these materials.[15]

Mechanical properties of fabrics for tarpaulins, tents and awnings

The material for lorry tarpaulins and tents, on which the joints made for the purposes of this chapter were tested, was joined on the machine TXPE PL-12, SN-3059, constructed in 2003, over a width of 20 mm and at a frequency of 27 MHz. Hot air welding reaches the plastic state at high temperatures which depend on the materials to be joined. Joining was performed on a Leister machine, Variant model, constructed in 1995. Test conditions were: temperature of hot air was 450°C and fabric transportation speed was 2 m/min. Joining tarpaulins and tents with sewn seams was carried out on the Pfaff sewing machine Model 1245, constructed in 2002. Joining awnings was performed on the two-needle sewing machine Garudan, model GF-230-446MH, constructed in 2006 (see Table 18.1, Figs 18.12 and 18.13).

According to the results obtained it can be seen that in the tarpaulins the hot air welded regions were the strongest, followed by the high-frequency welding, and the sewn seams were the weakest. The breaking forces of the awning materials without the joints were even lower than the hot air welded joints, while the other joints were weaker. Higher mass per unit area of the tarpaulin had higher breaking force for all samples. Elongation at break was the highest in the sewn seams followed by the seamless samples, and it was

Table 18.1 Mechanical properties of materials and sewing thread for tarpaulins, tents and awnings in seamed and seamless regions and basic technical data

	Mass (g/m^2)	Thickness (mm)		Seamless		High-frequency welding		Hot air welding		Joining with sewn seams	
				F_w (daN)	ε_w (%)	F_w (daN)	ε_w (%)	F_w (daN)	ε_w (%)	F_w (daN)	ε_w (%)
Tarpaulin and tent materials	640	0.52	1	231.34	17.58	229.76	14.91	238.94	17.29	148.14	20.77
			2	227.77	21.37	223.00	16.48	230.34	17.01	116.67	24.62
			3	242.10	13.28	289.07	7.33	280.02	9.00	202.33	7.00
		Woven fabric		Raw material composition: polyester, density: warp and weft 9 threads/cm; warp and weft count 1100 dtex, plain weave							
		Material		Double-sided PVC coating of the cloth and varnished surface, tear resistance: 250 warp/250 weft (N); adhesion: 110 (N/5 cm), cold resistance: –30°C; heat resistance: +70°C; colourfastness: min. 7, flame retardancy: burning rate <100 mm/min.							
	670	0.62	1	278.33	14.66	261.67	11.37	295.00	13.54	188.33	22.33
			2	256.67	18.92	221.94	16.73	277.33	16.29	120.77	24.60
			3	285.19	15.00	310.42	11.09	346.99	11.94	180.73	18.66
		Woven fabric		Raw material composition: polyester, density: warp and weft 9 threads/cm; warp and weft count 1100 dtex, plain weave							
		Material		Double-sided PVC coating of the cloth and varnished surface, tear resistance: 300 warp/300 weft (N); adhesion: 110 (N/5 cm), cold resistance: –30°C; heat resistance: +70°C; colourfastness: min. 7, flame retardancy: burning rate <100 mm/min.							
	900	0.75	1	396.75	18.58	393.39	14.85	415.00	16.37	303.08	25.90
			2	380.47	21.68	367.15	16.80	401.19	20.42	251.95	26.44
			3	427.90	14.55	430.09	12.02	437.61	10.82	331.05	18.90
		Woven fabric		Raw material composition: polyester, density: warp and weft 9 threads/cm; warp and weft count 1100 dtex, panama weave 2:2							

Table 18.1 Continued

	Mass (g/m^2)	Thickness (mm)		Seamless		High-frequency welding		Hot air welding		Joining with sewn seams	
				F_w (daN)	ε_w (%)	F_w (daN)	ε_w (%)	F_w (daN)	ε_w (%)	F_w (daN)	ε_w (%)
		Material		Double-sided PVC coating of the cloth and varnished surface, tear resistance: 600 warp/530 weft (N); adhesion: 120 (N/5 cm), cold resistance: –30°C; heat resistance: +70°C; colourfastness: min. 7, flame retardancy: burning rate <100 mm/min.							
	Sewing thread for tarpaulins			F_y = 8,554.70 cN; ε_y = 18.57 (%); ω = 34,174.90 (cN×cm); σ = 56.28 cN/tex; Twist = 310 (t/m); Tt = 50 × 3 (tex)							
Awning material	300	0.54	1	117.33	23.27					56.67	14.77
			2	72.67	20.01					48.93	16.82
			3	82.51	13.44					63.55	12.61
		Woven fabric		Raw material composition: polyacryl , density: warp 30 threads/cm, weft density 14 niti/cm; warp and weft count 30 × 2 tex; plain weave							
	350	0.55	1	120.33	21.98					59.33	9.37
			2	80.74	16.32					53.04	13.13
			3	93.11	11.53					64.55	10.44
		Woven fabric		Raw material composition: polyacryl; warp density 32 threads/cm, weft density 14.5 threads/cm; warp and weft count 30 × 2 tex; plain weave							
	400	0.56	1	141.33	18.70					67.67	8.25
			2	86.04	22.38					60.33	12.37
			3	98.47	15.69					80.24	12.99
		Woven fabric		Raw material composition: polyacryl; warp density 34 threads/cm, weft density 15 threads/cm; warp and weft count 30 × 2 tex; plain weave							
	Sewing thread for awnings			F_y = 3,066.13 cN; ε_y = 20.37 (%); ω = 13,708.67 (cN×cm); σ = 47.17 cN/tex; U = 540 (twists/m); Tt = 22 × 3 (tex)							

F_w/F_y – breaking force in accordance with ISO 13935-1/in accordance with ISO 2062 (daN/cN), $\varepsilon_w/\varepsilon_y$ – elongation at break (%), ω – work to rupture (cN×cm), σ – tensile strength (cN/tex), T – twist level (twists/m); Tt–thread count (tex),1 – warp direction. 2 – weft direction, 3 – bursting force and bursting elongation in accordance with ASTM D3787-89 (daN).

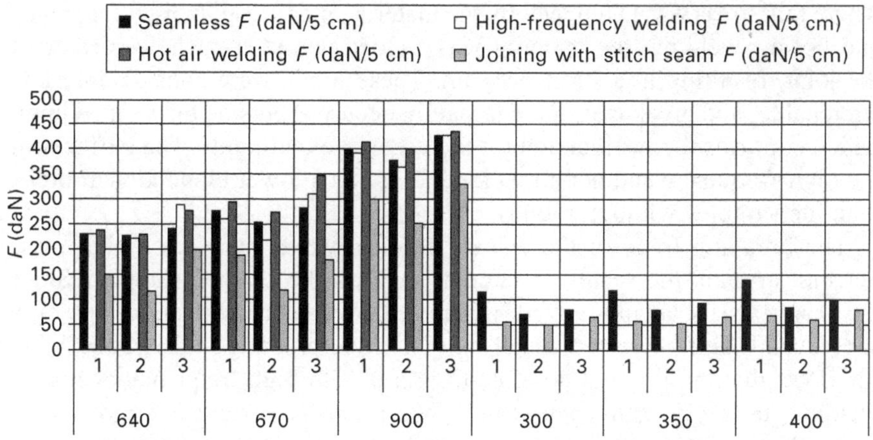

18.12 Breaking forces of tarpaulins, tents and awnings. F – breaking force (daN), 1 – warp direction, 2 – weft direction, 3 – bursting force.

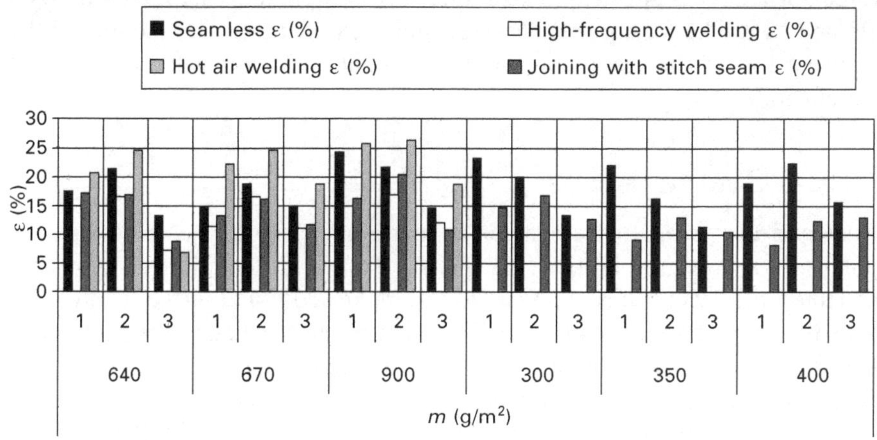

18.13 Elongation at break of tarpaulins, tents and awnings. ε – elongation at break (%), 1 – warp direction, 2 – weft direction, 3 – bursting elongation (%).

the weakest in the high-frequency welding. Higher mass per unit area of the tarpaulin had higher breaking forces for all samples. Elongation at break was the highest in the sewn seams followed by the seamless samples, and it was the weakest in high-frequency welding. The seamed samples burst in the seams and the thread elongation caused a higher elongation to rupture. In the sewn seam joint, the thread damaged the tarpaulin during stretching by cutting into the material until it broke in the stitch points. Therefore, the thread strength was satisfactory regardless of the weak seam, but the problem was that the materials were spot-welded together or only at the stitch points where this stress was taken over by the material and thread. In high frequency and hot air welding, the material was welded across the whole surface of the joint, resulting in a stronger seam. These joints were considerably less stretchable, and breakages of the tarpaulin occurred outside the welded joint, which confirms the justification for the use of welded joints. The stiffness of the high-frequency and hot air welded joint had a lower elongation at break than the samples without a joint.

The breaking force of the awning fabrics was higher at a higher mass per unit area on the seamless samples than on the seamed samples. Since the materials for awnings were not multilayered, but the fabric was only subsequently surface treated, usually only for waterproofing and impermeability for UV radiation, it is not possible to join it with high frequency or hot air welding, but only with sewn seams. Warp density is always almost twice as much as weft density, which caused higher breaking forces in the warp direction. The breaking force and ball burst elongation (Table 18.1 and Figs 18.12 and 18.13, bars marked '3') mostly follow the values of the breaking forces of a 20 × 5 cm strip (Table 18.1, Figs 18.12 and 18.13, bars marked '1' and '2'). The higher the mass per unit area of the fabric, the higher the bursting force. The results of bursting elongation have not led to a logical conclusion.

18.3.4 Boat tarpaulins

Boat and car tarpaulins are made of extremely strong, dense and heavy polyester or polyacryl fabrics (Fig. 18.14). To be able to satisfy all standards and thus utility value, it is necessary to undertake additional surface treatments. This is a protective layer to block UV radiation and leakage. The stability of the material after wetting, and resistance to stains are extremely important. Since boat tarpaulins usually have a small area, welded joints are mostly applied only at the edges of the finished product. Sometimes these covers have many folds, transparent inserted areas that imitate windows, welded belts, etc., which requires joining materials in folds and joining various materials. To join and cut awnings, tarpaulins, tents, etc., which have great cutting and joining lengths, many special machines have been developed by various manufacturers.[5]

18.3.5 Fabrics for sailboats and aircraft

Fabrics used for the outer parts of aircraft and sails are very similar in raw material composition, construction parameters and after-treatment. Fabrics for smaller aircraft are used for coating the wings, flaps and wing parts. Until the late 1950s, fabrics for that purpose were made of natural fibres, such as cotton or flax. Initially, joining was done by manual sewing using a strong

18.14 Boat tarpaulin (source: Krika).

and coarse sewing thread made of the same raw material as the fabric, while the stitch length as well as thread tension were uneven. Sewing was carried out directly on the aircraft or sailboat, in the wet state of the fabric which shrank during drying, and thus the creation of folds on the wings or sails was prevented. By further developing sewing machines with specialized additional devices, joining was then able to be carried out on sewing machines and parts were later built into the aircraft or the sailboat. This way of joining with finer threads and higher-quality seams allowed making higher quality and aesthetically more beautiful seams. Since no complete air tightness could be achieved in the seam area, sewing seams was increasingly abandoned in the areas in which air tightness is crucial. Today, cut parts are joined together and attached to other parts of the sailboat or the aircraft usually in combination with bonding with colourless nitrocellulose lacquer, welding and sewing. The raw material of the sewing thread must be very similar or the same as the fabric, so that the seam quality would be adequate.

Since these materials require complete air tightness, several parts (surface of the mast or surface of the aircraft wing) require welding. A sewn seam or any other form of fastening such as rings with ropes is still desirable for fastening materials to metal or wooden parts of the aircraft or the sail boat. Sewing thread, stitch type and length as well as sewing conditions must satisfy the quality and durability of the seam area.

Today, fabrics made from natural fibres are not used for that purpose, except for repair and renovation of old aircraft and boats. Fabrics made of natural fibres were replaced by synthetic fibers (polyamide, polyester, fiberglass and aramid fibres) which have low specific mass and high strength, which are resistant to decay, their water absorption is lower and resistance to solar radiation is higher. Woven aramid fabrics provide improved properties compared to polyester and glass, except in resistance to UV radiation. Fabrics of polyethylene fibres with ultra high modulus, as well as fabrics made from carbon fibers whose properties surpass any other are increasingly used.

Fabrics are woven in plain weave with a relatively high warp and weft density, low specific mass, resistance to shear stress and bending so that an exceptionally stability of the fabric is achieved.[16] Fabrics of various structural parameters, which are mainly distinguished by mass and physical-mechanical properties, are often used in manufacturing sail-boats. The fabric is coated on both sides with synthetic layers to a high degree of smoothness which provides resistance to solar radiation, flammability resistance, higher air and water tightness, and improved physical-mechanical properties. Synthetic coating is applied after weaving and before cutting, so that joining with sewn seams is not desirable because the material is permeable and has lower strength in this area. Today, efforts are made concerning the design of sails and aircraft wings that requires additional surface treatments and printing, but including the following requirements: low mass, high strength, air and water tightness, resistance to fire and radiation, structural stability, to keep the number of joints as small as possible, etc. Most aircraft today are made of composite materials that have relatively low mass, but high strength, stable shape, chemical stability, resistance to atmospheric conditions, large dynamic resistance, easy processing and sound dampening capability. Today's aircraft materials consist of several thin layers of Kevlar and carbon fibres joined together. The result is a very thin and extremely stable composite material. In this way it is possible to reduce the mass of the aircraft to only 350 kg. Due to small mass and high strength, composite materials have almost completely replaced the previously used materials (aluminium, wood, steel, etc.) in building light structures. In aircraft construction, new composites have found their full application primarily because they allow the creation of lightweight and solid parts of the aircraft with an incredibly smooth surface. Their disadvantages are still insufficient resistance to ultraviolet radiation, and complexity of repair in case of damage.

New fabrics from synthetic carbon and aramid fibres are more difficult to be welded because it is difficult to achieve the conditions of material softening that occurs at elevated temperatures and the movement of molecules through the material thickness. There are difficulties with sewn seams where needle penetration and material damage pose a problem, and as a consequence fabrics are often cut in one piece and joined only sideways, mostly with

Velcro straps. Taking account of their use, when cutting is performed, it is necessary to avoid joints which are sometimes the weakest link in the protection of humans because they represent potential areas of risk.

18.3.6 Protective clothing

The development of new materials for effective protection must follow the development of new harmful agents and media. Special programs and cost-effectiveness analysis of the finishing process for new materials are created, new solutions to improve the process of manufacturing materials and garments for protection are continually sought. Appropriate choice of materials, finishing treatment, appearance and size of garment, quality and method of joining cut parts allow the effective protection of the body from various external impacts harmful to the human body. In manufacturing protective clothing from multilayer laminated materials, because of its rigidity, we should respect the functionality of the design that must conform to the requirements of protection, wear, use and comfort. New laminated materials for protective clothing are usually watertight and windtight from the outside, whereas they are impermeable to perspiration in the opposite direction. In this way clothing achieves the complexity of thermophysiological performance and wear comfort, and this is why it is called 'breathable material or garment'.

Protective clothing must meet ergonomic criteria (corresponding cut adaptable to the body and must not restrain human movement) and physiological criteria (ensuring the balance of body temperature in a way that protects the body against cold and allows the flow of heat in the form of sweat and ensures physiological wear comfort). The development of new target materials allows complete body protection. Protection from chemicals, radiation and various microorganisms creates a fully impermeable joint on the protective clothing. In order to obtain a fully impermeable seam, it should be joined by high frequency or thermal welding, but a sewn seam is also possible in combination with a high-frequency or thermal weld. Since straight seam welding is desirable, garments have a simpler cut, such as overalls. Unfortunately, some materials can only be joined by a sewn seam, which poses a risk to liquid or media penetration through holes made by needle penetration. Due to their excellent protection, they are still used for single use items where a chain stitch seam provides sufficient protection (e.g. Tyvek overalls discussed in this chapter, see also Table 18.2).

Protective clothing can be made from two- or three-layer laminated fabrics of types Gore Tex, Sympatex, Pro-line, Windstopper, Polartec types, etc. Such laminates usually have a waterproof fabric on the outside, below which there is a membrane that allows the passage of water vapour (perspiration). The middle layer membrane is composed of tiny micropores that do not allow water droplets through, but at the same time they allow water vapour

Table 18.2 Average experimental values of protective garment materials

		Protective overalls 1			Protective overalls 2		
		F (N)	ε (%)	ω (N × cm)	*F* (N)	ε (%)	ω (N × cm)
Breaking force, elongation at break and work to rupture in accordance with the standard ISO 13935-1	Warp	244.06	15.71	461.00	83.13	18.57	199.36
	Weft	151.73	20.80	391.01	65.51	20.69	151.80
	Warp-sewn seam, 3 stitches/cm	77.43	8.21	78.45	73.22	10.44	112.47
	Weft-sewn seam, 3 stitches/cm	65.40	11.36	113.10	62.45	12.17	98.31
Bursting force and bursting elongation in accordance with the standard ASTM D3787-89		170.00	5.17	175.78	138.33	8.67	239.86
Materials		Woven fabric: raw material composition: polyamide, warp density: 30 threads/cm; weft density: 17 threads/cm; warp count: ? 8 tex; weft count: 7 tex, plain weave			Non-wovens raw material composition: polyethylene		
Sewing thread for sewing the protective overalls 1		*F* = 808.51 cN; ε = 15.82 (%); ω = 2,949.50 (cN×cm); *U* = 580 (twists/m); Tt = 34 tex					
Sewing thread for sewing the protective overalls 2		*F* = 1,286.89 cN; ε = 25.28 (%); ω = 9,084.12 (cN×cm); *U* = 80 (twists/m); Tt = 33 tex					

1 – Protective overalls for extreme weather conditions joined with sewn seam —⧺— with thermally welded tape on the back, 2 – PA/PVC protective overalls for chemical, radioactive and electrostatic protection (Tyvek) joined with chain stitches, *F* – breaking force (N), ε – elongation at break (%), ω – work to rupture (N × cm).

through. Since the size of micropores is around 0.2 μm, which is less than the size of bacteria, so the product provides both a barrier and antibacterial protection. This material is very lightweight and thin, and at the same time strong and durable.[17–20]

Ergonomic functionality of protective clothing implies comfort and adaptability to the body. Physiological criteria are: ensuring a balance of body temperature in a way that protects the body against cold and allows the flow of excessive heat in the form of sweat from the skin surface to the environment, ensuring cutaneous sensory comfort and satisfaction of physiological thermal comfort in wearing. Protective clothing with good thermal physiological characteristics must allow for thermal physiological balance of different climatic conditions and various physical activities of users, so that not only protection, but also optimal human body thermoregulation are provided. These materials are often characterized by a smaller mass of material that provides greater freedom of movement, as described in comfort properties.[17, 21–23]

Mechanical properties of fabrics for protective clothing

Protective clothing can protect from moving parts, the impact of cold, bad weather, chemicals, heat, contamination, etc. Complete protection for a specific purpose is essential, whereby very complex conditions with regard to safety, design and comfort of clothing are set. Figure 18.15 shows an example of the basic data protection and standards for overalls for chemical and radioactive contamination and electrostatic protection. For the purpose of this chapter, two protective overalls were chosen, namely for protection of people from extreme weather conditions and protection from chemical, radioactive and electrostatic effects (Du-Pont Tyvek). The mechanical properties of the material with and without joints were studied (Table 18.2, Figs 18.15 and 18.16). In the protective overalls for extreme conditions, cut parts were sewn with lock stitch seams; subsequently the sewn seam was protected using a thermally welded tape on the back.

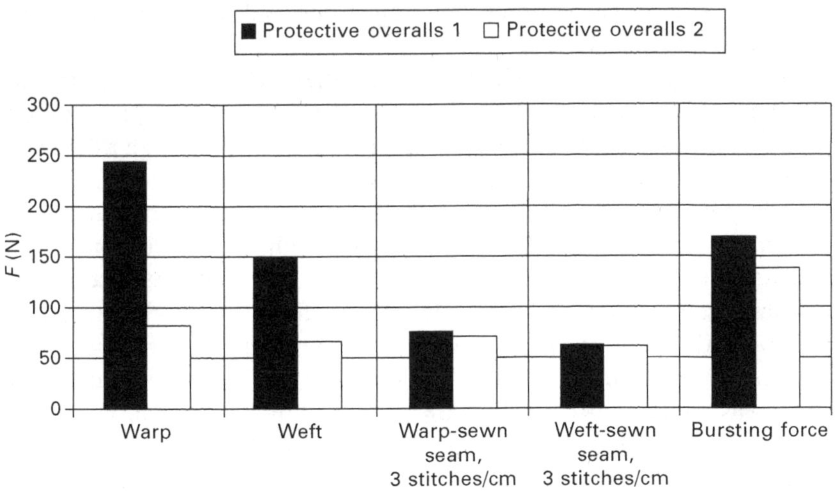

18.15 Breaking forces of the material selected for protective clothing.

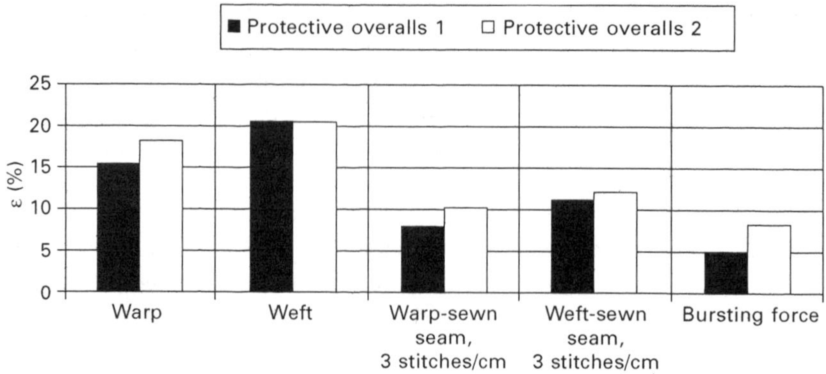

18.16 Elongation at break of the material selected for protective clothing.

Cut parts in overalls for chemical, radioactive and electrostatic protection were joined with chain stitches. According to the results obtained, it can be ascertained that the joints have a significantly lower breaking force and breaking elongation in both overalls. Since these are disposable overalls or used for a shorter time, the stress of the material and seams is not significant so that a relatively low strength of the seams does not pose a problem.

For testing materials intended for protective clothing, two multilayer materials known as artificial leather were selected. One material was a laminated fabric coated with a polyurethane coat, and the other was a laminated knitted fabric with the same polyurethane coat. Breaking force, elongation at break and work to rupture on seamless materials as well as on seamed materials with three different stitch lengths and high-frequency welding were tested. Furthermore, bursting force and bursting elongation were also tested (Table 18.3).

The artificial leather with textile fabric on the back with a seam length of 4 stitches/cm has the highest breaking force, while in case of the knitted fabric, the seam has a length of 5 stitches/cm (Table 18.3). The high frequency welded joints with an additional tape on the substrate with knitted and woven fabric had a lower breaking force than the sewn seams. The breaking force of the samples with knitted fabric was higher without an additional tape than in the samples with an additional tape, while it is the reverse in the woven fabric samples – with an additional tape they had a higher breaking force compared to the samples without an additional tape. The breaking force of the artificial leather with fabric on the back in the warp and weft direction was higher than the breaking force with knitted fabric on the back. This suggests that the strength of the artificial leather with woven fabric was higher in comparison to the knitted fabric. Since polyurethane coating and coating conditions for both the cotton fabric and the polyester

Table 18.3 Average experimental values of textile multilayer materials with polyurethane coat and sewing thread

	Polyurethane coat + woven fabric			Polyurethane coat + knitted fabric		
	F (N)	ε (%)	ω (N × cm)	*F* (N)	ε (%)	ω (N × cm)
Warp (wales)	393.03	21.89	300.49	251.35	135.67	766.16
Warp (wales)–sewn seam ╪═ 3 stitches/cm	368.92	18.94	286.51	216.96	109.03	629.00
Warp (wales)–sewn seam ╪═ 4 stitches/cm	392.41	20.54	298.00	242.47	120.07	723.44
Warp (wales)–sewn seam ╪═ 5 stitches/cm	380.06	19.87	279.44	257.68	128.99	770.99
Weft (courses)	311.74	26.76	472.04	179.05	123.95	1100.50
Weft (courses)–sewn seam ╪═ 3 stitches/cm	282.07	28.05	451.35	176.02	160.86	1144.63
Weft (courses)–sewn seam ╪═ 4 stitches/cm	305.37	32.26	493.86	178.23	175.89	1116.64
Weft (courses)–sewn seam ╪═ 5 stitches/cm	357.53	31.01	608.38	185.55	183.54	1228.58
Weft (courses)–high-frequency welded joint with a tape	224.23	38.48	471.49	175.05	260.61	1823.21
Weft (courses) – high-frequency welded joint without a tape	201.42	30.10	356.76	189.62	299.17	1961.12
Bursting force and bursting elongation seamless	253.33	8.17		391.67	12.04	
Bursting force and bursting elongation seamed – sewn seam ╪═ 3 stitches/cm	430.61	11.35		620	17.18	
Bursting force and bursting elongation seamed – sewn seam ╪═ 4 stitches/cm	482.51	12.47		663	18.57	
Bursting force and bursting elongation seamed – sewn seam ╪═ 5 stitches/cm	406.61	12.89		680	16.42	
Sewing thread	*F* = 1,581.63 cN; ε = 16.36 (%); ω = 5,612.15 (cN × cm); σ = 37.55 cN/tex; U = 650 (twists/m); Tt = 21 × 2 tex					

weft-knitted fabric were exactly identical, the knitted fabric and the woven fabric, respectively, affected the difference in the properties of the artificial leather. Viewing the break point of the samples with joints, it can be noted that the majority of breaks occurred at the sewn seam and the welded joint, respectively. This means that the joint is actually the weakest place. Bursting force was higher in the samples with knitted fabric, and it was different at the seams. Breaking elongation in the warp (wales) direction, in the weft (courses) direction and bursting elongation was significantly higher for the artificial leather with knitted fabric as compared with the artificial leather with woven fabric. Work to rupture was also greater for the artificial leather with knitted fabric.

18.4 Future trends

Since the materials and joints are limited by time and production conditions, it may happen that there are weak places that will not achieve their expected lifespan. By development of integrated monitoring systems based on the textile material, especially on the seams and welded joints, it is possible to monitor changes in material properties during use and replace them in time. This allows the evaluation of materials and the timely reaction, which is extremely important in products such as ropes and parachute materials.[24]

Since the joint is a potentially weak place on materials for tarpaulins, awnings and tents, weaving wider fabrics allows making fabrics with minimum lengths of joints used usually only at the edges of the finished product. In addition to the application of composite materials for aircraft and sailboats, different three-dimensional composites are being increasingly used. These structures consist of two or more composites, which are joined with a material with a low specific mass as filling material and shaped according to a predetermined form. The task of the filling material is to transfer the load from one composite to the other. The basic characteristic of these materials is their resistance to bending or twisting pressure under the action of tangential forces. Since there are no joints or only a minimum number, this construction has a maximum stability and long usage value. Some aircraft parts are manufactured in one piece in a three-dimensional shape so that there is almost no material joining. Reinforcement of composite materials in folds is desirable to obtain sufficient stability and strength. When outfitting the so-called smart house, it is possible to program automatic opening and closing of the awnings, meaning that they open in case of solar radiation or rain. Also, it is possible to develop an integrated system that opens in case of presence of higher humidity in the fabric and thus avoid the deterioration of materials.

18.5 Sources of further information and advice

A closer cooperation of scientific institutions and institutes with the producers of textiles and composites for target products is of great importance. The development of new materials to ensure effective protection and comfort requires an interdisciplinary approach, which is what material manufacturers need from scientific research institutions. It is possible to obtain the latest information and advice primarily at professional conferences, exhibitions of textile and other machinery for the production of new composite materials. However, such information is often insufficient and superficial in order to protect the knowledge and ideas of a new development. More comprehensive information based on experimental research can be found in the latest scientific literature as professional books and journals written by textile experts from scientific institutions and manufacturing companies, who are intensively dealing with these problems. New patented materials are published by the groups of experts who make the most successful products through teamwork.

18.6 Acknowledgement

The authors would like to thank the companies *Čateks*, Čakovec; *Belina*, Krapinske toplice; *Krika*, Nedelišće; *Du-Plast*, Zagreb and *Polimer*, Zagreb for their assistance in writing this chapter.

18.7 References

1. Rogale D, Ujevic D, First Rogale S and Hrastinski M, *Technology of Garment Manufacture with Work Study*, 2nd edn, Faculty of Technical Engineering, University of Bihać, 2005.
2. *News from the Ranges on Offer at the Techtextil 2001*, Messe Frankfurt, 2001.
3. Ujević D and Knez B, 'Stitch damages in the sewing process of Knitwear', *Tekstil*, 1991, 40, 465–470.
4. Sacevičienė V and Masteikaitė V, 'A study of the bending stiffness of laminates', *Tekstil*, 2006, 55, 458–466.
5. Pfaff, Brochures of Pfaff Sewing Machines – Welding Machine, Model 8310, 2008.
6. Nikolić G, Rogale D, Ujevic D and First Rogale S, 'Textile material joining and cutting machines with new technologies – IMB 2006', *Tekstil*, 2007, 56, 308–317.
7. http://www.twi.co.uk/content/altex_finalrpt.html, prosinac, 2010
8. Altas S and Yucel O, 'Examining the effect of lubrication rate on sewing thread friction coefficient and on seam slipppage', *Tekstil*, 2006, 55, 566–571.
9. Geršak J and Knez B, 'Determination of the strength of the sewed seams on clothing', *Tekstil*, 1991, 40, 361–368.
10. Mulack H, 'Fiber processing from the point of view of Pfaff Company', *Tekstil*, 1994, 43, 536–538.
11. Strmečki V, Koren T and Cerovec M, 'The impact of fabric construction on yarn slippage in the area of the sewing seam', *Tekstil*, 1999, 48, 71–76.
12. Dürkopp Adler, Machine program 2009.
13. Catalogue of the Stattler Company, 2010.
14. Tomljenović A, Cunko R, Pezelji E and Grgec S, 'Evaluation methodology of the quality of sunshade fabrics under application conditions', *Tekstil*, 2008, 57, 1–14.

15. The experimental part of this chapter results from the scientific programs Advanced Technical Textiles and Processes, code: 117-0000000-1376; and Anthropometric Measurements and Adaptation of Garment Size System, code: 117-1171879-1887, conducted with the support of the Ministry of Science, Education and Sports of the Republic of Croatia.
16. http://hr.wikipedia.org/wiki/ casement _ aeroplane # fabric, July, 2010.
17. Kirin S and Dragčević Z, 'State of the art and standardisation of proctective clothing', *Tekstil*, 2003, 52, 282–285.
18. Pezelj E and Pezelj D, 'Problems of care of garments containing functional membrane', *International Professional Conference TEXTILE AND GARMENT CARE – Dry Cleaning and Washing, Actual Problems*, Otočec ob Krki, Slovenia, 28–29 September, 2007.
19. Kirin S, Kovačević S and Vučinić J, 'Wear comfort of firefighting protective clothing', *4th International Ergonomics Conference*, Stubičke Toplice, Croatia, 30th June to 3rd July 2010.
20. Pezelj E, Andrassy M and Čunko R, 'Modern technical textiles – specific requirements on fibres', *Tekstil*, 2002, 51, 262–277.
21. Geršak J and Grujić D, 'Impact of clothing on thermal and physiological human comfort under different climatic stresses', *Tekstilec*, 2003, 46, 183–190.
22. Andrassy M and Domes Z, 'Do fibers contribute to garment comfort?', *Tekstil*, 1990, 9, 687–691.
23. Bockelmann R, 'Care of firefighting protective clothing according to special criteria', *Tekstil*, 1999, 48, 154–156.
24. Eichhoff J, Jungbecker N and Gries T, 'Research area of the technology of smart textiles', *Tekstil*, 2008, 57, 532–533.

19
Nonwoven materials and joining techniques

A. POURMOHAMMADI, University of Payame Noor, Iran

DOI: 10.1533/9780857093967.4.565

Abstract: Nonwoven fabrics have become an extremely important segment of the textile industry in recent years. The technical developments in polymers, nonwoven processing and fabric finishing have led to significant improvements in fabric physical and mechanical properties including fabric handling and drapability, tensile properties, abrasion resistance, pilling and washing stability, dyeing and printing that create prospects for nonwoven fabric applications in particular in apparel outerwear. This chapter briefly discusses the various nonwoven fabric production processes including web formation, web consolidation and finishing. An introduction to different joining techniques is also discussed and at the end future trends in the non woven market are outlined.

Key words: nonwovens, joining, web formation, web consolidation, finishing, apparel.

19.1 Introduction

Technical development in textiles, in particular nonwovens, provide the opportunity for the industry to penetrate into a wide range of applications from oceans to space. The smart fabrics incorporating electronic components, medical textiles, protective clothing and many other usages of nonwovens demonstrate the potential of this technology in the future of the textile industry. Whilst historically the penetration of nonwoven fabrics in garments has been limited, the integration of nonwoven fabrics into garments for functional purposes is well established. Among the various reasons behind this lack of success is the seaming or joining methods for nonwoven fabrics. Seamed parts are most commonly joined by stitching, a highly labor-intensive process which increases the production cost. It also makes holes in the fabric, which impairs the strength and sealing performance of the resulting seam. There have been intensive studies to look for automated processes with fast, clean and high quality seaming for nonwoven fabrics. Different welding methods and adhesive bonding have been demonstrated to be promising.

19.2 Principles of nonwovens

Nonwoven fabrics have become an extremely important segment of the textile industry in recent years. As a result of new technologies developed since the 1950s with the availability of synthetic fibers in the early 1970s, nonwoven fabrics competed with conventional textiles, i.e. woven and knitted structures.

Nonwovens had inherent versatility in terms of fiber composition, structure and performance. This facilitated the innovation of new concepts and opened up new market sectors such as geotextiles, civil engineering and construction materials, agriculture, filtration and aerospace. More recently nonwoven products have been widely developed for applications where conventional textile fabrics would not even be considered. Such products include: absorbents, protective clothing, disposable medical fabrics, and specialized filtration and insulation materials.

19.2.1 Definition and classification

The term nonwoven has been around for more than 800 years when nonwovens were regarded as a low price replacement for conventional textiles. All the definitions proposed so far describe what nonwovens are not, rather than what they are. EDANA (The European Disposable and Nonwovens Association) defines a nonwoven as 'a manufactured sheet, web or batt of directionally or randomly orientated fibers, bonded by friction, and/or cohesion and/or adhesion', but it excludes a number of materials such as paper, woven and knitted fabrics, tufted or stitchbonded, and felted by wet-milling. INDA (the Association of the Nonwoven Fabrics Industry) explains nonwoven fabrics as 'sheet or web structures bonded together by entangling fibers or filaments, by various mechanical, thermal and/or chemical processes. These are made directly from separated fibers or from molten polymers.'

Nonwovens are engineered structures that can form products with specific functions to fulfill various applications. Perhaps the most significant feature of nonwoven fabrics, is that the fabric is usually made directly from raw materials in a continuous production line, thus partially or completely eliminating conventional textile operations such as carding, roving, spinning, weaving, or knitting. This simplicity of fabric formation, combined with high productivity, allows nonwovens to compete favorably with woven and knitted fabrics on a performance per cost basis in many industrial applications. However, in fashion apparel fabrics, where drape properties, flexibility and strength are required, nonwovens need to be improved technologically, although there has been a continuous effort through innovative manufacturing processes to develop new nonwoven products with similar drape and flexibility to woven and knitted structures.

19.2.2 End-use application

Nonwovens are being used in numerous fields of application with new products appearing almost every day. Some remarkable aspects of nonwoven structures, such as tensile strength, elongation, stiffness, permeability, bulk, vapor and liquid absorption, chemical resistance, which can be engineered using different manufacturing processes, provide a spectrum of products used in industrial, engineering, consumer and healthcare goods. The range of applications is widening as the basic technology has become more sophisticated. The application of nonwoven fabrics can be classified as disposable and durable. Disposable nonwovens are those for short-life, single-use application, and durable nonwovens are those for long-life or multiple-use applications. Table 19.1 gives a list of these two classes of applications.

There are organizations in Europe and the United States such as EDANA and INDA that publish statistical data relating to the deliveries of produced nonwovens in each of the various sectors and applications. An example of this data is given in Table 19.2 showing the percentage for each end use in relation to the total weight of nonwoven deliveries. Hygiene is by far the largest of these categories, calling for over 33.3% of nonwoven production, followed by civil engineering/construction and building materials with 17.9%. The high growth in recent years of the use of wipes, both dry and wet, in a range of household and industrial applications is shown by about 15.8% of nonwoven deliveries.

19.2.3 Manufacturing processes

The manufacturing of nonwoven fabrics is very different from that of woven and knitted fabrics. Each nonwoven manufacturing method involves the following generic steps:

Table 19.1 Applications of nonwoven materials

Nonwovens in disposable applications	Nonwovens in durable applications
Hygiene Baby diapers, feminine care, etc.)	Garments (protective apparel, cleanroom garments)
Medical (drapes, gowns, dressings, etc.)	Interlinings
Consumer wipes (baby wipes, household cleaning wipes, personal care wipes)	Shoe and leather goods
Industrial wipes	Furnishing and bedding
	Floor coverings
	Air and liquid filtration
	Building and roofing
	Geotextiles and civil engineering
	Automotive
	Agriculture and horticulture
	Electronics and electrical (battery separator)

Table 19.2 Statistical data on deliveries of nonwovens in each application

Product application	2008 (1,000 T)	% of total
Hygiene	566.1	33.3
Medical/surgical	42.1	2.5
Wipes for personal care	163.2	9.6
Wipes others	105.7	6.2
Garments	13.3	0.8
Interlinings	30.9	1.8
Shoes/leather goods	24.1	1.4
Coating substrates	37.5	2.2
Upholstery/table linen/household	88.7	5.2
Floor coverings	43.0	2.5
Liquid filtration	60.8	3.6
Air and gas filtration	42.1	2.5
Building/roofing	195.5	11.5
Civil engineering/underground	108.6	6.4
Automotive	61.4	3.6
Agriculture	31.8	1.9
Others	71.0	4.1
Unidentified	16.1	0.9
Total	1,701.9	100

Source: EDANA.

- fiber/raw material selection
- web formation
- web bonding
- web finishing and converting.

The process of manufacturing nonwovens can be divided into three major areas: dry-laying, wet-laying, and extruded polymer-laying. Dry-laid materials have their origin in textiles, wet-laid materials in papermaking and extruded polymer-laid products in polymer extrusion and plastics. There is also a hybrid system which is a combination system or composite structure. An overview of nonwoven manufacturing technologies is given in Table 19.3.

Dry-laid or textile-based nonwoven fabrics are manufactured with machinery designed to manipulate textile fibers in a dry state. In general, this technology includes fiber openings, carding and/or aerodynamic forming of fibers into selectively oriented webs. This method provides maximum product versatility, since most textile fibers and bonding systems can be utilized.

Wet-laid or paper-based technology includes systems designed to use short synthetic fibers as well as wood pulps to form nonwoven structures by suspending the fibers in water. Short fiber air-laying may also be considered in this category since the principle of this web forming method originates from suspending the fibers in an air stream and collecting them in a perforated conveyor belt followed by different web consolidation methods. This technology provides the least product versatility with high capital investment;

Table 19.3 Overview of nonwoven manufacturing processes (courtesy of D.B. Brunnschweiler)

Paper technology Fibers 2–15 mm		*Nonwovens* *Textile technology* Fibers 10–200 mm **Web forming**					*Extrusion technology* Filments			
Wet-lay	Air-lay	Garnetts	Woolen cards	Cotton cards	Hybrid cards	Air-lay cards	Melt blown	Spun laid	Flash spun	Fibrillated film
		Web manipulation								
		Cross-laying	Drafting		Spreading	Scrambling		Crimping		
Bonding										
Adhesive		Heat	Needlepunching		Hydroentangling			Stitch bonding		Ultrasonics
Calender/mangle spray foam powder		Melt fiber (monofiber or bi-component) Powder Calendar (plain or embossed) Oven (drum or lattice)	Tacking Plain needling Multi-directional needling Textured needling		(spunlace)			With or without yarn		
					Finishing					
		Singeing	Coating		Printing	Embossing				Laminating

however, it yields outstanding product uniformity at very high production speeds.

Extrusion-based nonwovens are also referred to as spun-laid fabrics and include spun-bond, meltblown and flash spinning (a method developed by DuPont for their Tyvek product). This method of nonwoven manufacturing provides less versatility in product properties compared to dry-laid materials but they offer fabric structures with a high strength-to-weight ratio (spun-bonds), high surface area (meltblown) at a modest capital investment and relatively high production capacity.

19.3 Raw materials

Fibers are the basic element of a nonwoven structure and, therefore, the selection of the fibers mainly determines the properties of the final products. A wide range of fibers including natural and synthetic have been used in the production of nonwoven fabrics. According to the statistics the man-made fibers dominate nonwoven production accounting for over 90% of the total output. The use of specialized fibers is also helping to improve performance, productivity and efficiency during manufacturing. These are becoming vitally important in an industry which is using integrated systems running at very fast speeds with increasing raw material costs.

A convenient method for classifying fibers is based on their origin, whether natural or man-made. Natural fibers are obtained from the following natural polymers or inorganic materials:

- protein-based (i.e. wool and all other animal hairs such as goat, rabbit, cow, alpaca, camel, etc).
- cellulose-based
  - seed fibers such as cotton, kapok and milkweed
  - leaf fibers such as abaca, sisal, palm, pineapple, etc.
  - bast or stem fibers such as flax, jute, hemp and kenaf.
- mineral fibers which are rock based such as asbestos, mineral wool and basalt.

Man-made (synthetic) fibers are produced from synthetic polymers through linking their units together by either condensation or addition polymerization methods.

There are also regenerated polymers where the constituent polymer is obtained from natural sources but undergo chemical and/or mechanical modification in order to be formed as a usable fiber. Viscose rayon belongs to this category and is derived from naturally occurring cellulose (generally wood cellulose).

The most commercially used fibers in manufacturing nonwoven fabrics are:

- polypropylene
- polyester
- viscose rayon
- wood pulp
- polyamide
- other specialty fibers.

These days the natural and synthetic fiber producers are increasingly under pressure by consumers and retailers to offer environmentally friendly fibers for nonwovens. In particular, the producers of polyester and polypropylene are making efforts to launch sustainability initiatives, increasing recyclability, saving energy and water and reducing the landfill space. The imminent large-scale production of biodegradable polyester from corn-starch is a sign of this change of direction in fiber choice. For example, Foss Manufacturing is producing a polyester fiber made of 100% post consumer recycled (PCR) materials.[1] Freudenberg is launching Lutradur Eco, a PET spun-laid fabric made of PCR from drinking bottles used in a variety of applications such as automotive, carpet, furnishings, building and industrial.

Also cellulose-based fibers, made from wood pulp, are used as a sustainable fiber option in producing nonwovens. Wood is a natural resource that is sustainable, renewable and biodegradable. Examples of such commercial fibers are Tencel, Danufil, Galaxy, and viloft. In addition to cellulose, cotton has gained popularity in many nonwoven sectors. The advancement of spunlace technology parallel to the growth of the wipe industry and being natural with exceptional properties can be attributed to the growth of cotton usage in nonwovens. Using cotton fiber in its raw or virgin state, i.e. as it is harvested and ginned, is also being established in the nonwovens industry. Fiber in this form is naturally hydrophobic and oleophilic and is used in oil spill cleanup pads and booms.

Another bio-based and eco-friendly fiber is based on polylactic acid (PLA) that is made of plant sugars (e.g. cornstarch, sugar beet or wheat starch) and hence, it is an annually renewable plant-based material replacing conventional plastics like polypropylene and polyester. PLA has already found a place in nonwovens wipes, diapers and feminine hygiene products. It has very good inherent wicking properties as well as UV resistance, elastic recovery, and reduced flammability with low smoke and heat generation.[2] It has a melt-point in the same range of polypropylene. Fiberweb disclosed laminated nonwoven webs made of 100% PLA in 1997 and introduced a range of meltblown and spun-laid PLA fabrics.[3]

19.4 Web formation technology

Nonwoven fabrics are generally composed of fibrous materials. The web formation step of nonwoven manufacturing processes transforms the previously prepared fibers or filaments into layers of loosely held fiber networks. Various mechanical or fluid means are used to achieve preferred fiber orientation in the web through machinery adapted from the textile, paper or extrusion industry. The characteristics of the web determine the physical and mechanical properties of the final product. These characteristics largely depend on the web geometry which in turn is determined by web formation methods. In all nonwoven manufacturing systems, the fiber material is deposited or laid on a forming or conveying surface in a dry, wet or quenched molten state; hence the terms dry-laid, wet-laid and polymer-laid are used to describe different modes of web formation.

19.4.1 Dry-laid web formation

The dry-laid nonwoven sector employs carding, garneting and air-laying to convert staple fibers into a web or batt structure that is uniform in weight per unit area. Different steps in dry-laid manufacturing systems include:

- fiber selection
- fiber preparation
- web formation.

Applications of air-laid products

Depending on the fiber composition and the bonding method, a wide variety of air-laid products for different end-uses can be made which may be classified as below:

- Chemical bonded air-laid: napkins, table cloths and wipes.
- Thermal bonding: nappies (different components, i.e. acquisition layer, distribution layer, and absorption core), feminine hygiene/incontinence products and insulation.
- Spun-lacing: wet and dry wipes for domestic and industrial applications, medical textiles (including disposable gowns, curtains, wound-care dressings, bed sheets), filtration media.
- Needle-punching: interlinings and shoe linings, wadding, medical and hygiene products, geotextiles and roofing felts, insulation felts, automotive components, filters, wipes.

19.4.2 Wet-laid web formation

Wet-laid nonwovens technology is closely related to papermaking which itself goes back about 2,000 years with the craft of making paper by hand being developed in China. In 1975, Osborne made significant contributions to the development of a continuous process to manufacture long fiber papers and wet-laid nonwoven structures.[4] Traditional papermaking machinery uses a variety of woodpulps or other short (1–4 mm) cellulosic fibers with a dilution ratio of 0.5%. The incorporation of textile fibers including synthetic and high performance fibers such as PEEK, aramid, carbon and glass fibers is also possible with dispersion concentration of as low as 0.0025 fiber-to-water ratio.

Applications of wet-laid product

Wet-laid nonwoven fabrics in the range 5–2,000 g/m^2 can be produced using different machine designs. There are a wide range of application areas for wet-laid materials which are listed in Table 19.4.

19.4.3 Polymer-laid web formation

Polymer-laid, spun-laid or spun-melt technology is a one-step process for converting polymer resin directly into a nonwoven web. In this process, fabrics are produced by extrusion spinning in which filaments are directly collected to form a web and therefore the intermediate steps of conventional spinning (i.e., fiber production and preparation carding, cross-lapping, etc.), are eliminated resulting in a high production and the lowest cost method of web formation.

Historically, the concept of spun-laid fabric production was developed in the late 1950s and mid-1960s by adopting the melt-spinning techniques. It is believed that the term spun-bond was used by US-based company, DuPont which developed and commercialized the first spun-bond process in the late 1950s. They developed a polyester spun-bond product called Reemay® followed by a polypropylene product Typar® and the flash-spun polyethylene fabric Tyvek®. In Europe, there have been different spun-bond machinery manufacturers such as Freudenberg with Lutravil®, Lurgi with Docan®, Oerlikon Neumage and Reifenhauser with Reicofil® all in Germany, STP Impianti and Meccaniche Moderne SPA and ORV in Italy.

There are four simultaneous and integrated steps in the production of spun-bond fabrics, including filament extrusion, drawing, laydown and

Table 19.4 wet-laid product applications

Conventional applications	Technical applications
Surgical clothing and drapes	Glass fiber roofing substrate
Bed linen	Printed circuit boards
Table cloth and napkins	Filters
Towels	Wall coverings
Kitchen wipes	Insulation materials
Hygiene products	

bonding. In general, all the thermoplastic polymers including polyester (PET), polypropylene (PP), polyethylene (PE), thermoplastic polyurethane (TPU) and polyamide (PA) can be processed by spun-bonding to manufacture a uniform web. Also different types of rayon, i.e. viscose and cuprammonium rayon, have been processed into usable spun-bond product using wet-spinning. Also splittable bicomponent (two different polymer components combined into a single fiber that are arranged in sheath/core or segmented pie) fibers are used in spun-bond fabrics and have some interesting applications. One example is Evolon® fabric produced by Freudenberg that is split into microfibers during hydroentanglement of the spun-laid web. Additionally, bicomponent fibers with an eccentric sheath core arrangement are used to develop crimp in spun-laid fabrics by differential thermal shrinkage of the two polymer components. Spun-bonding usually requires a narrow molecular weight distribution (NMWD) and melt flow rate resins, typically of 25–42 (g/10min) MFR. For melt blowing, a narrow molecular weight distribution with a very high melt flow rate of 800–1,800 g/10 min is required.

Applications of spun-bond fabrics

Spun-bond fabrics are used in a variety of applications, namely:

- automotive (seat covers, interior door panels, trunkliners, etc.)
- civil engineering (roofings, geosynthetics, railway bed stabilization, revetment protection, erosion control, canal and reservoir lining protection)
- hygiene and medical
- packaging.

Applications of meltblown fabrics

Due to the inherently large fiber surface area in meltblown fabrics, applications are in filtration, insulation and liquid and oil absorption. The fine fiber network results in some characteristic properties:

- enhanced filtration efficiency (air filter, food and beverage filtration, surgical mask and respiratory filtration, water and liquid filtration)
- excellent barrier properties (used in insulation applications such as Thinsulate® from 3M Company)
- good wicking action (industrial wipes, oil sorbent pads and booms, also Coform® from KC Company).

19.5 Web bonding technology

The progress of the nonwovens industry has been the direct result of advancement in bonding technologies. The challenge is to take a formed web and give it the characteristics of a finished nonwoven product. At this stage the web is a flexible, soft, and porous material, lacking strength and durability that the individual fibers must somehow be connected. In response to all the demands being made by the wide range of nonwoven markets, several bonding processes are employed. The simplest way to categorize them is chemical, thermal and mechanical.

19.6 Nonwoven fabric finishing

Finishing of nonwoven fabrics is increasingly important since producers are looking to add value to the final product using different finishing methods. Principally, finishing can be classified as either wet finishing or dry finishing (or mechanical). Wet finishing includes washing (scouring), dyeing and coating, chemical applications such as antistatic agents, antimicrobial treatment, flame-retardant and hydrophobic or hydrophilic finishes and UV stabilizers. Dry or mechanical finishing processes include calendering, shearing and singeing, brushing or raising, embossing, creping, perforating, etc. A more detailed explanation of the above processes can be found in the various nonwovens textbooks.[5–7]

19.7 Techniques for joining nonwoven materials

Historically, seaming or joining methods for fabrics have centered on various types of stitching, a highly labor-intensive process which increases the production cost. Stitches also make holes in the fabric, which can impair the strength and sealing performance of the resulting seam. In searching for more efficient joining processes, companies have looked at welding and adhesive bonding, particularly for synthetic fabrics. The main joining techniques commonly used are briefly explained below. More detailed information can be obtained from Ref. 8.

19.7.1 Ultrasonic joining (welding)

Ultrasonic welding, one of the most widely used methods for joining thermoplastics, is a process that applies ultrasonic energy at high frequency (20–40 kHz) to produce low amplitude mechanical vibration (1–25 μm). The vibrations, produced by a welding sonotrode or horn, generate heat at the joint interface of the components being welded, resulting in softening or melting thermoplastic materials and weld formation after cooling. Ultrasonic bonding is the fastest known welding technique, with welding times typically between 0.1 and 1.0 seconds. Ultrasonic welding is fast, economical, and energy efficient, resulting in higher productivity at lower costs than other assembly methods. It is commonly used in the healthcare industry because it does not introduce contaminants to the weld that may affect biocompatibility of the medical device. Ultrasonic applications can be found in almost all major industries in which thermoplastic parts are assembled in high volumes. Some examples are: automotive, medical, and packaging.

Amorphous polymers, which gradually melt over a broad temperature range and also solidify gradually, transmit ultrasonic vibration efficiently and can be bonded with a broad range of process conditions. However, semi-crystalline plastics require high heat to disrupt the ordered molecular structure and melt it. The melting point is sharp and solidification occurs rapidly as the temperature drops slightly. In solid state this type of polymer is spring-like and absorbs a large part of the ultrasonic vibrations, instead of transmitting them to the joint interface. Therefore high amplitude is necessary to generate sufficient heat for joining.

Ultrasonic welding can be used to make seams very rapidly in nonwoven materials. Typically the fabrics are fed between two rollers, one of which is vibrated, generating melting through the full thickness of the fabrics. Filters or face masks are widely produced using this technique.

19.7.2 Laser welding

Laser welding is used mainly to attach fabrics to rigid components and also experimental work has demonstrated the feasibility of using this method for joining woven and nonwoven textiles made of synthetic fibers such as polyester, nylon, polypropylene and also coated or laminated fabrics with membrane layers.[9] Laser welding process consists of three stages including:

- application of laser absorber material which may be applied in the form of a low-viscosity liquid that dries rapidly to leave a very thin deposit on the surface of the textile
- assembly of the seam and application of clamping pressure
- irradiation of the seam with a near infrared laser beam at a wavelength of 940 nm to melt the material where the absorber has been applied and creates a permanent weld.

Some of the benefits of this method are it is potentially automated, it retains the external appearance (melting occurs at the interface only), it is clean and fast (20 m/min), and possesses good seam strength (40–100% of the material strength). The main limitation is that one part of joining should be transparent to the infrared laser.

19.7.3 Radio frequency welding

Radio frequency (RF) or high frequency welding (HF), also called dielectric welding, uses high frequency (normally 27.12 MHz) electromagnetic energy to generate heat in materials by virtue of their electrical properties. The material to be welded must be able to convert alternating electric field into heat. RF is normally used for joining films or thin sheets since a sufficiently strong electric field to make material melt can only be achieved when the gap between the electrode and the welded part is less than around 1.5 mm. RF welding is commonly used in sealing thin sheets and films of plasticized PVC for stationary products, medical items such as blood bags, disposable clothing, blood pressure cuffs and other types of packaging. The major disadvantage of RF welding is that it is only suitable for materials containing polar groups in the molecular structure such as PVC, TPU, nylons, PET, EVA, and PVDC. It is not possible to join standard PP and PE grades using this method.

19.7.4 Infrared welding

Infrared (IR) radiation is another noncontact method which can potentially be used for joining nonwoven fabrics made of thermoplastic polymers including polypropylene, polyamides, polycarbonate, and elastomers. The IR radiation is often supplied by heated metal (with wavelength of 1–15 μm) or IR lamps, which radiate at wavelength in the range of 1–3 μm. When the IR radiation is incident on a polymer surface it can be reflected, transmitted, or absorbed. It is the absorbed portion that leads to temperature increase and consequently leads to melting of the polymer.

Some of the advantages of IR welding include: speed and flexibility (this is a fast and economical method), the high welding quality leading to parts with excellent aesthetics, and there are no large heated platens as only the weld area is heated. IR welding is not suitable for clear materials unless an absorbing material is used at the interface. In addition to the wavelength of the IR source, the depth of penetration of IR radiation depends on many factors, including the polymer's absorption characteristics, which could vary strongly with minor changes in polymer formulation. Therefore, consistent IR welding requires close attention to be paid to batch polymer uniformity.

19.7.5 Hot wedge welding

Hot wedge welding was introduced as an alternative to sewing methods for seaming nonwovens, in particular geotextiles. Hot wedge welding depends on combination of heat, pressure, and speed to melt two materials together. The temperature settings depend on the fabric basis weight and thickness; for example, for nonwoven polypropylene fabrics with the basis weight of 200–700 g/m^2, the typical settings are reported to be 300–350°C. The wedge welding equipment runs automatically between two overlapped sheets of thermoplastic materials and heats the interface between them followed by pressing the materials together to form a seam line. According to the ASTM standards D-4884 and D-4595, the seams should provide more than 75% of the parent material strength. In an application of this method of welding it was shown that a seam strength as high as 95% of parent material strength can be achieved.[10] Wedge welding is most commonly performed on HDPE sheet materials; however, there are available technologies that can seam weld polypropylene, polyester, and nylon with the seam width ranging from 25 to 100 mm.

19.7.6 Heat sealing

Heat sealing is another welding process used to join thermoplastic nonwoven fabrics using a very precise heat and pressure. It is also a desirable method for fabric lamination which can provide high-strength joints for demanding applications. It was claimed that joints stronger than the base materials can be obtained. A method for heat sealing of nonwoven fabrics was explained in which a nonwoven fabric composed of thermoplastic materials was fed through a pair of rollers, which has sealing ridges of a predetermined pattern on its outer circumference.[11] As the fabric passes between the rollers, heating and pressing the fibrous materials at the sealing points (i.e. ridges) results in forming fused portions in the fabric. Nonwoven fabrics produced using this method, are used in cleaning sheets and industrial wipes.

19.8 Future trends in the nonwoven market

The nonwoven industry has passed through several phases of evolution since the 1950s when the industry was regarded as a supplier of low cost, low performance substitutes for woven and knitted materials. By the 1980s nonwovens were competing increasingly with conventional textiles in many market sectors offering similar level of performance at a lower cost. Until about the 1990s, most of the world's nonwoven industry was based in the areas where the process technology developed, mainly the United States, Europe, and Japan. The large-scale production facilities are highly capital intensive, making it too risky for small companies to set up production lines in particular for technologies such as spun-laid, wet-laid and spun-lace.

The industry is still seen as capital intensive, considering that, according to the latest estimate, 90% of the total nonwovens sales are made by some 40 companies. However, the impressive involvement of smaller companies in innovation and new product development is an indication that newer companies are taking an active leadership role in the future of the nonwoven industry.

According to the latest figures published by INDA and EDANA, the total production of nonwovens in North America, Europe and Asia is about 5.4

million tonnes.[12] On a comparative basis with the figures from 2007, the production of nonwovens in 2008 in Europe reached 1,722,400 tonnes with just about 1.4% growth. However, a growth of about 9.3% for the Asian producers was reported with nearly 60% of this produced in China (i.e. 1.347 million tonnes) (see Table 19.5).

The global sales of the nonwovens roll goods in 2008 were reported to reach to US$17.3 billion.[13] As shown in Table 19.6, the polymer-based nonwovens production, including online composites in Europe, has declined for the first time from 46% in 2007 to 44% in 2008. The share of other production processes is illustrated in Fig. 19.1.

19.8.1 Raw material consumption trends

Total fiber and polymer consumption in 2008 was 1,746,100 tonnes. In fiber-based processes, polyester was the most used fibers with 33.8% of the total fibers consumption. This is due mainly to increase in hydroentangled products, needled materials as well as substitution for viscose in chemical bonding. Polyester is followed by polypropylene (20% of total consumption) and woodpulp with 20.8%, with viscose in fourth position (14% consumption). In polymer-based processes, clearly polypropylene remains by far the most used polymer with consumption of 78.6% of the total production.

19.8.2 Outlook for nonwoven fabrics by leading applications

The main end-use for the nonwovens industry remains the hygiene market (see Table 19.7). Baby diaper is the major segment of the hygiene market and

Table 19.5 Nonwovens production in 2008

Regions	2007 (1,000 tonnes)	2008 (1,000 tonnes)	Growth rate (%)
Europe	1,699	1,722	1.4
North America	1,273	1,348	5.9
Asia	2,120	2,317	9.3

Table 19.6 Nonwovens production in Europe by webforming process

Process	2006		2007		2008	
	1,000 tonnes	Growth (%)	1,000 tonnes	Growth (%)	1,000 tonnes	Growth (%)
Spun-laid	710.6	9.4	776.5	9.3	758.7	−2.3
Dry-laid	623.8	5.8	659.6	5.7	685.7	4.0
Wet-laid	106.8	8.8	107.8	0.9	115.7	7.4
Air-laid	140.2	5.7	154.8	10.4	162.3	4.8
Total	1,581.4	7.6	1,698.7	7.4	1,722.4	1.4

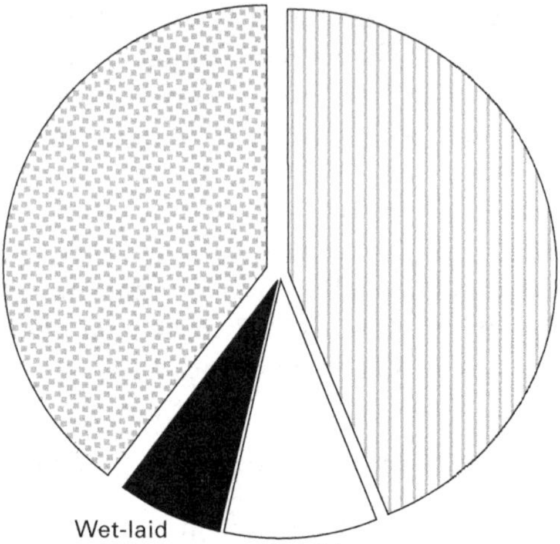

19.1 Nonwovens production in Europe by process in 2008.

Table 19.7 Outlook for nonwovens by leading applications in NAFTA, Europe and Asia Pacific (source: INDA, EDANA estimates)

Application	2007 (1,000 tonnes)	2012 (1,000 tonnes)	Growth (%/year)
Hygiene	1,410	2,144	8.7
Medical	250	331	5.8
Wipes	609	946	9.2
Apparel	274	374	6.4
Furnishing	459	570	4.4
Filtration	345	512	8.2
Automotive	289	415	7.5
Building/roofing	227	332	7.9
Other industries	657	923	7.0
Others	695	888	5.0
Total	5,219	7,435	7.3

had better fortune than the two other categories (incontinence and feminine hygiene). This is followed by wipes (personal, medical and others) and by building/roofing.

It can be concluded that there is increasing interest in nonwoven fabrics which combine the advantages of different textile structures including knitted and woven fabrics. This world growth is in part due to several industrializing economies and their increasing demand for nonwoven materials and key markets such as hygiene, consumer wipes, liquid and gas filtration, automotive, building and roofing and geotextiles. Further growth will derive from increased market penetration by new products with improved characteristics.

Customers will continue to put more demands on the nonwovens industry for products that not only perform but are sustainable and will not harm the environment.

19.9 Acknowledgements

The author would like to thank the contribution of the following organizations in preparing this chapter by kindly providing the technical literature and information: Trützschler Nonwovens, DiloGroup (Engineering for nonwovens), Andritz Küsters, Reicofil and EDANA. Special thanks are also due to Baftineh Company (Iran) for the opportunity that provided me to gain valuable experience and information during past five years.

19.10 References

1. Levy S (2010), 'Meet fibers, green dream team', *Nonwovens Industry*, Issue 6.
2. Woodings C, 'New developments in biodegradable nonwovens', available at: www.technica.net/NF/NF3/biodegradable.htm.
3. Ehret P, Guipouy P and Lahtenkorva K (1997), US Patent application 5702826, December 1997, assigned to Fiberweb.
4. Osborne F H (1975), *The History of Dexter's Long Fiber Paper Development*, C H Dexter, Windsor Locks, CT.
5. Rusell S J (ed.) (2007), *Handbook of Nonwovens*, Cambridge: Woodhead Publishing.
6. Albrecht W, Fuchs H and Kittelmann W (2002), *Nonwoven Fabrics: Raw Materials, Manufacture, Applications, Characteristics, Testing Processes*, Weinheim: Wiley-VCH.
7. Turbak A (1993), *Nonwovens: Theory, Process, Performance and Testing*, Atlanta, GA: TAPPI.
8. Troughton M J (ed.), (2008), *Handbook of Plastics Joining (a Practical Guide)*, Norwich, NY: William Andrew Publishing.
9. Jones I A and Wise R J (2005), 'Novel joining methods applicable to textiles and smart garments', *Wearable Future Conference*, University of Wales, September.
10. Cadwallader M W (1999), 'Hot wedge welding geotextiles', *GFR Magazine*, Vol. 17, No. 3.
11. Tanaka Y and Kenmoch Y (Uni-Charm Corporation) 2004, 'Heat sealing methods and apparatus for fiber sheet', US patent application 6713726B2, March 30, 2004.
12. Annual Statistics of EDANA, 2008.
13. 'The world nonwovens industry: profiles of ten leading producers', Textile Intelligence, December 2009.

20
Epilogue: joining textiles

I. JONES, TWI Ltd, UK and G. K. STYLIOS,
Heriot-Watt University, UK

This book shows that the joining process has no barriers between 'traditional' and 'technical' textiles and that the method depends only on the design and development of the product and its end use. When multidisciplinary techniques prevail, innovation thrives. Discipline fragmentation must be replaced by the imagination of the designer. This creates exciting new endeavours, but at the same time challenges and responsibilities. The main challenge is to continue to provide new knowledge and innovation. For example, the creation of a garment will not occur almost exclusively in the East, because economics will bring back manufacturing, rendering the robotisation of garment making and mass customisation a necessity. In order to avoid the mistakes of the past, however, this will only be possible through the underpinning of fabric mechanics in the machine/fabric/human interface, and by other methods of joining. It is essential to train the next generation of students with the necessary tools for optimising product design and performance through a knowledge of materials and processes. This philosophy is shown throughout the chapters in this book and we hope that the reader, whether lecturer, student, researcher or industrialist, will be able to apply it as best practice. The final word is to thank all those researchers and industrialists who, for many years, helped to shape this new knowledge so that it could finally be reported in the chapters in this book.

Index